Start-Up Creation

Related titles

Nonconventional and Vernacular Construction Materials: Characterisation, Properties and Applications

(ISBN 978-0-08-100871-3)

Biopolymers and Biotech Admixtures for Eco-Efficient Construction Materials

(ISBN 978-0-08-100214-8)

Acoustic Emission and Related Non-destructive Evaluation Techniques in the Fracture Mechanics of Concrete: Fundamentals and Applications

(ISBN 978-1-78242-327-0)

Woodhead Publishing Series in Civil and Structural Engineering: Number 66

Start-Up Creation

The Smart Eco-Efficient Built Environment

Edited by

Fernando Pacheco-Torgal, Erik Rasmussen, Claes-Göran Granqvist, Volodymyr Ivanov, Arturas Kaklauskas and Stephen Makonin

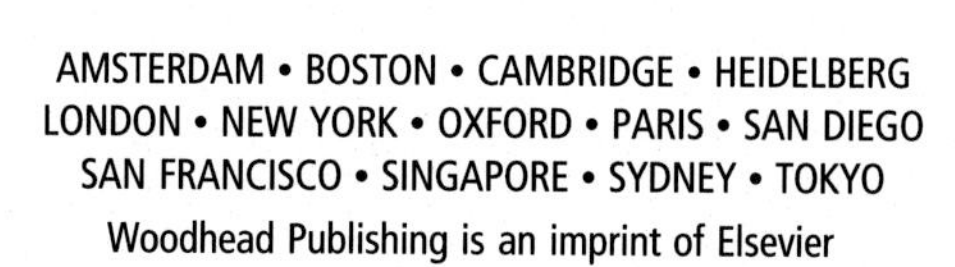
AMSTERDAM • BOSTON • CAMBRIDGE • HEIDELBERG
LONDON • NEW YORK • OXFORD • PARIS • SAN DIEGO
SAN FRANCISCO • SINGAPORE • SYDNEY • TOKYO
Woodhead Publishing is an imprint of Elsevier

Woodhead Publishing is an imprint of Elsevier
The Officers' Mess Business Centre, Royston Road, Duxford, CB22 4QH, UK
50 Hampshire Street, 5th Floor, Cambridge, MA 02139, USA
The Boulevard, Langford Lane, Kidlington, OX5 1GB, UK

British Library Cataloguing-in-Publication Data
A catalogue record for this book is available from the British Library

Library of Congress Cataloging-in-Publication Data
A catalog record for this book is available from the Library of Congress

ISBN: 978-0-08-100546-0 (print)
ISBN: 978-0-08-100549-1 (online)

For information on all Woodhead Publishing publications visit our website at https://www.elsevier.com/

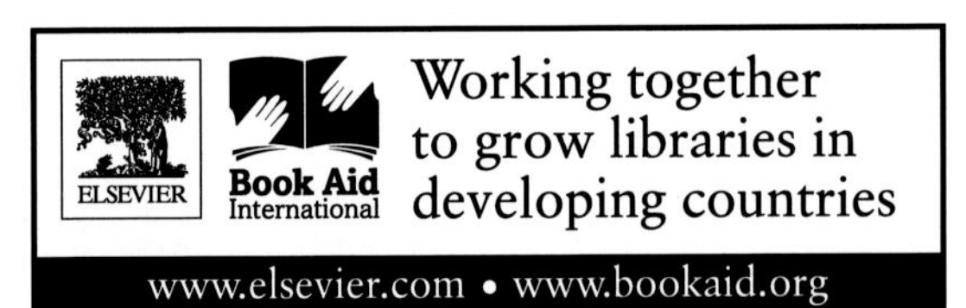

Publisher: Matthew Deans
Acquisition Editor: Gwen Jones
Editorial Project Manager: Charlotte Cockle
Production Project Manager: Debasish Ghosh
Designer: Greg Harris

Typeset by TNQ Books and Journals

Contents

List of contributors

Dr. K. Biswas Oak Ridge National Laboratory, Oak Ridge, TN, United States

A. Caplanova University of Economics in Bratislava, Bratislava, Slovakia

F. Cappelletti University Iuav of Venice, Venice, Italy

E. Carayannis Professor George Washington University

I. Chatzigiannakis Sapienza University of Rome, Rome, Italy

W.K. Chong Arizona State University, Tempe, AZ, United States

J.-S. Chou National Taiwan University of Science and Technology, Taipei, Taiwan; Arizona State University, Tempe, AZ, United States

C. Cristalli Loccioni Group, Angeli di Rosora, (An), Italy

A. Gasparella Free University of Bozen-Bolzano, Bolzano, Italy

G.E. Gibson Jr. Arizona State University, Tempe, AZ, United States

R. Gudauskas Vilnius Gediminas Technical University, Vilnius, Lithuania

M.R. Hammer University of Stuttgart, Stuttgart, Germany

K.R. Hansen University of Southern Denmark, Odense, Denmark

P. Harvard Professor EIGSI Engineering School France

Bjørn Petter Jelle Norwegian University of Science and Technology (NTNU), Trondheim, Norway; SINTEF Building and Infrastructure, Trondheim, Norway

A. Kaklauskas Vilnius Gediminas Technical University, Vilnius, Lithuania

J. Knippers University of Stuttgart, Stuttgart, Germany

C. Köhler-Hammer University of Stuttgart, Stuttgart, Germany

D. Kolokotsa Technical University of Crete, Chania, Greece

A. Köse Ege University, Izmir, Türkiye

L. Long University of Science and Technology of China, Hefei, PR China

S. Makonin Simon Fraser University, Burnaby, BC, Canada

N.-T. Ngo National Taiwan University of Science and Technology, Taipei, Taiwan

S.Ş. Öncel Ege University, Izmir, Türkiye

D.Ş. Öncel Dokuz Eylul University, Izmir, Türkiye

S.C. Oranburg Chicago-Kent College of Law, Chicago, IL, United States

F. Pacheco-Torgal University of Minho, Guimarães, Portugal

S. Papantoniou Technical University of Crete, Chania, Greece

P. Penna Free University of Bozen-Bolzano, Bolzano, Italy

A. Prada Free University of Bozen-Bolzano, Bolzano, Italy

E.S. Rasmusscn University of Southcrn Dcnmark, Odcnsc, Dcnmark

T. Shih Lund University, Lund, Sweden

G. Soreanu Technical University "Gheorghe Asachi" of Iasi, Faculty of Chemical Engineering and Environmental Protection, Department of Environmental Engineering and Management, Iasi, Romania

L. Standardi Loccioni Group, Angeli di Rosora, (An), Italy

S. Tanev University of Southern Denmark, Odense, Denmark

H. Ye University of Science and Technology of China, Hefei, PR China

Woodhead Publishing Series in Civil and Structural Engineering

1 **Finite element techniques in structural mechanics**
C. T. F. Ross
2 **Finite element programs in structural engineering and continuum mechanics**
C. T. F. Ross
3 **Macro-engineering**
F. P. Davidson, E. G. Frankl and C. L. Meador
4 **Macro-engineering and the earth**
U. W. Kitzinger and E. G. Frankel
5 **Strengthening of reinforced concrete structures**
Edited by L. C. Hollaway and M. Leeming
6 **Analysis of engineering structures**
B. Bedenik and C. B. Besant
7 **Mechanics of solids**
C. T. F. Ross
8 **Plasticity for engineers**
C. R. Calladine
9 **Elastic beams and frames**
J. D. Renton
10 **Introduction to structures**
W. R. Spillers
11 **Applied elasticity**
J. D. Renton
12 **Durability of engineering structures**
J. Bijen
13 **Advanced polymer composites for structural applications in construction**
Edited by L. C. Hollaway
14 **Corrosion in reinforced concrete structures**
Edited by H. Böhni
15 **The deformation and processing of structural materials**
Edited by Z. X. Guo
16 **Inspection and monitoring techniques for bridges and civil structures**
Edited by G. Fu
17 **Advanced civil infrastructure materials**
Edited by H. Wu
18 **Analysis and design of plated structures Volume 1: Stability**
Edited by E. Shanmugam and C. M. Wang

19 **Analysis and design of plated structures Volume 2: Dynamics**
Edited by E. Shanmugam and C. M. Wang
20 **Multiscale materials modelling**
Edited by Z. X. Guo
21 **Durability of concrete and cement composites**
Edited by C. L. Page and M. M. Page
22 **Durability of composites for civil structural applications**
Edited by V. M. Karbhari
23 **Design and optimization of metal structures**
J. Farkas and K. Jarmai
24 **Developments in the formulation and reinforcement of concrete**
Edited by S. Mindess
25 **Strengthening and rehabilitation of civil infrastructures using fibre-reinforced polymer (FRP) composites**
Edited by L. C. Hollaway and J. C. Teng
26 **Condition assessment of aged structures**
Edited by J. K. Paik and R. M. Melchers
27 **Sustainability of construction materials**
J. Khatib
28 **Structural dynamics of earthquake engineering**
S. Rajasekaran
29 **Geopolymers: Structures, processing, properties and industrial applications**
Edited by J. L. Provis and J. S. J. van Deventer
30 **Structural health monitoring of civil infrastructure systems**
Edited by V. M. Karbhari and F. Ansari
31 **Architectural glass to resist seismic and extreme climatic events**
Edited by R. A. Behr
32 **Failure, distress and repair of concrete structures**
Edited by N. Delatte
33 **Blast protection of civil infrastructures and vehicles using composites**
Edited by N. Uddin
34 **Non-destructive evaluation of reinforced concrete structures Volume 1: Deterioration processes**
Edited by C. Maierhofer, H.-W. Reinhardt and G. Dobmann
35 **Non-destructive evaluation of reinforced concrete structures Volume 2: Non-destructive testing methods**
Edited by C. Maierhofer, H.-W. Reinhardt and G. Dobmann
36 **Service life estimation and extension of civil engineering structures**
Edited by V. M. Karbhari and L. S. Lee
37 **Building decorative materials**
Edited by Y. Li and S. Ren
38 **Building materials in civil engineering**
Edited by H. Zhang
39 **Polymer modified bitumen**
Edited by T. McNally
40 **Understanding the rheology of concrete**
Edited by N. Roussel
41 **Toxicity of building materials**
Edited by F. Pacheco-Torgal, S. Jalali and A. Fucic

42 **Eco-efficient concrete**
Edited by F. Pacheco-Torgal, S. Jalali, J. Labrincha and V. M. John
43 **Nanotechnology in eco-efficient construction**
Edited by F. Pacheco-Torgal, M. V. Diamanti, A. Nazari and C. Goran-Granqvist
44 **Handbook of seismic risk analysis and management of civil infrastructure systems**
Edited by F. Tesfamariam and K. Goda
45 **Developments in fiber-reinforced polymer (FRP) composites for civil engineering**
Edited by N. Uddin
46 **Advanced fibre-reinforced polymer (FRP) composites for structural applications**
Edited by J. Bai
47 **Handbook of recycled concrete and demolition waste**
Edited by F. Pacheco-Torgal, V. W. Y. Tam, J. A. Labrincha, Y. Ding and J. de Brito
48 **Understanding the tensile properties of concrete**
Edited by J. Weerheijm
49 **Eco-efficient construction and building materials: Life cycle assessment (LCA), eco-labelling and case studies**
Edited by F. Pacheco-Torgal, L. F. Cabeza, J. Labrincha and A. de Magalhães
50 **Advanced composites in bridge construction and repair**
Edited by Y. J. Kim
51 **Rehabilitation of metallic civil infrastructure using fiber-reinforced polymer (FRP) composites**
Edited by V. Karbhari
52 **Rehabilitation of pipelines using fiber-reinforced polymer (FRP) composites**
Edited by V. Karbhari
53 **Transport properties of concrete: Measurement and applications**
P. A. Claisse
54 **Handbook of alkali-activated cements, mortars and concretes**
F. Pacheco-Torgal, J. A. Labrincha, C. Leonelli, A. Palomo and P. Chindaprasirt
55 **Eco-efficient masonry bricks and blocks: Design, properties and durability**
F. Pacheco-Torgal, P. B. Lourenço, J. A. Labrincha, S. Kumar and P. Chindaprasirt
56 **Advances in asphalt materials: Road and pavement construction**
Edited by S.-C. Huang and H. Di Benedetto
57 **Acoustic emission (AE) and related non-destructive evaluation (NDE) techniques in the fracture mechanics of concrete: Fundamentals and applications**
Edited by M. Ohtsu
58 **Nonconventional and vernacular construction materials: Characterisation, properties and applications**
Edited by K. A. Harries and B. Sharma
59 **Science and technology of concrete admixtures**
Edited by P.-C. Aïtcin and R. J. Flatt
60 **Textile fibre composites in civil engineering**
Edited by T. Triantafillou
61 **Corrosion of steel in concrete structures**
Edited by A. Poursaee
62 **Innovative developments of advanced multifunctional nanocomposites in civil and structural engineering**
Edited by K. J. Loh and S. Nagarajaiah
63 **Biopolymers and biotech admixtures for eco-efficient construction materials**
Edited by F. Pacheco-Torgal, V. Ivanov, N. Karak and H. Jonkers

64 **Marine concrete structures: Design, durability and performance**
Edited by M. Alexander

65 **Recent trends in cold-formed steel construction**
Edited by C. Yu

66 **Start-up creation: The smart eco-efficient built environment**
Edited by F. Pacheco-Torgal, E. Rasmussen, C.-G. Granqvist, V. Ivanov, A. Kaklauskas and S. Makonin

67 **Characteristics and uses of steel slag in building construction**
I. Barisic, I. Netinger, A. Fučić and S. Bansode

Foreword

Start-up companies are just one, but a valuable way towards progressing innovation so that people in the built environment may have a healthier place to live and work. The barriers to the pathway of innovation are many. For success, there needs to be a fruitful collaboration between academia and industry, but often this also depends on Government policies which can encourage cooperative ventures. Academics need to have entrepreneurship as part of their portfolio, but this needs time and perseverance besides communication skills and some either do not or can see this as time they need to concentrate on the research. Industry and commercial outlooks towards innovation vary a lot. Some industries are very conservative and tend to think more short term whereas other sectors take the long-term view. In 19 chapters, this book covers all the range of possibilities that need consideration when contemplating a start-up company besides describing some of the latest innovations which offer new opportunities for achieving energy-efficient buildings. The best ideas are those that start with a defined focus such as energy efficiency but then bring added value by, for example, improving the human conditions. It is important that academics produce convincing business cases in their proposals for seeking any financial investment. Often industry tends to look at capital cost whereas the most innovative ones are more likely to look at the value so balancing the benefits and whole life costs of any proposal. There are lessons to be learnt from forward looking across sectors to the likes of information technology, aeronautics and pharmaceuticals, for example. Start-up companies need to be lean, adaptable and open with a wide range of technical and business skills. This book is welcome as it fills a gap in the market for eco-efficient scientists who want to understand how their work can make an impact on the industry.

Derek Clements-Croome
Professor Emeritus in Architectural Engineering
Reading University
United Kingdom

Introduction to start-up creation for the smart eco-efficient built environment

1

F. Pacheco-Torgal
University of Minho, Guimarães, Portugal

1.1 A brief introduction to entrepreneurship and start-up creation

The paramount importance of entrepreneurs (and entrepreneurship) for economic development is mainly associated with the theoretical work of Joseph Schumpeter (1934). According to this economist entrepreneurs are key for the process of industrial mutation "that incessantly revolutionizes the economic structure from within, incessantly destroying the old one, incessantly creating a new one." For Schumpeter, innovations are disruptions that emanate from a pathological behavior, a social deviance from norms, from daring entrepreneurs (Louçã, 2014). However, and according to Leyden et al. (2014), the concept of the entrepreneur as an innovator precedes the work of Schumpeter (1934), dating back to the writings of Nicolas Baudeau in the 18th century (Baudeau, 1910) and the works of the economist Richard Cantillon as the first economist to recognize the importance of entrepreneurs to lead with uncertainty.

Kirchhoff (1989) emphasized the importance of entry and growth of new small firms as the sign of Schumpeter's "Creative Destruction" being the mark of the new entrepreneurial economy and the driving force underlying innovation and economic growth (Thurik et al., 2013). Start-up creation is especially important in the current knowledge-based economy in which knowledge production is shifting from universities to highly flexible multidisciplinary teams (Hsu et al., 2014). Despite that view, some still believe that in the next few years universities will continue to be the major sources of knowledge generation (Godin and Gingras, 2000). The truth is that its (indirect) role on the technology transfer process by providing highly qualified engineers to industry (as they did in the past) will no longer be considered enough. A European Union report (STAC, 2014) states that knowledge generation is no longer enough and emphasizes the need to translate knowledge into products and services. Universities will then have to face increased pressure to turn investigation budgets into profitable products and services (Kalar and Antoncic, 2015; Guerrero et al., 2015). According to Etzkowitz (2003) the universities' assumption of an entrepreneurial role constitutes the latest step in the evolution of a medieval institution from its original purpose of conservation of knowledge. This author points out that in

Start-Up Creation. http://dx.doi.org/10.1016/B978-0-08-100546-0.00001-7

US universities this evolution replaced the 19th century model of a single professor representing a discipline surrounded by a staff of assistants by a more democratic model in which, for instance, an assistant professor can set research directions if he or she can obtain outside research funding. This issue is especially important because around the world hundreds of universities still live by the outdated 19th century model. The interactions between universities, government, and industry (triple helix model) are and will be crucial for the development of the knowledge-based economy (Leydesdorff and Etzkowitz, 2001).

The germination of biomedical research in the 1970s, the passage of the Bayh-Dole act in 1980 (Mowery et al., 2001; Mowery and Ziedonis, 2002), and the increased financing of research by industry (Mowery et al., 2004) not only explain the increased rate in university spinoffs that occurred in the last decades but are the consequence of the triple helix model. However, only recently has the scientific community tried to explain why different universities show very different spinoff creation rates. In this respect Di Gregorio and Shane (2003) studied 101 US universities. Their hypothesis for the different spinoff generations encompassed intellectual eminence, the existence equity investment policies, and a low inventor's share of royalties. O'Shea et al. (2005) also studied US university-based spinoffs. For that they analyzed the spinoffs created in the top 20 US universities for the period 1980–2001 in which the Massachusetts Institute of Technology (MIT) had a lead position. These authors showed that spinoff creation is very dependent on university resource stock availability. These authors also confirmed the importance of intellectual eminence in faculty with critical expertise to create radical innovations that are essential for spinoff creation. Landry et al. (2006) analyzed a sample of 1554 Canadian researchers in natural sciences and engineering to understand the determinants of the creation of university spinoffs by Canadian researchers. They noticed that university laboratory assets are especially important for spinoff creation. They also noticed that the existence of experienced researchers and the degree of novelty of research knowledge have the largest marginal impact on the likelihood of university spinoff creation.

Krabel and Mueller (2009) state that scientists who hold a patent are four times more likely to be nascent entrepreneurs than those scientists without a patent. Astebro et al. (2012) reviewed three case studies known for their high percentage of student alumni that start new businesses. This included the case of MIT and two others from Swedish universities (Halmstad and Chalmers). These authors state that MIT is a unique case very hard to replicate because it combines an entrepreneurial culture with cutting edge research and a research budget that exceeds one billion dollars. The MIT exceptionality for spinoff creation was also highlighted by Roberts (2014). Astebro et al. (2012) pointed out the success of the Chalmers surrogate entrepreneur concept, where a student is chosen/hired specifically to develop the new venture. The reason for that has to do with the fact that the surrogate entrepreneur not only will add new entrepreneurial competence but also new network capability. This concept is based on a three-part division ownership rights. The university is entitled to one-third, the inventor to another third, and the remaining third to the surrogate entrepreneur. Lundqvist (2014) analyzed a total of 170 ventures; 35% were surrogate-based. The results show that the surrogate ventures outperformed nonsurrogate ventures both in terms of growth and revenue. The surrogate entrepreneurship concept is therefore a virtuous one because surrogate entrepreneurs will contribute to a more balanced distribution of expertise among the start-up team members, which is known to be a start-up success factor (Maidique and Zirger, 1984; Roure and Keeley, 1990).

The lack of knowledge of the commercialization part of the entrepreneurial process is recognized as a gap in faculty (Siegel et al., 2007). And the work of Visintin and Pittino (2014) carried out on a sample of 103 Italian spinoffs confirms the importance of the surrogate entrepreneur concept and of the proper balance between scientific and commercial expertise of the team members. Those authors also mention that team members' high profile differentiation could constitute a pressure toward separation, requiring that team members must share some common characteristics to counterbalance that pressure. Other authors (de Lemos, 2014) also confirm that interpersonal relationship problems is the most critical factor that leads to the failure of technology start-ups. Using a sample of 2304 entrepreneurs who have started new businesses, Cassar (2014) investigated the role of experience on entrepreneurs' forecast performance regarding new business growth, and found that entrepreneurs with greater industry experience have more realistic expectations. Still this finding says very little about the start-up success by experienced entrepreneurs.

Ouimet and Zarutskie (2014) state that start-up creation is dependent on the availability of young workers. And Teixeira and Coimbra (2014) recently showed that younger start-up members reveal higher levels of entrepreneurial spirit and entrepreneurial capabilities, being in a better position to internationalize earlier than older members. It is worth mentioning that the average start-up member funded by the Silicon Valley Y Combinator (YC) is around 29 years old—a typical Y-generation (millennial), known for having a high entrepreneurial spirit (Winograd and Hais, 2014). Founded by Paul Graham in March of 2005, YC was the first start-up accelerator and so far has funded over 800 start-ups with a combined value over $30 billion (YC, 2015).

Start-up accelerators are composed of four main features: a highly competitive application process (YC selects around 2% of applicant start-up (Stagars, 2014)); provision of preseed investment in exchange for equity; focus on small teams instead of individual founders; time-limited support comprising programmed events and cohorts or classes of start-ups rather than individual companies (Miller and Bound, 2011). In the last decade this new incubating technology variant has grown very rapidly, exceeding thousands of new accelerators across the globe (Cohen and Hochberg, 2014).

Another important aspect that may help boost start-up creation concerns crowdfunding. This is an innovative funding method in which start-ups raise capital from small contributions of a very large number of individuals. Crowdfunding is especially important in a context in which banks are much less prone to lend money than before the 2008 financial crisis, and because start-ups do not have a financial history that would make it harder to get bank funding. Is not without irony that this alternative financing scenario that matches enterprises and investors could turn out to be a much better and sustainable solution for funding the economy (Macauley, 2015).

Different crowdfunding business models are identified: donation, passive investment, and active investment (Schwienbacher and Larralde, 2012). So far crowdfunding has financed thousands of entrepreneurial ventures and the global crowdfunding market is expected to reach $93 billion by 2025 (Swart, 2013). In the United States, the Jump-start Our Business Start-ups (JOBS) Act signed into law on April 5, 2012, to legalize equity crowdfunding and enable entrepreneurs and small business owners to sell limited amounts of equity in their companies to a large number of investors via social networks and various Internet platforms (Stemler, 2013).

Kickstarter, the largest crowdfunding site, has already funded 48,526 US-based projects, amounting to $237 million (Mollick, 2014). This helps to explain why the United States dominates global crowdfunding with 72% whereas the shares of Europe and the rest of the world were 26% and 2%, respectively (Kshetri, 2015).

Distinguished Prof. Willian Baumol (2008) stated that promoting entrepreneurship and small firms would play a critical role for economic prosperity. Also in the current context of high graduate unemployment rates that will be more dramatic in the next decades (Biavaschi et al., 2015; Li et al., 2014; Roy, 2014; Schmid, 2015; Sadler, 2015; Min, 2015), a context in which tacit knowledge and formal education is recognized as not being enough (Lacy, 2011; Wagner, 2012; Agarwal and Shah, 2014; Thiel and Masters, 2014), start-up creation could become a way to solve this serious problem. Still much more effort is needed to bridge the gap between research and the entrepreneurial world (Allen and O'Shea, 2014; Stagars, 2014) in order to foster massive start-up creation. Since the right identification of market needs is of paramount importance to avoid start-up failure (da Silva et al., 2015), the following section tries to justify why the smart eco-efficient built environment is considered an important area for start-up creation.

1.2 Smart eco-efficient built environment: an untouched start-up pond?

Civil engineering is known as an area mainly concerned with directing the great sources of power in nature for the use and convenience of man through the construction of large and public infrastructures (bridges, dams, airports, highways, tunnels, etc.) by large construction companies. Never was this area known to be associated with high-tech start-up creation. This constitutes a sign of low innovation, which is confirmed by its low patenting level. In the United States the patenting level on civil engineering falls behind other areas (Rothe, 2006). According to Keefe (2012) very few civil engineers take their innovations to the US Patent and Trademark Office, in contrast to the considerable number of electrical and mechanical engineers who do so. This author gives data that shows that the patenting in the civil engineering area is 7 times less than in mechanical engineering and 10 times less than in electrical engineering. A worldwide study (Fisch et al., 2015) confirms the prone patenting nature of other more innovative areas than civil engineering. This low innovation level undermines the prestige of civil engineering and helps explain the reduction of undergraduate applications to civil engineering (Byfield, 2003; Lawless, 2005; Hubbard and Hubbard, 2009; Quapp and Holschemacher, 2013).

Nedhi (2002) stated that civil engineering is not traditionally viewed as high-tech engineering. Even in India this area is viewed as a low-tech one (Chakraborty et al., 2011). As a consequence low starting salaries are normal in this area (Hamill and Hodgkinson, 2003). The fact that construction enterprises have low productivity (Fulford and Standing, 2014) and have to compete for lower bids having lower and lower profit margins (Morby, 2014) and also have to face increasing and fierce Chinese competition already capable of building a dozen-story structure in just a few weeks (McKinsey, 2014; CWO, 2015) means that construction enterprises in the future will have less and less financial possibilities to offer high and attractive paychecks to civil engineers. Still, civil engineering has an important role to play given the environmental impact of the construction industry that will be exacerbated in the next decades due to the growth in world population. By 2050 urban population will almost double, increasing from approximately 3.4 billion in 2009 to 6.4 billion in 2050 (WHO, 2014). Recent estimates on urban expansion suggests that by 2030 a high probability exists (over 75%) that urban land cover will increase by 1.2 million km^2 (Seto et al., 2012). Since the global construction industry consumes more raw materials (about 3000 Mt/year, almost 50% by weight) than any other economic activity, the previously mentioned urban expansion will dramatically increase that consumption (Ashby, 2015). This not only will make it more difficult to reduce greenhouse gas emissions for which the built environment is a significant contributor (representing 30% of related emissions), but will also increase pressure on biodiversity loss, which is crucial for the survival of humanity (Wilson, 2003). It is worth remembering that humanity has already transgressed the planetary boundaries for climate change, rate of biodiversity loss, and changes to the global nitrogen cycle (Rockstrom et al., 2009). As a consequence, the role of civil engineering will be much more relevant if it was able to reinvent itself into an eco-efficient one with high added value.

The concept of eco-efficiency was first coined in the book, *Changing Course* (Schmidheiny, 1992), in the context of 1992 Earth Summit process. This concept includes "the development of products and services at competitive prices that meet the needs of humankind with quality of life, while progressively reducing their environmental impact and consumption of raw materials throughout their life cycle, to a level compatible with the capacity of the planet." Thus the eco-efficient built environment concerns reducing its environmental impact while enhancing the quality of life of its users. In this context the development of technologies for the smart eco-efficient built environment may provide the body of innovative knowledge that is known to be critical for entrepreneurs to transform innovative ideas into commercial products and services (Agarwal and Shah, 2014).

In the last decades nanotechnology became a hot area, crossing different scientific areas from electronics to life sciences; however, only in the last few years have the nanotech investigations for the construction industry begun to have enough expression justified by the published works on that particular field (Smith and Granqvist, 2011; Pacheco-Torgal et al., 2013a). A Scopus search of journal papers containing the terms "nanotechnology" and "eco-efficient construction" shows that a research shift from cement nanotech to nanotech energy-efficient materials has occured.

A high priority nanotech-related field concerns the development and production of cool materials incorporating new advanced nanomaterials (Santamouris et al., 2011). Cool materials have high solar reflectance, allowing for the reduction of energy cooling needs in summer. These materials are especially important for building energy efficiency because as a consequence of climate change, building cooling needs are expected to increase in the coming years. According to the IEA (2013), energy consumption for cooling is expected to increase sharply by 2050, by almost 150% globally, and by 300–600% in developing countries. The Cool-Coverings FP7 project (Escribano and Keraben Grupo, 2013) aimed at the development of a novel and cost-effective range of nanotech-improved coatings to substantially improve near-infrared reflective properties. The author's view is that this area could merit the formation of successful start-ups. Another important nanotech field concerns switchable glazing technology-based materials that make it possible to construct glazings whose throughput of visible light and solar energy can be switched to different levels depending on the application of a low DC voltage (electrochromics) or on the temperature (thermochromics), or even by using hydrogen (gasochromics). This technology has a large potential to minimize the energy use in buildings and allow for the nearly zero-energy building target (Granqvist, 2013; Pacheco-Torgal et al., 2013b; Favoino et al., 2015). Several commercial solutions are already available on the market (SAGE Electrochromics–USA, Econtrol Glas, Saint Gobain Sekurit, and Gesimat-Germany, among others) with a service life of 30 years and capable of 100,000 switching cycles, but their cost-efficiency is far from an optimum condition. ChromoGenics is a relevant company operating in this field that was established in 2003 as the outcome of over 20 years of research on electrochromic materials by Professor Claes-Göran Granqvist and his team at the Ångström Laboratory at Uppsala University in Sweden. Using a laminated electrochromic plastic foil, ConverLight™, rather than coating the glass itself, ChromoGenics has contributed to a more scalable and cost-effective smart-glass manufacturing. The most challenging point of smart windows at the moment is their higher cost compared to the other glazing technologies (Cuce and Riffat, 2015). Hee et al. (2015) states that due to the higher costs of dynamic glazing, it is more suitable to be installed in the building that needs high performance in terms of day-lighting and energy savings such as commercial buildings.

Biotechnology is one of the world's fastest growing industries that could constitute a hot area allowing for radical changes in the eco-efficiency of construction materials and technologies. Since this area is one of the six key enabling technologies that will be funded under the EU Framework Programme Horizon 2020 (Pacheco-Torgal, 2014) this can also foster the development of start-ups for the eco-efficient built environment. The use of biotechnology for indoor air purification is also a crucial biotech innovation of major significance for the eco-efficient built environment with high marketable potential. US expenditures for indoor air quality (IAQ) are currently $23 billion per year, or, accounting for uncertainty, between $18 and $30 billion per year (Mudarri, 2014). IAQ is a main issue for researchers motivated by the time that humans spend indoors, the wide range of pollutants present in indoor air, their concentration and toxicity, and the higher indoor concentrations with respect to outdoor ones. Building ventilation is a simple and efficient measure to improve IAQ (provided that the building is not located in a polluted city or near polluted areas like high-traffic roads). However, high ventilation rates are associated with high energy costs. Wang and Zhang (2011) developed an active biofiltration system based on carbon as a hydroponic substrate for indoor plants reporting a reduction in ventilation energy costs.

Currently some biofilters are already on the market, like for instance the active modular phytoremediation systems developed by CASE, cohosted by Rensselaer Polytechnic Institute and Skidmore, Owings & Merrill LLP (Torpy et al., 2014). Still the development of improved biofilters could merit the creation of new start-ups. Another important biotech feature concerns the production of bioenergy through microalgae photobioreactors (PBRs) integrated as facades or roofs. This technique seems to have high potential for start-up creation. Microalgae has a high oil content and most importantly, shows an extremely rapid growth. It doubles its biomass within 24 hours, being the fastest growing organism in the world. This is about 100–200% higher than any other energy crop (Chisti, 2007). The major constraint to the commercial-scale algae farming for energy production is the cost factor. But since microalgae have the ability to assimilate nutrients like nitrogen and phosphates (which are present in wastewaters) into the cells for its growth, the application of microalgae for wastewater treatment can be an interesting option to enhance its economic value and at the same time to solve environmental problems related to wastewater management. As for the use of architectural PBRs, their synergy generated by summation profits can turn the architecture into something iconic, environmentally didactic, active energetically, surface-saving, and environmentally friendly (Cervera and Pioz, 2014).

Smart-home solutions are another important area (especially for users with special needs) that may unleash a lot of business opportunities for built environment professionals. The investigation on smart homes began in the 1990s with the MIT pioneering work "Smart rooms" (Pentland, 1996). A case study that also took place in the 1990s of an adaptive house that used neural networks to control air heating, lighting, ventilation, and water heating without previous programming by the residents was described by Chan et al. (2008). This system, termed ACHE (**a**daptive **c**ontrol of **h**ome **e**nvironments), attempted to economize energy resources while respecting the lifestyle and desires of its inhabitants. De Silva et al. (2012) defines it as a "home-like environment that possesses ambient intelligence and automatic control, which allow it to respond to the behavior of residents and provide them with various facilities." They also mention that currently there are three major application categories. The first category aims at providing services to the residents and includes smart homes that provide elder care, smart homes that provide health care, and smart homes that provide child care. The second category aims at storing and retrieving multimedia captured within the smart home, in different levels from photos to experiences. The third area is about surveillance, where the data captured in the environment are processed to obtain information that can help to raise alarms in order to protect the home and the residents from burglaries, theft, and natural disasters like flood, and so on.

Other authors (Wong et al., 2005) mentioned that intelligent buildings started three decades ago. Still it was only in 2006 that Derek Clements-Croome (2013) formed the Intelligent Buildings Group of the Chartered Institution Services Engineers. For this author intelligent buildings are not only responsive to the occupants' needs but at the same time are sustainable in terms of energy and water consumption and maintain a minimal impact on the environment in terms of emissions and waste including the use of self-healing and smart-materials technology (Clements-Croome, 2011). Several terms are used in this respect including self-aware and sentient buildings (Mahdavi, 2008); however, this concept has not been widely used. The concept of smart buildings has been associated with a more advanced grouping (Buckman et al., 2014) that integrates and accounts for intelligence, enterprise, control, and materials and construction as an entire building system, with adaptability, not reactivity, at its core, in order to meet the drivers for building progression: energy and efficiency, longevity, and comfort and satisfaction.

Apart from the discussion between the intelligent/smart/sentient concepts the important thing to retain is that the overall objective relies on the development of housing to be healthier, safer, and comfortable (GhaffarianHoseini et al., 2013). In the next few years three major disruptive drivers (big data/Internet of Things (IoT)/cloud computing) will radically change smart homes. The data generated from thousands of home sensors and home appliances that are able to connect to each other, to send data, and to be managed from cloud network services will boost smart home advantages (Kirkham et al., 2014). Thanks to IoT, the largest software companies will make a shift to the physical world as did Google, which acquired a company producing thermostats to enter its trademarks in the smart-home world (Borgia, 2014). This highlights the importance of building energy efficiency. This importance is also shared by some works on the IoT area (Moreno et al., 2014) and is especially needed to address ambitious energy consumption targets for instance like the Zürich 2000 Watt Society (Zurich, 2011). More on the role of the energy-efficient built environment to European smart cities can be found in Kylili and Fokaides (2015). Smart homes will be able to assess an occupant's satisfaction, which is one of the main shortcomings of built environment, even in green buildings that surprisingly are not as occupant-friendly as previously alleged. In a large-scale occupant survey Guo et al. (2013) found that in some green buildings lower satisfaction and comfort were reported. Hirning et al. (2014) reported discomfort glare in five green buildings in Brisbane, Australia. Altomonte and Schiavon (2013) found that LEED-certified buildings show no significant influence on occupant satisfaction. In a postoccupancy study of a LEED Platinum building some occupants mentioned thermal discomfort (Hua et al., 2014).

The assessment of the occupant's feedback in smart homes will trigger interactive actions to adapt homes' performance accordingly. This is a leap from neutral comfort (absence of discomfort) into a new one in which the well-being of occupants is at the heart of the smart-home concept (Clements-Croome, 2014). Older people constitute an important group of users with special needs that could benefit from smart-home features. In the next decades this group will increase dramatically. The global population of people over the age of 65 is expected to more than double from 375 million in 1990 to 761 million by 2025 (Dishman, 2004). By 2040 it is expected to reach 1300 million (Kinsella and He, 2009). Between 2100 and 2300 the proportion of the world population in the 65 or over age group (the retirement age in most countries) is estimated to increase by 24–32%, and the 80 or over age group will double from 8.5% to 17% (UN, 2004). Elderly people prefer to live in their own house rather than in hospitals, which means that is important that homes can be studied and adapted to enhance elderly users' satisfaction. For instance home sensors can be used to balance daylight exposure and artificial light in order to guarantee enough light to maintain circadian rhythmicity, or else to warn elderly occupants on heat waves and high UV exposure. Sensors can also be used to detect air pollutants like volatile organic compounds and trigger ventilation to reduce its concentration. This means that the design of a built environment to help elderly occupants being independent in their homes is a crucial issue, to be addressed by built environment professionals integrating multidisciplinary teams.

Redirecting the focus of civil engineering from construction and rehabilitation of grand infrastructures to smart eco-efficient built environment-related areas and the needs of individual home users will enlarge the number of future clients. Different user problems will require different tailored solutions and this may represent a wide market of millions of clients that may foster high-tech start-up creation.

Other books have already been written about start-ups. However, as far as the author is concerned, none was published concerning civil engineering-based start-ups or start-up creation for the built environment. Parts 2 and 3 of this book cover a wide range of innovative technologies (ideas) that could generate start-ups. However, although innovative ideas are crucial they are not enough for start-up creation—the ability to put those ideas to work is. That is why the first part of this book assembles an important group of issues that are crucial for those who need to set their start-up in motion.

1.3 Outline of the book

This book provides an updated state-of-the-art review on the start-up creation for the smart eco-efficient built environment.

The first part encompasses an overview on business plans, start-up financing, marketing, creativity, and intellectual property (Chapters 2–8). Chapter 2 concerns business plan basics for engineers. It discusses the unique characteristics and challenges of technology-driven business environments, and describes the two key components of the business planning process: the articulation and the development of a viable business model, and managing the scaling up and the growth of the business.

Chapter 3 addresses the concept of the lean start-up approach as a way of reducing the risk of starting new firms (or launching new products) and enhancing the chances for success by validating the products and services in the market with customers before launching it in full scale.

Chapter 4 discusses the pro and cons of different start-up financing options. These include debt financing, equity financing, convertible debt financing, and crowdfunding.

Marketing for start-ups is the subject of Chapter 5. It describes how start-ups interact, how their networks are built, and what contributions various actors have in terms of how they coinfluence each other and add to the possibility of the start-up to develop in the long-term.

In Chapter 6, a representative and pertinent survey is presented, covering research literature about measuring and defining entrepreneurship, and more especially, entrepreneurial creativity. It discusses a minimalist model for measuring entrepreneurial creativity based on three criteria: timing, cognitive capacity, and quantifiable changes. An application of the model to information about the career of three well-known entrepreneurs is made.

Chapter 7 reviews intellectual property-related issues. It includes forms of intellectual property rights, trademarks, industrial designs, patents and utility models, copyrights, and trade secrets. A review of the historical development of the intellectual property protection is made and the regulatory aspects of the intellectual property protection are discussed.

Nano- and biotechnologies for eco-efficient buildings are the subject of Part II (Chapters 8−13).

Chapter 8 is concerned with nano-based thermal insulation for energy-efficient buildings. It starts with a review on the advantages and disadvantages of traditional building thermal insulation materials. A special focus is given to nano-based thermal insulation materials. Comments on start-up creation for manufacturing nano-based thermal insulation for energy-efficient buildings are made.

Chapter 9 is related to nano-based thermal storage technologies for building energy efficiency. Synthesis and characterization of the heat transfer and thermal storage properties of nano phase change material (nanoPCM) are included. A review of nanoPCM applications and their potential energy benefits is performed. The chapter also discusses whether the higher conductivity of nanoPCM is desirable in all applications and if the property enhancements are worth the cost and effort needed to create nanoPCMs.

Chapter 10 covers nano-based chromogenic technologies for building energy efficiency, especially thermochromic and electrochromic windows. Application performances were demonstrated through both experiments and simulations. A guidance on the performance improvement was also discussed.

Chapter 11 analyzes façade-integrated PBRs for building energy efficiency. A review on microalgae and the different type of PBRs is included. Design and scale-up parameters are discussed. The role of PBRs in building, particularly as building facades, is also discussed. This chapter is closed with relevant comments on start-up creation to the development of innovative PBRs for the built environment.

Biotechnologies for improving indoor air quality is the subject of Chapter 12. It reviews the different indoor air pollutants and current air cleaning methods. It addresses the theoretical basics on biotechnologies for air cleaning, the types of bioreactors, and the evaluation of bioreactors performance. Also discussed are the opportunities and challenges of using bioreactors for indoor air cleaning. The removal of specific indoor air pollutants are also covered as well as future trends in this field.

Chapter 13 addresses the use of biobased plastics for building facades. Comments on feedstocks, resource efficiency, and recycling are included. Requirements for use of biobased plastics as building components are addressed. Performance of some biobased plastics concerning fire resistance, heat stability, and weathering resistance is also addressed. A case study of a biobased plastic façade is included.

Finally Part III (Chapters 14–19) deals with algorithms, big data, and IoT for eco-efficient and smart buildings.

Chapter 14 is concerned with the development of algorithms for building retrofit. It contains an overview of different methodologies to deal with multiobjective projects, and methods to assist and define the retrofit interventions is described.

Chapter 15 looks at the use of control algorithms in lighting systems for high energy savings and for the fulfillment of lighting requirements. This chapter introduces light control algorithms as enabler of differentiation, which is a key requirement for a successful start-up rollout. Moreover, the proposed control lighting systems are customized and implemented in three real operational environments: two hospitals and one office building.

Chapter 16 is concerned with the use of big data and cloud computing for building energy efficiency. This chapter presents the framework of a smart-decision support system (SDSS) that integrates smart-grid big data analytics and cloud computing for building energy efficiency. A real-world smart metering infrastructure was installed in a residential building for the experiment. The SDSS accurately identified the building energy consumption patterns and forecasted future energy usage.

Chapter 17 addresses the case of intelligent-decision support systems and the IoT for the smart built environment. This chapter outlines the general theory of the IoT in the built environment. An analysis of possibilities to integrate intelligent decision support systems with IoT in the built environment is carried out. The main trends and the future of IoT in the built environment are discussed.

Chapter 18 is concerned with app programming and its use for smart-building management systems. An overview of different issues to consider when developing apps is included. A discussion about app types and how they are used is also included.

Chapter 19 deals with the usage of smart-home technologies for home security, the available networking technologies, and their benefits and vulnerabilities. A number of existing products are presented, in terms of the features provided for making a home secure, along with their advantages and potential disadvantages.

References

Allen, T., O'Shea, R., 2014. Building Technology Transfer within Research Universities: An Entrepreneurial Approach. Cambridge University Press.

Agarwal, R., Shah, S., 2014. Knowledge sources of entrepreneurship: firm formation by academic, user and employee innovators. Research Policy 43, 1109–1133.

Altomonte, S., Schiavon, S., 2013. Occupant satisfaction in LEED and Non-LEED certified buildings. Building and Environment 68, 66–76.

Ashby, F., 2015. Materials and Sustainable Development, first ed. Butterworth-Heinemann, Elsevier, Oxford, UK.

Astebro, T., Bazzazian, N., Braguinsky, S., 2012. Startups by recent university graduates and their faculty: implications for university entrepreneurship policy. Research Policy 41, 663–677.

Baudeau, N., 1910. In: Dubois, A. (Ed.), Première introduction à la philosophie Économique. P. Geuthner, Paris.

Baumol, W.J., 2008. Small enterprises, large firms, productivity growth and wages. Journal of Policy Modeling 30 (4), 575–589.

Biavaschi, C., Eichhorst, W., Giulietti, C., Kendzia, M., Muravyev, A., Pieters, J., Rodriguez-Planas, N., Schmidl, R., Zimmerman, K., 2015. Youth Unemployment and Vocational Training. IZA Policy Paper No. 6890. Institute for the Study of Labor, Bonn, Germany.

Byfield, M.P., 2003. British civil engineering skills: defusing the time bomb. Proceedings of the Institution of Civil Engineers: Civil Engineering 156 (4), 183–186.

Borgia, E., 1 December 2014. The internet of things vision: key features, applications and open issues. Computer Communications 54, 1–31.

Buckman, A., Mayfield, M., Beck, S., 2014. What is a smart building? Smart and Sustainable Built Environment 3 (2), 92–109.

Cassar, G., 2014. Industry and startup experience on entrepreneur forecast performance in new firms. Journal of Business Venturing 29, 137–151.

Cervera, R., Pioz, J., 2014. Architectural bio-photo reactors: harvesting microalgae on the surface of architecture. In: Pacheco-Torgal, F., Labrincha, J., Diamanti, M., Yu, C.-P., Lee, H.K. (Eds.), Biotechnologies and Biomimetics for Civil Engineering. Springer, London, pp. 163–180.

Chakraborty, S., Iyer, N., Krishna, P., Thakkar, S., 2011. Assessment of Civil Engineering Inputs for Infrastructure Development. Indian National Academy of Engineering.

Chan, M., Esteve, D., Escriba, C., Campo, E., 2008. A review of smart homes – present state and future challenges. Computer Methods and Programs in Biomedecine 91, 55–81.

Chisti, Y., 2007. Biodiesel from microalgae. Biotechnology Advances 25, 294–306.

Clements-Croome, D., 2014. Sustainable Intelligent Buildings for Better Health, Confort and Well-being. Report for Denzero project.

Clements-Croome, D., 2013. Intelligent Buildings: An Introduction. Earthscan, Routledge.

Clements-Croome, D., 2011. Sustainable intelligent buildings for people: a review. Intelligent Buildings International 3, 67–86.

Cohen, S., Hochberg, Y.V., March 30, 2014. Accelerating Startups: The Seed Accelerator Phenomenon. Available at SSRN: http://ssrn.com/abstract=2418000 or http://dx.doi.org/10.2139/ssrn.2418000.

Construction Week Online-CWO, 2015. Chinese Firm Builds World's Tallest Prefab Tower. http://www.constructionweekonline.com/article-32916-chinese-firm-builds-worlds-tallest-prefab-tower/.

Cuce, E., Riffat, S., 2015. A state-of-the-art review on innovative glazing technologies. Renewable and Sustainable Energy Reviews 41, 695–714.

Dishman, E., 2004. Inventing wellness system for aging in place. IEEE Computer Magazine 37 (5), 34–41.

da Silva, F., Romulo, F., Pinto, R., Galegale, N., Akabane, G., 2015. Why technology-based startups fail? An it management approach. In: Production and Operations Management Society, POMS 26th Annual Conference,Washington. http://www.pomsmeetings.org/ConfProceedings/060/Full%20Papers/final_full_paper.htm.

De Silva, L., Morikawa, C., Petra, I., 2012. State of the art of smart homes. Engineering Applications of Artificial Intelligence 25, 1313–1321.
de Lemos, H., 2014. The failure of early-stage technology startups in Brazil: a study about the contributing factors to the early-death of startups between 2009 and 2014 in Southeastern Brazil. Master Dissertation. Fundação Getulio Vargas. http://bibliotecadigital.fgv.br/dspace/handle/10438/13098 (accessed on 3 of September of 2015).
Di Gregorio, D., Shane, S., 2003. Why some universities generate more start-ups than others. Research Policy 32, 209–227.
Escribano, M.A.B., Keraben Grupo, S.A., 2013. Development of a Novel and Cost-effective Range of Nanotech Improved Coatings to Substantially Improve NIR (Near Infrared Reflective) Properties of the Building Envelope. FP7 Project. http://cordis.europa.eu/project/rcn/94644_en.html.
Etzkowitz, H., 2003. Research groups as 'quasi-firms': the invention of the entrepreneurial university. Research Policy 32, 109–121.
Favoino, F., Overend, M., Jin, Q., 2015. The optimal thermo-optical properties and energy saving potential of adaptive glazing technologies. Applied Energy 156, 1–15.
Fisch, C., Hassel, T., Sandner, P., Block, J., 2015. University patenting: a comparison of 300 leading universities worldwide. Journal of Technology Transfer 40, 318–345.
Fulford, R., Standing, C., 2014. Construction industry productivity and the potential for colaborative practice. International Journal of Project Management 32 (2), 315–326.
GhaffarianHoseini, A., Dahlan, N., Berardi, U., GhaffarianHoseini, A., Makaremi, N., 2013. The essence of future smart houses: from embedding ict to adapting to sustainability principles. Renewable and Sustainable Energy Reviews 24, 593–607.
Guerrero, M., Cunningham, J., Urbano, D., 2015. Economic impact of entrepreneurial universities' activities: an exploratory study of the United Kingdom. Research Policy 44, 748–764.
Godin, B., Gingras, Y., 2000. The place of universities in the system of knowledge production. Research Policy 29, 273–278.
Guo, Z., Prasad, D., Lau, S., 2013. Are green buildings more satisfactory and comfortable? Habitat International 39, 156–161.
Granqvist, C.-G., 2013. Switchable glazing technology for eco-efficient construction. In: Pacheco-Torgal, F., Diamanti, V., Nazari, A., Granqvist, C.G. (Eds.), Nanotechnology in Eco-efficient Construction. Materials, Processes and Applications. Woodhead Publishing Limited Abington Hall, Cambridge, UK, pp. 236–269.
Hamill, L., Hodgkinson, L., 2003. Civil engineering's image in schools - and how to change it. Proceedings of the ICE - Civil Engineering 156 (2), 78–85.
Hee, W., Alghoul, M., Bakhtyar, B., Elayeb, O., Shameri, M., Alrubaih, M., Sopian, K., 2015. The Role of window glazing on daylighting and energy saving in buildings. Renewable and Sustainable Energy Reviews 42, 323–343.
Hirning, M.B., Isoardi, G.L., Cowling, I., 2014. Discomfort glare in open plan green buildings. Energy and Buildings 70, 427–440.
Hsu, A., Shen, Y.-C., Yuan, B., Chou, C., 2014. Toward successful commercialization of university technology: performance drivers of university technology transfer in Taiwan. Technological Forecasting & Social Change 92, 25–39.
Hua, Y., Goçer, O., Gocer, K., 2014. Spatial mapping of occupant satisfaction and indoor environment quality in a LEED platinum campus building. Building and Environment 79, 124–137.
Hubbard, B.J., Hubbard, S.M., 2009. Activities to enhance civil engineering recruitment and coordination with industry. Transportation Research Record 22–30.

IEA, 2013. Technology Roadmap: Energy Efficient Building Envelopes. International Energy Agency, Paris.
Kalar, B., Antoncic, B., 2015. The entrepreneurial university, academic activities and technology and knowledge transfer in four european countries. Technovation 36–37, 1–11.
Keefe, S., 2012. Patent eligibility: an open fieled for civil engineering. Civil Engineering 70–73.
Kinsella, K., He, W., 2009. An Aging World: 2008. US Census Bureau, Washington DC.
Kirchhoff, B.A., 1989. Creative destruction among industrial firms in the United States. Small Business Economics 1 (3), 161–173.
Kirkham, T., Armstrong, D., Djemame, K., Jiang, M., 2014. Risk driven smart home resource management using cloud services. Future Generation Computer Systems 38, 13–22.
Kylili, A., Fokaides, P., 2015. European smart citties: the role of zero energy buildings. Sustainable Citties and Society 15, 86–95.
Krabel, S., Mueller, P., 2009. What drives scientists to start their own company? An empirical investigation of Max Planck Society Scientists. Research Policy 38, 947–956.
Kshetri, N., 2015. Success of crowd-based online technology in fundraising: an institutional perspective. Journal of International Management 21, 100–116.
Lacy, S., 2011. Peter Thiel: We're in a Bubble and It's Not the Internet. It's Higher Education. Techcrunch.
Landry, R., Amara, N., Rherrad, I., 2006. Why are some university researchers more likely to create spin-offs than others? Evidence from Canadian universities. Research Policy 35, 1599–1615.
Lawless, A., 2005. A wake up call to address the capacity crisis in SA civil engineering. Civil Engineering/Siviele Ingenieurswese 13 (10), 40–43.
Leyden, D., Link, A., Siegel, D., 2014. A Theoretical analysis of the role of social networks in entrepreneurship. Research Policy 43, 1157–1163.
Leydesdorff, L., Etzkowitz, H., 2001. The transformation of university–industry–government relations. Electronic Journal of Sociology 5 (4).
Li, S., Whalley, J., Xing, C., 2014. China's higher education expansion and unemployment of college graduates. China Economic Review 30, 567–582.
Louçã, F., 2014. The elusive concept of innovation for Schumpeter, Marschak and the early econometricians. Research Policy 43, 1442–1449.
Lundqvist, M., 2014. The importance of surrogate entrepreneurship for incubated Swedish technology ventures. Technovation 34, 93–100.
Macauley, C., 2015. Capitalism's renaissance? The potential of repositioning the financial 'meta-economy'. Futures 68, 5–18.
Mahdavi, A., 2008. Anatomy of a cogitative building. In: EWork and EBusiness in Architecture, Engineering and Construction – Zarl & Scherer. Taylor and Francis.
Maidique, M.A., Zirger, B.J., 1984. A study of success and failure in product innovation: the case of the US electronics industry. IEEE Transactions on Engineering Management 4, 192–203.
McKinsey, 2014. How to Build a Skyscraper in Two Weeks. Insigths & Publications. http://www.mckinsey.com/insights/engineering_construction/how_to_build_a_skyscraper_in_two_weeks.
Miller, P., Bound, K., 2011. The Startup Factories: The Rise of Accelerator Programmes to Support New Technology Ventures. NESTA, London.
Min, W., 2015. The challenge facing Chinese higher education in the next two decades. International Higher Education 80, 11–12.
Mollick, E., 2014. The dynamics of crowdfunding: an exploratory study. Journal of Business Venturing 29, 1–16.

Mowery, D., Ziedonis, A., 2002. Academic patent quality and quantity before and after the Bayh–Dole act in the United States. Research Policy 31, 399–418.

Mowery, D., Nelson, R., Sampat, B., Ziedonis, A., 2001. The growth of patenting and licensing by U.S. universities: an assessment of the effects of the Bayh–Dole Act of 1980. Research Policy 30, 99–119.

Mowery, D.C., Nelson, R.R., Sampat, B.N., Zeidonis, A.A., 2004. Ivory Tower and Industrial Innovation. Stanford University Press, Stanford.

Morby, A., February 10, 2014. Cost Rises Squeeze Builders Profit Margins. Construction Enquirer.

Moreno, M.V., Úbeda, B., Skarmeta, A., Zamora, M., 2014. How can we tackle energy efficiency in IoT based smart buildings? Sensors 14, 9582–9614.

Mudarri, D., 2014. National expenditures, jobs, and economic growth associated with indoor air quality in the United States. Journal of Environmental Health 26–31.

Nehdi, M., 2002. Crisis of civil engineering education in information technology age: analysis and prospects. Journal of Professional Issues Engineering Education and Practice 128 (3), 131–137.

O'Shea, R.P., Allen, T.J., Chevalier, A., Roche, F., 2005. Entrepreneurial orientation, technology transfer and spinoff performance of U.S. universities. Research Policy 34 (7), 994–1009.

Ouimet, P., Zarutskie, R., 2014. Who works for startups? The relation between firm age, employee age, and growth. Journal of Financial Economics 112, 386–407.

Pacheco-Torgal, F., 2014. Eco-efficient construction and building materials research under the EU Framework Programme Horizon 2020. Construction and Building Materials 51, 151–162.

Pacheco-Torgal, F., Mistretta, M., Kaklauskas, A., Granqvist, C.G., Cabeza, L.F., 2013b. Nearly Zero Energy Building Refurbishment: A Multidisciplinary Approach. Springer-Verlag London Ltd.

Pacheco-Torgal, F., Diamanti, M.V., Nazari, A., Granqvist, C.G., 2013a. Nanotechnology in Eco-efficient Construction. Woodhead, Cambridge, UK.

Pentland, A., 1996. Smart Rooms. Scientific American, pp. 54–62.

Quapp, U., Holschemacher, K., 2013. Efforts to reduce the drop-out rate in civil and structural engineering programs. In: Research and Applications in Structural Engineering, Mechanics and Computation. Proceedings of the 5th International Conference on Structural Engineering, Mechanics and Computation, SEMC 2013, pp. 2545–2548.

Roberts, E., 2014. In foreword. Allen, T. and ÓShea, R. (2014) Building Technology Transfer within Research Universities: An Entrepreneurial Approach. Cambridge University Press.

Rockström, J., Steffen, W., Noone, K., Persson, Å., Chapin III, F.S., Lambin, E., Lenton, T.M., Scheffer, M., Folke, C., Schellnhuber, H., Nykvist, B., De Wit, C.A., Hughes, T., van der Leeuw, S., Rodhe, H., Sörlin, S., Snyder, P.K., Costanza, R., Svedin, U., Falkenmark, M., Karlberg, L., Corell, R.W., Fabry, V.J., Hansen, J., Walker, B., Liverman, D., Richardson, K., Crutzen, P., Foley, J., 2009. Planetary boundaries: exploring the safe operating space for humanity. Ecology and Society 14 (2), 32.

Roy, S., 2014. Reengineering our vision: breaking through the paradoxical crisis of unemployment. International Journal of Human Resource Management 1, 11–17.

Rothe, C., 2006. Using patents to advance the civil engineering profession. Civil Engineering 67–73.

Roure, J.B., Keeley, R.H., 1990. Predictors of success in new technology based ventures. Journal of Business Venturing 5 (4), 201–220.

Sadler, D., 2015. The Challenges Facing Chinese Higher Education. And Why They Matter. The Observatory of Borderless Higher Education.

Santamouris, M., Synnefa, A., Karlessi, T., 2011. Using advanced cool materials in the urban built environment to mitigate heat islands and improve thermal comfort conditions. Solar Energy 85, 3085–3102.

Schmid, G., 2015. Youth Unemployment in India: From a European and Transitional Labour Market Point of View. IZA Policy Paper No. 95. Institute for the Study of Labor, Bonn, Germany.

Schumpeter, J., 1934. The Theory of Economic Development (R. Opie, Trans.). Harvard University Press, Cambridge.

Schwienbacher, A., Larralde, B., 2012. Crowdfunding of small entrepreneurial ventures. In: Cumming, D. (Ed.), The Oxford Handbook of Entrepreneurial Finance. Oxford University Press, New York.

Scto, K.C., Buneralp, B., Hutyra, L.R., 2012. Global forecasts of urban expansion to 2030 and impacts on biodiversity and carbon pools. PNAS 17–21.

Siegel, D.S., Veugelers, R., Wright, M., 2007. Technology transfer offices and commercialization of university intellectual property: performance and policyimplications. Oxford Review of Economic Policy 23, 640–660.

Smith, G., Granqvist, C., 2011. Green Nanotechnology: Solutions for Sustainability and Energy in the Built Environment. Taylor & Francis Group.

Stagars, M., 2014. University Startups and Spinoffs. Guide for Entrepreneurs in Academia. Springer.

STAC, 2014. The Future of Europe Is Science. A report of the President's Science and Technology Advisory Council (STAC). https://ec.europa.eu/programmes/horizon2020/en/news/report-pres-barrosos-science-and-technology-advisory-council-stac-future-europe-science-oct2014.

Stephan Schmidheiny with BCSD, 1992. Changing Course: A Global Perspective on Development and the Environment. MIT Press, Cambridge, MA.

Stemler, A., 2013. The JOBS Act and crowdfunding: harnessing the power—and money—of the masses. Business Horizons 56, 271–275.

Swart, R., December 10, 2013. World Bank: Crowdfunding Investment Market to Hit $93 Billion by 2025. http://www.pbs.org/mediashift/2013/12/worldbankcrowdfunding investmentmarkettohit93billionby2025/.

Teixeira, A., Coimbra, C., 2014. The determinants of the internationalization of Portuguese university spin-offs: an empirical investigation. Journal of International Entrepreneurship 12, 270–308.

Thiel, P., Masters, B., 2014. Zero to One: Notes on Startups or How to Build the Future. Crown Business, New York.

Thurik, A., Stam, E., Audretsch, D., 2013. The rise of the entrepreneurial economy and the future of dynamic capitalism. Technovation 33, 302–310.

Torpy, F., Irga, P., Burchett, M., 2014. Reducing indoor air pollutants through biotechnology. In: Pacheco-Torgal, F., Labrincha, J., Diamanti, M., Yu, C.-P., Lee, H.K. (Eds.), Biotechnologies and Biomimetics for Civil Engineering. Springer, London, pp. 181–210.

UN Department of Economic and Social Affairs, 2004. Population Division. World Population in 2300. United Nations, New York.

Visintin, F., Pittino, D., 2014. Founding team composition and early performance of university—based spin-off companies. Technovation 34, 31–43.

Wagner, T., 2012. Creating Innovators: The Making of Young People Who Will Change the World. Scribner/Simon & Schuster.

Wang, Z., Zhang, J.S., 2011. Characterization and performance evaluation of a full-scale activated carbon-based dynamic botanical air filtration system for improving indoor air quality. Build Environ 46, 758–768.

Winograd, M., Hais, M., 2014. How Millenials Could Upend Wall Street and Corporate America. Governance Studies and Brooking, pp. 1–19.

Wilson, E.O., 2003. The Future of Life. Vintage, New York, US.

WHO, 2014. Urban Population Growth. Global health observatory. http://www.who.int/gho/urban_health/situation_trends/urban_population_growth_text/en/.

Wong, J.K.W., Li, H., Wang, S.W., 2005. Intelligent building research: a review. Automation in Construction 14 (1), 143–159.

YC, 2015. Y Combinator. https://www.ycombinator.com/atyc/.

Zurich, 2011. On the Way to the 2000-watt Society. www.stadt-zuerich.ch/2000watt.

Part One

Business plans, start-up financing, marketing, creativity and intellectual property marketing

Business plan basics for engineers

2

S. Tanev, E.S. Rasmussen, K.R. Hansen
University of Southern Denmark, Odense, Denmark

2.1 Introduction

2.1.1 What makes business planning for engineers so unique?

What would be the context that would require discussing the nature of business planning activities with a specific focus on engineers? One of the possible contexts is the one of engineering firms interested in refreshing or refining their existing business plans. There is also the entrepreneurial context, when engineers become entrepreneurs in a technology-driven business environment aiming at creating and growing a business focusing on the development of new technological products and services. Engineers who have become entrepreneurs typically develop a business that reflects their expertise or previous research interests. Thus the product or service they are willing to commercialize has often been part of their research and development (R&D) passion for many years. They would therefore often ramp up their business creation from an R&D perspective, approaching the entrepreneurial process with a "technology in search of a marketplace" mindset as opposed to a "need in search of a solution" mindset (Servo, 2005). The "need in search of a solution" mindset has its starting point in the marketplace where the outcome consists of incremental innovations whereas the "technology in search of a marketplace" is usually developed with a focus on the new technology itself being more disposed to result in radical or disruptive innovations. Quite often in such situations the specific product or service developed by the engineers is new to both the firm and to the world, thus requiring the exploitation of new or emergent markets.

This chapter focuses on the nature of business planning activities from an entrepreneurial perspective. This perspective, however, is not limited to technology start-ups or newly created engineering firms. It is equally relevant for established firms investing in projects that assemble and deploy highly qualified human resources and heterogeneous assets that are intricately related to advances in scientific, engineering, and technological knowledge for the purpose of creating and capturing value for the firm (Bailetti, 2012). We can speak, therefore, of a business planning approach that could be applied to a broader entre-/intrapreneurial context of technology-driven business environments. The adoption of such perspective implies multiple challenges for engineering professionals, which are related to the various uncertainties and risks they are forced to deal with.

Start-Up Creation. http://dx.doi.org/10.1016/B978-0-08-100546-0.00002-9

2.1.1.1 Uncertainties and risks typical of technological business environments

There are multiples ways of conceptualizing and categorizing uncertainty and risk. It should be pointed out, however, that we should start by making a distinction between uncertainty and risk (Schmidt and Keil, 2012). In a situation of risk, the future is known in terms of statistical probabilities expressed in the form of means and distributions. Uncertainty, on the other hand, characterizes a situation where neither means nor distributions can be known. In an environment characterized by uncertainty, it is not possible to have more accurate information in a strict sense. What really matters is the interpretation of information. Firms and managers use pieces of information to classify future states subjectively and form beliefs about these states or estimates of them, essentially transforming the situation of uncertainty into one of risk (Schmidt and Keil, 2013). Entrepreneurs use their personal judgment, their ability to integrate many bits of information, to view objectively the various aspects of their particular situation and conceptualize alternative feasible futures. The emphasis on the ability of managers to transform situations of uncertainty into situations of risk is particularly relevant within the context of technology-driven or engineering businesses.

Maurya (2012) adopted a product development perspective on defining uncertainty and risk. According to him, building a successful product is fundamentally about risk mitigation. Technology start-ups, new engineering firms, or new R&D-based product development projects in existing firms are a risky business, and the real job of the managers is to systematically de-risk the key business activities over time. The biggest risk of all for a new firm developing a new product is building something that nobody wants. Maurya refers to Douglas Hubbard (Hubbard, 2014) in order to make a clear distinction between uncertainty and risk. Uncertainty is the lack of compete certainty; that is, the existence of more than one possibility. Risk is a state of uncertainty where some of the possibilities involve a loss, catastrophe, or other undesirable outcome (Maurya, 2012, p. 49). This is why the way to quantify risk when developing a new product and designing a new business model is by quantifying the probabilities of a specific outcome along with quantifying the associated loss of being wrong about something associated with that outcome.

Recent publications have explicitly discussed the nature of uncertainty in technology-driven business environments (Tanev et al., 2015). There are, for example, technological, market, competitive (Allen, 2010; Yadav et al., 2006), and resource uncertainties (Arteaga and Hyland, 2013). The technological uncertainty is associated with issues such as whether or not the product will function as promised; the delivery timetable will be met; the vendor will give high-quality service; there will be side effects of the product; or the new technology will make the existing technology obsolete. The market uncertainty is associated with issues such as the kind of needs that are supposed to be met by the new technological product and how these needs would change in the future; whether or not the market would adopt industry standards; how fast the innovation will spread and how large a potential market niche is. The uncertainty of the competitive environment is associated with the inherent competitive volatility of high-technology markets. It refers to changes in the competitive landscape such as newly emerging competitors, their product offerings, or the tools they use to compete as well as new defensive intellectual property (IP) protection moves by competitors. Resource uncertainty is associated with the lack of information about the availability of funding, specific types of human resources, and competency gaps about product commercialization in a specific global context; innovation talent and relevant technology development and commercialization partners at a specific global location; and so on (Arteaga and Hyland, 2013, p. 51).

Furr and Dyer (2014) have pointed out that greater degrees of uncertainty in the overall global business environment has created the need to change the way most organizations and especially start-ups are managed. According to Furr and Dryer there are three types of uncertainty that influence a firm's ability to create customers: demand uncertainty; technological uncertainty; and environmental uncertainty, which is associated with the overall macroeconomic environment and government policy. Sarasvathy et al. (2014) focus on the uncertainties associated with running a business internationally or on a global scale. According to them there are at least three characteristics of conducting cross-border business activities. The first one is the need to explicitly deal with cross-border uncertainty. The second is the need to leverage limited resources. Operating by leveraging limited resources within a context involving political, economic, and sociocultural risks is particularly challenging for new technology-based firms. The third characteristic is related to the challenges associated with taking into account network dynamics since "creating, maintaining, growing, and managing networks, whether at the individual, organizational, or inter-organizational level, becomes more challenging across borders because of geographic and cultural distance" (Sarasvathy et al., 2014, p. 76).

The different types of risk could be categorized in different ways. It should be admitted that the different categorization schemes could sometimes be overlapping but still useful as an exploratory lens helping the risk management process. For example, Maurya (2012) suggests that risks in a start-up company could be divided into three general categories: product risk (getting the product right), customer risk (building a path to customers), and market risk (building a viable business). Adner (2012) points out, however, that an overemphasis on the risk associated with the internal execution of a firm creates an innovation blind spot. By adopting an ecosystem perspective, Adner suggests using two additional types of risk: coinnovation and adoption chain risks. Coinnovation risk is the extent to which the success of an innovation depends on the successful commercialization of other innovations. Adoption chain risk is the extent to which partners will need to adopt a firm's innovation before end-consumers have a chance to assess the full value proposition. Interestingly, Girotra and Netessine (2014) have suggested a categorization of risk as part of a business model development and innovation framework that could be related to the categorization suggested by Adner (2013). According to them the key choices executive managers or entrepreneurs make in designing a business model either increase or reduce two characteristic types of risk: information risk, when an entrepreneur makes strategic operational decisions without enough information, and incentive-alignment risk, when entrepreneurs need to make assumptions about the expected incentives of all the relevant stakeholders involved in the company value creation network. Girotra and Netessine (2014) are fully aware that there are other types of risk such as financial and technological, but believe that by mitigating information and incentive-alignment risk firms can improve their ability to tolerate all other risk categories. The reason to mention the approach suggested by Girotra and Netessine is twofold. First, their framework relates the management of risk and business model development, which is quite relevant for the context of early-stage technology-based or engineering businesses. Second, the explicit articulation of incentive-alignment risk provides additional support to the categorization of risk in the wide-lens approach to innovation suggested by Adler (2012), which specifies the need for the alignment of the relative benefits of all potential stakeholders involved in technology-driven business ecosystems.

2.1.1.2 Three primary challenges: financing, sizing markets, and intellectual property management

The interplay between these uncertainties and risks in new or existing engineering firms could result in many unique issues that could be structured under three primary challenges associated with financing, sizing markets, and IP management (Servo, 2005).

The challenge of financing

From a financial perspective, the initial R&D phase of a technology-driven or engineering business is costly and risky. There might be a lot of uncertainty associated with the development of the technology and its specific functionality. Furthermore the market adoption rate will be difficult to determine when the technology is new. The combination of high cost and high risk is not a formula that attracts investors. The high amount of technological and market uncertainty keeps private sector investors away simply because they prefer that risks are reduced before providing capitalization. Capitalization during the initial R&D phase can therefore be challenging for engineers and requires a focus on risk mitigation during the development phase. Engineers should be very careful, therefore, when considering the value of raising venture capital at the early stages of their businesses. Once the R&D phase is complete, they must find partners and funding mechanisms to support engineering, manufacturing, marketing, sales, and distribution. This leads to a number of choices that require familiarity with the pros and cons of a wide range of equity and debt financing methods (Servo, 2005).

The challenge of sizing markets

The R&D context of technology-driven and engineering businesses forces their managers to deal with not only new technologies but also new or emergent markets. Dealing with new technologies and new or emergent markets results in difficulties when the market size and adoption process is to be determined. Unlike entrepreneurs who are providing a solution to an existing market where relevant market data can be easily found, technology entrepreneurs have to build a hypothetical construct to determine the market size for their products and the rate of growth.

The rate of acceptance depends not only on nailing the problem/solution fit, but also on legislations, breakthroughs in other related technologies, and amount of training the customers need before using the product. These factors together create market risk in terms of adoption chain risk and coinnovation risk, which needs to be assessed. The new technology that is being commercialized might constitute a high reward opportunity, but at the same time it is also a high-risk business environment where blind spots are lurking around every corner because of the high amount of uncertainty. The truth is that even the most brilliant innovation cannot succeed when its value creation depends on other innovations or if you do not reach your customers through the right channels. The global aspect of a business complicates even further the marketing perspective when cross-border uncertainties become quite relevant (Sarasvathy et al., 2014).

The challenge of intellectual property management

Creating a business from an R&D perspective might result in a technology or product that is patentable. Patents together with other forms of IP such as trade secrets, copyright, and trademarks are valuable assets not only for the engineering entrepreneurs, but also for potential investors. IP management depends on several factors such as technology or industry sector, size and maturity of the business, technology lifecycle, and the business and market environment (Wilton, 2011, p. 5). The IP strategy should be aligned with the business strategy from the beginning since it might constitute an important source for generating increased returns on R&D investments and added business value (Wilton, 2011). At the same time, the risks adopted by engineers during the commercialization of a patent are generally poorly understood.

There is a high degree of personal risk, which is a function of the chance that the project will fail, and the amount of resources invested by the inventors. The inventors should therefore try to understand risk from an objective perspective by making realistic assessments of the likelihood of success. Technology risk on the other hand is a function of the specific scientific or engineering field and the extent of novelty. In addition, there is commercial risk, which is described by the possibility that despite overcoming its technical milestones, the invention does not become a success due to factors related to the market environment or competition. Commercial risks can be moderated to a certain extent by the right management skills. Inventors and investors must be aware that IP risk management is a continuous effort and does not end with the success of an invention or patent (Levy, 2006). It is therefore important that the engineers engaged in entrepreneurial activities become familiar with issues regarding IP protection in order to develop a strategy for appropriately protecting their innovations from early on. However some forms of IP protection are expensive and involve a complicated process that could be associated with additional financial and human resource management and decision-making risks (Servo, 2005).

2.2 How to approach business planning for engineers

Given the specific characteristics of engineering entre-/intrapreneurial context the business planning process could be articulated in terms of two major components or phases: (1) the articulation and shaping of a viable business model, and (2) scaling and growth of the business.

Any entrepreneur can tell you about what the technology or the product is, about the potential customers, through which channels it could be sold, and what the costs would be. Business planning, therefore, always starts with the plan the entrepreneur has in his or her head. The initial plan usually contains a myriad of assumptions because of little initial data collection. Adding more empirical data to the initial plan most often starts to shift the initial idea in order to better fit the emerging market needs. Business planning is therefore iterative in nature and should be approached this way.

While established companies seem to focus on executing their business models according to their initial plan, start-ups are operating in a search mode, looking for a scalable business model around which to develop their plan (Blank and Dorf, 2012). This is what lean start-ups are greatest at (see chapter: Startup financing). By combining hypothesis-driven product development, agile engineering, and customer development they focus on testing their initial hypothesis about the product, running experiments and learning until the product/market fit has been reached. Product/market fit is where a start-up meets the early adopters' needs and expectations, by solving a highly valuable problem.

Scaling happens after achieving the product/market fit, when the technology or product has achieved enough traction. Traction is a measure of your product's engagement with its market. Investors care about traction over everything else (Maurya, 2012). This is where there is a visible transition from early adopters to mainstream customers. In *Crossing the Chasm*, Geoffrey Moore (1991) emphasizes that this is a challenging point and the chasm should be crossed in order to scale. Once the mainstream market is reached, everything changes.

Today there are several approaches to establishing start-ups; however there seems to be a dearth of tools that could help start-ups to scale. As Daniel Isenberg asks, "Which is more important, giving birth or raising children?"[1] He further claims that current entrepreneurship policy favors quantity of start-ups at the expense of scale-ups. Focusing on the challenges associated with the transformation from being a start-up to a scale-up, therefore, constitutes a critical part of the business planning process.

2.3 Developing and articulating the business model: the lean canvas approach

The lean canvas approach has become the most popular tool for the articulation and refining of business models in both start-ups and established firms (Blank and Dorf, 2012; Osterwalder and Pigneur, 2010). It is especially popular in the case of

[1] https://hbr.org/2012/11/focus-entrepreneurship-policy#.

Table 2.1 The Business Model Canvas together with the questions that could help clarifying the key components of a business plan at the early stages of a business

Key partners	Key activities	Value proposition	Customer relationships	Customer segments
Who are our key partners? Who are our key suppliers? Which key resources are we acquiring from our partners? Which key activities do partners perform?	What key activities does our value proposition require? Our distribution channels? Customer relationships? Revenue streams? **Key resources** What key resources does our value proposition require? Our distribution channels? Customer relationships? Revenue streams?	What value do we deliver to the customer? Which one of our customers' problems are we helping to solve? What bundles of products and services are we offering to each segment? Which customer needs are we satisfying? What is the minimum viable product?	How do we get, keep and grow customers? Which customer relationships have we established? How are they integrated with the rest of our business model? How costly are they? **Channels** Through which channels does our customer segment want to be reached? How do other companies reach them now? Which ones work best? Which ones are most cost-efficient? How are we integrating them with customer routines?	For whom are we creating value? Who are our most important customers? What are the customer archetypes?
Cost structure What are the most important costs inherent to our business model? Which key resources are most expensive? Which key activities are most expensive?			**Revenue streams** For what value are our customers really willing to pay? For what do they currently pay? What is the revenue model? What are the pricing tactics?	

Osterwalder, A., Pigneur, Y., 2010. Business Model Generation: A Handbook for Visionaries, Game Changers, and Challengers. John Wiley & Sons; Blank (2013).

start-ups or new product development projects where the managers need a tool for framing their initial hypothesis and documenting their learning, as they iterate through the early stages of their business. The lean canvas approach is more flexible and rather holistic as compared to the direct focus on writing a business plan. Rather than engaging months of planning and research, managers accept that all they have on day 1 is a series of untested hypotheses and, instead of writing an intricate business plan, entrepreneurs summarize their hypothesis on the canvas. As the hypothesis get tested and validated the canvas is used as a way to evolve the business rather than to come up with a fixed solution.

The traditional and most popular canvas is the business model canvas (BMC) developed by Alexander Osterwalder (Osterwalder and Pigneur, 2010). Table 2.1 shows its nine building blocks together with a number of questions that could help in the process of their formulation. The BMC provides the main structure of a business plan with a focus on ease of use, flexibility, and transparency, including the key drivers of a business: customer segments, value propositions, channels, customer relationships, revenue streams, key activities, key resources, key partnerships, and cost structure.

There are two important practical points that should be mentioned with respect to the use of the BMC. First, the BMC is just a tool that helps the initial formulation and the continuous refinement of the business model as a key component of the business plan. It is not the filling up of the building blocks on the canvas that will make a business successful, but the proper managerial actions and activities corresponding to them. Second, the BMC is not a dogmatic framework but just a starting point that could be modified or refined depending on the specific business context, technological solution, or customer base. This is why there are publications suggesting modified versions of the canvas that could be better adapted to specific business circumstances. Examples of such modified versions are the lean canvas (Maurya, 2012) and the business model snapshot (Furr and Dyer, 2014). Both of them focus on providing more systematic tools to mitigate risk in new product, service, and business development. The lean canvas approach proposed by Ash Maurya (Table 2.2) appears to be more intuitive and better suited to address the multiple uncertainties and risks (Section 2.1.1.1) that are typical of the context of new technology start-ups and engineering professionals in technology-based businesses. The main objective behind its introduction was to make it as actionable as possible while staying as close as possible to the entrepreneurial context. The way to making the canvas actionable was to focus its intended use on capturing which was most uncertain and most risky.[2]

The lean canvas is shown in Table 2.2. It helps in deconstructing the business model into nine distinct subparts that are then systematically tested, starting from the highest and moving to the lowest risks. Following the road map proposed in the lean canvas approach, it is important to emphasize one of its key assumptions: your product is not the technological solution you are providing; your product is the business model. Ash

[2] http://leanstack.com/why-lean-canvas/.

Table 2.2 The lean canvas (the numbering indicates the order in which the different building blocks are usually addressed)

<table>
<tr>
<td rowspan="2">1. Problem
Top 3 problems</td>
<td>4. Solution
Top 3 features</td>
<td colspan="2" rowspan="2">3. Unique value proposition
Single, clear compelling message that states why you are different and worth paying attention</td>
<td>5. Unfair advantage
Can't be easily copied or bought</td>
<td rowspan="2">2. Customer segments
Target customers</td>
</tr>
<tr>
<td>8. Key metrics
Key activities you measure</td>
<td>9. Channels
Path to customers</td>
</tr>
<tr>
<td colspan="3">7. Cost structure
Customer acquisition costs
Distribution costs
Hosting
People etc.</td>
<td colspan="3">6. Revenue streams
Revenue model
Life time value
Revenue
Gross margin</td>
</tr>
</table>

Maurya, A., 2012. Running Lean: Iterate from Plan a to a Plan that Works, O'Reilly Media, Incorporated, Sebastopol, CA.

Maurya has built the road map based on the three key stages of a start-up: problem/solution fit, product/market fit, and scaling. Every start-up runs through each of these stages where risk mitigation through each stage should be the main focus of its activities. In the problem/solution fit stage the focus is on finding out if there is a problem worth solving. The first stage can be navigated without even building a product. Instead a demo can be developed (a screen shot, video, or a physical prototype) that could engage customers in sharing their vision about how to solve the problem. The demo should help the customer visualize the solution by demonstrating the unique value proposition (UVP). During this stage, entrepreneurs attempt to answer the key question (the existence of a problem worth solving) by using a combination of qualitative customer observation and interviewing techniques on the basis of which they derive the minimum feature set to address the right set of problems, which is also known as the minimum viable product (MVP). The problem/solution fit is validated when you repeatedly get the customers to accept the UVP.

During the second stage (product/market fit) the key question is whether the company has built something that people want. Once there is a problem worth solving and an MVP has been built, we need to test how well the solution solves the problem. The first significant milestone is achieving market traction. At this stage, the initial plan should start working—there are customers who are signing up, the company retains them, and gets paid.

The third stage (scaling) focuses on acceleration, growth, and scaling the business model. Before product/market fit, the focus of a start-up centers on learning and pivots (substantial changes in the initial idea). After product/market fit, the focus shifts toward growth and optimizations. In this sense, achieving product/market fit is the first significant milestone of a start-up, which greatly influences both its strategy and tactics.

The overall framework is organized around three metaprinciples: (1) documenting your plan A; (2) identifying the riskiest part of your plan; and (3) systematically testing your plan. Documenting plan A is a snapshot of the initial plan. Entrepreneurs start by documenting their plan A, focusing on the identification of a potential customer segment and then continue sketching out their first guess at the business model for each customer segment, potential solution, or customer channel. When this first step is done, an entrepreneur will have multiple versions of plan A, which is ready to be prioritized in terms of risk in the next step. Having developed multiple canvases, the focus in the next step is on prioritizing where to start. The prioritizing should be undertaken by focusing on ranking the business models in terms of lowest risk. The objective is to find a business model with a big enough market that can be reached, with customers who need the product, around which a viable business can be built (Maurya, 2012).

The weighting criteria for prioritizing the risk are the customer pain level (the problem), the extent of reach (the channels), the price/gross margin (revenue streams and cost structure), market size (the customer segment), and the technical feasibility (the solution). These criteria are evaluated against the three types of risks suggested by Maurya (2012): product risk—getting the product right; customer risk—building a path to customers; and market risk—building a viable business. The lean canvas automatically captures uncertainties that are related to risk in terms of loss of opportunity costs and real costs. The final result is a lean canvas that captures the key business components that could be further tested and validated. Systematically test the initial plan is the third step, where all assumptions made during the articulation of the business model are transformed into hypotheses that can be either validated or invalided through running experiments with customers.

2.4 Scaling up the business

Once the business model has been developed and validated it is time to grow the business. Following a method such as the lean canvas approach the managers will get past the early stages of growth. However, the transition from focusing on exploration to execution of a business model will change the company in fundamental ways, which involves new challenges.

One of the main challenges for managers of new technology-based or engineering firms is managing the transition to growth. At this stage a start-up should have nailed the product and the business model and many unknowns should have been clarified. In this sense, the amount of uncertainty declines and so does the reason for applying purely entrepreneurial management practices. The focus shifts on applying more traditional management principles focusing on execution, value capture, and optimization. However the start-up may enter a transitional phase where neither entrepreneurial nor traditional management alone is appropriate. In order to master the scaling process the start-up needs to blend in the two management practices as it transitions to a mature business (Furr and Dyer, 2014).

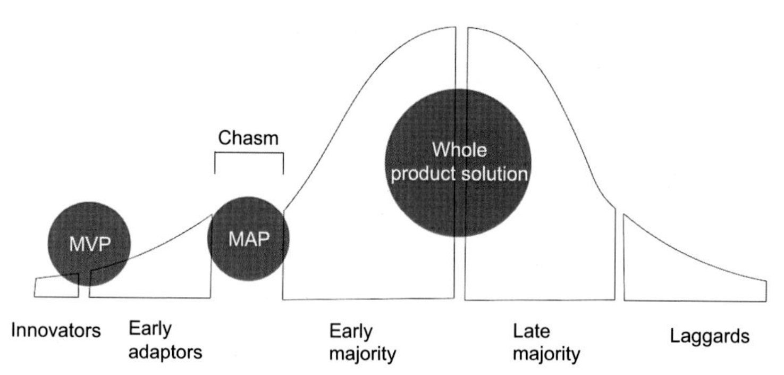

Figure 2.1 Minimum viable product (MVP) and minimum awesome product (MAP) versus whole product solution across the technology/product adoption life cycle.
Adapted from: Furr and Dyer (2014).

The transition from a start-up to a mature business requires going through several stages, leading to fundamental changes in the way it operates. While there are a lot of tools available on how to develop a business, there are very few tools dealing with how to manage the scaling process in a systematic way. There is, however, an agreement among practitioners that the growth and scaling process can be organized around three key areas: market, processes, and team transitions.

2.4.1 Market scaling

The main challenge for new technology firms is crossing the chasm. Geoffrey Moore (1991) argues that the gap between early adopters and early majority provides a significant challenge for companies because these groups are quite different and require completely different marketing strategies. While early adopters are willing to try something entirely new, the early and late majority wants a product solution that is error-free and full-featured. The only way to cross the chasm is to put all your eggs in one basket, meaning that the strategy should be to identify a niche segment among the early majority and focus all efforts on developing the whole product solution by serving this particular segment. When this particular customer niche has adopted the solution, the firm can focus its effort on a second customer niche. The key to getting a foothold

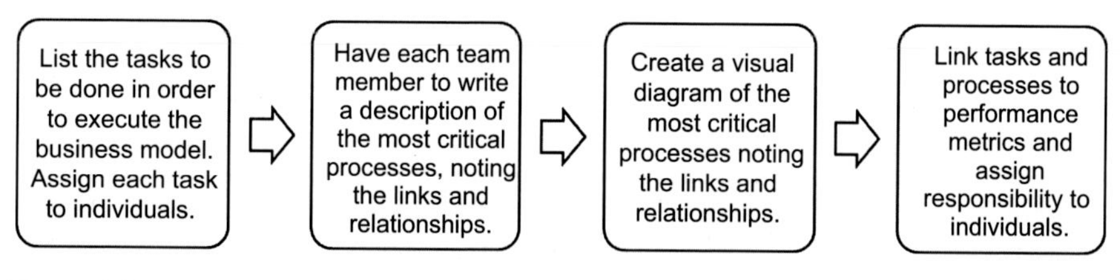

Figure 2.2 Scaling the process.
Adapted from: Furr and Dyer (2014).

in the mass market is to use the initial customer segment as reference customers. Thereby the firm can start shaping all marketing communications to position itself as a market leader in order to derisk mass-market adoption. The key to successfully redefining the market or create a market leader position is to choose an unoccupied space where there is a legitimate market need (Furr and Dryer, 2014) (Moore, 1991).

In order to attract the first customer niche among the early majority, the firm should focus on developing a minimum awesome product (MAP) (Furr and Dryer, 2014). While the MVP is used for validating the core assumptions during the initial stages among the innovators and early adaptors, the MAP is a solution that is extraordinary on the dimensions that customers value the most (Fig. 2.1). The point is to use the MVP to improve the key attributes of the solution that can evoke positive emotions and thereby turn it into a MAP by focusing on the functional, social, and emotional dimensions of the solution. While using the MAP to get the solution adopted by the early majority, the firm can move to a second customer niche. In parallel to that the MAP is further developed into the whole solution that could address the needs of the main market.

2.4.2 Process and team scaling

As a newly created technology firm grows it will begin to see the same types of problems cropping up again and again. These issues indicate a need for standardization of its processes and workflow. As the market grows, tasks and workflow need to be standardized in order to continually deliver a quality product (Furr and Dyer, 2014). In order to introduce scalable and standardized processes in an organization, a simple four-step road map can be used as a guiding tool (Fig. 2.2). The process starts by creating a list of tasks to be done in order to execute the business model. Each of these tasks is assigned to an individual. The objective of the next step is to create a common understanding among the employees. Each team member has to write a job description for the assigned task and by reviewing each of the descriptions together in the team, people agree on how to perform certain tasks and who is responsible for what. The third step is about visually mapping out the most critical processes in order to detect critical linkages. The diagram will help establish a common understanding about the most critical aspects of the processes as well as make sure that someone is assigned the responsibility for each of the key processes. The last step is to establish key metrics for the tasks and the processes and make sure someone is accountable for those metrics.

It is critically important for the new firm to shift the performance metrics when scaling. While the discovery phase should be using "love metrics" such as activation, retention, and payment, the focus should now shift to using growth metrics. Growth metrics are focused on determining whether the firm delivers a reliable solution with increasing economics of scale and can include more detailed measures of users in terms of acquisition and referral (Maurya, 2012), measures of the efficiency of the processes and the revenue growth (Furr and Dyer, 2014). By measuring the performance around the right metrics and reporting the results, the firm can improve its processes significantly.

During the start-up phase the team most often consists of people who possess good discovery skills. However as the start-up scales up, there is a need for expertise profiles that can execute the business model. There could be a need therefore for the talent pool to change. Building the team during the growth phase requires a mix of discovery-oriented people and experts.

As the team grows, it is important to focus on creating a working culture around which communication processes and activities are shaped. As the team grows it is furthermore important to plan and structure how meetings should be organized in order to secure that everyone is moving in the same direction.

2.4.3 The danger of getting things wrong

Even though new technology-based firms have a great potential for wealth, value, and job creation, there is enough evidence showing that, on average, 90% of them fail. What is the main cause of such high percentage of failures? According to the experts at Startup Genome (https://startupgenome.co/), the failures could be related to issues associated with premature scaling. Furr and Ahlstrom (2011) define this problem as "spending money beyond the essentials on growing the business (e.g., hiring sales personnel, expensive marketing, perfecting the product, leasing offices, etc.) before nailing the product/market fit." In addition, start-ups "are doing good things but doing them out of order. In other words, they are doing things that seem to make sense, like investing to build the product, hiring good people to help them sell it, developing marketing materials, and essentially doing all the kinds of things that big companies with lots of resources do when they are executing on a known opportunity" (Furr and Ahlstrom, 2011). The problem is that the risk associated with these investments could be justified only by extensive preexisting market research or sales data. Instead of assessing the risks and opportunities objectively and scaling those investments accordingly, start-ups tend to rely on guesswork without really looking into the real facts. It is true that many start-ups bring products that are new to the market; that is, they lack substantial market research and sales data. This is, however, exactly why they need to manage the scaling process in a more structured way. Another study has reported the top 20 most common reasons for start-ups to fail (CB Insights, 2014). Some of them are as follows: there was no market need; the firm ran out of cash; it did not have the right team; it got outcompeted; it got the pricing/cost wrong; the product design was poor; there was a need/lack of a business model; the marketing was poor; the customers were simply ignored; and so on. Interestingly, all of these issues could be related to the first stage (articulation and shaping of the business model) or the second stage (the proper scaling up of the business) of the business planning process.

2.5 A business plan template

The business plan serves as the executable plan for the start-up (Faley, 2015). While the phrase "business plan" conjures images of 60 pages of documents full of dense charts and diagrams, the business plans of today have become a lot shorter (Blumberg, 2013; Anthony, 2014; Faley, 2015). From 60-page documents they now often come in the form of slide decks or shorter executive summaries. Furthermore, the key of a successful business plan in today's rapidly changing environment is to consider the plan as the definite pivot point of the company, which means that the business plan should be treated as a dynamic tool by updating it as new learning emerges (Faley, 2015).

Besides being an executable plan for a new firm another fundamental aspect of the business plan is to keep everyone involved on the same page. For a start-up there are two primary stakeholders to articulate the business plan to: investors and employees. While investors focus on the size of the market opportunity and the likely revenue, employees are more interested in knowing the shape of their future work life (Blumberg, 2013). Two versions of the business plan can therefore be articulated: a version for investors focusing on the likely revenue of the opportunity and a version for employees focusing on articulating the overall vision and strategy that guide all employees toward the same goal during the initial phase.

2.5.1 A mini business plan for investors

Articulating a business plan to investors can be done either in a written document or by conducting an oral investor presentation using slide decks (Blumberg, 2013). There are several ways to structure the content of the business plan; however at this early stage the start-up should at least be able to articulate the size of the opportunity, the competitive advantages, current status and road map from today, and the strengths of the team (Blumberg, 2013).

Anthony (2014) proposes a structure for organizing a mini business plan, which articulates the elements of the business model and the aspects of scaling. A mini business plan template can be downloaded at www.innosight.com/first-mile/index.cfm. It contains the following components:

- An executive summary or a pitch if presented orally to investors
- The target customers and their problems
- The proposed solution
- The key business model elements
- The plan to scale the idea
- The thumbnail financials
- The critical assumptions
- The proposed testing plan

The insights from the early discovery-based stages as well as the required scaling activities could be easily articulated by using this template.

2.5.2 Key points in the business plan for employees

The challenges for a company in the process of scaling up are well known. Furr and Dyer (2014) recommended using the so-called V2MOM (vision, values, methods, obstacles and measures) scaling tool to simplify the process. It is a strategic planning tool that allows a start-up or an established company to define goals and organize ways to execute them. The V2MOM tool ensures that everyone is moving toward the same goal, regardless of the size of the company. The tool is therefore especially well suited for companies placed in a rapidly changing environment where there is a need to evolve continuously.

The articulation of a vision helps define what the company wants to do. The values refer to the principles and beliefs that guide the company toward its vision. The methods illustrate the actions and steps that everyone needs to take to achieve the vision. The obstacles identify the challenges and issues that have to be overcome to achieve the vision. The measures define the results aimed to achieve. V2MOM therefore helps the company to define where it wants to go, what things are important, what it will do to get there, what would prevent it from getting there, and how it knows whether it is successful or not (Furr and Dyer, 2014).

Salesforce.com used V2MOM in their early days as part of their business plan and is still using it to guide the overall organizational goal of today (Furr and Dyer, 2014). The structure of the tool can therefore be used in all phases of the life cycle of an organization.

2.6 Conclusion

This chapter focused on describing the basics of the business planning process in the context of engineering professionals. One of its key contributions is the detailed discussion of the unique characteristics and challenges of technology-driven business environments that are typical of engineering professionals. One of the key characteristics of such environments is the combination of multiple uncertainties and risks that could potentially affect the initiation and the evolution of a newly created technology-based or engineering business. The main challenge in addressing the multiple uncertainties and risks is that they have completely different sources and nature. For example, there is interplay between the uncertainties associated with the degree of newness of the technology and the early stage of a new technology firm. Fortunately, the recent growth of interest in the lean start-up approach, agile technology/product development, and hypothesis-driven technology entrepreneurship has resulted in some key publications, frameworks, and models focusing on the implementation of a rigorous scientific approach to uncertainty management.

This chapter reflects this trend and hopes to offer a brief introduction of how it could be implemented within the context of the business planning process. The business planning process consists of two major parts: articulation and development of a viable business model, and managing the growth and scaling up of the business. The development of a viable business model is described through the lean canvas approach suggested by Maurya (2012). The lean canvas approach is specifically designed to address the unique uncertainties and risks associated with the development and the introduction of new products by newly created firms. It works with a simple risk categorization focusing on product, customer, and market risks. It allows us, however, to address any other types of risks such as the ones that are discussed here and typical of technology-driven and engineering business environments.

The growth and scaling up of the business is described as consisting of three major components: market scale-up, team scale-up, and process scale-up. Our description refers to a practical tool suggested by Furr and Dyer (2014), which should help the operationalization of the business scale-up process in the context of engineering businesses. Without pretending to offer an exhaustive picture of the complex process of business development and planning, the chapter is an expression of a vision that emphasizes the fact that business development and planning knowledge and skills should become part of the culture of present-day engineering professionals.

References

Adner, R., 2012. The Wide Lens: What Successful Innovators See That Others Miss. Penguin Group.

Allen, K., 2010. Entrepreneurship for Scientists and Engineers. Prentice Hall.

Anthony, S., 2014. The First Mile: A Launch Manual for Getting Great Ideas into the Market. Harvard Business Review Press.

Arteaga, R., Hyland, J., 2013. Pivot: How Top Entrepreneurs Adapt and Change Course to Find Ultimate Success. John Wiley & Sons.

Bailetti, T., February 2012. Technology entrepreneurship: overview, definition, and distinctive aspects. Technology Innovation Management Review 5−12.

Blank, S., Dorf, B., 2012. The Start-up Owner's Manual: The Step-by-step Guide for Building a Great Company. K&S Ranch Incorporated, Pescadero, CA.

Blumberg, M., 2013. Start-up CEO: A Field Guide to Scaling Up Your Business. Wiley.

CB Insights, 2014. The Top 20 Reasons Startups Fail. Available at: https://www.cbinsights.com/blog/startup-failure-post-mortem (accessed 3.10.15.).

Faley, T., 2015. The Entrepreneurial Arch: A Strategic Framework for Discovering, Developing and Renewing Firms. Cambridge University Press.

Furr, N., Ahlstrom, P., 2011. Nail It Then Scale It: The Entrepreneur's Guide to Creating and Managing Breakthrough Innovation. NISI Institute.

Furr, N., Dyer, J., 2014. The Innovator's Method: Bringing the Lean Start-up into Your Organization. Harvard Business Review Press.

Girotra, K., Netessine, S., 2014. The Risk-Driven Business Model. Harvard Business Review.

Hubbard, D., 2014. How to Measure Anything: Finding the Value of Intangibles in Business, third ed. Wiley.

Levy, A., 2006. Starting up and financing your venture. In: Junghans, C., Levy, A. (Eds.), Intellectual Property Management: A Guide for Scientists, Engineers, Financiers, and Managers: A Guide for Scientists, Engineers, Financiers, and Managers. Wiley-VCH Verlag, pp. 119−142.

Maurya, A., 2012. Running Lean: Iterate from Plan a to a Plan That Works. O'Reilly Media, Incorporated, Sebastopol, CA.

Moore, G., 1991. Crossing the Chasm: Marketing and Selling High-Tech Products to Mainstream Customers. HarperBusiness (revised 1999, 2006 and 2014).

Osterwalder, A., Pigneur, Y., 2010. Business Model Generation: A Handbook for Visionaries, Game Changers, and Challengers. John Wiley & Sons.
Sarasvathy, S., Kumar, K., York, J.G., Bhagavatula, S., 2014. An effectual approach to international entrepreneurship: Overlaps, challenges, and provocative possibilities. Entrepreneurship Theory and Practice 38 (1), 71–93.
Schmidt, J., Keil, T., 2012. What makes a resource valuable? Identifying the drivers of firm-idiosyncratic resource value. Academy of Management Review 38 (2), 206–228.
Servo, J., 2005. Business Planning for Scientists & Engineers, fourth ed. Dawnbreaker.
Tanev, S., Rasmussen, E., Zijdemans, E., Lemminger, R., Limkilde, L., June 2015. Lean and global technology start-ups: linking the two research streams. International Journal of Innovation Management 19 (3), 41. http://dx.doi.org/10.1142/S1363919615400083.
Wilton, A., December 2011. Patent value: a business perspective for technology start-ups. Technology Innovation Management Review 5–11.
Yadav, N., Swami, S., Pal, P., 2006. High technology marketing: conceptualization and case study. Vikalpa 31 (2), 57–74.

Lean start-up: making the start-up more successful

E.S. Rasmussen, S. Tanev
University of Southern Denmark, Odense, Denmark

3.1 Introduction

In the 2000s a new type of literature emerged with Steven Blank and Eric Ries in front claiming that it is possible to reduce the risk of launching new products. In this chapter the literature will be grouped under the term lean start-up (LS). The core of the LS principles is to reduce waste by not using resources on hypotheses about the product or marketplace that have not been validated by the customer (Ries, 2011). It is thus extremely important to learn from the potential customers early in the process, thereby producing a solution based on customer needs and wants. All too often, entrepreneurs right from the start fall in love with their product or technology, ignoring negative feedback from customers and in the end spending years building a product based on a vision that no one else shares (Furr and Ahlstrom, 2011). To avoid this LS suggests an approach of going through an iterative process where problem, products, and customer hypotheses are developed and validated by the customers.

Jim Collins and Morten T. Hansen (2011) in their book, *Great By Choice*, state that top-performing companies "fire bullets before firing cannonballs." This can be seen as the difference between traditional business planning and the LS approach. LS is about continuously firing small bullets—testing different hypotheses about the product and business model to find out if the customers would validate them or not. First, when every aspect of the business model has been validated, the cannonball is fired and the business moves from an exploration to an execution mode and focuses on scaling. The very same argument is also at the core of the book, *Nail It Then Scale It*, by Furr and Ahlstrom (2011). Here the argument is that it is beneficial for a company to implement lean thinking in their launch process. Instead of using large amounts of both resources and money to go through the traditional business planning process, which results in the cannonball being fired on launch day, they support the idea of making small inexpensive tests about the business model with the actual customers—firing bullets. This creates an iterative process where parts of the business model are adjusted from the feedback of the tests until every part of the business model reaches a stage where it has been validated by the customers; that is, the company has nailed it.

3.1.1 How to be a successful start-up

Starting a new business—as a new firm or inside an existing firm—has always been difficult and with a high risk of failure, but these days when you often have to start

Start-Up Creation. http://dx.doi.org/10.1016/B978-0-08-100546-0.00003-0

on a global scale it is even more complicated. The main idea of the emerging LS paradigm is to provide new businesses with some tools for a more successful start-up. From a number of studies and extensive research it has been well documented that most start-ups fail. The risk of starting a new enterprise is high, especially if it is combined with the development of a new technological product that has to be commercialized and scaled after the final business model has been developed. The lean approach can help start-ups test new ideas and get customer feedback in a way that enhances the market success of their new products (Blank et al., 2013; Blank, 2007, 2011, 2013; Ries, 2011).

When launching new products or starting new companies a common piece of advice taught in business schools around the world is to write a solid business plan. This process has been considered as one of the best methods for securing a successful launch. The traditional business plan describes relevant information to the planning process such as resources, assets, strategy, competitors, and market-related factors. The plan is intended to cover all available information that can contribute to making the right choices for the business model as well as estimating the future revenue. In short, business planning typically starts with the identification of an opportunity, followed by the development of specifications, building the product, and finally selling it. The reasoning logic behind this model is that it is possible to make reasonable predictions about the customers' wants and thereby understand how a given product will be positioned in the customers' mind as well as how the new product will perform in the marketplace. Often this approach follows what is called a waterfall process; assuming that it is possible to make precise predictions, the product development is conducted as a step-by-step process leading to the final product launch.

Even though business plans have become a tool that is deeply rooted in entrepreneurial practices, there is little empirical evidence concluding that writing a business plan increases the chances of start-up success. The business planning process requires a deep understanding of the marketplace as well as the customers in order to be able to make precise predictions that can be applied in the planning. For an established and mature company with large amounts of historical data on how past products have performed as well as extensive knowledge about their target customers, the writing of business plans can be beneficial. With a plan based on solid data the odds of having made reasonable predictions increase and the company can easily allocate the right resources to the different steps in the plan, effectively leading to an efficient and controlled process (Furr and Ahlstrom, 2011).

Planning and forecasting, however, are only accurate when based on a long, stable operating history and a relatively stable environment, which is not the case of a start-up. Most often start-ups do not yet know who their customer is or what their product should be. Combined with the fact that the world becomes more uncertain, it gets harder and harder to predict the future and the traditional management techniques are no longer up to the task (Ries, 2011).

These facts have ignited the discussion of whether business planning is the optimal tool for navigating a start-up. Steve Blank (2013) criticizes the use of business planning for start-ups because "start-ups are not just a small version of large companies." While mature companies execute on a proven business model, start-ups are searching for a viable business model as entrepreneurs start out with a guess at the right problem and solution. This means that instead of executing, entrepreneurs must search for the right problem and solution.

Start-ups face high uncertainty and the absence of a business plan can be economically reasonable. The start-up does not have to spend extensive resources on planning and does not risk spending resources on some part of the plan that is based on assumptions that might prove wrong. In the best scenario the resources will be wasted but even worse the plan could result in cognitive limitations, making the entrepreneur too focused on executing the wrong plan instead of searching for other solutions.

3.1.2 What is lean in a lean startup?

Eisenmann et al. (2012) defined, from the work of Ries (2011), an LS as a firm that follows a hypothesis-driven approach to evaluate an entrepreneurial opportunity and develop a new product for a specific market niche. The LS methodology focuses on translating a specific entrepreneurial vision into falsifiable hypotheses regarding the new product as part of an emerging business model that is going to be used to deliver it. The hypotheses are then tested using a series of well-thought-out prototypes that are designed to rigorously validate specific product features or business model specifications. In this context, the entrepreneurial opportunity is based on shaping the new solution in a way that could solve a specific customer problem. The uniqueness of the methodology consists of its ability to explicitly take into account the numerous uncertainties regarding the suitability of a given solution toward a specific customer problem.

In the recent years a wide array of authors has contributed to the method's evolution by giving their take on the matter. Originally the methodology was developed with high-tech companies in mind but has since been expanded to apply to a broader category of companies looking to introduce new products to the market.

Steve Blank's (2007) introduction to the customer development process launched the LS movement. He was the first to describe how entrepreneurs should test and refine business hypotheses through customer validation. *The Startup Owner's Manual* (Blank and Dorf, 2012) describes a step-by-step process for managing the search for a new business model and provides entrepreneurs with a path from idea to a scalable business model.

Eric Ries is a former student of Blank. After being involved in several start-ups Eric Ries started to wonder why they were failing despite doing everything right in the traditional ways. He decided to try a different approach inspired by Steve Blank's Customer Development and the core idea was that the business and marketing functions of a start-up should be considered as important as engineering and product development, and therefore deserve an equally rigorous methodology (Ries, 2011). Measured against the traditional theories on product development, Ries's new ideas did not make sense, although it seemed they had a very positive influence on the performance of the start-ups. To describe his new ideas he used the term "lean" from lean manufacturing in order to emphasize the core idea behind the methodology—to eliminate the waste, the non-value-creating efforts, that he saw in start-ups around him building products that nobody wanted. After refining and developing his theories with other start-ups, writers, and thinkers, Ries published his book, *The Lean Startup* in 2011.

Two other prominent contributors to the LS methodology are Nathan Furr and Paul Ahlstrom (2011) with their book, *Nail It Then Scale It*. Both authors have been involved in multiple start-ups. By observing both failures and successes they started to see a pattern, which came to serve as the foundation of their book. They suggest a three-step process where the entrepreneur starts with a hypothesis about the customer pain and then tests it. Once the customer pain has been identified and validated, a hypothesis about the minimum feature-set necessary to drive a customer purchase should be made. From there, a series of gradually more advanced prototypes should be built, while discussing and validating with customers each step of the way. Eventually, the solution to the customer pain will be nailed and the start-up can start developing a go-to-market strategy and scaling the business.

Other authors have been publishing their refinements of the original methodology by focusing on two different aspects. The first aspect is the operationalization of the LS approach with a focus on start-ups. The most valuable example in this direction is the book, *Running Lean*, by Ash Maurya (2012), which has received a lot of attention. The second aspect is the extension of the LS methodology to a broader context including the management of new product ideation, design, development, and commercialization in established firms. Examples of books focusing on this aspect are Scott Anthony's (2014), *The First Mile: A Launch Manual for Getting Great Ideas into the Market*, Nathan Furr et al. (2014), *The Innovator's Method: Bringing the Lean Startup Into Your Organization*, and Remy Arteaga and Joanne Hyland's (2013), *Pivot: How Top Entrepreneurs Adapt and Change Course to Find Ultimate Success*.

3.1.3 The link to the business model idea

The articulation of the LS approach was complemented by the adoption of the business model canvas (BMC)[1] approach (Osterwalder and Pigneur, 2010) to form the basis for its conceptual status quo. The BMC is a visual chart including nine elements (building blocks) describing a firm's value proposition, infrastructure, customers, and finances. It assists firms in aligning their activities by illustrating all potential trade-offs and evolving their initial vision into a refined operating business model. The BMC was not developed specifically for start-ups but was later adopted by the LS community as a key reference model. The purpose of the model is not to fix or codify the initial entrepreneurial vision but to provide a tool for its continuous refinement. One of the key benefits of the model is the possibility to be adapted to a specific business context, industry sector, technological domain, and particular firm's circumstances. It is not by accident that the BMC was creatively modified by other authors, resulting in several different versions: the Lean Canvas[2] suggested by Maurya (2012) as an

[1] http://www.businessmodelgeneration.com/canvas/bmc.
[2] http://leanstack.com/lean-canvas/.

adaptation of the BMC to the context of LSs, the Business Model Snapshot suggested by Furr et al. (2014) as a simpler and more intuitive version of the BMC, and the Big Idea Canvas suggested by Paul Ahlstrom as a practical tool helping the adoption of the nail-it and scale-it process (Furr and Ahlstrom, 2011).[3] More details about some of the canvas approaches can be found in chapter "Business plan basics for engineers."

3.2 The main elements of lean start-ups

3.2.1 Overview of key elements

The LS process of validation was described initially by Steve Blank (Blank, 2007, 2011, 2012, 2013; Blank et al., 2013; Blank and Dorf, 2012) through the introduction of a customer development model (CDM). It was later popularized by Ries (2011) through the articulation of several key paradigmatic principles as part of a build-measure-learn (BML) loop framework, which he described in his book, *The Lean Startup*.

The emergence of the LS approach is based on Blank's and Ries's study of successful entrepreneurs who tended to follow the CDM model in new product development and commercialization instead of a purely product-centric development model. According to Blank (2013), one of the key starting points is to emphasize that a start-up is not a smaller version of a large company, but "a temporary organization designed to search for a repeatable and scalable business model." Eric Ries (2011) pointed to another important aspect of the LS by defining it as "a human institution designed to create new products and services under conditions of extreme uncertainty." The LS approach favors experimentation over planning, customer feedback over intuition, and iterative design over traditional business planning (Blank, 2013; Blank and Dorf, 2012). The focus on experimentation as a source of customer knowledge is associated with the concept of minimum viable product (MVP). This is a product or service consisting of a minimum set of features that is used first as a tactic to reduce wasted engineering hours and financial resources; second, as a specific commercialization strategy bringing the product to the hands of early and visionary customers as soon as possible; and third, as a specific approach to product codevelopment with customers by looking for quick adjustments of the initial product feature set in order to align in real time with specific customer needs. The MVP approach seeks to validate as many assumptions as possible about the viability of the final product before using extensive financial resources. In addition, the new venture may adjust its course in a way that may involve pivoting from the original agenda. The MVP concept is the basis for another difference of LSs as compared to traditional ones—the need for the adoption of success metrics tolerating experimentation and productive failure.

[3] http://www.nailthenscale.com/the-big-idea-canvas/.

3.2.2 Customer feedback

One of the most emphasized principles of LS is to get out of the building to learn from the potential customers. As Blank and Dorf (2012) state, "there are no facts inside the building, so get the heck outside," implying that the facts a start-up needs to gather about customers, markets, suppliers, and channels exist only "outside the building." According to Furr and Ahlstrom (2011), 90% of businesses fail because the start-up could not get anyone to buy it, not because they could not build it. A deep understanding of customers is thus important in the development of new products and services and in the establishment of a new business model for a start-up.

The way LS measures progress is through validated learning. Validated learning is the process of demonstrating empirically that the start-up has discovered valuable truths about its present and future customers. Ries (2011) states that it is much more accurate and faster than traditional market forecasting or traditional business planning. It is the answer to the problem of achieving failure by successfully executing a plan that leads nowhere. The entrepreneur should develop an attitude to learning that enables the start-up to spot new opportunities and understand how a different business model might bring more value to the customers. In the words of Furr and Ahlstrom (2011), the entrepreneurs should "maintain a seed of doubt that they may be wrong."

3.2.3 Big design or iterative design: pivot or persevere

If the assumptions tested with the customers turn out to be incorrect, the entrepreneur should be ready to make a fundamental pivot. Ries (2011) describes the pivot as "a structured course correction designed to test a new fundamental hypothesis about the product, strategy, and engine of growth." The point of the pivot is to realize when the initial assumptions about some part of the business model are wrong in order to avoid spending excess resources on moving the company in a wrong direction (Blank and Dorf, 2012).

The MVPs a start-up builds can be seen as experiments to learn about how to build a sustainable business. Reframing the purpose of the start-up to be learning what the customer wants rather than proving that the traditional business plan holds true is the first step to shred the learning traps holding many entrepreneurs back. Ries (2011) suggests a tool to facilitate this learning process described by the BML feedback loop. Through testing initial MVPs with the customers, their feedback should result in changes that steer the start-up in the right direction (Blank and Dorf, 2012). By continuously going through the loop and iterating rapidly the start-up is making incremental progress in their business model to target their customers in the right ways, thereby increasing the odds of success. The entrepreneur will at the same time be facing the most difficult question: whether to pivot the original strategy or stick to the original strategy. The answer to this question will, in part, be gauged by the metrics used by the start-up to evaluate the customer response.

3.2.4 Business planning or hypothesis testing

Standard accounting is not helpful in evaluating start-ups. Start-ups are too unpredictable for forecasts and milestones but should instead be evaluated on other measures. Startup metrics focus on tracking the start-up's progress in converting guesses and hypotheses into incontrovertible facts rather than measuring the execution of a static business plan. It is critical that management continuously test and measure each hypothesis until the entire business model is worth scaling into a company (Blank and Dorf, 2012). The first MVP should be used to establish a baseline for the different assumptions. When choosing which metrics to focus on, the ones best describing the riskiest assumptions should be chosen (Ries, 2011). By focusing on the right metrics the entrepreneur will be able to cut through all the noise involved with launching a new product (Furr and Ahlstrom, 2011).

After the baseline metrics have been chosen they should then be used to evaluate new changes in the business model. Once the baseline has been established, the start-up can work toward the second learning milestone by targeting every product development, marketing, or other initiatives at improving one of these metrics.

Premature scaling is thought of as one of the major causes of start-up failure. Premature scaling means "turning on the engine of growth" by hiring sales people, setting up production facilities, or building offices before the business model has been validated in the marketplace. Furr and Ahlstrom (2011) argue that before the start-up has proven the sustainability of their business model, defined as reaching the product/market fit, it should stay in the iterative process of improving and testing the business model.

3.3 The key concepts of lean start-ups

Although the authors chosen to represent the LS methodology share their views to a large extent, there are differences in the process they suggest start-ups should adopt. The chosen authors all agree that the first phase of a start-up should focus on understanding the problems the customers are facing. From this understanding an MVP should be built based on the customer requirements. Ries (2011) emphasizes building an MVP targeted at early-adopters and then going through the BML loop to refine the product until the start-up metrics suggest the business is ready to be scaled. Blank and Dorf (2012) propose a more rigid approach inspired by the book, *Business Model Generation*, by Osterwalder and Pigneur (2010). They define a business model by nine building blocks. Following the LS methodology the most critical hypothesis for success of the business should be tested first. As the product is validated and refined by customer feedback the later and less critical building blocks are tested until the business model is fully validated and can move to the phase of scaling. Blank and Dorf (2012) and Furr and Ahlstrom (2011) also follow a step-by-step process. After the first phase of interviewing customers about the problem a virtual prototype should be developed to test the solution hypothesis. When these have been validated and refined the start-up should build a prototype to test price point. From there the start-up should launch the product to test the remaining parts of the business model. Once every part of the business model has been validated and indicate that the start-up has found a sustainable business model it should move to the phase of scaling the company.

3.3.1 Minimum viable products: do we have a problem worth solving?

As seen in Fig. 3.1 the LS process begins with the formulation of working hypotheses that later will be tested through conversations with customers. The first phase of the process includes the creation of problem and solution hypotheses, contacting customers and scheduling interviews, validating hypotheses, and an exploration of the market attractiveness.

The start-up must at first figure out which customers to listen to as well as finding specific questions to investigate. Although using different terms, Blank and Dorf (2012), Furr and Ahlstrom (2011), and Ries (2011) agree that the initial hypotheses should seek to investigate the problem the customers are facing and then test the proposed solution to this problem.

This can be seen as the problem and the solution hypotheses. The problem hypothesis should be created to determine whether a problem worth solving exists, identify early-adopters, and learn how customers currently solve these problems.

After understanding the problem the start-up should develop a solution hypothesis stating the problem should be solved. This hypothesis can be tested by simply describing it to potential customers and asking whether something like that would solve their problem. This could be a virtual prototype or a PowerPoint presentation of the solution used to qualitatively validate the hypothesis, or it could be setting up a web page for quantitative validation. Measuring metrics from this should give some indication of whether the right solution to the problem has been found.

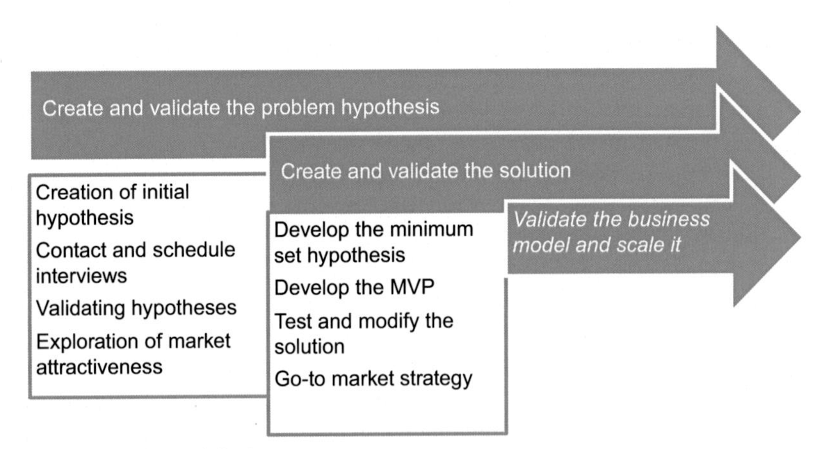

Figure 3.1 Overview of the lean startup process.

Although the initial hypotheses should seek to investigate whether a market for a given product is present at all and gain a deep understanding of the way the customers perceive the problem, the start-up should also have the big idea hypothesis in mind (Ries, 2011). The big idea hypothesis represents an idea of how a possible solution to the customer pain should look and how it should be delivered. However, in the first phase this idea should be left on the paper to make sure that the entrepreneur keeps an open state of mind and is able to accurately listen to the customers, thereby establishing the right solution for the future product.

Once the problem and solution hypotheses have been formulated it is time to test them. Before they are validated in the marketplace they are nothing but an educated guess. Again at this step it should be remembered not to waste too much time and resources only to discover that the assumptions were wrong. The goal is therefore to quickly get in contact with customers to test the hypothesis, measure the results, and objectively determine if they were right.

Of course, it is important to contact the right customers. The customers should in some way feel the pain the solution is trying to solve. Different segments might have different perception of the intensity of pain they have in the given area. In the first round of customer contact the segment with the highest pain level should be chosen. Methods of contacting customers include cold calling, people within the founders' network, or leads collected from the web page.

When in contact with first customers it is important not to just ask the customers what they want but to gain a deep understanding of their motivations, needs, and the problem they want solved (Furr and Ahlstrom, 2011). The focus should be on listening and learning and not trying to sell a product. Instead of presenting the hypothesis and asking the customers if they agree, the customers should be asked open questions. From their answers it should be possible to evaluate the hypothesis qualitatively, remembering not to draw conclusions from single customers and considering the type of customer who answers. To decrease the probability of making wrong conclusions the process should accurately capture the data in the interview by continuously taking notes or recording the conversations.

In addition to qualitative customer feedback the start-up could use a web page as a way of testing the hypothesis quantitatively. By describing the problem as well as how the entrepreneur intends to solve it on the web page the response could be used to evaluate the hypotheses. Furr and Ahlstrom (2011) set a cut-off point at 50%; if 50% respond positively to having the problem and the purposed solution the start-up can move to the next phase of creating an MVP. However, if the response rate is lower the entrepreneur should first consider if the right customer segment was chosen as well as whether the hypothesis was stated correctly. If none of these are the case the entrepreneur should use the cues from the customer feedback to reformulate the hypothesis and test again.

In addition to primary research with customers, secondary material such as reports, analyses, and other published material should be included to gain a deeper understanding of the customer pain. This will help evaluate the competitive environment and the health of the industry the start-up is trying to enter. Also it could provide valuable information on areas that need further testing.

Often customers do have significant pains or desires but if the market is not large enough or has too many entrenched competitors the chances of launching a new product is problematic. From talking to the potential customers the entrepreneur should have an idea of which segments and which markets the product is targeting. Blank and Dorf (2012) and Furr and Ahlstrom (2011) argue for the importance of retrieving market information. The identified segments and markets should be analyzed to understand the dynamics and competition as well as the potential of the product. Only when a big enough customer base is evident to justify the needed investments should the start-up move to the next phase. Ries (2011), however, warns about using too many resources on market research in this early phase, as answers to other questions are more important.

The company is ready to move to the next phase once there is a clear understanding of which problem the start up is trying to solve as well as which customers face the problem and how high they perceive it on the pain scale. It should also be known how customers currently deal with the problem, understand the competitive dynamics, and have well-documented reasons to believe that the solution is attractive enough to make a viable business in the long run.

3.3.2 Pivoting: have we built something people want?

After completing the first phase the start-up ought to have a deep understanding of the problem it is trying to solve and some ideas of which customers are facing this problem. In other words the problem and solution hypotheses should have been validated and there should be reason to believe that a viable market exists for the product. Where the previous phase tested the customer problem or need and explored the customers' passion for it, the next phase tests whether the solution to the problem—the value proposition—gets the customers enthusiastic enough to buy and use it.

To get more detailed information about the solution the next step is to create a minimum feature set prior to building the first MVP. The first MVP should contain only the minimum features required to drive customer purchase. To identify these, the start-up should review the feedback from the first phase and look for the features repeatedly mentioned as must-haves during the customer interviews. By focusing on a simple product the start-up both makes it easier for the customer to evaluate the core value proposition and makes the start-up able to move much faster with less resources. This makes it easier to get to the market fast and gain new feedback that in turn gives information and time to refine the product.

An MVP helps start-ups to start the process of learning as early as possible. Contrary to traditional product development, which usually involves a long process and strives for product perfection, the goal of the MVP is to begin the process of learning. Also the point is to find an inexpensive MVP because "no business model survives the first contact with customers" (Blank and Dorf, 2012). By not spending excess resources on the MVP the start-up should have enough money left over to try their second idea (Furr and Ahlstrom, 2011).

As long as the entrepreneur has nothing but an educated guess about the customer the focus should be on rapid, inexpensive, and simple experiments. This focus should be directly applied to the development of MVPs. The first MVP should only be complex enough to be able to test the initial hypotheses about the customer.

By reviewing the collected customer feedback coupled with the industry and competitive research the solution and business model should be discussed. Combined with the chosen minimum feature set the entrepreneur should arrive at new hypotheses about the needed product specs, customer segments, channels, pricing, and revenue model (Blank and Dorf, 2012). These hypotheses should lay the foundation for the first MVP. While Ries (2011) suggests starting with an actual physical product, Blank and Dorf (2012) and Furr and Ahlstrom (2011) argue that even simpler alternatives could be used to gain feedback before building an actual product. These include setting up a web page to test customer interest or a virtual presentation of the proposed solution.

Unlike when testing a traditional prototype, the idea is not just to test the product's design or technical questions but rather to test the fundamental business hypothesis. As long as the company does not know who the customer is and what the customer needs, they are not able to define the right quality. This implies that the MVP might have flaws and sometimes be perceived as a low quality product by the test customers. But, the point is not to build a perfect product from the beginning but rather to learn which attributes the customers care about and thereby provide a solid empirical foundation on which to build future products. The initial MVP should thus focus on finding a dominant position with the early-adopters before targeting the mainstream market.

Building an MVP is not without risks. The start-up has to be aware of patent protection as well as the danger of established companies stealing the idea. However, as Ries (2011) states, "if a competitor can out-execute a start-up once the idea is known, the start-up is doomed anyway." The exact idea behind the iterative MVP process is to be able to accelerate faster than anyone else, making what the competitors know insignificant. Furr and Ahlstrom (2011) state it in this way: "Pursuing a rapid experiment and finding out where you were wrong and changing direction is not failure. It is the road to success."

Another concern is that a poorly designed MVP can damage the brand of the start-up and result in negative word-of-mouth. However, this should not be a big concern. But new product releases in early start-ups rarely draw much attention without a simultaneous marketing campaign. Furthermore, the first MVP is not designed to satisfy the mainstream customer. No start-up can afford to build a product with every feature the mainstream customer needs all at once. Instead, the successful start-up focuses on building a product aimed at a small group of customers, early-adopters, who have bought into the start-up's vision. These early-adopters are characterized by having a problem high on the pain scale and searching actively for a solution as well as having the budget to try new solutions that might aid their problem. But, if the start-up fears negative brand perception the solution could be to release the product or service under another name.

By specifically targeting the early-adopters with a simple product it should be possible to go through a series of iterations in partnership with the customers to perfect the product until the business model has been validated and mainstream customers can be targeted.

3.3.3 Agile development together with the customers

Once the MVP has been made it is time to test it with real customers. Almost all LS authors suggest using iterative processes to test and further refine their MVPs. Ries (2011) illustrates the iterative process with his BML loop (Fig. 3.2):

In the first phase, build, the MVP is developed on the basis of the problem and solution hypotheses. In the next phase, measure, the entrepreneur seeks customer feedback, which is then analyzed in the last phase, learn, and used to refine the solution in the subsequent build phase.

As a start there should be some idea about the market and the applications and where the customers see a problem that the start-up can solve. These customers often represent more than one group or segment. To figure out which group to target first the profile of each segment should be considered. Although a given segment might suggest having the widest reach and biggest potential it should only be targeted in the early stages of development if the customers in the segment have the characteristics of early-adopters.

When these customers have had the opportunity to test the product they should be directly engaged in dialogue with the start-up. During this dialogue the start-up should try to look at the product from the customers' perspective and make them feel that their feedback is truly appreciated. Again the focus should not be on selling because that might distort the ability to pick up on cues the customers are providing.

The LS authors advocate different methods of evaluating the customer responses. Blank and Dorf (2012) and Furr and Ahlstrom (2011) argue that the entrepreneur should first seek to validate the solution qualitatively and then verify it quantitatively. The reasoning for this sequence is that qualitative customer feedback is superior when little is known about the perception of the solution in the customer's mind. A dialogue

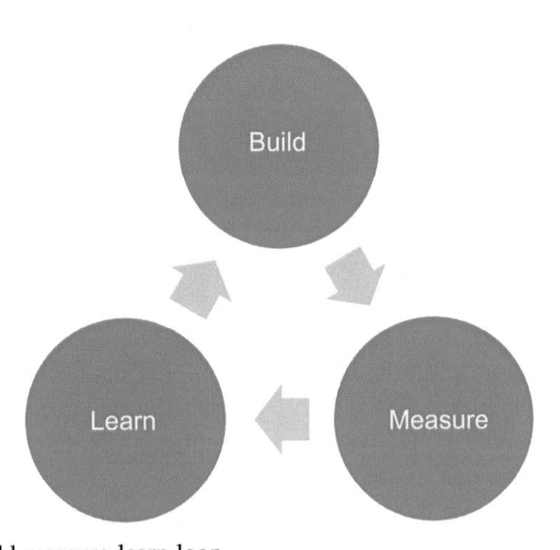

Figure 3.2 The build-measure-learn loop.
Adapted from Ries, E., 2011. The Lean Startup: How Today's Entrepreneurs Use Continuous Innovation to Create Radically Successful Businesses, Crown Business, New York.

with the potential customers about their experience with the proposed solution makes it possible for the start-up to answer the pivot question. If the feedback is sufficiently positive the entrepreneur should refine the solution based on the feedback and test it again. Once the customers start to validate the solution quantitative measures should be implemented to try to verify the customer hypotheses on a larger scale. These quantitative measures are termed start-up metrics as they differ from more traditional accounting metrics. Income statements, balance sheets, and cash flows are great at monitoring a company's financial health when executing on a proven business model. However, they do not provide the insights on whether the chosen business model is viable. Startup metrics can include cost of customer acquisition and retention rate. By analyzing such metrics the start-up will be able to make decisions about whether the current business model will prove viable.

In contrast to this, Ries (2011) proposes to implement quantitative measures as early as possible in addition to the qualitative feedback. Ries not only uses start-up metrics to answer questions concerning the viability of the business model but also to evaluate each incremental refinement of the solution and make decisions on whether to pivot or proceed. He argues that these metrics should seek to establish a baseline for the riskiest assumptions that the business model resides upon. After analyzing the metrics along with the qualitative feedback in the learning phase the entrepreneur should make a decision about whether to pivot or proceed. If the refinement of the solution in the building phase does not result in a satisfying improvement on the chosen metrics the entrepreneur should consider a pivot. If the entrepreneur decides to pivot it is important for the entrepreneur to use the experience received in previous steps when finding a new approach to the problem. The start-up should strive to reuse the validated learning from the customers and try to change.

Before moving to the next phase management should be confident that they have found the right solution to an important problem, which a large enough number of customers are willing to pay for. Also they should understand the demographics and archetypes of the target customers and know enough about their behavior to reach them cost-effectively. At last they should have validated the revenue model, including market-size estimates, production costs, and customer-acquisition costs.

3.3.4 Searching for a business plan: do we have the right business model?

After having verified the solution on a small scale the next step in the process is to launch the product. This is the most critical phase of the LS, where the start-up determines whether there is a scalable, profitable business model ahead. It is time to evaluate if the company is ready to start spending money to scale and whether the result will be a great profitable company. Often it becomes clear for the start-up company that there is a market but that this market is global and that the product launch has to be in a number of markets at the same time (Tanev et al., 2015).

The entire business model is now tested, not just its individual components, as in the prior phases. This does not mean going into full execution mode as the viability of the business model is not yet validated and a large number of aspects still need to be defined, such as a global strategy. The start-up to this point has done a great deal to uncover and test the initial hypotheses about the problem, the solution, potential marketing channels, and the general understanding of the market. The company is now acting on data about the customers. It is now time to launch the product and test the remaining hypotheses about the business model in a vigorous way. Especially, financial metrics have to be validated while seeking to improve the baseline metrics established in an earlier phase. The financial model will of course be one of the final hypothesis-testing activities before launching the business in full force. The financial model includes metrics such as fixed versus variable costs, margins, customer acquisition costs, customer lifetime value, and break-even. Although this process focuses on collecting the relevant quantitative data about the financial model of the start-up this does not entail ending the iterative process of refining the current solution, and the work on the most critical areas of the business model will continue. This iterative process of refining the solution will continue until all the data indicate that the business model is viable and ready to be scaled into a larger company.

Another important point to consider is break-even or the point where the revenue matches the expenses. Combined with the current cash-burn rate the entrepreneur will know how many months' worth of cash are left. Furr and Ahlstrom (2011) suggest that the start-up should have at least double the amount of cash that the entrepreneur estimates to use before reaching break-even. Prior to this the positioning and unique value proposition has been tailored from the customer feedback without much consideration given to competition or market type. Before the start-up can start scaling the company they must address these areas strategically. The market type and competitive environment have a large influence on both the investment needed to enter the market as well as the chances of success. If facing a market with a clear market leader a superior product will not be enough to win the battle but a marketing budget multiple times that of the marketing leader will be needed. Most start-ups do not have access to those financial resources. Therefore, the entrepreneur should consider resegmenting the market or creating a new market where the product can gain a unique and substantially different position.

The strategic decisions made by the start-up are interrelated with the evaluation of the financial metrics. Targeting a large market obviously entails a larger potential compared to targeting a small niche market. However, if a few strong players dominate the large market the cost of customer acquisition will in many instances be very high. Such a choice will only be possible if the start-up has the adequate funding to support the large investment needed to reach break-even.

3.3.5 How to find or create the next customers: scaling

The start-up will now have found a scalable and repeatable business model that predictably generates revenue and should be moving full scale into growing the business. As a company grows, a shift occurs away from facing an unknown problem/solution, which requires the iterative search-oriented approach, to facing a known problem/solution that requires execution. In other words the company's focus changes from customer learning to more traditional measures such as deadlines and quality standards. As a result the start-up must transition the way it operates fundamentally to reach the next level. Overall Furr and Ahlstrom (2011) argue that three areas of the company should be addressed: (1) market, (2) process, and (3) team transition.

As a start-up begins to scale, they often have early success, followed by a period of stagnant growth. This process is often described by the technology adoption life cycle model suggested by Moore (2014/1991) (Fig. 3.3).

This adoption gap emerging between the visionaries and the pragmatists is often named the chasm. The start-up will reach this period once the early-adopters have started to use the product but the mainstream customers are still waiting to adopt. If the start-up is not prepared to handle this phase in advance they risk continuing to burn cash on a strategy that does not target the mainstream customers properly. In short, to cross the chasm, the start-up must adjust its value proposition not only to meet the needs of the early-adopters but to reach the additional needs of the mainstream customers. In the words of the LS this implies moving away from the simple MVP to a full-product solution.

One of the major tasks now for the start-up is to begin scaling the company's processes. The customer base is growing and the focus has to change to serve these customers, too. The company must therefore turn most of its activities into repeatable processes such as an automated logistical setup to ensure satisfying and consistent delivery times.

As the start-up starts growing it will need to hire new employees to handle the increasing workload. This will entail new requirements for the organization and individual team members. A more formal organizational chart will be needed to ensure effective communication as well as an increased focus of how to manage the communication, and more specific areas of responsibility will be needed for each employee. All of these changes will require a different skill-set from the employees. Not only will

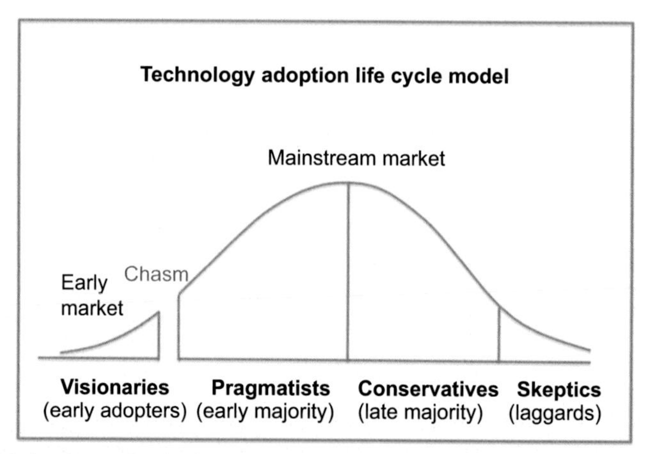

Figure 3.3 Technology adoption life cycle.
Adapted from Moore, G.A., 2014/1991. Crossing the Chasm: Marketing and Selling Disruptive Products to Mainstream Customers, HarperCollins.

more specialized talent be hired for specific areas such as production optimization or sales; the management will also need abilities to facilitate these changes and manage a large organization. In many cases this will imply that the founder is no longer the best-suited candidate to lead the company (Furr and Ahlstrom, 2011).

3.4 Some examples of lean processes

The reader may find several examples of LSs on Eric Ries's website, http://theleanstartup.com/casestudies, including Dropbox, Wealthfront, Grockit, Imvu, Votizen, and Aardvark. Just as an example, once Drew Houston, the CEO and founder of Dropbox discovered Eric Ries's Lean Startup blog, the company started iterating their product much faster in order to test what customers really wanted, early and often. Using LS principles, in just 15 months, Dropbox went from 100,000 registered users to over 4,000,000.[4]

3.5 Conclusion and future trends

Many of the LS companies face problems other than developing a new product or a new service. In many cases this has to be done on a global scale, with all the problems a fast internationalization may entail. In the blog of Steve Blank this problem was introduced with the slogan "Born global or die local." A large number of the LS firms are facing the problem of very small home markets and thus have to see market opportunities in larger markets all over the world right from the beginning.

3.5.1 *Lean and global*

Being a global company right from the foundation is often described as being born global (BG) or as being an international new venture (INV). Combining BG and INV with the LS approach leads to a type of company that has been labeled a lean and global start-up (LGS) (Tanev, 2012; Tanev et al., 2015). This type of firm is a new technology start-up that has to deal with business development, innovation, and early internationalization, not as processes separated in time or in firm functions but as one process leading to an achievable business model that operates on a global scale. The complexity, uncertainty, and the risk of being highly innovative and global at the same time is of course high but as shown in the first empirical studies companies deal with these problems by being very disciplined in their development process by following the LS approach, for example (Tanev et al., 2015).

The analysis in Tanev et al. (2015) showed that it is important to distinguish between the upstream and downstream global resources—a distinction without which

[4] http://theleanstartup.com/casestudies#dropbox.

it is impossible to conceptualize the early internationalization of the LGS type of firms. LGS firms have had problems dealing with the complexity, uncertainties, and risks of being innovative on a global scale. Some of the specific ways of addressing these problems include a disciplined knowledge and intellectual property protection strategy, the efficient use of business support and public funding mechanisms, and pivoting around the ways of delivering a value proposition and not around the value proposition itself. In addition, all the firms have managed the different types of uncertainties by moving one step at a time in a way that they could maximize the value of newly emerging relationships.

As markets are getting more and more global together with both downstream and upstream relations and activities the LGS firm must be expected to become a common new firm type in the years to come. It will thus be important for new firms to be able to deal with innovation and internationalization at the same time as part of one integrated process.

3.5.2 Further reading and links

The classic books and articles about LSs are Blank (2013), Blank et al. (2013), Hart (2012), and Ries (2011), which can be supplemented with Blank (2007), Blank and Dorf (2012), Furr and Ahlstrom (2011), Hart (2012), Miski (2014), and Moogk (2012).

Web resources

The classic start: http://theleanstartup.com/
An LS site addressing a larger community: http://www.leanstartupcircle.com/
Steve Blank's blog: http://steveblank.com/
Ash Mauria's website: http://leanstack.com/
Disruptive entrepreneurs: An interview with Eric Ries, McKinsey & Company: http://www.mckinsey.com/Insights/High_Tech_Telecoms_Internet/Disruptive_entrepreneurs_An_interview_with_Eric_Ries

References

Anthony, S.D., 2014. The First Mile: A Launch Manual for Getting Great Ideas into the Market. Harvard Business Review Press, Boston, Massachusetts.
Arteaga, R., Hyland, J., 2013. Pivot: How Top Entrepreneurs Adapt and Change Course to Find Ultimate Success. Wiley.
Blank, S., 2007. The Four Steps to the Epiphany – Successful Strategies for Products that Win.
Blank, S., 2011. Embrace failure to start up success. Nature 477, 133.
Blank, S., 2012. Where the Next Big Thing Lives in Our Nation's Research Labs. Hard Part: Turning Scientists into Entrepreneurs. Mansueto Ventures LLC on behalf of Inc.
Blank, S., 2013. Why the Lean Start-up Changes Everything. Harvard Business Review, pp. 1–9.
Blank, S., Benjamin, S., Turner, E., Eisenberg, I., Warren, H., Telleen-Lawton, D., Guido Hassin, B., 2013. "Lean" Is Shaking up the Entrepreneurial Landscape: Interaction, 91. Harvard Business Review, pp. 14–15.

Blank, S.G., Dorf, B., 2012. The Startup Owner's Manual: The Step-by-step Guide for Building a Great Company. K&S Ranch Incorporated, Pescadero, CA.
Collins, J., Hansen, M.T., 2011. Great by Choice: Uncertainty, Chaos and Luck - Why Some Thrive Despite Them All. Random House.
Eisenmann, T., Ries, E., Dillard, S., 2012. Hypothesis-driven Entrepreneurship: The Lean Startup. Harvard Business School Entrepreneurial Management Case.
Furr, N., Ahlstrom, P., 2011. Nail it Then Scale it: The Entrepreneur's Guide to Creating and Managing Breaththrough Innovation. Nathan Furr and Paul Ahlstrom.
Furr, N., Dyer, J., Christensen, C.M., 2014. The Innovator's Method: Bringing the Lean Startup into Your Organization. Harvard Business Review Press.
Hart, M.A., 2012. The lean startup: how today's entrepreneurs use continuous innovation to create radically successful businesses. Journal of Product Innovation Management 29, 508–509.
Maurya, A., 2012. Running Lean: Iterate From Plan a to a Plan That Works. O'Reilly Media, Incorporated, Sebastopol, CA.
Miski, A., 2014. Development of a mobile application using a lean startup methodology. International Journal of Scientific & Engineering Research 5, 1743–1748.
Moogk, D.R., 2012. Minimum Viable Product and the Importance of Experimentation in Technology Startups. Technology Innovation Management Review.
Moore, G.A., 2014/1991. Crossing the Chasm: Marketing and Selling Disruptive Products to Mainstream Customers. HarperCollins.
Osterwalder, A., Pigneur, Y., 2010. Business Model Generation: A Handbook for Visionaries, Game Changers, and Challengers. Wiley-Blackwell, Hoboken, NJ.
Ries, E., 2011. The Lean Startup: How Today's Entrepreneurs Use Continuous Innovation to Create Radically Successful Businesses. Crown Business, New York.
Tanev, S., March 2012. Global From the Start: The Characteristics of Born-global Firms in the Technology Sector. Technology Innovation Management Review, pp. 5–8.
Tanev, S., Rasmussen, E.S., Zijdemans, E., Lemminger, R., Svendsen, L.L., 2015. Lean and global technology start-ups: linking the two research streams. International Journal of Innovation Management 19, 1–41.

Start-up financing

4

S.C. Oranburg
Chicago-Kent College of Law, Chicago, IL, United States

4.1 Introduction

Start-ups are innovative, high-risk, high-growth business ventures that often require a significant amount of outside financing (Cable, 2010). Most start-ups "bootstrap," meaning self-fund, in their earliest stages (Cole, 2009, p. 477). Start-ups may also receive informal investment from friends and family or other insiders (Alden, 2011). Fig. 4.1 shows various sources of spin-off funding analyzed by Soetanto and Van Geenhuizen (2015) in a study covering 100 firms.

This chapter will focus on the ways in which start-ups are formally financed by outside investors to grow rapidly. Professional investors fund start-ups primarily through the use of debt, equity, or convertible debt (a hybrid of debt and equity) (Deeb, 2014). Crowdfunding is emerging as a new way for start-ups to get outside investments. At the outset it is important to note that all of these methods may involve the issuance of securities, which are subject to various laws. Engage a competent attorney in any securities issuance because compliance can be tricky and penalties can be harsh.

Before a company can issue securities, it must be formed or incorporated. Start-up formation involves two formative decisions that affect financing and have repercussions for the life of the company: In what state should you form or incorporate the company? What type of entity do you want to form there? There are many excellent

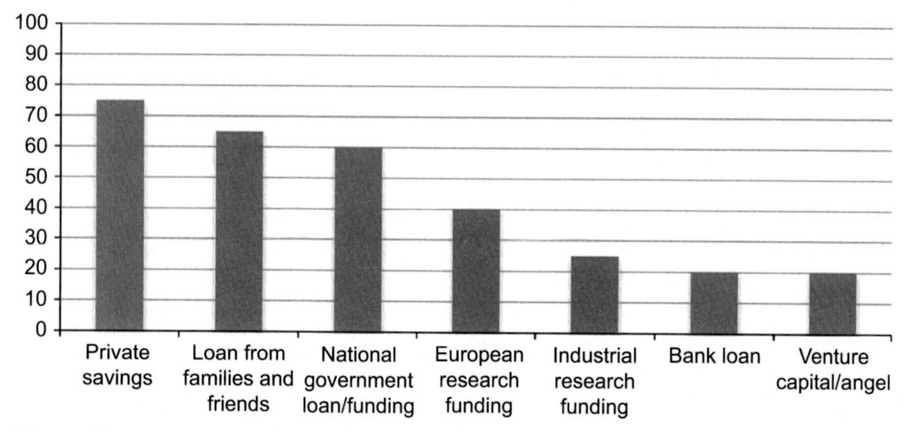

Figure 4.1 Various sources of funding for innovation (Soetanto and Van Geenhuizen, 2015).

Start-Up Creation. http://dx.doi.org/10.1016/B978-0-08-100546-0.00004-2

resources about how to make this important decision, but most American entrepreneurs choose to form a Delaware corporation. This chapter will therefore focus on start-up financing for a Delaware corporation, although it may be helpful to understand the reasons why many entrepreneurs make this choice.

Corporations are the most common entity choice for a start-up because the corporate form is well understood, separates ownership and control, limits liability for shareholders and directors, simplifies accounting, can be created quickly and easily, and has many standard characteristics that make it possible to use standard forms to issue securities (Hyman, 2014, §1:1). This is not, however, the only choice. In recent years many states have authorized the creation of new entity types. The most popular new entity type is the Limited Liability Company, which is lauded for its flexibility and tax benefits (Sargent and Schwidetzky, 2014, §1:1). Flexibility is not always a benefit to start-ups seeking financing because complexity and customization quickly increases legal costs, and investors have to review more information to understand investors' rights in flexible entities. The tax benefits may also be illusory, as most start-ups lose money for a period of years.

The state of incorporation matters a great deal because the corporate law of the state of incorporation governs the corporation's internal affairs (such as voting rules and other shareholder protections), and corporations may be sued in their state of incorporation. Most start-ups incorporate in Delaware because Delaware law is generally considered to be the most predictable and reasonable (Fisch, 2000). The Delaware courts are also reputed to be the best in resolving corporate issues quickly and fairly. Some start-ups choose to incorporate in the state in which they do business because this saves costs, but outside investors from other states may be hesitant to invest in entities that could be subject to quirky state laws. Accordingly, this chapter assumes the entrepreneur will form a Delaware corporation, which is widely regarded as the safest choice for a start-up that seeks outside investment, even though there are many good reasons to make other choices. Consult a legal advisor to determine if another entity choice of state of incorporation is better for you.

4.2 Debt financing

4.2.1 Introduction

Many entrepreneurs are already familiar with debt financing from their personal life. While debt agreements to borrow money can be very complicated, they have several common and straightforward characteristics. The most important characteristics of debt financing to many entrepreneurs are, first, that lenders typically do not have rights to tell the entrepreneur how to run the business and, second, failure to make regularly scheduled debt payment can put the start-up into bankruptcy.

4.2.2 Pros and cons

Perhaps the biggest advantage of debt financing is the end of the relationship, when the debt is repaid and lenders do not have the right to tell the entrepreneur how to run the business in the meantime (unless such rights are given by contract, which is unusual). This is a pro for the entrepreneur who wants to maintain total control of the start-up. So long as the debt is repaid on time and as scheduled, the entrepreneur does not have to answer to investors. Doing so can even raise the start-up's credit rating, making it easier to borrow money in the future.

But early stage start-ups often have dips in cash flows that make it difficult to make debt payments every month. Moreover, the entrepreneur may prefer to reinvest profits in growing the business, rather than making debt payments. Entrepreneurs may not realize that corporate debt requires regular repayment; severe repercussions can result when payments are missed.

When a start-up fails to make a debt payment when it is due, the loan immediately goes into default. But default for nonpayment is only one of many ways in which a start-up can default on a loan. Commercial debt agreements also have a number of covenants, which are conditions that must be met while the loan is outstanding. Financial covenants can include a requirement to maintain a minimum level of assets or not to take on more debt. Another con associated with these covenants is that the entrepreneur will have to provide the bank with balance sheets, income statements, and cash flow documents on a regular basis so the bank can confirm that the financial covenants are not violated (Booth, 2014, Chapter 6).

Banks may also ask start-ups to agree to operating activity covenants. These can be unobjectionable, like a requirement to pay taxes and comply with laws and regulations, or they can be onerous, like a prohibition against using company money for certain purposes—such as leasing equipment or real estate, changing management, selling assets, or paying dividends—without bank approval (Booth, 2014, § 6:13). It may feel like there is a banker in the boardroom while the debt is outstanding.

Fortunately, not all loans are so onerous. In fact, another advantage of loans is that there are many products from which to choose. For example, the Small Business Association helps start-ups get commercial loans up to $5 million at reasonably good terms (SBA loans, n.d.). Such loans can have fixed interest rates and long terms, so the monthly payment is low and certain. There is also a new lending industry called peer-to-peer (P2P) lending or crowdfunding, which does not use a traditional bank at all (see Section 4.5).

These alternative (nonbank) lenders are important for start-ups because many banks are unlikely to fund a start-up that does not have sufficient collateral to guarantee a loan. Collateral is property of the borrower that can be seized if the borrower does not repay the loan. This is often called a security interest, and it can include cash, real property, and intellectual property. Posting security often helps the borrower obtain a lower interest rate because it makes the loan less risky for the lender, but many early-stage start-ups (and some late-stage ones) lack sufficient collateral for a traditional loan (Bradford, 2012).

4.2.3 Issues

There are a number of issues with start-up debt financing, but perhaps the largest issue is that currently it is hard for a small business in America to get a loan, especially without collateral. One study showed that small business owners have difficulty obtaining loans (Chakraborty and Mallick, 2012). This is particularly problematic for the busy entrepreneur. Filling out each loan application can take hours, and the bank may require days' worth of follow-up information before denying the loan application. Studies show that the trend away from small business bank lending is unlikely to reverse (Peirce et al., 2014).

Fortunately, there are new sources of debt financing that put fewer onerous requirements on start-ups. The new industry of P2P lending has emerged to provide personal and small business loans. Lending Club, Prosper, Realty Mogul, and other companies provide a P2P platform where borrowers and lenders can transact without going through a traditional bank. Each platform has its own methods, but generally the P2P lending works on a reverse-auction model. The start-up creates a profile and submits financial information to the platform, which assigns a credit rating. Then the start-up posts a request on the platform's website for funds. Lenders view potential borrowers and contribute some fraction of the requested amount. Depending on the attractiveness of the borrower, it may take some time for enough lenders to syndicate and fund the entire loan amount. Typically the cash is not available to the borrower until the loan is fully funded, so this creates an issue if money is needed urgently.

Another issue with P2P lending is that the request for a loan is visible to thousands of people, not just a few banks. Some entrepreneurs may not want to disclose their business model and financial situation so broadly. An analogy may be posting a request for a personal loan to a Facebook page. This publicity may not be desirable in all situations.

Finally, loan documents, especially ones with security or collateral agreements, are complicated, technical, and hard to read. The loan agreement may set forth a great number of restrictive covenants, and it can be hard to track when they are tripped. The risks of default not only for nonpayment of principal and interest but also for technical noncompliance may require the entrepreneur to dedicate substantial efforts to servicing the loan. It is critical to carefully read and fully understand any loan agreement before signing. If the terms of the loan are so onerous that they will prevent the business from running properly, it may be best to look for another source of financing.

4.3 Equity financing

4.3.1 Introduction

Equity is a share of a business. In fact, equity in corporations is actually called shares, or shares of stock. An entrepreneur who raises money through equity financing effectively sells pieces of the company in return for outside investment. The outside investors thereby become inside shareholders, who have certain legal rights and privileges that lenders do not enjoy.

Professional equity investors include angels and venture capitalists (Oranburg, 2015). Angels form groups that collectively invest an average of $350,000 of their own money in a very early-stage start-up, while venture capitalists form funds that invest an average of about $7.3 million in a more mature start-up company (Oranburg, 2015). These independent venture capital (IVC) investors (which may also include private corporations that have venture funds) contribute the vast majority of start-up capital in America and the world, but some countries have established government venture capital (GVC) funds and mixed public-private partnership funds. Public investors provide capital to entrepreneurs that IVC investors may not because GVCs are under political pressure to pursue domestic employment and other nonfinancial goals (Cumming et al., 2014).

4.3.2 Pros and cons

The overwhelming advantage of equity financing for start-ups is that there is no loan that must be paid back. With equity financing, there is no risk of bankruptcy as a result of failing to repay investors. This gives start-ups the flexibility to deploy invested capital in growing the business. Investors do not expect to be repaid until the business becomes profitable, and even then equity investors may be willing to reinvest the profits to continue growing the business (Booth, 2014, §1:10).

The reason equity investors are so flexible about being repaid is that equity owners are entitled to a share of the total value of the start-up. The investor joins the entrepreneur in sharing business risks, and both have similar incentives to grow the business as large as possible before cashing out (Dent, 1992). This creates a partnership dynamic between the entrepreneur and the equity investor, which may be welcome in some situations but which can also create some frustrations for the entrepreneur.

As a coowner in the business, the equity investor is entitled to vote on fundamental transactions, such as a merger or a sale of substantially all the company's assets. Once equity investors purchase stock in a start-up, the entrepreneur may lose control over when to liquidate the business and exit the market. In fact, many equity investors will push for an exit within eight to ten years of their initial investment. While the entrepreneur frequently gets paid to operate the business, the equity investors typically get a return on their investment only when it liquidates. Some equity investment contracts even contain a provision where the investors can force the company to go public (Smith, 2005).

Another con about equity financing is that the investors typically receive preferred stock, whereas the founders and employees typically receive common stock. Preferred stock is so named because it has certain preferences, including the right to be paid first in a merger or liquidation, the right to receive dividends, and the right to block certain transactions such as another debt or equity financing. Founders need to realize that equity financing not only gives up a share of their business to investors, but it also subordinates founders' equity position to investors. The common stockholders usually have financial rights only in the residual, which is the amount that remains after the preferred stockholders are fully paid for their initial investment, dividends they have accumulated, and any liquidation preference they are owed. Preferred stockholders may even have the right to participate, which means that after they receive their preference, the preferred stock converts to common and receives a percentage of the residual as well (Walther, 2014, p. 167).

Preferred stockholders lack the affirmative covenants often found in bank loans, but they need a way to make sure that management is acting in the investors' best interests, so they often bargain for management rights or even a seat on the company's board. Giving an investor one seat out of three may not seem like a big deal at first, but remember that start-ups often raise multiple rounds of outside investment. Each fund-raising round may result in giving up another board seat, so that by the third round the equity investors may outnumber the founders on the board by three to two. At this point, the investors can fire the CEO and replace management. Only the strongest start-up founders are able to raise multiple rounds of equity financing without eventually giving up a majority of board seats to investors. Most founders who raise equity financing should expect to be at the mercy of outside investors at some point in their start-up's life cycle (Wasserman, 2012).

But founders are often overly worried about maintaining control. Equity investors are in the business of selecting and overseeing start-ups in which to invest, not in the business of running them. A good CEO will find his position secure, and quite frankly it may be in everyone's best interest to replace an underperforming CEO with someone who can really build the business. The original founding team should be vested in most or all of their common stock (meaning they have the right to keep that stock even if they are fired) by the time equity investors control the board, so they can actually profit financially from such a change of control (Empey, n.d.). Sometimes the founding CEO is a visionary who can build something new, but not an administrator who can run a large and successful organization. This can be hard for a CEO to admit, but successful repeat entrepreneurs learn their own strengths and weaknesses.

The close and interdependent relationships between entrepreneurs and investors that arise from equity financing agreements are not for every start-up founder. Those founders who demand complete control may be frustrated when each financing round slowly drains power from their grasp. But founders who are open-minded about the business model and their role in the company may actually find that equity investors have valuable experience, connections, and skills that can be employed to dramatically improve the business. The key thing to remember when sorting out the pros and cons of equity financing is this: equity investors are more than money. Equity investors are partners. Some analogize the founder/investor relationship to a marriage. Choose someone you trust and want to work with for many years.

4.3.3 Key issues

The biggest issue with equity financing is that it may not be available during certain stages of start-up development. There are currently two main types of equity financing investors, angels and venture capitalists. Angels typically fund less than $1,000,000 and venture capitalists typically fund more than $5,000,000, so start-ups trying to raise $3,000,000 often have trouble attracting investors (Oranburg, 2016). Moreover, these investors prefer business models that can scale quickly, which is why they often invest in software start-ups. Start-ups that make physical goods or that require a large input of human capital may find it difficult to attract equity investors. Fortunately, there are more resources than ever to find angels and venture capitalists who are in specific sectors. Web sites like Angel List connect start-ups to equity investors, incubators and accelerators have "demo day" to highlight emerging companies, and many large corporations have established venture financing groups to help grow start-ups in their sector (Incubator, n.d.). When equity crowdfunding becomes legal, that will introduce yet another way to fundraise. But until then, start-ups should plan their business model around the realities of equity financing by planning to seek financing at points in the start-up's life cycle where investment will be at the highest levels.

Another issue is that raising money through equity financing requires a public disclosure of the securities issuance. There are web sites like Crunch Base, whose business is to look for filings about such securities issuances and to post that information online (www.crunchbase.com). Therefore, one key issue with equity financing is that it effectively takes the start-up out of stealth mode and subjects it to public scrutiny. The solution to this issue is to treat the equity financing like a PR campaign and use the press to the company's advantage through effective communication and good timing.

Once a start-up raises money through equity financing, the company effectively has a value as of a certain point in time. The value is determined simply by dividing the amount of money invested in the start-up by the percentage of equity received from that investment (eg, if investors pay $1,000,000 for 10%, the start-up has a value of $10,000,000). This has many implications. For example, start-ups often incentivize employees to work long hours for less pay by offering stock options. The value of a stock option is the difference between the price that the employee must pay to get the stock (this strike price is set by contract and does not change) and the value of the stock (the market price, which changes frequently). Clearly, the employees prefer to get stock options when the strike price is low, but there are laws about setting the strike price based on the value of the company. Once the company has a value determined by outside investment, the strike price typically increases dramatically, making stock options less valuable. Therefore, it may be a good idea to grant stock options before seeking an equity investor (Casserly, 2013).

Another issue is the expense of doing an equity financing. Expect legal fees to cost at least $25,000 for a standard equity financing. That cost can increase dramatically if the start-up has done a poor job of maintaining records and complying with corporate formalities before the financing. The equity investors will require the start-up to pay lawyers to clean up the books and records before closing the investment. This process can also require the entrepreneurs to spend a lot of time on due diligence, the process by which the start-up discloses and the investors review the business. During this process, some issues like lawsuits, patent infringement, and disgruntled employees may come out and be memorialized indefinitely in the stock purchase agreement. Start-ups that have the financial means to do so would be well advised to ensure their corporation is compliant with the law before seeking financing (Gartner et al., 2012).

Finally, the stock purchase agreement and other equity financing documents are technical and complicated, but in a different and more dangerous way than bank financing documents. Whereas debt-financing documents can contain numerous restrictive covenants and severe penalties for late payment, a founder can easily recognize the important economic terms such as interest rate and loan term. In equity financing, however, the economic terms are much more complex. For example, simply changing the liquidating preference multiple from one to two means that, in the event of an acquisition or liquidation, the preferred stockholders get twice their investment before the common stockholders get anything. A sophisticated entrepreneur can learn about these terms by studying a term sheet and reviewing the annotated version of these forms (available online at nvca.org), but a good lawyer is truly worth hiring in this context, where a seemingly tiny detail can result in millions of dollars.

4.4 Convertible debt financing

4.4.1 Introduction

Convertible debt is a hybrid between debt and equity. Convertible debt is technically a loan, typically with a very low or nominal interest rate. The interest rate can be low because the point of the loan is not to earn money on interest but to convert the debt to equity upon a triggering event, such as another financing or the expiration of a period of time. Convertible debt was traditionally used for bridge loans, which are loans between two rounds of financing to carry the start-up through a brief but tough time. Nowadays, start-ups frequently receive their first financing (also called seed financing) through convertible debt because this seed note method is relatively quick, cheap, does not require the company to be firmly valued, and may not require the start-up to make the public disclosures associated with stock issuances (Werner, n.d.).

4.4.2 Pros and cons

The main advantage of convertible debt is that this method is quick and inexpensive. Whereas an equity-financing round can easily cost $25,000 or more, a seed note round can be accomplished for $5000 or less. The reason for the low cost is that seed notes are very simple instruments that effectively put off discussions regarding valuation, board seats, protective provisions, liquidation preferences, and many other complex terms until a triggering date or event in the future (Werner, n.d.).

Until the seed note converts to equity, the noteholder is not a stockholder. Noteholders are not entitled to shareholder rights such as voting on mergers, and they almost never receive protective provisions. Seed notes also generally lack the strict affirmative covenants found in conventional debt agreements. As a result, entrepreneurs are generally free from virtually any investor influence during the preconversion period (Kramer and Levine, 2012).

Seed notes also do not require the entrepreneur and the investor to agree on a valuation of the company. The notes do not translate into a percentage of equity until the next equity financing. This eliminates what is often the most contentious aspect of start-up financing: valuing a new and unique company in its very early stages. The flip side of this kick-the-can-down-the-road method is there is a great amount of uncertainly as to what investors will ultimately receive. To understand why requires a brief discussion of how seed notes work. There are essentially only four material terms in a seed note agreement. First, how much will be invested as debt? Second, at what interest rate should the debt accrue? Third, when does the debt investment convert into equity? Fourth, does the debt investor get a discount (relative to later investors) when the note converts at the next financing? There are some other material terms such as what happens in the event of a preconversion acquisition and whether the investor shall receive financial statements during the preconversion period, but the focus is generally on the economics of conversion in the next financing (Werner, n.d.).

The seed notes will likely stipulate that the noteholder gets equal or better rights than any new investor in the next equity financing, and the noteholder will pay the same or less than any new investor in that next round (Werner, n.d.). But when the note is formed, neither party has any way to know what the rights nor price of the next financing will be. There are a few techniques employed to address this uncertainty, such as a valuation cap, which reflects the highest price per share that the noteholder will pay upon conversion, but the uncertainty regarding final terms about control, preference, and voting rights is the unavoidable consequence of not negotiating these terms up front.

4.4.3 Key issues

Negotiations over seed notes typically focus on two main economic terms, the cap and the discount. First consider a situation where there is no cap and no discount. The start-up issues 1,000,000 shares of common stock to founders. Then a seed note investor purchases a $100,000 note with 0% interest. Later, an equity investor values the 1,000,000 shares at $1,000,000 (this is called the premoney valuation), or $1.00 per share, and agrees to purchase 200,000 shares of preferred stock for $200,000. Upon the closing of that stock purchase, the seed note investor receives 100,000 shares of preferred stock, too. After the equity financing, the founders hold 77% of the equity (Werner, n.d.).

Now take these same facts, except that the note has a 20% discount. That means the noteholder pays $0.80 per share and receives 125,000 shares of preferred stock. After the equity financing, the founders hold only 75% of the equity.

Instead of a discount, assume the note has a $500,000 valuation cap. That means that the noteholder will never pay more than $0.50 per share (which is derived from a $500,000 valuation for 1,000,000 shares). Now the noteholder receives 200,000 shares of preferred stock, and after the equity financing, the founders hold only 71% of the equity.

Finally, consider a note that has a $500,000 cap and a 20% discount. Typically, a noteholder gets the better of the two options, but not both. Under the previous facts, the cap is more valuable than the discount. However, if the new equity investor valued the company at less than $625,000, the discount would instead be applied because it is more valuable to the noteholder.

As this hypothetical situation should make clear, the number of shares that a seed note investor ultimately receives cannot be determined when the note is issued. Rather, the noteholder receives a number of shares that is dependent on the valuation by the next equity investor. This can create a conflict of interest between the noteholder and the founder. The holder of a note that does not have a cap may want the company to close its next equity financing when the stock value is as low as possible as to get the highest number of shares, whereas the company is best served by holding out for a higher valuation. On the other hand, if the note is capped, both the note holder and the founders may want to receive a high valuation in the next round, but if the valuation is too high then the note holder will effectively get a massive discount vis-à-vis the next equity investor. This can seem unfair and thus discourage an equity investor from investing in the start-up.

Another issue with seed-note financing is that the noteholders eventually become equity holders, so many of the same issues discussed earlier apply here. Founders should consider how such investors will act as stockholders before accepting their investment. In the early stages of a closely held corporation, stockholders can influence management and prevent certain corporate actions. An investor who contributes $100,000 in a seed note round may need to consent to a sale of the company for $10,000,000 several years later. Before accepting funds, remember that equity investment is a long-term relationship, and that seed note holders will become equity holders.

4.5 Crowdfunding

4.5.1 Introduction

Crowdfunding is "the practice of funding a project or venture by raising many small amounts of money from a large number of people, typically via the Internet" (Prive, 2012). New laws allowing crowdfunding have been promoted by the Obama administration because "crowdfunding offers real promise for underserved business entrepreneurs and may allow the organizations that serve them the ability to reach even deeper into the entrepreneurial community" (Rand, 2012). There are high hopes for this new fundraising model. Business scholars have even suggested that crowdfunding can change the very nature of capitalism by dramatically lowering the cost and difficulty of raising funds (Macaulay, 2015).

But crowdfunding is not entirely new. Crowdfunding is a blanket term that covers a range of different models for raising capital, which can be categorized into five types: (1) donations, (2) rewards, (3) prepurchases, (4) lending, and (5) equity crowdfunding. Entrepreneurs can combine these models for specific fundraising purposes.

4.5.1.1 Donations

The donation model of crowdfunding does not involve the sale of securities (Sheik, 2013). Those who donate to crowdfunding projects do not receive any financial interest in the product (Sheik, 2013). This appeals to donors who want to foster the development of intangible common goods and social welfare projects. Donors for these projects do not receive a direct, private benefit, although they may appreciate an indirect, public benefit to donors and nondonors alike. This altruistic donation model of crowdfunding has proven popular for charities due to its transparency and personal touch (National Overview, n.d.), but it is not a common way that for-profit start-ups raise money (Belleflamme et al., 2011). It is more common for entrepreneurs to seek financing with an approach that eschews the donation model or combines it with one or more of the other models described next (Belleflamme et al., 2011).

4.5.1.2 Rewards

The rewards model of crowdfunding allows a supporter to invest in an organization or cause in return for a reward such as "recognition, a show ticket, [or] a t-shirt." (Manderson, 2012 as cited in Sumners, 2012, p. 40). The reward model is often combined with the donation model. For example, large donors to *The Canyons* movie received cameos, script reviews, and even the money clip that Robert DeNiro gave director Burt Schrader on the set of the movie *Taxi Driver* (Rodrick, 2013).

Kickstarter is a crowdfunding website that encourages the use of the rewards model (Rewards, n.d.). Rewards include copies of the thing produced, limited editions of the product, collaborations with the donor in the production, experiences with the producers, and mementos of the project (Rewards, n.d.). Being rewarded with the final product is similar to engaging in the prepurchase model of crowdfunding described later, but the distinction is that rewards are technically considered gifts that are not offered for sale (Terms of Use, n.d.). Start-ups that produce consumer goods find that the rewards model is an effective way to raise capital, although more money can be raised more easily with the prepurchase model described next, and start-ups that cannot easily produce an entertaining reward have trouble using the rewards model of crowdfunding.

4.5.1.3 Prepurchase

The prepurchase model is the most common type of crowdfunding (Belleflamme et al., 2011). Under the prepurchase model, the consumer pays in advance for the product. If the start-up launches the product, the consumer typically receives that product for a lower price than regular customers who purchase the product after it is already on the market (Belleflamme et al., 2011), and the consumer receives nothing if the start-up fails (Manderson, 2012).

4.5.1.4 Lending

The lending model leverages technological and financial innovation to connect lenders and borrowers in what is called peer-to-peer (P2P) lending, as exemplified by Lending Club (Verstein, 2011). On Lending Club (the first P2P lender to register with the SEC), customers interested in a loan complete web-based applications (FAQ for New Investors, n.d.). The applicant is evaluated by Lending Club for creditworthiness and assigned an interest rate (FAQ for New Investors, n.d.). Investors select loans in which to invest (FAQ for New Investors, n.d.). The minimum required to invest is only $25, so broad diversification is possible for Lending Club investors with limited capital (Earn Solid Returns, n.d.). The term of loans on Lending Club are generally 3 or 5 years, at a fixed interest rate and straight-line amortization (Lending Club, 2014).

The lending model works best for start-ups that deal with tangible goods and have some sort of collateral. The risk of nonrepayment for unsecured business loans is very high, so investors require a commensurately high return on investment to compensate for this risk. To receive such a high rate of return through the lending model of crowdfunding, interest rates on the loans may be so high as to be illegally usurious. By posting collateral, producing valuable goods, or holding real property, start-ups can reduce their risk of nonpayment and obtain "crowdlending" on more favorable terms.

4.5.1.5 Equity crowdfunding

The Jumpstart Our Business Startups Act of 2012 (the JOBS Act) amends Section 4(a)(6) of the Securities Act of 1933, as amended (15 U.S.C. § 77d(a)(6) (2015)) (the Securities Act), to allow a private corporation to offer and sell up to $1 million worth of equity securities (stock) in a 12-month period to the general public without registering the securities with the SEC. This new exemption to registration under the Securities Act is generally called crowdfunding, although this federal law is more specifically called equity crowdfunding. (Securities and Exchange Commission (SEC), 2012). There are also of dozens of intrastate equity crowdfunding laws that are available where companies and investors are located in the same state, but these varied and nascent laws are beyond the scope of this chapter.

The SEC issued its final rules on equity crowdfunding on October 30, 2015, and they will go into effect to allow Internet funding portal registration starting on January 29, 2016, and to allow stock issuance starting 180 days after the rules are published in the Federal Register (probably May 2016) (SEC, 2015). The size of this financing market, the role of the funding portals, and many other details will take time to emerge. However, some of the key crowdfunding rules are provided in the JOBS Act itself. Individuals who have between $100,000 and $1 million in annual income or net worth may invest 10% of it each year in start-ups through crowdfunding (15 U.S.C. § 77d(a)(6)(B)(ii) (2015)). Individuals who have or annually earn less than $100,000 may invest the greater of $2000 or 5% of their annual income each year in start-ups (15 U.S.C. § 77d(a)(6)(B)(ii) (2015)). At this time, however, federal equity crowdfunding is not yet operating in the United States, so it is not possible to describe exactly how it will function.

4.5.2 Pros and cons

Crowdfunding can be very useful for start-ups with a primary purpose of producing a social benefit because crowds may be inspired to fund projects that create a public good. The rewards model of crowdfunding in particular has been used by nonprofit organizations like the Smithsonian Institute to fund public-work projects like the Reboot the Suit campaign (Reboot the Suit, n.d.). This may extend to eco-friendly projects and socially responsible investing (Chamberlain, 2013).

On the other hand, successful crowdfunding campaigns on leading platforms like Kickstarter and Indiegogo generally regard consumer products. Only one of the top 15 crowdfunding campaigns in 2014 was not for a consumer product, although it was for Reading Rainbow, a social benefit project (15 most funded, 2015). Business-to-business (B2B) crowdfunding has not taken off with a donation, reward, or prepurchase model. B2B companies and service companies are probably limited to the lending and equity crowdfunding models, unless their business focuses primarily on providing a compelling social benefit.

A benefit of non-equity crowdfunding is that entrepreneurs maintain control of the business. Backers obtain no control rights under the donation, reward, or prepurchase model, although start-ups that raise money in these ways may feel obligated to keep backers informed about progress. Recently the Federal Trade Commission has sued start-ups that sought money on Kickstarter and failed to deliver the product, so even the non-equity crowdfunding create legal risks for entrepreneurs and companies (Federal Trade Commission (FTC), 2015). Under lending crowdfunding models, the lender may obtain some contractual rights to the start-up's books and records, and the start-up may be forbidden from selling more equity or taking on more debt while the crowdfunded loan is outstanding. The restrictive covenants on crowdfunded loans vary widely depending on the platform, loan amount, and other facts.

On the other hand, raising money through equity crowdfunding requires making a number of disclosures. Start-ups must file a Form C with the SEC, keep various records, and produce audited financial statements before raising more than $500,000 (Lingam, n.d.; Crowdfunding, 2013). If a start-up raises a substantial amount of money through equity crowdfunding, it may have to get stockholder approval from the crowdfunding investors, which may be a diverse and hard-to-contact group. That can slow down or even prevent management from engaging in corporate finance and acquisition activities.

4.5.3 Key issues

The donation, reward, and prepurchase crowdfunding models do not involve the sale of any securities under federal law (Bradford, 2012). The touchstone of the Supreme Court's test for what defines a security is whether the investment is premised on a reasonable expectation of profits (United Housing Foundation, 1975). But start-up investors are generally not interested predominately in nonfinancial rewards. Investors generally want to earn a return on their investment, so donations, prepurchases, and nonfinancial rewards can only attract a limited number of investors (Bradford, 2012). To offer investors an opportunity to earn business profits, entrepreneurs will have to comply with securities regulations including Title III of the JOBS Act of 2012.

As discussed earlier, the biggest issue with donation, rewards, and prepurchase crowdfunding is that there are very few examples of B2B or service companies using those models successfully. That means lending and equity crowdfunding may be the only options. Crowdlending has many of the pros and cons discussed earlier, although it may be easier for a start-up to get a loan through crowdfunding than through a conventional bank.

The biggest issue with equity crowdfunding that it is not yet operational, so entrepreneurs cannot be sure how it will work. Congress is working on JOBS Act 2.0, so there is additional regulatory uncertainty around equity crowdfunding because Congress may change the statute (Clifford, 2015). When equity crowdfunding does go online, many academics are concerned that many companies will use it to commit fraud, which may cause a backlash against equity crowdfunding that potentially shuts down the entire model (Hazen, 2012). At this stage in the development of equity crowdfunding, there are many political, legal, business, and societal uncertainties that make this new model both exciting and risky.

4.6 Conclusions and future trends

Many start-ups are eager to get funded so they can begin operations. Entrepreneurs should first consider bootstrapping, or self-funding, but obtaining outside financing from professional investors is often a necessary step toward achieving business goals. Equity financing remains the gold standard for venture capital investment, and seed-note financing has become quite common for smaller round and angel investment due to its quick speed and low cost. But new approaches to start-up financing are emerging. For example, in December 2013, the start-up incubator Y Combinator announced the Simple Agreement for Future Equity as "a replacement for convertible notes," although this form has not yet become widely adopted (Graham, 2013). Series Seed is an open-source project to simplify equity financing documents and thus make equity financing quicker and cheaper. Series seed is simpler in part because it puts off negotiations on some material terms until the next equity financing round, thus creating some of the same uncertainty problems that arise in seed notes (although series seed is a true equity issuance, in which the company is valued, whereas seed notes do not impute a value to the company). Crowdlending web sites like Lending Club, Prosper, and Realty Mogul now offer debt financing without banks. And crowdfunding legislation may create new portals for equity investment that do not involve angel investors or venture capitalists. Yet despite all the novelty and complexity, start-up financing remains rooted in the fundamental principles of debt and equity as set forth in this chapter.

References

15 Most-Funded Crowdfunding Projects on Kickstarter and Indiegogo, February 16, 2015. Forbes. Retrieved from: http://www.forbes.com/sites/gilpress/2015/02/16/15-most-funded-crowdfunding-projects-on-kickstarter-and-indiegogo/4/.

Alden, E., 2011. Primum Non Nocere: the impact of Dodd-Frank on Silicon Valley. Berkeley Business Law Journal 8, 107−127.

Angel Capital Association. FAQs about angel groups. Retrieved August 14, 2015 from: http://www.angelcapitalassociation.org/press-center/angel-group-faq/.

Angel List. Incubator – Company types. Retrieved August 14, 2015 from: https://angel.co/incubators.

Belleflamme, P., Lambert, T., Schwienbacher, A., 2011. Crowdfunding: tapping the right crowd. Journal of Business Venturing 29 (5), 585–722.

Booth, R.A., 2014. Financing the Corporation. Thomson Reuters.

Bradford, C.S., 2012. Crowdfunding and the federal securities laws. Columbia Business Law Review 1–150.

Cable, A.J.B., 2010. Fending for themselves: why securities regulations should encourage angel groups. University of Pennsylvania Journal of Business Law 13, 107–172.

Casserly, M., March 8, 2013. Understanding Employee Equity: Every Startup's Secret Weapon. Forbes. Retrieved from: http://www.forbes.com/sites/meghancasserly/2013/03/08/understanding-employee-equity-bill-harris-sxsw/.

Chakraborty, A., Mallick, R., 2012. Credit gap in small businesses. International Journal of Business 17, 65–80.

Chamberlain, M., April 24, 2013. Social Responsible Investing: What You Need to Know. Forbes. Retrieved from: http://www.forbes.com/sites/feeonlyplanner/2013/04/24/socially-responsible-investing-what-you-need-to-know/.

Clifford, C., February 2, 2015. Déjà Vu 2012: A Zombie-Frankenstein JOBS Act 2.0 Is in the Works. Entrepreneur. Retrieved from: http://www.entrepreneur.com/article/242442.

Cole, C.W., 2009. Financing an entrepreneurial venture: navigating the maze of corporate, securities, and tax law. University of Missouri-Kansas City Law Review 78, 473–502.

Cumming, D., Grilli, L., Murtini, S., 2014. Governmental and independent venture capital investments in Europe: a firm-level performance analysis. Journal of Corporate Finance. http://dx.doi.org/10.1016/j.jcorpfin.2014.10.016.

Crowdfunding, November 5, 2013. 78 Federal Register 66,428 proposed.

Deeb, G., March 19, 2014. Comparing equity, debt and convertibles for startup financings. Forbes. Retrieved from: http://www.forbes.com/sites/georgedeeb/2014/03/19/comparing-equity-vs-debt-vs-convertibles-for-startup-financings/.

Dent Jr., G.W., 1992. Venture capital and the future of corporate finance. Washington University Law Quarterly 70, 1029–1085.

DonorsChoose.org. National overview. Retrieved August 18, 2015 from: http://www.donorschoose.org/docs/DonorsChoose_org-NationalOverview.pdf.

Empey, G. Protecting Your Founder Equity. Cooley Go. Retrieved August 14, 2015 from: https://www.cooleygo.com/protecting-founder-equity/.

Federal Trade Commission, 2015. Crowdfunding Project Creator Settles FTC Charges of Deception. Retrieved August 19, 2015 from: https://www.ftc.gov/news-events/press-releases/2015/06/crowdfunding-project-creator-settles-ftc-charges-deception.

Fisch, J.E., 2000. The peculiar role of the Delaware courts in the competition for corporate charters. University of Cincinnati Law Review 68, 1061–1100.

Gartner, W.B., Frid, C.J., Alexander, J.C., 2012. Financing the emerging firm. Small Business Economics 39, 745–761.

Graham, P., December 6, 2013. Announcing the Safe, a Replacement of Convertible Notes. Y Combinator Posthaven. Retrieved from: http://blog.ycombinator.com/announcing-the-safe-a-replacement-for-convertible-notes.

Hazen, T.L., 2012. Crowdfunding or fraudfunding? Social networks and the securities—why the specially tailored exemption must be conditioned on meaningful disclosure. North Carolina Law Review 90, 1735–1769.

Hyman, M., 2014. Corporation Forms. Thomson Reuters, Eagan, MN.

Indiegogo, Inc. Terms of use. Retrieved August 19, 2015 from: https://www.indiegogo.com/about/terms.

Jumpstart Our Business Startups Act, 2012. Pub. L. No. 112-106, § 303(b). 126 Stat. 306.

Kickstarter. Reboot the Suit: Bring Back Neil Armstrong's Spacesuit. Retrieved August 19, 2015 from: https://www.kickstarter.com/projects/smithsonian/reboot-the-suit-bring-back-neil-armstrongs-spacesu.

Kramer, B.J., Levine, S.S., 2012. Seed Finance Survey 2012: Internet/Digital Media and Software Industries. n.p.. Fenwick & West LLP. Retrieved from: https://www.fenwick.com/FenwickDocuments/2012_Seed_Survey_Report.pdf.

LendingClub. FAQ for new investors. Retrieved August 19, 2015 from: https://www.lendingclub.com/public/investing-faq.action.

Lending Club, 2014. Prospectus. Lending Club, San Francisco. Retrieved from: https://www.lendingclub.com/fileDownload.action?file=Clean_As_Filed_20140822.pdf&type=docs.

LendingClub. Earn solid returns. Retrieved August 19, 2015 from: https://www.lendingclub.com/public/steady-returns.action.

Lingam, K. Equity Crowdfunding Rules: The Good, the Bad, & the Ugly. SeedInvest.com. Retrieved August 19, 2015 from: http://www.seedinvest.com/blog/equity-crowdfunding-rules-good-bad-ugly-part-ii/.

Manderson, C., April 28, 2012. The JOBS Act, Part 4: Will Crowdfunding Live up to the Hype? PE HUB. Retrieved from: https://www.pehub.com/2012/04/will-crowdfunding-live-up-to-the-hype-part-4-of-a-4-part-series-on-the-jobs-act/.

Macaulay, C., 2015. Capitalism's renaissance? The potential of repositioning the financial 'meta-economy'. Futures 68 (Melbourne, Australia. GI Think tank).

Oranburg, S., April 6, 2015. The Law and Economics of the Series A Gap. The CLS Blue Sky Blog. Retrieved from: http://clsbluesky.law.columbia.edu/2015/04/06/the-law-economics-of-the-series-a-gap/.

Oranburg, S., 2016. Bridgefunding: crowdfunding for startups across the Series A gap. Cornell Journal of Law & Public Policy 25, 695.

Peirce, H., Robinson, I., Stratmann, T., 2014. How Are Small Banks Faring under Dodd-Frank? Mercatus Center at George Mason University, Arlington, VA. Retrieved from: https://www.stlouisfed.org/~/media/Files/PDFs/Banking/CBRC-2014/SESSION3_Peirce_Robinson_Stratmann.pdf.

Prive, T., November 27, 2012. What Is Crowdfunding and How Does it Benefit the Economy. Forbes. Retrieved from: http://www.forbes.com/sites/tanyaprive/2012/11/27/what-is-crowdfunding-and-how-does-it-benefit-the-economy/.

Rand, D., June 28, 2012. The Promise of Crowdfunding for Social Enterprise. Message posted to https://www.whitehouse.gov/blog/.

"Rewards." Creator Handbook. Kickstarter. Retrieved August 19, 2015 from: https://www.kickstarter.com/help/handbook/rewards.

Rodrick, S., January 10, 2013. Here Is what Happens when You Cast Lindsay Lohan in Your Movie. New York Times. Retrieved from: http://www.nytimes.com/2013/01/13/magazine/here-is-what-happens-when-you-cast-lindsay-lohan-in-your-movie.html?_r=0.

Sargent, M.A., Schwidetzky, W.D., 2014. Limited Liability Company Handbook. Thomson Reuters, Eagan, MN.

SBA loans: A Primer. Entrepreneur. Retrieved August 13, 2015 from: http://www.entrepreneur.com/article/217372.

Securities Act of 1933, 2015, 15 U.S.C. § 77d(a)(6).

Sheik, S., May 2013. Fast forward on crowd funding. Los Angeles Lawyer 36, 37.

Smith, D.G., 2005. The exit structure of venture capital. University of California-Los Angeles Law Review 53, 315–356.
Soetanto, D., van Geehuizen, M., 2015. Getting the right balance: University networks' influence on spin-offs' attraction of funding for innovation. Technovation 36–37, 26–38.
Sumners, P.C., 2012. IV. Crowdfunding America's small businesses after the JOBS Act of 2012. Review of Banking and Finance Law 32, 38–49.
United Housing Foundation, Inc., 1975 v. Forman, 421 U.S. 837, 852–53.
United States Securities and Exchange Commission, 2012. Jumpstart Our Business Act: Frequently Asked Questions about Crowdfunding Intermediaries. Retrieved from: http://www.sec.gov/divisions/marketreg/tmjobsact-crowdfundingintermediariesfaq.htm.
United States Securities and Exchange Commission, 2015. SEC Adopts Rules to Permit Crowdfunding. Retrieved from: http://www.sec.gov/news/pressrelease/2015-249.html.
Verstein, A., 2011. The misregulation of person-to-person lending. University of California-Davis Law Review 45, 445–530.
Walther, B., 2014. The peril and promise of preferred stock. Delaware Journal of Corporate Law 39, 161–209.
Wasserman, N., 2012. The Founder's Dilemmas: Anticipating and Avoiding the Pitfalls that Can Sink a Startup. Princeton University Press, Princeton, NJ.
Werner, P. Primer on Convertible Debt. Cooley Go. Retrieved August 14, 2015 from: https://www.cooleygo.com/convertible-debt/.

Marketing for start-ups

T. Shih
Lund University, Lund, Sweden

5.1 Introduction

Science- and technology-based start-up firms have received considerable attention in academic, policy, and business circles. Academic studies note that start-ups contribute to industrial renewal, innovation, and job creation (Almus and Nerlinger, 1999). As such, the support to start-up firms has also been extensive as illustrated by the increased financial capital available, emergence of innovation support structures, and so forth (Autio et al., 2014).

However, the challenges to turn new discoveries and ideas into products and commercialize these are considerable. Research and investigations have documented the high failure rates of technology- and science-based start-ups with concern to the commercialization of ideas (Zahra et al., 2007). Failures are explained, for example, by researchers not being interested in actually running businesses (Visintin and Pittino, 2014) or that the ideas are too radical or disruptive (Almus and Nerlinger, 1999; Sandberg and Aarikka-Stenroos, 2014). The latter infers that the ideas do not fit well with current solutions or expectations on the market. Thus, Lindelöf and Löfsten (2006) note how the start-ups often lack connections to business parties. The business component is raised as particularly important by Bathelt et al. (2010), who identify technology-based start-up firms' customers as the most important component for stimulating innovation.

With this understanding, the notion of marketing becomes important. Marketing is defined as bridging gaps between market actors (Agndal and Axelsson, 2012). This general definition includes the gaps between the idea provider and the end-user/consumer. It also refers to relational activities within the broader network in which products are developed, produced, and used in an industrial setting (Håkansson and Waluszewski, 2007). The latter particularly focuses on the inter-organizational relationships that are formed and developed between the start-up firm and its network partners, both business and nonbusiness actors (Aaboen et al., 2011). This chapter discusses the nature of marketing, defined as bridging gaps between market actors, and the contextual setting in which marketing occurs. The aim is to investigate the value-creating network and how the start-up utilizes form and uses others to reach the market. To that end the chapter presents two cases of how start-up firms have built networks, the challenges they face, and how they manage the process.

Start-Up Creation. http://dx.doi.org/10.1016/B978-0-08-100546-0.00005-4

5.2 Conceptual framework

5.2.1 Science- and technology-based start-ups

Science- and technology-based start-ups are firms that develop and seek to commercialize new science- and/or technology-based solutions (Mian, 2011). The origins of these firms can be universities, research institutes, or established firms. They have been acknowledged as important for the renewal of industries, economic growth, innovation, and job creation (Almus and Nerlinger, 1999). With this backdrop science- and technology-based start-ups have received considerable attention from universities, policymakers, and investors (Di Gregorio and Shane, 2003).

However, the challenges to turn new technological discoveries and ideas into products and commercialize these are considerable (Waluszewski et al., 2009). Scholars have documented the high failure rates of technology- and science-based start-ups (Zahra et al., 2007). Common explanations of failures include researchers not being interested in actually running businesses (Visintin and Pittino, 2014) or that the ideas are too radical or disruptive (Almus and Nerlinger, 1999; Sandberg and Aarikka-Stenroos, 2014). Perez and Sanchez (2003) note that start-ups are often founded, not on market opportunities, but on the basis of technological advancements. This infers that the ideas might not fit well with current solutions or expectations in the market. Accordingly, Lindelöf and Löfsten (2006) describe how start-ups often have weak connections to business parties. Moreover, Feinleib (2012) identifies the weak marketing skills of start-up managers as a reason for failure. Table 5.1 summarizes some common reasons for failure among start-ups.

The identification of the perceived weaknesses and reasons behind start-up failure, in conjunction with the liability of newness, referring to start-ups' lack of resources and business relationships has led to the emergence of commercialization support structures (Malerba, 2010). For example, much has been written about the incubators (Clarysse et al., 2005; Mian, 1996; Salvador and Rolfo, 2011) or the technology transfer offices (Di Gregorio and Shane, 2003) and their supportive functions. In studies of incubators and technology transfer offices the focus is primarily on the strategies that these organizations apply to aid start-ups, which can entail the provision of resources or the establishment of organizational links (Rothaermel and Thursby, 2005). The attention thus aims at the commercialization mechanisms that the support organizations help to create (Baraldi and Waluszewski, 2011). The understanding of such environments is important as commercialization mechanisms embed in complex interorganizational interactions. This chapter departs from the industrial marketing perspective (Håkansson, 1987; Håkansson et al., 2009).

5.2.2 Navigating in an innovation context and building a network: an industrial marketing perspective

From an industrial marketing perspective the understanding is that firms are embedded in networks of relationships where actors and organizations are interdependent (Håkansson and Waluszewski, 2002). Industrial marketing research illustrates that it

Table 5.1 **Reasons to why start-ups fail**

Reasons for failure	Description
Founders have no interest in running businesses	The increased attention of research policy to stimulate the commercialization of academic research has led to the increase in number of academic entrepreneurs. This group of entrepreneurs has been driven by various logics, such as starting companies because of advancement opportunities in academia, the infusion of public money to form new firms, or academic researchers seeing firms as alternative channels to develop technologies or scientific ideas. Hence academic entrepreneurs start firms for various reasons but do not always seek to become business managers.
Solutions are disruptive	The fit of a certain solution into an existing structure of already made investments in human resources, technical facilities and so forth is an important aspect of whether a solution is adopted in a specific context. The more disruption the solution creates for an existing system the more likely that it will encounter resistance to use.
Focus on technological novelty rather than existing market demand	The focus on the newness of scientific and/or technological characteristics and not on whether there is a market demand has been observed as an obstacle to business development. To develop a prototype for the market requires the development to stop, and this has been noted as problematic in technology/science-based start-ups where founders mainly view their role as researchers and not as business developers.
Weak connections to market actors	Firms started based on existing customer relationships have a higher success rate than firms that starts based on a business idea without established market ties. Thus weak connections to market actors associate with longer time to market, less knowledge of market demand and so forth.
Poor marketing skills	Failure of business development due to poor marketing skills often relate to the lack of a functional business model, including the promotion of the product. Here some of the challenges include defining the customer and establishing ties.

is within established business relationships where innovation usually occurs, while only a minor part of the firms' most important development relationships are with external research and development units (Gadde and Håkansson, 2001). Thus, most innovations occur as the result of ongoing improvements in established business structures. The reason is that interaction between two parties leads to adaptations and learning, which in turn infers higher efficiency and innovation (Håkansson and Snehota, 1995). The notion of established business structures is therefore important when studying innovation. Of integral concern here is how an innovation can create value not only for already existing resources and investments such as physical equipment, production processes, and technologies in use, but also for established relationships (Linné and Shih, 2013). In this vein innovation does not need to be based on novel discoveries, but can be old solutions introduced into new contexts or new solutions incrementally developed in existing structures, and so forth.

However, start-ups commonly originate from outside these established business structures. Hence as the new firm starts out with a new product or idea it needs to establish legitimacy and garner resources from a heterogeneous network context (La Rocca and Snehota, 2014; Zimmermann and Zeitz, 2002). Having network relationships associates heavily with the access to resources (Snehota, 2011). This situation has been well described in the entrepreneurship and start-up literature concerning how new firms suffer from a liability of newness and lack established business relationships (see, eg, La Rocca and Snehota, 2013). The predicament is particularly obvious for firms that are formed based on novel technology (Clarysse et al., 2011). These firms are often well embedded in research networks, but less so in business networks of suppliers, distributors, and customers (Aaboen et al., 2013). Hence an issue for start-ups is how such firms can form and develop relationships with other actors and navigate within networks that encompass actors within research, production, and business.

5.2.3 Marketing for start-ups

A major goal of the start-up is to seek to establish relationships relating to the search for resources, such as financial, personnel, social capital, and so forth (Andersson et al., 2010; Lechner and Dowling, 2003). It can also be to gain legitimacy for its products and ideas, by testing them with potential customers and users (Öberg, 2010). As Kotler et al. (2010) note, an important aspect of marketing is for the company to convey its values and deepen relationships with partners. Hence, the start-ups need to be able to handle a number of challenges related to bridging gaps between market actors. Of these, gaps to the market is considered a major challenge and associates with the observation that ideas and technologies that researchers are trying to commercialize are too far away from the existing solutions on the market (Sandberg and Aarikka-Stenroos, 2014). This infers a considerable conceptual, financial, and technological distance to bridge in order to attract users and customers. The distance can relate to the lack of established interorganizational relationships to market actors (Aaboen et al., 2013; Cantú et al., 2012). The start-up firms often start with a nonexistent network, and need to establish themselves in business networks that are willing to test and adopt the solutions that they offer (Ciabuschi et al., 2012).

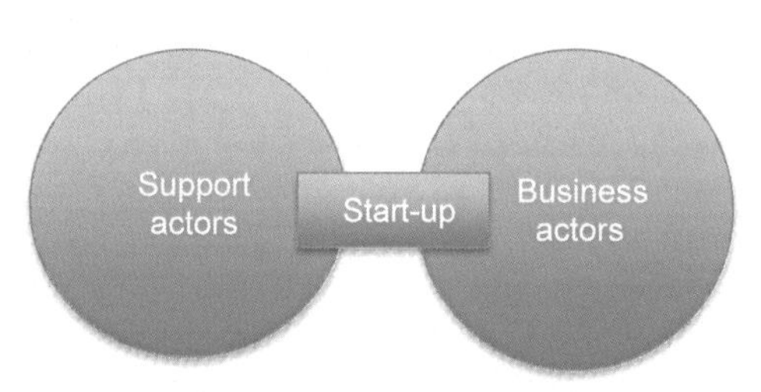

Figure 5.1 Analytical framework.

Research has noted that the failure of start-ups to thrive has been due to their inability to form meaningful business relationships with established business actors (Aaboen et al., 2011). Several reasons in the extant literature are offered to explain this inability. Some examples include the novelty of product offerings (Chandy and Tellis, 1998), lack of already established business contacts (Aaboen et al., 2013), or differences in logics (Öberg and Shih, 2014). Due to the observed difficulties, start-ups have received help from supportive actors that function as intermediaries to help start-ups to embed within business networks. For example universities have been described as integral in facilitating entrepreneurship (Rasmussen and Borch, 2010). Organizations that are related to the state and policy are also frequently argued as important (Audretsch et al., 2007). These policy actors contribute, with helpful start-up capital, in the early stages.

A major challenge for the early years is often how the start-up can remain afloat by gaining sufficient capital and resources to continue with development-related and market-promotion activities. Hence to become self-sustaining companies, start-ups require the support from actors that can provide financial resources and sometimes also network support (Clarysse et al., 2007). But generally start-ups need to progress and move from academic- and policy-supported networks into networks of business actors (Clarysse et al., 2014). This move can mean a change in logics and strategies required in order to handle other kinds of goals prevalent in the business world (Öberg and Shih, 2014). In order to describe the logics of different actors, Fig. 5.1 illustrates two different kinds of interrelated spheres of actors that often make up the environment for the start-up. On the left-hand side, there are the actors that associate with the support functions to start-ups and on the right-hand side are the business actors.

To illustrate the processes of forming and developing relationships for the purpose of commercializing products, the chapter presents two case studies of how start-ups manage in the two spheres and the broader network.

5.3 Case studies

To illustrate marketing activities in terms of market bridging actions in which start-ups engage, two cases are presented. They have been anonymized by using fictional names, and are based on empirical data collected primarily through 10 semistructured in-depth interviews (each lasting between 1 and 2 h). The aim was to catch from a network perspective how the start-ups had developed and key themes have revolved around development of business relationships and the product offerings. The respondents, some of whom have been interviewed on several occasions, consist of key individuals such as the CEO, founder, marketing manager, or scientist. Complementary data has been collected from secondary sources such as websites, press releases, and annual reports and sometimes more informal meetings or lunches with managers in the start-ups. They have provided basic information about the firm as well as more specific information on financial performance and events. The accuracy of case descriptions (current or previous more lengthy versions) has been checked with interviewees.

5.3.1 Alpha

Alpha was founded in 2010, by four participants of a business development course organized by a nonprofit organization focused on entrepreneurship and innovation. During an open innovation project, the group came together and decided to start a company. One of the founders was a biologist and professor in chemical ecology at Lund University. Her research focused on the communication of bed bugs.

The other came from engineering and/or business backgrounds. All had been working for a number of years and felt that they needed a change in the working life. With their different backgrounds and competencies, they formed a shared vision for building a company based on the biologist's research on bed bugs (their chemical communication and behavior). For the first year the company was located in a tech-based incubator in Lund. Early on the start-up the company did engage in discussion with potential customers, for example the two largest pest control firms in Sweden. While it did not result in any direct customer relationships being formed, customers were being informed about the existence of the company and also served as a way to educate the market.

The incubator was a way to locate in an environment with other entrepreneurs. This functioned as an environment where experiences could be shared with other entrepreneurs. The incubator did not provide any laboratory space and the business support was limited. In 2012, the company instead moved into a life science-based incubator in the same area. Here the company gained laboratory space and was able to build up its own research-and-development environment. This was necessary as the company needed its own research facility as it hired nonuniversity-affiliated researchers.

During the time in the first incubator the firm had been encouraged to participate in local and national business plan competitions, to test its competitiveness, and also create some media exposure. The company was being acknowledged as having good potential in competition, as well as attracting new capital from private and governmental backers, securing a Research & Grow grant from Vinnova, a governmental innovation agency, and receiving coverage by regional and national media in many forms.

The company was thus forming and establishing relationships with main actors in the innovation support system, such as the incubators, government agencies, municipality, and so forth. In parallel the discussion with potential customers were ongoing. In 2013 the firm reached agreements with their first customers, the pest control companies and the Swedish migration bureau. Fig. 5.2 illustrates the network of Alpha.

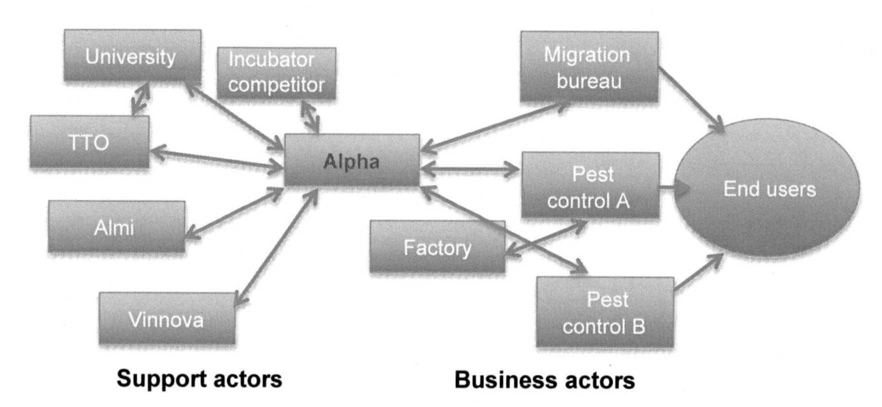

Figure 5.2 Alpha's network.

5.3.2 Beta

Beta is a firm founded in 2011, which develops engineering solutions with the aim to optimize the use of resources and provide an environmentally friendly lighting system. The start-up develops lighting-control hardware that can be used as an independent system on existing installations, or as a component in a lighting unit. Smaller amounts and prototypes can be produced at the firm's own facilities, but small and medium commercial volumes are outsourced to a larger manufacturing site.

The idea to start a firm was based on research conducted at Lund University's faculty of Engineering (LTH) in Sweden. When writing his master's project the firm's cofounder and CEO realized that it was possible to turn his research findings into a real product. He teamed up with a graduate of business development and marketing from Lund University, and his academic supervisor, a professor in telecommunications and the other cofounder of the start-up provided the technical knowledge. Expertise from the industry came from external advisors, such as a former managing director of Ericsson Mobile Platforms, as well as industry consortia.

During the first year, the firm located in a student incubator with free office space and was run with modest means. Initial funding was allocated through ALMI, a state-owned investment and advisory company, and was used to pay the founder's salary. Research was conducted in the university laboratories with the former academic supervisor, and there were few other expenses. Early discussions were also made with the university technology transfer office (TTO), which also provided an accelerator grant that was used to develop a prototype. To further the scientific base of the project, students and master thesis projects were encouraged. This enabled the project and the technological idea to further develop into a product.

Beta was able to stay in the student incubator for a year and a half, but had to leave in late 2013. Meetings with other incubators in the Science Park were held to understand where the company could relocate. Three incubators were of interest, a life science incubator, a tech-based incubator, and a service-based incubator. Finally the company decided to locate in a tech-based incubator. The incubator had no laboratories of its own, but provided office space, business coaching, and an interactive

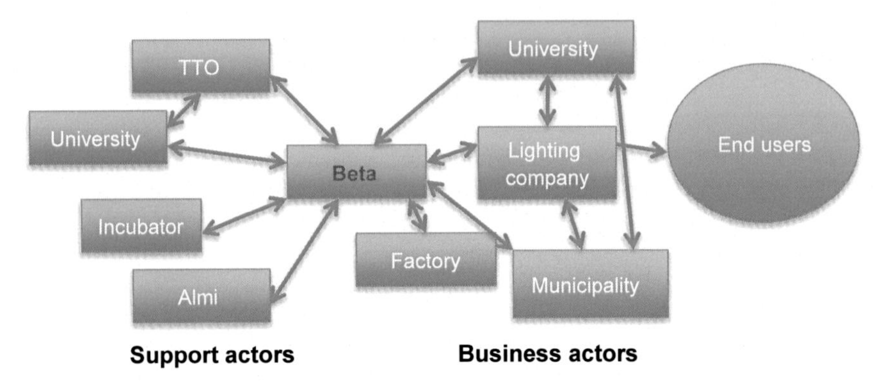

Figure 5.3 Beta's network.

environment. The benefits of joining the incubator were low rent, access to meetings with investors and other entrepreneurs, legitimacy and the brand name of the incubator, and so forth. However there were also activities that did not work toward business development. The incubator receives most of its funding from public sources, the municipality, the regional government, and the national government. The metrics used to evaluate the effect of the incubator services can take form in measuring the number of relationships formed with local actors (eg, in R&D and business), number of patents, female board members, and so forth. To some extent such pressures are not always conducive for business development. Beta thus felt that the business development support from the incubator was limited.

However, the brand name and network of the incubator provided exposure of the incubated firms to investors and potential business partners. While no investors have invested in the Beta, the exposure has led the firm to reflect on what services and products that it is offering, which has been helpful in the discussions with potential customers. The first customer came in 2014, and the firm is also in negotiation with municipalities and lighting companies concerning introducing its solutions in street lighting. In 2015 a national clean tech organization described the company as belonging among the top 25 most interesting clean-tech companies in Sweden.

The subsequent section discusses the network relationships of the two start-up firms and how they are built, with a particular emphasis on the relationships with business actors, customers, and the innovation support actors. Fig. 5.3 illustrates the network of Beta.

5.3.3 Relationships with innovation support actors

Both companies established themselves in the local innovation support structure early on. Considering the firms' lack of financial resources, it was convenient and advantageous to locate within the local science park and solicit the help of university support structures to academic entrepreneurs. This included establishing relationships with incubators, university TTO, and the regional public funding agencies.

The initial seed funding came from the public funding agencies and the TTO. The requirements to meet in order to be granted funding are often more lenient than if the start-ups go directly to private investors. The grants were small, but no share was taken by the funding organizations. This allowed the start-ups to work on their ideas for the first few months, and to test whether the ideas were viable in a business context. The role of the incubators is to offer services and support entrepreneurs in the business development process. In the two cases the incubators have offered limited business development. However they have provided cheap office space and infrastructural support.

Early on both start-ups did encounter conflicting logics between the requirements/advice from the innovation support actors and the market-based actors. This included, for example, the case of the incubators that need to report back to funders that have been the regional policy organization, national innovation agencies, and the municipality. These funding actors have been asking to view results, for instance, in terms of the number of relationships being formed between actors regionally and nationally, the amount of women on firm boards, and the number of patents. Hence there has been a deviation sometimes between goals of advancing business development and pursuing metrics sought after from the innovation funding agencies and other business support actors such as the TTO and incubators.

5.3.4 Relationships with business actors and customers

Alpha engaged with potential industrial customers from the start. This provided the company with indications for product development. Beta first developed a product and thereafter approached the customers. While these are two different approaches to customer engagement, early customer contact has benefits. For example customers are the main sources of innovation. To quickly gain revenue from customers also demonstrates the viability of the business, eases the validation of the firm to external actors, and enhances credibility.

Although building customer relationships take time, the necessity to discuss with potential customers early has been described as imperative (Ries, 2011). For Alpha the time from initiation of contact to actually signing contracts was 2 years. It took this time for Alpha to learn of the specific needs of the large customer and to also convince the customer of the reliability of the product, its production, and its delivery. There was also during this time the build-up of trust and personal connections. Lessons learned from a specific customer relationship were of course also internalized in the development of new customer ties.

For Beta, the solution and product is engineering-based and required a certain level of refinement of the prototype before approaching customers. The use of the solution is multifaceted and a number of different customers were attractive for the start-up firm. These included other companies, the university, and municipalities. The development of the product was based on current standards, the direction of change, and the regulatory changes that were occurring in the area of smart lighting, including requirements for power efficiency, quality, and so forth. The first customer was a research unit at Lund University that was doing commissioned research on lighting. The business contract, including the purchase of a lighting product as well consulting services allowed Beta to verify the applicability of its product.

As the product complied with industry standards and the efficiency of its product in relation to alternatives was demonstrated, the firm was able to initiate discussion with customers. Several customer contacts came through an industry consortia of which the firm was part; the help of senior advisors on the board were also of integral importance to create contacts. Nonetheless the solution needs to be tested in the context of the technological and social systems into which it is to be embedded. Beta is currently in the process of negotiating possible contracts with municipalities and lighting companies.

5.4 Concluding discussion

This chapter has focused on the marketing of start-ups. While traditionally this is understood as analyzing the customer needs and how to bridge that gap, for example, through segmentation and branding, the chapter has focused on bridging the gaps between industrial actors in order to build networks. The rationale of this focus relates to the fact that science- and tech-based start-ups often do not sell directly to the end consumer. Instead marketing activities are often targeted toward industrial actors and the formation of network relationships. Such network building activities can be particularly troublesome for newly formed firms as they often lack resources and business legitimacy. This chapter has discussed how start-ups interact, how their networks are built, and what contributions various actors have in terms of how they coinfluence each other and add to the possibility of the start-up development in the long term.

Some implications for start-up firms on marketing can be derived from the cases. First, the initial relationship with customers can play an important role in the business direction of the start-up and the service and/or product offerings. Handling customer relationships entails long-term commitment and the opportunity to create value by embedding new solutions in existing structures.

There are also various conflicting logistics that start-up firms have to handle. For example, in the above cases, the innovation support actors had requirements that were often incongruent with efficient business practice and consequently nonconducive for start-up development. The metrics used to measure performance by the incubators cause unnecessary disturbance to the start-up's business development. Nonetheless, these relationships might be necessary for the survival of the start-up at an early stage.

This chapter suggests that understanding the nature of a start-up firm's relationship with marketing relates to how the start-up manages in the network, including the different goals and logistics that the network partners represent.

5.4.1 *Managerial implications*

A critical issue for a start-up firm is whether there is a demand for its product. The start-up also often starts without customers or established relationships, and few resources. Given this conundrum, one of the main concerns is how the start-up can build a network of partners that can support the company in gaining resources to eventually commercialize its products. The implications for this study, as illustrated by the cases, suggest that the supportive environment plays an important role in providing resources related to office space, closeness to research intensive actors, and some minor funding opportunities. Start-up firms, however, need to carefully assess the requirements of the supportive system and make a sound judgment about how the support actually drives forward business development and the formation of customer relationships. As deeper relationships are being created with support actors, stronger interdependencies are being created, which can influence the business development plans.

To interact early on with potential customers can provide that information window to start-ups to understand the demand. Here, it can be difficult to try and understand who is a relevant customer. In the case of a start-up, the product is often also created through the interaction with the potential customer and be customized to such an extent that it cannot be directly sold to other customers. In this case the business model is also evolving, and cocreated with the demands from market actors. This study suggests that a start-up needs to interact early on with the two spheres analyzed in the cases. It is imperative that start-up firms seek to integrate and work within the two spheres in parallel from the earliest.

5.4.2 *Limitations*

The two cases embed in the same support structure and have followed similar paths in the early stages of business development with respect to how they interact with the support system actors and the resources they offer. The support system will appear differently in different contexts, with respect to actors, resources, and so forth. The exact strategies that the start-up firms used might not be relevant for firms in other contexts. However, irrespective of environment, start-up firms would arguably need to interact with potential customers from the beginning, in order to close gaps to the market. Often there are also structures that can support the formation of networks in early stage ventures. In this case the findings from this study go beyond idiosyncratic conditions and would be general to start-ups in most science and high-tech regions in economically developed nations.

References

Aaboen, L., Dubois, A., Lind, F., 2011. Start-ups staring up: firms looking for a network. IMP Journal 5 (1), 42–58.

Aaboen, L., Dubois, A., Lind, F., 2013. Strategizing as networking for new ventures. Industrial Marketing Management 47 (7), 1033–1041.

Agndal, H., Axelsson, B., 2012. Professional Marketing. Studentlitteratur, Lund.

Almus, M., Nerlinger, E.A., 1999. Growth of new technology based firms: which factors matter? Small Business Economics 13 (2), 141–154.

Anderson, A.R., Dodd, S.D., Jack, S., 2010. Network practices and entrepreneurial growth. Scandinavian Journal of Management 26 (2), 121–133.

Audretsch, D., Grilo, I., Thurik, A., 2007. Handbook of Research on Entrepreneurship Policy. Edward Elgar, Cheltenham.

Autio, E., Kenney, M., Mustar, P., Siegel, D., Wright, M., 2014. Entrepreneurial innovation: the importance of context. Research Policy 43 (7), 1097–1108.

Baraldi, E., Waluszewski, A., 2011. Betting on science or muddling through the network two universities and one innovation commission. The IMP Journal 5 (2), 1–21.

Bathelt, H., Kogler, D., Munro, A., 2010. A knowledge based-based typology of university spin-offs in the context of regional development. Technovation 30 (9), 519–532.

Cantú, C., Corsaro, D., Snehota, I., 2012. Roles of actors in combining resources into complex solutions. Journal of Business Research 65 (2), 139–150.

Chandy, R., Tellis, G., 1998. Organizing for radical product innovation: the overlooked role of willingness to cannibalize. Journal of Marketing Research 35 (4), 474–488.

Ciabuschi, F., Perna, A., Snehota, I., 2012. Assembling resources when forming a new business. Journal of Business Research 65, 220–229.

Clarysse, B., Wright, M., Lockett, A., Van de Velde, E., Vobora, A., 2005. Spinning out new ventures: a typology of incubator strategies from European research institutions. Journal of Business Venturing 20 (2), 183–216.

Clarysse, B., Wright, M., Lockett, A., Mustar, P., Knockaert, M., 2007. Academic spin-offs, formal technology transfer and capital raising. Industrial and Corporate Change 16 (4), 609–640.

Clarysse, B., Wright, M., Van de Velde, E., 2011. Entrepreneurial origin, technological knowledge and the growth of spinoff companies. Journal of Management Studies 48 (6), 1420–1442.

Clarysse, B., Wright, M., Bruneel, J., Mahajan, A., 2014. Creating value in ecosystems: crossing the chasm between knowledge and business ecosystems. Research Policy 43 (7), 1164–1176.

Di Gregorio, D.D., Shane, S., 2003. Why do some universities generate more start-ups than others? Research Policy 32 (2), 209–227.

Feninleib, D., 2012. Why Startups Fail: And How Yours Can Succeed. Apress, New York.

Gadde, L.-E., Håkansson, H., 2001. Supply Network Strategies. John Wiley, Chichester.

Håkansson, H., Snehota, I., 1995. Developing Relationships in Business Networks. International Thomson Press, Boston.

Håkansson, H., Waluszewski, A., 2002. Managing Technological Development. Routledge, London.

Håkansson, H., Ford, D., Gadde, L.-E., Snehota, I., Waluszewski, A., 2009. Business in Networks. John Wiley & Sons, Chichester.

Håkansson, H., 1987. Industrial Technological Development a Network Approach. Routledge, London.

Kotler, P., Kartayaja, H., Setiawan, I., 2010. Marketing 3.0: From Products to Customers to the Human Spirit. Wiley, New Jersey.

La Rocca, A., Snehota, I., 2014. Relating in business networks: innovation in practice. Industrial Marketing Management 43 (3), 441–447.

La Rocca, A., Ford, D., Snehota, I., 2013. Initial relationship development in new business ventures. Industrial Marketing Management 42 (7), 1025–1032.

Lechner, C., Dowling, M., 2003. Firm networks: external relationships as sources for the growth and competitiveness for entrepreneurial firms. Entrepreneurship & Regional Development 15 (1), 1–26.

Lindelöf, P., Löfsten, H., 2006. Environmental hostility and firm behavior – an empirical examination of new technology-based firms on science parks. Journal of Small Business Management 44 (3), 386–406.

Linné, Å., Shih, T., 2013. The political embeddedness of business networks in a Chinese context: The case of a biopharmaceutical business network. IMP Journal 7 (3), 180–187.

Malerba, F. (Ed.), 2010. Knowledge-Intensive Entrepreneurship and Innovation Systems. Evidence From Europe. Routledge, London.

Mian, S.A., 1996. Assessing value-added contributions of university technology business incubators to tenant firms. Research Policy 25 (3), 325–335.

Mian, S.A., 2011. Science and Technology Based Regional Entrepreneurship: Global Experience in Policy and Program Development. Edward Elgar, Cheltenham.

Öberg, C., Shih, T.T., 2014. Divergent and convergent logic of firms: barriers and enablers for development and commercialization of innovations. Industrial Marketing Management 43 (3), 419–428.

Öberg, C., 2010. Customer roles in innovations. International Journal of Innovation Management 14 (6), 989–1011.

Perez, M., Sanchez, A., 2003. The development of university spin-offs: early dynamics of technology transfer and networking. Technovation 23 (10), 823–831.

Rasmussen, E., Borch, O., 2010. University capabilities in facilitating entrepreneurship: a longitudinal study of spin-off ventures at mid-range universities. Research Policy 39 (5), 602–612.

Ries, E., 2011. The Lean Startup: How Today's Entrepreneurs Use Continuous Innovation to Create Radically Successful Businesses. Crown Publishing, New York.

Rothaermel, F.T., Thursby, M., 2005. Incubator firm failure or graduation? The role of university linkages. Research Policy 34 (7), 1076–1090.

Salvador, E., Rolfo, S., 2011. Are incubators and science parks effective for research spin-offs? Evidence from Italy. Science and Public Policy 38 (3), 170–184.

Sandberg, B., Aarikka-Stenroos, L., 2014. What makes it so difficult? A systematic review on barriers to radical innovation. Industrial Marketing Management 43, 1293–1305.

Snehota, I., 2011. New business formation in networks. IMP Journal 5 (1), 1–8.

Visintin, F., Pittino, D., 2014. Founding team composition and early performance of university-based spin-off companies. Technovation 34 (1), 31–43.

Waluszewski, A., Baraldi, E., Shih, T., Linné, Å., 2009. Resource interfaces telling other stories about the commercial use of new technology: the embedding of biotech solutions in US, China and Taiwan. IMP Journal 3 (2), 86–123.

Zahra, S., Van de Velde, E., Larraneta, B., 2007. Knowledge conversion capability and the growth of corporate and university spinoffs. Industrial and Corporate Change 16 (4), 569–608.

Zimmerman, M., Zeitz, G.J., 2002. Beyond survival: Achieving new venture growth by building legitimacy. Academy of Management Journal 27 (3), 414–431.

A minimalist model for measuring entrepreneurial creativity in eco-systems

E. Carayannis [1], *P. Harvard* [2]
[1]Professor George Washington University; [2]Professor EIGSI Engineering School France

This presentation is an *esquisse* of entrepreneurial creativity beginning and ending with Irishmen. To begin with we shall refer to an 18th century Irish banker in France, Richard Cantillon, who was one of the first to define the theoretical importance of the role of entrepreneurs [44]. We shall also later refer again to the lucky Irish when examining some pertinent empirical experiences from the life of a more recent 21st century Irishman and entrepreneur, Tony Ryan of Ryanair [1,28]. Cantillon referred to entrepreneurs as sources of new wealth or new money circulating in the economy, thus contributing to economic development. His idea of new money or new sources of wealth as an integral part of economic growth is the Cantillon effect similar to William and McGuire's drivers of growth [21]. The Office of Commercial and Economic Development (OCED) claims it was not until the 1990s when entrepreneurship became a worldwide economic buzzword [35]. After all, since definitions are always formulated with words, whether keywords or buzzwords, it seems only logical to take a moment to examine the importance of words that can be chosen to define, in our case, entrepreneurial creativity. Researchers like Dell'Era, Buganza, Speranzini, and Bacot give great importance to speech, semantics, and words in the innovation process. Their language brokering process insists on the importance of word choice to stimulate entrepreneurial creativity [4]. In this light, it is only logical to take the time for very brief etymological study comparing interesting words pertinent when defining entrepreneurial creativity. For example, the etymological origin of the English word pirate is the Greek word *peirates* [23]. But the root of *peirates* is *peiran* and is translated in French as meaning *celui qui entreprend*, or he who undertakes [23]. French for undertake is *entreprendre* and *entreprendre* is the root of entrepreneurship in English. The French used *entreprandre* (another form of the same word) in the Middle Ages to mean to seize, surprise attack, and aggress [23]. The French connection between Greek and English can help explain why adventure and risk are so often associated with entrepreneurship. To complete our etymological word search it is important to realize, the French word *entreprendre* is actually two words combined into one word: *entre* and *prendre*. In French *entre* means "between" and *prendre* means "to take." Therefore, we can induce the French word means to successfully achieve specific objectives. This is how entrepreneurs see tomorrow today. In short, entrepreneurship is a pattern of preplanned intermediary actions leading to foreseen and hoped-for results or what the OCED refers to as performance [35]. Considering these two French words, we could

Start-Up Creation. http://dx.doi.org/10.1016/B978-0-08-100546-0.00006-6

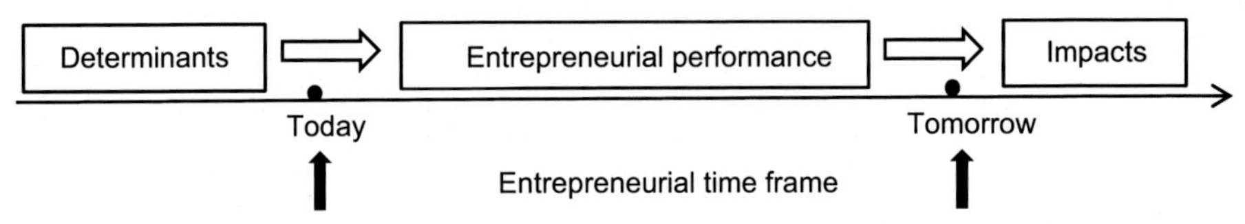

Figure 6.1 Entrepreneurship is what to do between two points in time.

say entrepreneurship is a point of view of what to do between two points in time—today and tomorrow—thus offering a more creative vision of future potential. These two points can be situated between the three main phases of OCED's model (see Fig. 6.1).

The OCED, together with Eurostart, has contributed richly to international efforts and leads the way for measuring entrepreneurship. Their research is carried out in their industrialized member countries: Australia, Austria, Belgium, Canada, Czech Republic, Denmark, Finland, France, Germany, Greece, Hungary, Iceland, Italy, Japan, Korea, Luxembourg, Mexico, Netherlands, New Zealand, Norway, Portugal, Slovak Republic, Spain, Sweden, Switzerland, Turkey, the United Kingdom, and the United States [35]. The OCED leads the world in proclaiming how much measuring to define entrepreneurship requires favorable government policies everywhere around the world and the great need to expand such measurements beyond self-owned and medium-sized companies. This is why the authors prefer to choose to examine a trio of worldwide entrepreneurs with some of their empirical global experiences. We read in the OCED's literature a scale of companies considered in their research, which were categorized into five sizes based on the number of employees: micro ($\geq$10 employees that employ 30+% work force in their member countries); small ($\leq$10 but $\geq$50 employees), medium-sized ($\leq$50 but $\geq$250 employees), and large ($\geq$250 employing 40+% work force in the same countries) [35]. They put into place a model for measuring entrepreneurship based on reference points or criteria called entrepreneurial indicators, which are categorized in three phases: determinants, performance, and impacts. These three phases are composed of indicators such as regulatory authorities, R&D and technology, culture, capabilities, market conditions (the determinants); plus action, firms, employment, wealth (the performance); and finally monetary or nonmonetary value, job creation, economic growth, and poverty reduction (the impacts). According to the OCED, entrepreneurship is "…not just creation of financial wealth-…increasing employment, tackling inequalities, or indeed…environmental issues…." The OCED see entrepreneurs as people who "…generate value through creation or expansion…by identifying …new products, processes or markets…" [35]. We shall later refer to the Schumpetrian-Kirznerian duality of entrepreneurial opportunities.

Creativity in almost all professional fields of activity is always greatly influenced by language, culture, and history, which are essential elements in any research. Leaving behind an incomplete and very brief etymological word search, we now move on to culture—one of OCED's leading determinants for measuring entrepreneurship [35]. A comparative analysis about different aspects of entrepreneur eco-systems between cultures offers interesting, and at the same time, important considerations. This can

Pioneer heritage / self-reliance or D-Y-I	Latin heritage / *patronus* & *pater* / guilds / specialists
management by objectives for profit	paternalistic management / legal strikes / employee rights
wealth is Heaven's blessing for hard work	rich are selfish and dishonest / rich are mean & poor are good
admire winners / want to be the best	admire survivors / being #1 is pretentious while #2 is more authentic
overview from top down / American eagle symbol	details from bottom up / French rooster symbol
material progress improves life	social progress for all
private financing through tax deductions	government funding through high taxes
Bill Gate's 70 hour work week	Martine Aubry's 35 hour work week
$10+ registration fee to create a company	required capital to create a company (≈ 7500–30000 Euros)
< $50 copyright fee / ≤ $500 patent fee	≥ 30000 Euros patent fee
(lasts lifetime of creator or inventor + 50 years)	(fees to renew patents)
American freedom = financial autonomy	French *noblesse oblige* = civic responsibilities & duties
take risks	analyze risk
competitive	collaborative
team players & team spirit to create together	isolated individual creativity
for / positive	against / criticize
We can - so why not?	because of......

Figure 6.2 Some cultural determinants influencing American and French entrepreneur eco-systems.

be easily done by comparing the point of view of an American, the second of our two authors, Harvard, who has been working and researching in France, his adopted culture, for over 30 years. The following Franco-American exercise allows an empirical comparative analysis of American and French entrepreneurial creativity in different eco-systems (see Fig. 6.2) [8,11].

At this point it is important to expand our very brief survey of research literature, beyond that of the OCED, to include other known prominent experts and researchers. For example, in the *International Journal of Entrepreneurship and Small Businesses*, experts like Justo, De Castro, and Maydeau-Olivares reinforce the need for better measuring to better define entrepreneurship and refer to "...difficulties when it comes to cross-country comparisons regarding entrepreneurial activity..." [31]. They promote global entrepreneurship monitor (GEM) as "...a unique attempt to both provide homogeneous cross-country measures of entrepreneurship activity and ascertain the relationship between entrepreneurship and economic development..." [31]. Their approach measures the likelihood of someone engaging in entrepreneurial activities and focuses on criteria 1, entrepreneurial behavior (eg, what is his or her current job?) as well as criteria 2, social environment of the entrepreneur (eg, does he or she know any business angels?) [31]. Interestingly enough, according to the OCED/Eurostart, there are nearly twice as many business angels in France than in Germany and Spain while Spain and Germany have twice as many as the United Kingdom [35]. Research literature of the Kaufman Foundation says there over 75,000 business angels throughout Europe investing around 15 million Euros while there are over 250,000 in the United States investing over $17 billion [34]. The OCED literature refers to young high-growth, high-tech dynamic entrepreneurs as gazelles, and an entrepreneurial time frame, where these gazelles have a lower survival rate in their second year of doing business compared to their first year of doing business [35]. Expanding Kaufman's research model focuses on two indicators or criteria for measuring entrepreneur eco-systems, vibrancy and outcomes, which offer a very rich, almost universal, framework to examine four aspects of profiles and skills of entrepreneurs. The first is density (which is more present in certain sectors of activity, etc.). The second is fluidity (including population, market growth, labor market, relocating, etc.). The third is connectivity (meaning high-tech networking, mobility, etc.), and the fourth is diversity (or spin offs, deal-making, etc.) [33]. According to experts like Jonathan Ortmans, there are many other institutions acting as information-gathering data banks about entrepreneurship such as Global Entrepreneurship and Development Index (GEDI) [35]. Justo, De Castro, and Maydeu-Olivares conclude by confirming "...entrepreneurship is indeed a multidimensional concept...depending on...very distinct...realities..." [31]. In the *Journal of Developmental Entrepreneurship*, Robichaud, McGraw, and Roger (not unlike Amabile) emphasize the importance of entrepreneurial motivation as something that must happen [20,45], before but leading up to, entrepreneurial behavior (as in Justo, De Castro, and Maydeau-Olivares); then it gives birth to entrepreneurial performance (as in OCED/Eurostart) [35]. We cannot ignore the very pertinent research of Amabile, who stresses the importance of an individual's intrinsic motivation as the key to progress, as a dominating principle of creativity [2,30].

The writings of the first of our authors, Carayannis [6,7], and his colleague Dubina, in the *Journal of Knowledge Economy*, confirm the importance of understanding "...innovative entrepreneurial behavior..." of entrepreneurs as "economic actors who are engaged, or want and can be engaged in innovative actions..." [40]. Kirzner's prolific research and writings are rich and exhaustive but contrasting him to Schumpeter offers insights into entrepreneurs also as economic actors. Schumpeter and Kirzner offer opposing definitions of entrepreneurs as economic actors and see entrepreneurial behavior and roles in precise yet different ways [13−17,24−36]. Schumpeter says entrepreneurs are economic actors who are highly motivated and dynamic innovative individuals catalyzing creativity in structures of rapidly growing markets. Kirzner sees entrepreneurs as those who creatively respond with innovative solutions to fulfill the needs of the day in markets where there is strong incumbent competition [13−17,24,36]. Basadur, Pringle, Speranzani, and Belitski also contribute richly in trying to better understand the behavior and roles of entrepreneurs as actors on various economic levels. Their quadruple collaboration resulted in giving priority to a collaborative win−win model increasing creativity and innovation between heterogeneous cognitive styles for greater employee satisfaction in team-related performance [4,41]. Ward also links creativity to cognitive capacities in the *Journal of Business Venturing* [20,33]. He wrote about a "...cognitive approach to creativity...generating novel and useful ideas...." In the *International Handbook of Innovation*, Carayannis and Gonzales contribute to the defining of entrepreneurial creativity and suggest creativity is thinking "...out of the box..." and a "...capacity to imagine....to perceive future potentials...based on current conditions..." [25]. The Basadur, Pringle, Speranzani, and Belitski model of entrepreneurial behavior goes further and defines four main axes or roles: the implementer, the generator, the optimizer, and the conceptualizer, who gravitate between binary poles of creativity including experiencing–thinking and evaluation–ideation [4,22,41]. They attribute a great deal of importance to and emphasize problem-defining as a key element of entrepreneurial creativity as well as the nurturing of tools in a given workplace favoring creativity going out of and beyond the corporate work box structure [4,22,41]. In short, entrepreneurs can be either short-term corporate creativity stars shining in stormy markets, or long-term, low-key innovators with a strong stable presence in highly competitive markets.

Whichever of the myriads of models chosen to measure and define entrepreneurship it is necessary to understand what to measure. Murphy, Traiter, and Hill confirm, in the *Journal of Business Research*, the existence of an enormous diversity of models and measures plus the overly numerous performance variables being proposed as essential to consider when defining entrepreneurial creativity [32]. Carayannis and Gonzales even proclaim "...imagination is more important than knowledge..." [25]. Yet it is knowledge that reshapes, increases, and changes the limits of one's work-box structure because knowledge is organized information put into action through decisions [25]. Only when information being raw data put into a logical order make it useful and usable. Knowledge, which includes sociocultural knowledge and semantics of the aforementioned Dell'Era, Buganza, Speranzani, and Bacot [10,37], can make almost any work-box structure evolve technologically and therefore become commercially and economically more competitive on the market. The Creativity Theory of Knowledge Spillover found in the research and writings of Audretsch and Belitski confirms how entrepreneurial creativity generates and increases knowledge within the limits of a given economic framework where an entrepreneur is present and active [3,29]. Carayannis and Gonzalez have the same logic describing sources of creativity as originating from a personal view point or vision, seeing things from product-oriented or process-oriented points of view, and finally, being empirically inspired by particularities of a given environment or culture [25]. An example of cultural and environmental considerations about entrepreneurial creativity in two different eco-systems are presented in Fig. 6.2.

Carayannis and Gonzalez team up with Wetter again in the *International Handbook of Innovation* to go further and describe how innovation is either evolutionary or revolutionary and can thus result in discontinuous or continuous creativity disrupting (or not) in OCED's work box structures full of employees [25]. The echoes of Kirzner–Schumpeter's duality is once again recognizable [13–17,24,36]. In the *Journal of Knowledge Economy*, Carayannis teams up with Hens and Nicolopoulon-Stamati to elaborate ideas about creativity with regard to the work-box structure syndrome [40]. They explain evolutionary differences between just thinking outside the box to thinking beyond the box and even transdisciplinary thinking out of the box. Here it is easy to discern a sort of Maslowian progressive evolution from lower to higher levels. Just as in the different levels on Maslow's pyramid of employee needs, for Carayannis and his colleagues, the issue is not a judgmental one of better or worse or of superior or inferior, but progressing and evolving to improve and develop employee competencies to be more creative in his or her given work-box structure [40]. Going back to the 1990s, research literature clearly demonstrates laudable efforts to propose models of measuring cognitive behavioral capacities of entrepreneurs. Theresa Kline and Flora Stormer, both from the University of Calgary in Canada, proposed the General Enterprising Tendency Test with a 54-item questionnaire to measure the capacity of being an entrepreneur [27,42]. Still in the 1990s, Holly Buttner and Dorothy Moore, from South Carolina, focused on the motivation of women to succeed as entrepreneurs [39].

In more up-to-date literature in 2015, we read in the Harvard Business Review what Brynjolfsson and MacAfee of the Massachusetts Institute of Technology wrote about today's second machine age of digital technologies, as opposed to the first machine age of mechanical technologies during the industrial revolution, and the resulting disruptive decoupling [30]. Such a disruptive differentiating gulf between previously statistically coupled measures now separates productivity from income and employment. Brynjolfsson clearly declares "...we need more entrepreneurship...." He emphasizes a more pertinent and very decisive role of creativity in future work-box structures. He says "...it takes a lot more creativity...to do something never done before...that will be more valuable in the marketplace..." [30]. Therefore the importance of creativity for entrepreneurs has an even greater importance in the 21st century than ever before. Apparently, it seems Carayannis and Gonzales were right in saying "...imagination is more important than knowledge..." [25].

Winter and Govindarajan offer a pristine empirical example of true entrepreneurial creativity in a recent issue of Harvard Business Review. They explained the dilemma of the long-term advantage or utility of the PlayPump project. It was viable only on a short-term basis. It consisted of African children playing and pushing a merry-go-round, in isolated desert regions, creating energy to pump needed water from a source in the ground up into an elevated storage tank for a nearby village. Unfortunately it has been abandoned since 2010 because it does not offer long-term feasibility. It would require up to 200 children playing 10 min a day to provide 3000 L of water for a village of 1000 people. Biais and Perotti, from the *Rand Journal of Economics*, would be right to bring up entrepreneurial considerations like technical feasibility, legalities, and patentability about the PlayPump case [38]. In research published in the *Journal of Business Venturing*, Gartner gave great importance to, even a priority to, the vital importance of time frames, specific time periods and sustainability over time of entrepreneurial actions [33]. In the case of PlayPump, entrepreneurial creativity clearly was timely but only over a specific period of time before the time frame expired. In other words, entrepreneurs must be able to see, size-up, then seize chances to create within a time frame.

The preceding, but all too succinct, small sampling of research literature paves the way for an audacious attempt to propose later a minimalist model of three criteria for measuring, hence better defining, entrepreneurial creativity. But first it is important to clarify two very pertinent concepts: measuring and defining. Often when doing surveys or analysis or research, links between concepts or ideas seem to occur as if in a pattern of cycles. Such cyclic linking of ideas surface regularly in many research pursuits and therefore seem appropriate when measuring entrepreneurship in order to better define entrepreneurial creativity. Therefore, it is only logical that the outcomes of models for measuring entrepreneurship can only enrich and greatly contribute to updating and revising a definition of entrepreneurial creativity. In fact in order to define it, it is also necessary to establish criteria or indicators (as OCED/Eurostart) or reference points. This means there is already a degree of some kind of measuring going on when defining. We can hence surmise definitions as being theoretical, yet all the while, based upon empirical measurements providing criteria indicating reference points and indicating what and determining how to measure. We can also say the nonuniversality of existing measurement models is due to their complexity and to the fact they seek to conform to measure a vast number of overly theoretical variables rather than empirical experiences lived by real entrepreneurs. Whatever it be the research criteria, indicators, variables, or standard, the purpose of any measurement (theoretical or empirical) is to offer a better definition. Once success has been achieved, the intermediary creative actions must be replaced by permanent more structured ones to guarantee on-going long-term success.

Entrepreneurs carrying out successful intermediate actions often have a hard time adapting themselves to being managers who have a more day-to-day vision necessary for long-term stability. The spontaneous high-level energy of an entrepreneur's risk-taking creative vision of the future is not always transformable into continuous on-going day-to-day energy to manage and maintain the structure he or she set up and put into place. In other words, should entrepreneurs be able to be both Schumpeterian and Kirznerian? Therefore it is appropriate to insist how spontaneity, so typical of entrepreneurial creativity, is usually short-term or intermediary, and not easily integrated into long-term management actions. Creativity remains a very important star word among the main key words of the OCED/Eurostart's research about entrepreneurship cited. This means no matter which work-box structure, or company where you may find yourself as an employee on OCED's structural scale of companies (micro, small, medium-sized, or large) entrepreneurial creativity can exist!

A few poignant examples from various professional activities, like education or the music industry or the medical industry, clearly demonstrate a unique aspect of entrepreneurial creativity: a high-risk survival factor. As has already been mentioned, this risky pirate syndrome of entrepreneurs goes back to the Greek *peirates*, origins of the French word *entreprendre*. If we consider a university lecturer in higher education, such as the second of our authors, Harvard, his pedagogical creativity to redo a module and put it online does not necessarily change his salary, yet he might receive a bonus or other employee incentives to be more creative in his teaching responsibilities or tasks. For him there is relatively little chance of a risky pirate syndrome threatening his present stability for future actions. When it comes to the music industry, we can consider the well-known examples of Michael Jackson and Madonna, who contributed richly and creatively to the music industry markets over decades. Their professional creativity benefited from a comfortable safety net called fame, due to acquired client loyalty and repeated client purchase of music they produced. The risk factor for similar famous entertainers is diluted and reduced by a risk-free celebrity status framework favoring creativity that defines, even initiates, the creation of music market trends.

This kind of creativity stabilizes, hence minimalizes, the risk factor, almost eliminating entirely the risky pirate syndrome. Medical researchers in pharmaceutical companies, such as Harvard's oldest daughter (a recent med school grad), benefit from a comfort zone of stability when creating under the protection of a salary, a stipend, a sponsor, and other forms of pre-agreed-upon financing for a medical development project. For her too, the presence of any risky pirate syndrome is relatively nonexistent in a global corporate career path scheme. In this light, may we propose the following about entrepreneurial creativity as opposed to educational creativity, musical creativity, and medical creativity: entrepreneurial creativity includes a high-risk survival factor, not only because of a high-level energy so typical of entrepreneurs, but also because of a high level of spontaneity.

From a minimalist point of view there seems to be at least three reoccurring elements in the research literature considered previously about measuring entrepreneurship to better define entrepreneurial creativity. Above all, the first criteria of the minimalist model for entrepreneurial creativity must foresee the importance of detecting timely opportunities where others only see risks, and organize risk-taking into a profitable pattern of actions at the right time and place. This means being able to see what to do between two points in time, today and tomorrow, in order to accomplish a future objective during a foreseen time frame (see Fig. 6.1). This first criteria of a more universal model might be able to provide some kind of a common working ground attempting to put into harmony ideas proposed by certain researchers, institutes, and experts cited (Audretsch, Belitski, Brynjolfsson, Gartner, Govindarajan, Kaufman, Kirzner, MacFee, OCED, Schumpeter, Winter, etc.). Second in our minimalist model, we must consider the fact the engaged behavior of entrepreneurs favors the cognitive process of coming up with what is different and new. Such creative ideas permit them to take the risk of judging, choosing, and deciding how to think out of and beyond their work-box structure or even on a transdisciplinary level, all the while taking into consideration sociocultural knowledge and semantics of their in-house innovation processes. These different thinking levels of entrepreneurial creativity provide the insight and foresight to discern what needs to be done now to opens doors to feasible future changes. This second criteria of a more universal model should be able to pave the way for another common working ground, this time allowing the harmonizing of concepts proposed by the some of the experts and researchers cited (Amabile, Bacot, Basadur, Biais, Buganza, Buttner, Carayannis, De Castro, Dell'Era, Dubina, Fecchio, Gonzales, Hens, Justo, Kaufman, Kline, Maydeau-Olivares, McGraw, Moore, Nicopoulon-Stamati, Perotti, Pringle, Robichaud, Roger, Speranzini, Stormer, Vergaati, Ward, Wetter, etc.). Third, a final criteria of a minimalist model for entrepreneurial creativity should be likened to a red carpet of ideas across a bridge through time leading up to quantifiable future changes. Such quantifying requires statistics in order to numerically analyze and interpret the long-term viability of changes resulting from entrepreneurial actions. This third criteria of a more universal model opens the door to a common working ground, permitting probable harmonizing of principles researched by several experts, institutes, and researchers cited (Audretsch, Belitski, Eurostart, GEDI, GEM, Hill, Kaufman, Murphy, OCED, Traiter, etc.).

So the question now is how to measure in order to better define entrepreneurial creativity. Here we adopt the French Cartesian approach based upon Descartes's idea that everything in the universe has its opposite to consider when trying to understand it completely. May we dare offer a definition of entrepreneurial creativity by taking into consideration the empirical opposite of the myriad of theoretical complex models traced by researchers, institutions, and experts? Allow us to give priority to a model based on empirical experiences lived by real entrepreneurs. In other words, we dare propose a more empirically based minimalist model articulating around the aforementioned three criteria: (1) timing including timeliness and time frame, (2) engaged behavior patterns favoring a cognitive capacity to think outside of work-box structures, and (3) quantifiable changes, resulting from entrepreneurial actions. A probable experimental application can be attempted by comparing the careers of three entrepreneurs of the British Commonwealth, still the world's oldest existing eco-system; a savvy Scotsman: Roy Thomson (1894–1976), an impossible Irishman: Tony Ryan (1934–2007), and an eccentric Englishman: Richard Branson (1950–) spanning an accumulated 100 years of entrepreneurship between the three of them. We adhere to Carayannis and Gonzalez, previously mentioned, when they express the same preference as OCED, to examine aspects of entrepreneurship in an empirical business and industrial context and not examine "...artists, scientists, entrepreneurs as obsessed manias and clairvoyant oracles..." [25].

The following very short keyhole retrospectives of Thomson and Ryan and Branson as entrepreneurs offer empirical sources of critical information to consider when attempting to measure and hence better define entrepreneurial creativity. Would Amabile be pleased with our chosen trio of entrepreneurs? Would she feel they all three demonstrate individual intrinsic motivation? How would Basadur attribute to them his cognitive profiles? Would Thomson be as more an implementer, or Ryan a generator, or Branson a conceptualizer? What about being capable of thinking outside the limits of a given work-box structure to see new work-box structures? Thomson was capable of going from one box to another and Ryan certainly did see beyond a certain airline work-box structure and saw market potential to create a new airline work-box structure. Branson's vivacious zest for diversity undoubtedly puts him in the transdisciplinary thinking-out-of-the-box category (see Fig. 6.3). Measuring by quantifying changes on various kinds and levels of empirical entrepreneurial performance can include quantifying changes in production cost, market price, profit margin, job creation, employee performance, as well as unemployment rates, different kinds of income like employee salaries, and work-box structure company turnover. Changes in turnover may seem a cruel crude commonsense approach, even perhaps too empirically primitive or overly simplistic and very materialistic for certain researchers, institutes, and experts.

Out of all the models considered, the simplicity of the Kaufman Model places it near the top of the list of candidates for being the most universal model [34]. Simplicity is universality. Kaufman's first indicator, or criteria, for measuring is vibrancy and corresponds to cognitive capacity, which is the second criteria of the minimalist model [34]. Outcome is the second indicator, or criteria, for measuring in the Kaufman model and corresponds to the third indicator of the minimalist model.

In...	His creative project was...	In...	His creative project was...
1966	Student magazine	1996	Second record label
1969	Mail order-company	2004	Space travel launchers
1971	Record shop	2006	Cable, broadband, telephone
1972	Recording studio	2007	Health bank
1973	1st artist on his record label	2008	Health care
1980	Travel agency	2009	Formula one
1983	Radio station and Virgin Megastores	2010	Hot air balloons
1984	Airline	2012	Test launch of a spaceship
1992	Financial difficulties (he cried when he sold Virgin Megastores for $1 billion.)		
1993	Railways	2014	Drones

Figure 6.3 Branson's entrepreneurial creativity.

Outcomes change things and what changes is measurable numerically. After all, when it comes to outcome, we cannot ignore the hard cold economic and commercial fact that in industry, engineers do not consider inventions to be an innovation until if and when they become quantifiable profits. Was is not the entrepreneur's new wealth or new money contributing to economic growth that makes Contillon's effect famous and accepted for centuries? [45] The limited career elements considered here as empirical experiences from the lives of Thomson, Ryan, and Branson are nonexhaustive therefore not representative. But they are still indicative of long-term profitability, as plausible and appropriate, quantifiable change for measuring the outcome of entrepreneurial creativity (criteria 3 of the minimalist model).

6.1 A savvy Scotsman

One of Roy Thomson's first newspapers was *The Scotsman* [18,45]. We respect his ferocious fidelity to his origins and present him as being Scottish because his hereditary peerage as a Baron attributed in 1962 cost him his Canadian citizenship. Son of a barber, this Canadian-born entrepreneur was a prominent 20th century newspaper magnate and media entrepreneur. His family was among the first settlers of Scarborough. He tried to follow the tenant farmer family tradition of his Scottish ancestors of Wester Kirk, Dumfriesshires, but he failed at farming in Manitoba, Ontario. During World War I, he attended a business college but was rejected by the Army due to his poor eyesight. After the war, he returned to the big city of Toronto trying different jobs until he finally ended up selling radios in Northern Ontario. Early in his career, his entrepreneur mindset allowed him to see, size up, then seize the chance to buy the rights to a radio frequency and in 1931 set up a radio station so his customers would have something to listen to in North Bay, Ontario. His creativity gave him the insight to understand and know what needed to be done. In 1934 he made a down payment to purchase a newspaper, the *Timmins Daily Press*, in Ontario [18,45]. By 1949 his entrepreneurial creativity had expanded to include fitted kitchen manufacturing, ice cream cones, and ladies hairstyling shops. He was not afraid of what was new or different and cognitively capable of thinking outside one box to invest in another one (criteria 2 of the minimalist model). He returned to his ancestral country, purchasing *The Scotsman* in 1952 and Scottish television in 1957. In 1959 he purchased a group of British newspapers including *The Sunday Times* and later bought the legendary *The Times* in 1966.

Roy Thomson could see an opportunity where others saw only a risk. He even had some joint projects with John Paul Getty in oil ventures [18,45]. Even if the Thomsons might tease each other about being descendants of Scottish pirates, it may or may not confirm the Anglo–Franco–Greco etymological origins of the word entrepreneur. At the same time it goes without saying that Roy Thomson did have the entrepreneurial savvy to organize his risk-taking profitable actions into a chronological time frame with a pattern of precalculated, but at the same time very creative, approaches to risk management. For example, chronologically speaking, there was a 3-year period between his creating his radio in North Bay, Ontario and purchasing the *Timmins Daily Press*, a 2-year period between *The Scotsman* and Scottish Television, another 2-year period between Scottish Television and *The Sunday Times*, but a 9-year period before acquiring *The Times*. Thomson's preplanned entrepreneurial time frame (criteria 1 of the minimalist model) was a pattern of intermediary actions leading to foreseen and hoped-for results. Thomson understood that the spontaneity of creativity typical of entrepreneurs is usually short term or intermediary, thus difficult to transform into long-term management actions.

We might say the crowning act of Roy Thomson's entrepreneurial creativity was expressed for his posterity in his last will and testament requiring (male) heirs to assume their entrepreneurial responsibility in the running of family affairs. His son, Kenneth Thomson (1923–2006) did follow in his father's footsteps. Ken's gentlemanly entrepreneur style probably explains his kind coaching through a mutual correspondence with Harvard [12]. At one time, Harvard asked him advice about a conflict with some California partners, who were, above all, Harvard's very close friends. Harvard was having trouble thinking beyond his work-box structure. The conflict concerned Harvard's own personal business ethics and his partners' way of managing a trust fund for clients. Ken Thomson warned Harvard to be very careful in doing business with friends because what seems to be the right thing to do at a given moment, under pressure or stress, seldom is in the long run. Ken Thomson's advice proved providential and Harvard eventually left Laguna Beach, California for La Rochelle, France where he began a second career in higher education as a university lecturer and researcher. Ken's son, David (1957–), began assuming his Thomson heritage of an entrepreneur role at the head of the family enterprises well before Ken's death. David's sister has also followed the family entrepreneur tradition by trying out her own entrepreneurial creativity in Hollywood's Entertainment Industry arenas. Today the Thomson Organization is a multifaceted global structure worth over 25 billion dollars, making them one of the planet's wealthiest families (criteria 3 of the minimalist model) [18,45].

6.2 An impossible Irishman

The historian, Professor Richard Aldous, refers in his writings to Tony Ryan as the "...epitome of what it means to be an entrepreneur..." and claims Ryan was "the Great Gatsby of Ireland" [1,28]. Tony Ryan was as proud of being Irish as Thomson was of being Scottish. T.R.'s life was the cold hard reality of a working man's self-made dream come true, not an impossible Gatsby-style fantasy. But how did the impossible dream become possible for this lucky Irishman? Did he see clearly enough between points in time: today's risks and tomorrow's results? Tony was born in Ireland in a railway man's cottage in Thurles, County Tipperary and died of pancreatic cancer in Celbridge, County Kildare [1,28]. His entrepreneurial creativity survived the stress of losing one fortune (Guinness Peat Aviation) and the building of another (Ryanair). But why did he invest millions in a Georgian version of Scarlett's O'Hara's Tara? Had fictitious dreams become Tony Ryan's reality? What about foresight and creative vision? After T.R.'s success was achieved it seems his intermediary creative entrepreneur actions were not completely replaced by permanent, more structured ones. Although he and Roy Thomson shared the same do-or-die work ethic mind-set, unfortunately, in comparison to Thomson and his heirs, T.R.'s entrepreneurial creativity was not transformed into continuous on-going day-to-day energy to maintain the structure he set up and put into place. He and his heirs did not adapt to being managers. Hence his spontaneous high-level energy necessary for his risk-taking creative vision of the future was short term and never converted into long-term management actions by himself or his heirs as is the case of the Thomson family.

It is also very interesting to consider and compare further information researched about Tony Ryan as an entrepreneur. His company GPA ordered very expensive new aircraft in 1992, the same year he tried to go public with GPA, and it turned into a disaster. He mismanaged his entrepreneurial time frame (criteria 1 of the minimalist model). He owed Merrill Lynch millions, and even though his properties in Mexico and Spain plus his art plus antiques along with his bank accounts were worth millions, his GPA shares once estimated at $234 million were now worthless [1,28]. But T.R. had his way out! With the luck of the Irish, he had foreseen a very creative approach to risk-taking. With a friend from Air Lingus, he had also created his renegade ruthless Ryanair as a no-frills airline [1,28]. Here we see his entrepreneurial creativity at its best when he was inspired to create Ryanair in the names of his three children; Cathal, Declan, and Shane. He made the impossible possible [1,28]. Ryanair was T.R.'s safety exit out of his GPA crisis. Ryanair corresponds to the growing need of low-cost transportation systems and networks in a global economy for work-at-home no-collar creative yuppies (as opposed to the traditional white- and blue-collar employees) in their cozy comfy suburbs. The New York/London/Tokyo mega-trio no longer dominates banks, business, and industry as they did at the end of the 20th century. Even more than before, websites and e-mails have become the new offices of Internet start-ups and e-commerce@companies far away from the stressful corporate lifestyle of megacities [8,11]. T.R. saw beyond the box of traditional airline service (criteria 2 of the minimalist model). T.R. could detect an opportunity where others only detected risk. His entrepreneurial creativity definitely still matches global economy needs and favors a formerly unimaginable lifestyle for the working man. For example, Harvard is personally acquainted with an English engineer, Nicolas Long, a fellow Ryanair passenger who works for the Tube in London. During the week he was in London and flew home on Ryanair to La Rochelle on weekends. His wife and daughters were living in a restored French country house in the village of La Jarne near the Château du Buzay just outside of La Rochelle. At this time, Long was spending weekends with his family enjoying an almost priceless quality lifestyle far beyond anything imaginable or comparable to the megacity corporate jungle of London. T.R.'s entrepreneurial creativity of different ideas was his red carpet leading up to, and consequently, the opening up of future airline markets. Ryanair went public in 1997, becoming a multimillion dollar airline venture. T.R.'s impossible dream became reality. It is important to note that his multimillionaire protégés, Denis O'Brien and Michael O'Neal, attribute their own success to T.R.'s hard-nosed coaching style [1,28]. At his death Tony Ryan was worth 1.5 billion Euros and considered to be one of the wealthiest Irishmen to have ever lived. His entrepreneurial creativity lives on through Ryanair as Europe's largest airline worth over 10 billion Euros (criteria 3 of the minimalist model) [1,28].

6.3 An eccentric Englishman

Richard Branson was born in London, resides in the Virgin Islands, and is a multi-media-, website-, blog-, Twitter-, Facebook-, LinkedIn-, and e-mail-connected entrepreneur. He founded the $5 billion Virgin Group (criteria 3 of the minimalist model) [5,43]. His global business empire, in more than 30 countries, is governed by a complex structure of offshore trusts. His father was a barrister and his grandfather was a judge of the High Court Justice. Dyslexic, he attended different schools until 16 [5,43]. A headmaster once told him he would end up either in a prison or as a millionaire. He believes in the decriminalization of drugs, is pro-European, is antinuclear, is a minister of the Universal Life Church Monastery, has already written his own autobiography, was married twice, and had three children. Even though Her Majesty the Queen named him Knight Bachelor for "services to entrepreneurship," it was Prince Charles who actually knighted him in a ceremony at Buckingham Palace [5,43]. It was the fundraising, fun-loving Solihull dinner party friends of the late Austrian Baroness Notburga Tilt Van Hann (one of Schindler's World War II Austrian resistance courier/party girls) [19] who tried to convince Harvard, but failed to make him an ardent fan of Sir Richard. They portrayed Sir Richard as a sincere philanthropist, not just another big-city carpetbagger. It is true he is involved in a myriad of humanitarian causes such as The Elders (to resolve global conflicts) as well as admirable efforts for missing and exploited children, or his own school of entrepreneurship plus his nonprofit foundation. His career clearly reflects an entrepreneurial time frame (criteria 1 of the minimalist model) with a logical year-to-year astounding creative pattern as highlighted in the following impressive yet incomplete listing (see Fig. 6.3)) [5,43]. Sir Richard is an enigmatic transdisciplinary entrepreneur who became a celebrity media star. He incarnates the entrepreneur who creatively undertook a wide range of entrepreneurial actions ranging from entertainment to transportation to health care (criteria 2 of the minimalist model). He was not afraid of different new ideas leading to the opening of future markets by perceiving opportunities where others were overwhelmed by risks. He successfully replaced his intermediary creative actions by permanent, more structured ones to guarantee his on-going long-term success and stability. Sir Richard's track record proves his capability to adapt and transform his spontaneous high-level energy as an entrepreneur into continuous stable on-going day-to-day energy to manage and maintain structures he set up and put into place. The spontaneity of his creativity has lasted nearly half a century since 1966. There are those who will claim Sir Richard is the entrepreneur's entrepreneur [5,43].

6.4 Conclusion

In conclusion, a very brief analysis, limited to the information cited, reveals things in common between our trio of Scottish, Irish, and English British Commonwealth eco-system entrepreneurs through empirical experiences from the careers of Roy Thomson, Tony Ryan, and Richard Branson. Indeed, what they have in common does contribute to a better empirical understanding of entrepreneurial creativity in an eco-system by offering some very practical down-to-earth measurements to help define it (see Fig. 6.4).

These three empirical elements—timing, cognitive capacity, and quantifiable changes—seem to be plausible universal measurements that not only indicate, but also help define and determine, what is entrepreneurial creativity in an eco-system. God bless us everyone; espcially creative entrepreneurs [9] !

Criteria #1: timing

All 3 carried out a successful pattern of entrepreneurial actions within a specific timeframe.

Criteria #2: cognitive capacity

All 3 possessed a cognitive capacity to think outside and beyond their own boxes and become involved in different new entrepreneurial actions.

Criteria #3: quantifiable changes

All 3 experienced evolutionary profitable changes which are numerically quantifiable on a long term basis.

Figure 6.4 The minimalist model for measuring entrepreneurial creativity in an eco-system.

References

[1] Aldous R. Tony ryan: Ireland's advisor. Dublin: Gill & Macmillan; 2013.
[2] Amabile TM. How to kill creativity. Harv Bus Rev 1998;76(5):76–87.
[3] Audretsch DB, Belitski M. The missing pillar: the creativity theory of knowledge spillover entrepreneurship. Small Bus Econ 2013;41(4):819–36.
[4] Basadur M, Pringle P, Speranzini G, Bacot M. Collaborative problem solving through creativity in problem definition: expanding the pie. Creativity Innovation Manag 2000; 9(1):54–76.
[5] Bronson R. Losing my virginity: how I survived, had fun, and made a fortune doing busniess my way. (London): Virgin Books; 2005 (hardback 1998/paperback 2002/update 2005).
[6] Carayannis E, editor. Encyclopedia of creativity, invention, innovation, and entrepreneurship. (New York): Springer; 2013.
[7] Carayannis E, editor. Leading and managing creators, inventors, and innovators. West Port (Connecticut): Praeger; 2007.
[8] Carayannis E, editor. The story of managing projects. West Port (Connecticut): Praeger; 2005.
[9] Dickens C. A christmas carole. (New York): Barnes & Noble; 2009.
[10] Dell'Era C, Buganza T, Fecchio C, Verganti R. Language brokering: stimulating creativity during the concept development phase. Creativity Innovation Manag 2011;20(1): 36–48.
[11] Harvard P. Franco-American cultural differences in entrepreneurship, leadership & teamwork. (France): Post graduate Module in the Department of Organizing & Managing Enterprises, EIGSI General Engineering School of La Rochelle; 2010.
[12] Harvard P. Personal correspondence. 1980–2006.
[13] Kirzner I. The alert and creative entrepreneur: a clarification. Small Bus Econ 2009;32(2): 145–52.
[14] Kirzner IM. Competition and entrepreneurship. Chicago (IL): University of Chicago Press; 1973.
[15] Kirzner IM. Entrepreneurial discovery and the competitive market process: an Austrian approach. J Econ Literature 1997;35(1):60–85.
[16] Kirzner IM. How markets work: disequilibrium, entrepreneurship and discovery. (London): Institute of Economic Affairs; 1997.
[17] Kirzner IM. Perception, opportunity and profit. Chicago: University of Chicago; 1979.
[18] Thomson R. After I was sixty. (London): Hamish Hamilton Ltd.; 1975.
[19] Tilt N. The strongest weapon. (Devon): Arthur H. Stockwell Ltd; 1972.
[20] Ward TB. Cognition, creativity, and entrepreneurship. J Bus Ventur 2004;19(2):173–88.
[21] Williams LK, McGuire SJ. Economic creativity and innovation implementation: the entrepreneurial drivers of growth? Evidence from 63 countries. Small Bus Econ 2010; 34(4):391–412.
[22] www.basadur.com.
[23] www.bard.edu.
[24] www.druid8.sit.auu.dk.
[25] www.elevier.com/international-handbook-of-innovation.
[26] www.figaro.com.
[27] www.get2test.net.
[28] www.gillmacmillan.ie.
[29] www.henley.ac.uk.

[30] www.hbr.org/magazine (Harvard Business Review).
[31] www.inderscience.com/jhome (International Journal of Entrepreneurship and Small Businesses).
[32] www.elevier.com/journal-of-business-research.
[33] www.journals.elsevier.com/journal-of-business-venturing.
[34] www.kaufman.org.
[35] www.oecd.org.
[36] www.ondernemerscap.nl.
[37] www.onlinelibrary.wiley.com.
[38] www.rje.org (Rand Journal of Economy).
[39] www.sc.edu (University of South Carolina).
[40] www.springer.com (Journal of Knowledge Economy).
[41] www.train4creativity.eu.
[42] www.ucalgary.ca (University of Calgary).
[43] www.virgin.com.
[44] www.whitman.syr.edu/programs-and-academics (Journal of Developmental Entrepreneurship).
[45] www.wikipedia.com.

Intellectual property

7

A. Caplanova
University of Economics in Bratislava, Bratislava, Slovakia

7.1 Introduction

Start-up businesses are frequently very proinnovation businesses. Because of the specific nature of innovation it is important that they paid adequate attention to the protection of their intellectual property. This chapter provides an overview of different forms of intellectual property and of the framework for its protection. We also look at the historical development of the intellectual property rights protection. We pay special attention to the current aspects of the intellectual property protection, which is influenced in the first place by rapid technological development and globalization processes. Also, we consider some macroeconomic, microeconomic, and business-related implications of the intellectual property protection.

In the current period of rapid spread of information it is of utmost importance that new products, technologies, and other forms of innovation are protected in an appropriate way, using either formal or informal forms of intellectual property protection. If the issue of their protection is not sufficiently considered, the firm may lose its competitive advantage and be ousted from the market. This is especially important in the case of small start-up businesses, which may have limited resources and experience with protecting their intellectual property. Large, well-established firms usually have designated departments or employees who are responsible for considering the intellectual property issues companywide. Frequently, they also outsource these services to professional firms, or use a combination of both methods.

Among other issues related to launching a new business, start-ups frequently face the complexity of different aspects to be considered when starting a business and may fail to recognize the necessity to properly consider intellectual property rights. This is relevant especially if a start-up's creation is based around new innovation(s), which the start-up aims to bring to the market, and its key activities are related to it. If in this situation their intellectual property is not adequately protected, it may jeopardize the possibilities for survival of the start-up business itself. We discuss the key aspects of which an entrepreneur should be aware when considering appropriate ways to protect his innovations, which as we will see can take different forms. The form of product or service in which the innovation is materialized also crucially influences the optimal choice of the intellectual property protection.

Start-Up Creation. http://dx.doi.org/10.1016/B978-0-08-100546-0.00007-8

7.2 Forms of intellectual property rights

Intellectual discoveries, which are transformed into innovations, take the form of new products, services, academic papers, new designs, production methods, or artistic works. Their original nature and innovative character raise the need for their protection. At the beginning of the innovation process there is usually an idea. However, to become intellectual property, ideas have to be transformed into innovations. Thus, in general, intellectual property represents ideas that are encoded in innovative goods. When ideas are expressed in a specific form, contained in a specific good, or recorded in a specific media, they become innovation, which is the subject of intellectual protection.

The process of coming up with new ideas and transforming them into innovations is a costly process, which requires substantial investment. Adequate protection of intellectual discoveries is to ensure reasonable return from resources invested into research and development. On the other hand, from the perspective of an innovating firm, also the protection of intellectual property carries positive costs; from the perspective of other firms it limits their access to these innovations. This from the point of view of an economy as such restricts the transfer of new innovations and innovation dynamics in the economy, since innovations and technological progress are an important factor of economic growth and development.[1]

If innovations lead to the production of products that are difficult to imitate, the problem of intellectual property protection is less pronounced. On the other hand, if it is easy to imitate the technology, the barriers of firms trying to adapt the innovation without paying for it are low, and the innovation needs to be protected in an alternative manner. It is also important to note that frequently consumers of the innovative products and services become their secondary producers, since they can distribute them further without the previous consent of the original producer and the need to invest additional inputs. In the case of some innovative products, it is easy to extract the content of the intellectual good by looking at them and studying them, which is sometimes referred to as reversed engineering. In this context there is a competition between a producer of intellectual property and potential free riders. If free riders succeed in their effort, they can sell the new technology at a lower price than the original seller, because they did not invest in producing the underlying innovation.[2] This then leads to the decreased revenue of an innovating firm. In case of some goods based, for example, on a secret recipe, this method is not efficient (eg, the secret recipe for Pepsi Cola precludes efficiently the possibility of reproducing it). However, in such cases it is important to ensure that the trade secret does not leak.

If the possibility of free unpaid transmission of innovative products can be precluded, the intellectual property can be sold in the same way as private goods (eg, in the case of printed books). In such situations owners of innovations can successfully protect their intellectual property informally without seeking formal protection

[1] See: Helpman (1992) for the discussion of the welfare impact of tighter intellectual property right protection.

[2] See: Lemley (2005) for a more detailed discussion.

(Hall et al., 2012). Also, internal norms within specific professional circles can help to protect the innovations within this specific professional group (eg, the ice cream producers sell original ice cream recipes among themselves and stick to the norm not to reveal recipes to outsiders).

If the methods of informal protection of intellectual property are not effective, the formal protection should be considered. The nature of the intellectual property is then a core to decide among the alternative ways of protection. In the next parts of this section we look at alternative forms of the protection of intellectual property rights.

7.2.1 Trademarks

Trademarks represent words, phrases, names, symbols, logos, images, or their combination, which are or will be used for commercial purposes to differentiate goods or services produced by one producer from the goods or services produced by others. Trademarks are also called brands or logos. As an intellectual property, trademarks can be sold, depending on specific legislation in individual countries, either independently from the underlying product, or together with the related product or service. The trademark helps the customers to identify the good or service with the specific producer and serves also as a guarantee of its origin, standards, and quality. It also provides an incentive to producers of the good or service to keep the standards of their production so as to ensure the quality associated with the brand and to sustain their reputation.

Producers having a registered trademark for their products have the right for exclusive use of their trademark and the right to prevent its unauthorized use. When addressing the case of illegal use of a trademark, it is always considered whether a consumer is confused with regard to the origin of the product or service using a similar trademark by the competition. In this context it is important to realize that the trademark can be only registered if it has distinctive characteristics and will not confuse consumers in relation to the origin and quality of the products.

Most countries require formal registration of trademarks, and the fact that the trademark was formally registered is crucial if there is the case of the trademark infringement. However, in some countries (eg, the United States, Canada, Germany), unregistered trademarks are also recognized and protected. The recognition of unregistered trademarks is normally based on the market share of a given brand on the sale of the specific type of goods. Such protection is related to the business reputation or goodwill. This also applies to situations, when a specific firm used the trademark for a long time and its business activities can be adversely affected if a similar trademark is used by the competition.

Trademarks are awarded territorially, thus they are fully enforced only on a specific territory, where they were registered. Some protection can be conferred also beyond the limits of this territory, but the nature of this protection should be carefully checked and not to be relied upon. Trademarks cannot be registered globally, thus, businesses have to decide on which territory they seek the protection and submit the application for registration accordingly. International registration of trademarks is facilitated by the Madrid system, which is a centralized system to protect the trademark on the territory of member countries of the system and allows for the registration of a trademark in multiple signatory countries.

As will be explained in more detail later, The Agreement on Trade-Related Aspects of Intellectual Property Rights (TRIPS) provides the guidelines to facilitate the compatibility of the trademark legislation applied in different countries, or regions.[3]

In the United States the protection of the trademarks is indicated by two symbols: trademarks "TM" for the trademark, and "®" for the registered trademark. In the European Union trademarks are registered by the Office for Harmonization in the Internal Market (OHIM).[4] The community trademark provides to its owner the protection in all member countries. The application for the community trademark is to be filed just once at the OHIM. The Office also has tools for searching the trademark databases for the availability of a trademark a business would like to use. A prerequirement for maintaining the trademark is its active use.

7.2.2 *Industrial designs*

Industrial design is related to aesthetic aspects of a product, to its specific shape, its color, pattern, or other visual characteristics. Thus, the industrial design protects visual characteristics and appearance of the article. The protection of industrial design aims to protect new, distinctively looking products. Thus, the designs, which are eligible for the registration, must be original and cannot closely resemble designs, which have already been registered. Industrial designs are sometimes referred to as design patents. The development of an innovative design of a product requires investment and it aims to attract consumers and differentiate the product from products produced by competitors. If this strategy is successful, it represents for the firm valuable intellectual property and the source of a competitive advantage, which needs to be protected. The owner of the industrial design, which was registered, has the right to prevent others from producing and selling products containing features of the protected design. In terms of product categories the industrial design protection is broad, ranging from jewelry to electronic devices.

Industrial design has to be registered with an industrial design office on the territory, where the protection is sought. In some countries industrial design laws provide limited protection for unregistered industrial designs without the need for their registration. Industrial designs may also be protected as art works under the copyright law.

International protection of industrial design is provided by the Hague Agreement and allows for the possibility to have a design protected in several countries by filing one application with the International Bureau of the World Intellectual Property Organization (WIPO).[5] In the European Union the industrial design protection is provided by means of the registration with OHIM, through which a registered community design can be obtained. The European legislation also recognizes the unregistered community design. Even though both registered and unregistered designs offer similar protection, the scope of the protection is different. The registered design protection is initially valid for the period of 5 years from the date of filing and can be repeatedly renewed for 5 years, for

[3] http://www.wipo.int/madrid/en/.
[4] https://oami.europa.eu/.
[5] http://www.wipo.int/hague/en/.

up to 25 years. An unregistered community design is protected for 3 years from the date on which it became available to the public on the European Union territory. After this period the protection cannot be extended. However, the owner of the unregistered industrial design can apply for the design registration within 1 year of its disclosure.

7.2.3 Patents and utility models

Patent protection is granted for an invention, a product, or a process, which brings a new technical solution. The invention, which is to be protected by a patent, must be new, useful, functional, and innovative; that is, the solution for which the patent protection is sought should not be an obvious one. Patent protection is usually granted for new innovative products, their composition and technology. The prevailing majority of patent applications are made to patent an improvement of previously existing patented inventions. After the patent was awarded, the patent owner has an exclusive right to prevent others from the commercial use of the patented invention.[6]

Since frequently it is difficult to identify which inventions have commercial potential at the point when they appear, firms may have the tendency to use patenting excessively, which can limit innovation activities of their competitors. Thus, on one hand, patent protection aims to encourage innovation by facilitating that innovative businesses achieve adequate return from their innovations. On the other hand, the over-patenting can also have adverse effects. The cost of patenting is relatively low (especially in case of larger, well-established firms) and potential losses from insufficient protection of innovations are big. Rivera (2000) provides specific examples of companies, which due to inefficient intellectual property protection experienced the loss of their competitive advantage. Due to this businesses have an incentive to file a patent application at an early stage of the innovation process and also to file several patent applications related to the same technology to ensure that alternative commercial applications of their innovations are protected and that they will obtain a patent on the technology, which will ultimately become commercially successful.

It is important to note that patents represent territorial rights granted in a country or a region in which a patent has been awarded. Patent protection is granted for a limited period of time only, usually 20 years from the date, when the application is filed. Thus, if the firm wants to market patented product in different countries, separate patent protection is to be sought in all countries where the business is to be conducted. This contributes to the higher cost of the patent protection. Except for initial fees related to the patent application, patent holders are also to cover maintenance fees, which are to be paid periodically to "maintain" the patent protection. The owner of the patent can grant the right to use the patent to other entities.

International cooperation in the area of the patent protection of intellectual property is granted by the Patent Cooperation Treaty,[7] which was signed in 1970 and has 148

[6] The information on patented processes has to be provided by patent applicants in the patent specification and this way it enters the public domain. However, in many countries there are limited possibilities for free use of patented inventions (limited eg to the use for research purposes).

[7] http://www.wipo.int/pct/en/texts/.

signatories. The treaty aimed to harmonize procedures for patent application in the signatory states. Even though the international patent application can be filed under this treaty, it does not result in granting an international patent and has to be followed by the national patent application. In Europe the European Patent Organisation[8] is an intergovernmental organization established in 1977 on the basis of the European Patent Convention signed in 1973. The European Patent Office (EPO) as its executive body examines and grants European patents, which, however, have to be validated separately in each European country, where the patent protection is sought, and the documentation has to be translated into respective national language. There is the effort to provide the unified patent protection in Europe as described in more detail later, but so far the relevant legislation has not been ratified.

In some countries[9] specific forms of patents—utility models—are also granted. These are also known as innovation patents, petty patents, short-term patents, or functional designs. The period for which a utility model is granted is shorter (on average between 6 and 10 years) and the procedure for granting it is less complicated. Usually, the utility models are used for the protection of less substantial inventions.

Even though the cost of patenting may not be high for established firms, for start-ups and new businesses it may represent an important cost item to consider. That relates not only to the fees associated with filing patents, but also to the lawyers and other professionals who are hired to carry out the patent search and prepare the documents required to file the application. After the patent is granted, maintenance fees represent an additional cost of the patent protection. Also, it is important to plan well the global patent strategy to provide adequate territorial protection of the innovation.

An alternative to the patenting is the defensive publication, which is used to prevent other parties from obtaining patent protection on a patentable innovation. This strategy is sometimes used by businesses for which the cost of patent protection is excessively high. It allows the inventor and others to use the innovation without any limitation. However, inventors sacrifice revenues they could obtain from commercial exploitation of the patented invention.

7.2.4 Copyrights

Copyrights are related to legal rights of authors on their published and unpublished literary or artistic original works. They limit the rights for their use and distribution. As in the case of other forms of intellectual property, copyrights protect the tangible form of the original work, not the underlying ideas. Tangibles protected by copyrights include books, music, paintings, and also computer programs, databases, and technical drawings. It is possible for two authors to obtain copyrights on similar work if their work was produced independently. Copyright protection establishes the authorship of the work and allows the owner to obtain revenue from the utilization of copyrighted material by other entities. In case of the coauthorship the copyright is shared by all

[8] http://www.epo.org/about-us/organisation.html.

[9] The list of countries, where the utility models can be acquired is available here, http://www.wipo.int/sme/en/ip_business/utility_models/where.htm.

coauthors. The owner of a copyright has also the right to authorize or prevent the reproduction of the copyrighted material. If the copyrighted work was created within the contract, if not stipulated otherwise in the employment contract, the copyright is owned by the employer. To avoid any dispute, it is suggested that the ownership of the copyright is determined at the point, when the employment contract is concluded. This is also important because the owner of the copyright is responsible for enforcing the rights, which in case of the infringement can be accompanied by substantial costs.

Copyrighted material can be provided without exercising the rights of the copyright owner in specific situations justified by the fair use. Miceli and Adelstein (2006) formulated the formal model of the efficient legal standard of fair use and pointed to the impact of technology in this respect. Even though the specification of fair use may vary across countries, in general, the term refers to cases promoting social welfare. In the European Union the fair use of copyrighted materials is based on the EU directive,[10] and it is related to their reproduction by libraries, museums, or archives, or reproductions for the benefit of people with disabilities and for noncommercial research purposes. Copyrights represent territorial rights and provide the copyright protection on a specific territory.

The copyright laws have been standardized in the Berne Convention and in the Universal Copyright Convention,[11] which have been ratified by most countries. According to the Berne Convention[12] copyright protection is obtained automatically and special registration of copyrights is not needed. Copyright protection is provided for the lifetime of the author plus subsequent decades (depending on a country ranging between 50 and 100 years). The WIPO adopted the Copyright Treaty[13] in 1996 to provide additional copyright protection, which became necessary due to the technological development (eg, with regard to the spread of computer software).

7.2.5 Trade secrets

Trade secrets represent any confidential information that is the source of a competitive business advantage. These can be related to manufacturing methods and secret recipes, but also to the information about customers or suppliers of the company. The unauthorized use of the confidential information (by external or internal parties) is considered to be the violation of the trade secret. The protection of trade secrets is not based on their formal registration; it is usually an integral part of the protection against unfair competition, or of the protection of confidential information.

According to Art. 39 of the Agreement on TRIPS,[14] specific conditions must be met for the information to be considered a trade secret. The information must be secret

[10] The Directive 2001/29/EC of the European Parliament and of the Council of 22 May 2001 on the harmonisation of certain aspects of copyright and related rights in the information society.

[11] http://portal.unesco.org/en/ev.php-URL_ID=15381&URL_DO=DO_TOPIC&URL_SECTION=201.html.

[12] Berne Convention for the Protection of Literary and Artistic Works of September 9, 1886, http://www.wipo.int/treaties/en/text.jsp?file_id=283698.

[13] http://www.wipo.int/treaties/en/text.jsp?file_id=295166.

[14] https://www.wto.org/english/tratop_e/trips_e/t_agm3_e.htm.

(ie, not generally known or easily accessible), it must have commercial value, and the owner of the secret must take steps to protect the confidential character of the information. Unlike patents the protection of intellectual property in the form of trade secrets does not require the disclosure of related information to relevant authorities. Trade secret protection has an advantage of not being limited in time and may continue indefinitely as long as the secret is not revealed to the public. As shown below trade secrets also have some disadvantages and thus, should not be considered as a straightforward alternative to the patent protection.

If the intellectual property of the company is protected using the trade secret, the company should make sure that only a small number of people have access to confidential information. These people should also be made aware of its confidentiality and agree to comply with its protection. This can be done, for example, in the form of confidentiality clauses included in the contracts of employees and business partners having access to the confidential information. It is also important to note that a trade secret does not provide an exclusive right for its commercial use. If it is embedded in the product, but the competition can "break" the secret and consequently use the information commercially. Thus, if the trade secret becomes publicly known, anybody can use it for commercial or any other purpose. Even though the degree of protection extended to trade secrets varies across different countries, generally it is lower than the patent protection.

When deciding on the appropriate form of intellectual property protection, small and medium-sized enterprises frequently rely on the use of trade secrets. However, as pointed out earlier, this is not always the best strategy to follow. When deciding between the trade secret and the patent protection the advantages and disadvantages of each of these should be carefully considered. Businesses should decide whether the innovation meets the criteria of its patentability. Some information, which might have the characteristics of the trade secret (eg, the list of business customers), is not suitable for patenting. If the information is patentable, then the choice of an appropriate alternative should be carefully considered. As stated earlier, trade secrets have an advantage that they do not require formal registration and thus, this form of protection is less costly. This advantage may represent an important reason for the choice of this alternative by small firms. Also, unlike patents the trade secret type of protection becomes effective immediately, whereas granting the patent right takes a relatively long period of time.

7.3 Historical development of the intellectual property protection

Individual forms of intellectual property have been protected for several centuries. In this section we briefly look at the historical development of main forms of intellectual property and its protection, which helps us better comprehend issues related to intellectual property protection, and the development of intellectual property itself.

7.3.1 Patents

Even though there are indications that the roots of the patent-like protection can be traced back to ancient Greece, the first patent systems were introduced in the middle ages. The word patent comes from the Latin *litterae patentes*, which means a patent letter. In the middle ages such letters were used by monarchs to award exclusive rights to individuals to produce specific goods or services. Such examples can be found in Italy, France, and England.

The first system of the patent protection related to the technology was introduced in Italy in the middle of 15th century, when patents were granted to Venetian glassmakers to protect their innovative production methods, which were used to produce unique Venetian glassware. It is generally agreed that the Venetian Republic introduced the first patent system in Europe in 1474 known as the Venetian Patent Statute. According to the statute, patents could be granted for new and useful products. As Venetian glassmakers exported their glassware, or moved to other countries, they also sought a similar type of protection there. At the beginning of the 18th century the patent law was further elaborated in England during the reign of Queen Anne, which required that the specification of the innovation was provided in a written form. In addition, since then patents could be granted not only on new products, but also on the improvements of already existing products. This system became the basis for the development of the patent legislation in other countries, in the first place those being under the rule of England. But in the 18th century the patent legislation was also introduced in France, the United States, and other countries. During this time the industry was developing dynamically, numerous innovations were introduced, and thus, the new patent legislation met the requirements of the developing industrial society.

In the 19th century the Paris Convention provided the basis for the international patent protection and thus, the international standards of patent protection replaced patent rules fragmented across individual countries.

7.3.2 Trademarks

The history of trademarks goes back to the ancient period, when different stamps and seals were used to stamp some goods, which can be considered as predecessors of current trademarks. Thus, looking at the development of forms of intellectual property from an historical perspective, trademarks can be considered its oldest form. On one hand, seals and stamps were considered as an indication of quality; on the other hand, they informed people about the origin of the product if there was a problem with its quality. Long ago farmers also marked their animals so their property could be easily identified. During the period of the Roman Empire local blacksmiths identified the origin of their products on the swords, which they produced. The formation of guilds in the middle ages led to the spread of trademarks on the products produced in guilds (eg, bells, paper, etc.). As guilds built up their reputation, their marks became the way to communicate the quality to their customers.

The first trademark legislation was passed in the 13th century in England, which required bakers to use a distinctive mark for their bread. In the 14th century silversmiths and also other producers were required to mark their products. More comprehensive trademark legislation was only introduced in England at the end of the 19th century. Subsequently, it was adopted in France, England, and the United States. The legislation established the procedure for the trademark application and the rights of trademark holders. Since then the use of the trademarks spread to virtually any industry and territory and became an important factor of the business competitiveness. In the course of the development of trademarks their legislative protection relied in the first place on the common law, even though in specific countries special trademark legislation was also introduced (eg, England, the United States, Japan, France). The Paris Convention set international standards of the trademark protection.

Counterfeit products experienced the boom with the increasing popularity of specific brands. They represent large industry and they force producers of the original brand products to sacrifice a large fraction of their revenues. Even though counterfeiting is legally prosecuted, so far this problem has not been sufficiently dealt with. The increase of the production of counterfeited goods led to the effort to introduce a comprehensive international legislation aimed specifically at the protection of trademark owners against counterfeiting. As explained later, the complex nature of the issue so far has not led to the resolution of this problem in the global context.

7.3.3 Copyrights

Before the development of printing machines the protection of intellectual property embedded in books did not represent a serious problem. The multiplication of books required their rewriting, which was very time-consuming. The invention of print in the 15th century brought forward the need to develop an appropriate form of the copyright protection of printed works, since the development of printing methods provided an opportunity for their uncontrolled multiplication. On the other hand, it also facilitated the access of printed materials to a much larger audience. The copyright protection introduced during this period was based on the procedure of authors providing to publishers exclusive rights to copy their work for a limited time period. After the expiration of this period, the copyrighted material became freely accessible.

The first comprehensive copyright law was the Statute of Anne introduced in England in 1709. On the basis of this act, authors, not publishers, were recognized as the owners of the copyrights of their work, and the act also specified the conditions of the copyright's protection. The authors were required to deposit their published works in copyright libraries and register them. The unpublished works were not protected by this legislation. After the period of the copyright protection expired, the copyrighted material became a part of the public domain. The statute provided the basis for the development of copyright legislation in other countries and it is still referred to today.

The purpose of the introduction of the copyrights was to maximize the utility of authors and at the same time to meet the social objectives. On one hand copyrights were aimed to provide incentives to produce intellectual property; on the other hand, they were to ensure the penetration of the works among the public. Limited duration of copyright protection aimed to address the contradiction between the incentives to produce intellectual property and the effort to ensure its widest possible distribution.

Throughout the development of the copyright legislation the issue of fair use of copyrighted material was considered. Under specific circumstances (eg, for the public purpose) free use of protected material was advocated.

The Berne Convention (1886) brought an internationally coordinated approach to the copyright legislation. The convention stipulates that it is not necessary to register works protected by copyrights in individual countries, provides the basis for mutual recognition of copyrights across countries, and extends the protection to unpublished work. This regulation was adopted by the majority of countries. It has provided the basis for the international copyright law to date.

The creation of the new photocopying technology facilitated further multiplication of printed materials and created the need for the new regulation. It also put pressure on publishers threatening that they would lose large parts of the market if illegal copying of printed works was not dealt with. Technological development also made the copyright protection of intellectual property embedded in audio and video carriers more challenging. The development of VCRs facilitated the copying of video production. In the area of video sharing the development of MP3s facilitated the file sharing and the copyright owners did not have adequate tools to fight against the breach of copyrights. In this situation at the beginning there was an effort to stop the production of the new carrier products. However, it was soon realized that such a solution was not possible and it was the illegal behavior that had to be limited. Even though new technologies did not lead to the destruction of the underlying intellectual property protection, they contributed to substantially higher prices of the original products—carriers of this kind of intellectual property. The owners expected that after their initial sale, subsequent copying will decrease their sales and revenues and tried to mitigate this impact by higher prices of the original products.[15]

7.4 Regulatory aspects of intellectual property protection

The objective of the legislation aimed to protect intellectual property rights is to create incentives for the investment in R&D, to promote innovation and creativity, and to ensure adequate return from resources invested into the innovation process for innovating firms (Besen and Raskind, 1991). On the other hand, however, the protection of intellectual property rights potentially creates economic inefficiencies by limiting excess to information, the spread of which does not require additional cost and creates the monopoly related to its use.

The infringement of intellectual property rights refers to their violation and may be prosecuted in line with the civil or criminal law, depending on the type of intellectual property concerned and the local legislation. In case of trade secrets, misappropriation is related to the violation of the confidentiality of information. On the other hand, in case of patents and other forms of formal intellectual property, the information on

[15] See a more detailed discussion of the contextual aspects of the historical development of intellectual property rights in, eg, Sell (2004).

the characteristics of innovation is registered and publicly available. Patent infringement can take place if a patented invention is used or sold without the permission of the patent holder. However, in many countries (with the exception of the United States) the use of patented inventions for research purposes is permitted.

In the subsequent part of this section, we discuss international regulatory framework, which has brought harmonization into the fragmented national regulation of intellectual property rights and in some instances also allowed for their international or regional protection. In today's globalized world when seeking to protect intellectual property it is not sufficient for businesses to concentrate on one country only. There is the need to be aware of the global regulatory framework, since many firms operate across national borders. This is even more pronounced in the case of intellectual property rights protection, since new information technologies interconnecting the continents allowed spreading of information at minimum cost and with problematic ways to protect accessing it. In the second part of the section we pay attention to the key regulatory framework in Europe.

7.4.1 International framework of the protection of intellectual property rights

The national legislation related to intellectual property protection varies in individual countries. However, a number of international intellectual property agreements and conventions facilitate the harmonization of the legislation in the international context (Franklin, 2013). In these processes the WIPO plays an important role, especially with regard to the harmonization of individual forms of the protection of intellectual property. The increasing globalization of trade contributes to an increasing role of the World Trade Organization (WTO) of intellectual property protection regarding the regulatory framework in the area of different forms of intellectual property protection.[16] The most comprehensive international agreement on intellectual property rights is the Agreement on TRIPS.[17] It came into effect on January 1, 1995. TRIPS sets the minimum standards for different forms of intellectual property protection (copyrights and related rights, industrial designs, trademarks and trade secrets, and also new plants). The agreement was agreed upon within the round of discussions of the General Agreement on Tariffs and Trade in 1994.

TRIPS is based upon previously introduced international agreements related to the intellectual property protection (eg, Berne convention). It introduces further specifications of types of intellectual property and of the enforcement procedures. The TRIPS Agreement requires that the stipulations of Paris and of Berne conventions were complied with. On one hand, the aim of TRIPS is to promote innovation; on the other hand, the aim is to create conditions facilitating its transfer and dissemination. The TRIPS Agreement sets out minimum standards of protection, which have to be provided by each of its signatories. Thus, adhering to the minimum requirements specified by TRIPS, signatory countries can decide to implement even larger protection of the

[16] For a more detailed discussion of global aspects of intellectual property protection, see Maskus, 2000.
[17] http://www.wto.org/english/docs_e/legal_e/legal_e.htm#TRIPs.

intellectual property rights on their territory. The agreement also specifies core principles applicable with regard to the intellectual property rights enforcement. It stipulates that the disputes related to the TRIPS Agreement should be subject to the WTO's dispute settlement procedures. The stipulations of the agreement apply equally to all signatory countries, but developing countries have been allowed a longer period for adjustment. Also, each country can decide itself on the ways the provisions of the agreement are to be implemented in its territory.

Historically, the Paris Convention for the Protection of Industrial Property,[18] which was adopted in 1883, was the first major agreement to ensure the international protection of different intellectual property rights (including patents, utility models, trademarks, and industrial designs). Subsequently, it was revised periodically. The last revision took place in 1967 and was amended in 1979. The convention stipulates that each signatory state must grant the same protection to citizens of another signatory state as it grants to its own citizens. It also introduced the right of priority, when individuals, after submitting the application in one state, have 6 or 12 months to seek the protection in other countries. Thus, it is not necessary to file the application at the same time in all countries, where the intellectual property protection is sought. The convention also specifies common rules related to individual types of intellectual property, which each signatory state has to adhere to.

In 1886, the Berne Convention for the Protection of Literary and Artistic Works[19] was concluded. Since then it has undergone several revisions, the last of them being agreed upon in Paris in 1971 and amended in 1979. The convention provides authors, musicians, poets, and painters with the right to control the use of their works. It stipulates main principles of the protection and the authors should be provided the same protection in every signatory state. If the work originates from a signatory state of the agreement the protection is provided automatically. The protection must be provided for a minimum of 50 years after the death of the author. However, in the case of applied art and photographic works the minimum term of protection is 25 years from the creation of the work. Individual countries frequently provide longer copyright protection than stipulated by the Berne Convention.

The Berne Convention also specifies provisions for the free use of the types of intellectual property it covers. Some countries still require the registration of copyrights, since the registration facilitates the investigation in case of any arising disputes related to the copyright protection. The Intellectual Property Rights Office (IPRO) was established to facilitate the international registration of copyrights. Its Copyright Registration Service serves as a depositary of unpublished work originating in signatory countries of the Berne Convention.

A specific agreement under the Berne Convention is the WIPO Copyright Treaty[20] of 1996, which entered into force in 2002. It deals with the protection of intellectual property in the digital form including computer programs and databases.

[18] http://www.wipo.int/treaties/en/text.jsp?file_id=288514.

[19] http://www.wipo.int/treaties/en/text.jsp?file_id=283698.

[20] http://www.wipo.int/treaties/en/text.jsp?file_id=295166.

Next to the rights of intellectual property owners as specified by the Berne Convention, it grants the authors the rights to distribute originals and copies of their works through the transfer of ownership and specifies conditions for their commercial rental. Also, it provides owners the right to communicate the works to the public, so that they can be accessed by the public individually, using different channels, including the Internet. The minimum duration of the protection is set at 50 years.

In the area of the copyright protection another principal regulation was adopted within UNESCO in 1952. It is the Universal Copyright Convention, which came to force in 1955, was revised in 1971, and since then the revised version of the convention has been applied. The aim of the convention was to extend the international copyright protection to all countries. The convention requires that equal protection of copyrights is provided to all authors regardless of whether they come from the same or other signatory countries. It also requires identifying the copyrighted work with the symbol ©, with the name of the copyright owner, and with the year when the work was first published. It stipulates that the minimum term of the copyright protection in any signatory country must be 25 years after the death of the author; in case of photographic works and applied art the protection is specified to be 10 years after the death of the author. Except for the Berne Convention the stipulations of the Universal Copyright Convention have priority over other copyright legislation. However, most countries are signatories of the Berne Convention, which limits the applicability of the Universal Copyright Convention. The revisions of the Universal Copyright Convention and of the Berne Convention in 1971 were prepared so that special needs of developing countries were taken into consideration with regard to copyrighted works.

International regulation also pays attention to the problem of counterfeit goods. Even though the problem has gained more attention with the spread of branding in recent decades it had already been addressed internationally by the end of the 19th century. The Madrid Agreement for the Repression of False or Deceptive Indications of Source on Goods[21] was concluded in 1891. Similarly to other earlier agreements, it has since underwent several revisions with the last one taking place in 1967. According to this agreement those goods that bear a false or deceptive indication of their source by which one of the signatory countries is directly or indirectly indicated as being the country of origin, must be confiscated at the import point or their import should be prohibited.

Recently, The Anti-Counterfeiting Trade Agreement (ACTA) was prepared as the legal act aimed at fighting the infringement of intellectual property rights related to the counterfeiting and piracy in the global context (Mercurio, 2012). Building upon the minimum standards introduced by TRIPS, ACTA establishes strengthened standards for the protection in this area. The agreement was signed by signatory parties in 2011. However, for the agreement to enter into force, it is required that the parties include into their legislation relevant penalties for the infringement of intellectual property rights concerned. In the European Union the implementation of ACTA was rejected by the European Parliament in 2012. The agreement led to substantial

[21] http://www.wipo.int/treaties/en/text.jsp?file_id=286779.

criticism from different stakeholder groups including civil society in Europe and in other parts of the world. It was pointed out that foreseen benefits of the agreement are outweighed by the potential related cost, especially with regard to their adverse impact on civil rights, including freedom of expression and privacy in communication.

Building upon previously introduced international regulation, the patent protection is specifically concerned by the Patent Law Treaty,[22] which was adopted in 2000 and came into force in 2005. Its aim is to harmonize and streamline formal procedures with respect to dealing with the national and regional patents and with patent applications. With the exception of filing date requirements, which are not regulated by the treaty, the treaty stipulates maximum requirements, which can be applied by signatory countries. Based on the specifications provided by the treaty, the procedures used by patent offices were simplified. This contributed to the decreased costs for patent applicants. The treaty also aims to facilitate the electronic filing of patent applications, but in specified cases the paper submissions are also requested to be accepted.

The international protection of intellectual property rights in the area of trademarks is stipulated by the Trademark Law Treaty,[23] which was concluded in 1994. The standards introduced by this treaty are related to the application for the trademark registration, to the changes in the trademarks after they were registered, and also to the renewal of the registration. The requirements for each of these aspects are set so as to contribute to the simplification of these procedures.

The Hague Agreement[24] concerning the international deposit of industrial design represents a specific agreement related to the establishment of the Hague system of international protection of the industrial design. It was first adopted in 1925; subsequently it was revised several times. Currently, its 1960 and 1999 revisions are applied. The industrial design protection is based on the registration of the industrial design using the application with the WIPO. Thus, the formalities for the international protection of the industrial design are reduced, not only for the first design registration, but also for the renewal of the registration and the recording of possible changes. The protection is provided for 5 years, and it is renewable for a minimum of one 5-year period (according to the 1960 act), or two 5-year periods (according to the 1999 act).

7.4.2 Intellectual property protection in the European Union

The previously mentioned international property rights regulations guarantee a degree of harmonization of the legislation in the area of intellectual property in the regional and global context. In the European Union there are two designated organizations, which have specific responsibility for addressing different aspects of the intellectual property protection. These are the OHIM[25] and the EPO.[26] These organizations are responsible for the progress of the harmonization in the area

[22] http://www.wipo.int/treaties/en/text.jsp?file_id=288996.

[23] http://www.wipo.int/treaties/en/text.jsp?file_id=294357.

[24] http://www.wipo.int/hague/en/legal_texts/.

[25] https://oami.europa.eu/ohimportal/en/.

[26] https://www.epo.org/index.html.

of the intellectual property rights protection in the European Union and the implementation of relevant directives and regulations.

In the European Union several directives were introduced to facilitate the processes of strengthening the intellectual property protection. The Directive 2001/29/EC of the European Parliament and of the Council of May 22, 2001[27] is related to the harmonization of certain aspects of copyright and related rights in the information society in the framework of the internal market. The Directive 2004/48/EC of the European Parliament and of the Council of April 29, 2004 on the enforcement of intellectual property rights[28] aims at establishing the procedures to ensure the enforcement of intellectual property rights including rights related to the industrial property.

The harmonization in the European Union has also progressed with regard to the trademark registration, when by submitting a single application European firms not only can obtain the protection of their trademarks throughout the European Union, but also in countries that are signatory of the Madrid Protocol.[29] Also, the entities having the international registration under the Madrid Protocol may apply for the protection of their trademarks in the European Union under the European trademark system. Thus, the firms carrying out their business in the European Union can use the Community trademark, which allows them to make their products distinctive on the European Union territory. The European trademark system provides uniform protection, which allows the owner to prevent another entity from using the trademark for the same products or services. Also, the use of the trademark in case of similar products is to be forbidden if a confusion could possibly arise. The Community trademark is registered for 10 years and is renewable.

The industrial property is protected by the Directive 98/71 of the European Parliament and of the Council of October 13, 1998 on the legal protection of designs.[30] The Council Regulation No 6/2002 of 2001[31] establishes the basis for the protection of industrial design using the Community design, which provides unified protection on the territory of the European Union.

The EPO, which is not bound to the European Union, but has broader European character, allows for a single application for the European patent. However, after the European patent is granted, it has to be separately validated in each member state, where the protection is sought. That is linked to further costs related, for example, to the need to translate the application documents into relevant national languages and other costs related to filing the patent application at national level. Thus, there has been an effort to progress toward the unified patent, which would provide protection on the European Union territory and allow reducing cost of patent protection in Europe. In December 2012 the European Parliament approved two regulations

[27] http://eur-lex.europa.eu/legal-content/EN/TXT/?uri=celex:32001L0029.

[28] http://eur-lex.europa.eu/legal-content/EN/ALL/?uri=CELEX:32004L0048R(01).

[29] The text of the Protocol is available at: http://eur-lex.europa.eu/LexUriServ/LexUriServ.do?uri=CELEX:22003A1114(01):EN:HTML and the related decision of the European Council is available at: http://eur-lex.europa.eu/LexUriServ/LexUriServ.do?uri=CELEX:22003A1114(01):EN:HTML.

[30] http://old.eur-lex.europa.eu/LexUriServ/LexUriServ.do?uri=CELEX:31998L0071:EN:HTML.

[31] http://eur-lex.europa.eu/legal-content/EN/TXT/?uri=URISERV:l26033.

(EU 1257/2012 and EU 1260/2012) that allow for the unitary patent protection. The protection can be obtained so it covers all European Union member states except for Italy, Spain, and Croatia, without the need for further national validation. These regulations of the European Parliament entered into force in January 2013. However, the unitary patents will be applied only from the date when the Agreement on a Unified Patent Court enters into force. The agreement was signed by 25 European Union member states in February 2013. To enter into force, it needs to be ratified by at least 13 countries, including France, Germany, and the United Kingdom.[32] Unitary patents will be accepted in English, French, or German and no further translation will be required after they are granted. Unitary patent protection is to make the patent system in the European Union easier, less costly, and legally secure. It is to be available to entities that want to protect their innovations in the countries of the European Union, regardless of their nationality or their place of residence.

The European Union also pays attention to the protection of biotechnological inventions (Directive 98/44/EC of the European Parliament and of the Council of July 6, 1998 on the Legal Protection of biotechnological inventions; Regulation No. 1610/96 of the European Parliament and of the Council of July 23, 1996 concerning the creation of a supplementary protection certificate for plant protection products), which have become a more pressing issue with the rapid development of biotechnologies that has taken place in recent years.

In case of start-ups and their effort to protect their intellectual property rights the financial and human resources can represent a problem. There is the scope for the public sector institutions to contribute to building up capacities by providing them the intellectual property consultancy and training. In Europe the creation of innovative start-ups has been supported from European and regional schemes, such as Horizon 2020.

7.5 Some considerations of the intellectual property protection for start-up businesses

In this section we reflect upon some of the recent developments in the area of intellectual property protection that are relevant for start-up businesses. We concentrate on the patent protection and licensing, not only because it frequently concerns start-ups, but some of the recent research in the field also sheds new light on the pros and cons of the existing system of the patent protection.

The results of the Berkley Patent Survey carried out in 2008 among top managers of more than 15,000 US entrepreneurial companies pointed out that a substantial number of surveyed companies opted for the patent protection (Sichelman, 2012). However, when trying to identify motives behind the use of patenting, it was revealed that firms did not seek the patent protection so as to secure high profit, but to strengthen their

[32] By September 2015 the Agreement was ratified by eight countries. From among Germany, United Kingdom and France, so far only France ratified the agreement. For the update on its ratification see: http://www.consilium.europa.eu/en/documents-publications/agreementsconventions/agreement/?aid=2013001.

negotiating position with investors. It was revealed that investors place high value on companies with patents and also strive to secure the competitive advantage brought by the new technology. On the other hand, the Berkley survey did not confirm that patents would provide strong incentives for increased innovation among businesses. Also, with respect to start-up businesses high cost of the patent enforcement and of the protection against infringement were identified as factors diminishing their incentives to seek the patent protection.

The cross-sectorial study carried out among more than 9000 start-ups in the United States (Lerman, 2015) confirmed the positive correlation between the venture funding raised by start-ups and the number of patents filed by them. Thus, patents held by start-ups represent one of the factors that influence investors' decisions about which start-up they would invest in. The study also pointed to a very active patent behavior of start-ups at the beginning of their existence and before they acquired any venture funding. This may reflect more intensive innovation activities at early stages of the start-up existence, but possibly also the realization that the ownership of patents serves as a signal to potential investors. On the other hand, the anecdotal evidence indicates that investors may also perceive an early filing of patent applications by start-ups as a negative feature, and consider it as spending scarce resources, which at an early stage of the existence of the business should be directed to other activities (Lerman, 2015). Lerman also looked at the territorial and sectorial distribution of patenting among US start-ups and concluded that Californian start-ups tended to patent more than those in other US regions. As for the sectorial differences she indicated that start-ups in biotechnology, hardware, and medical industry were prone to patent significantly more than those in the software industry.

Next to seeking the patent protection, the tendency for firms to provide their patents free of charge has been observed more frequently (Ziegler et al., 2014). It became common in the software industry. In many cases it was recognized that firms were motivated by economic reasons, since when providing an open access to software, they were able to appropriate profit by the sale of complementary products. Also, free access to software protected by the copyrights was recognized to facilitate its further improvement and development initiated by the comments from the software users. The donation of patents may also be motivated by the feeling of moral obligation, by seeking the tax deductions, or other cost benefits. The establishment of the patent pools allowing the members the free use of patents that are in the pool, has appeared as a new form of research and technological cooperation among entities. Examples of such cooperation initiatives are Eco-patent Commons, the Medicines Patent Pool, or WIPO Re:Search initiative (Ziegler et al., 2014).

Some studies indicate that the patent protection is not vital for start-up companies during the first years of their existence. Landers (2015) pointed out that, for example, Facebook and Microsoft increased their patent holdings only after achieving commercial success. To address the disadvantages of the patent system for start-up businesses, Landers suggested the implementation of the antipatent system for small businesses of the start-up nature, which would allow the owner to opt out of the patent system for a minimum period of 20 years. The system would consist of two fundamental components: first, the participants would obtain immunity from the third-party patent infringement assertions and second, during the immunity time period the start-up would not acquire any liability for nonwillful infringement of any patents. The system could be based on the system of an application submitted to the relevant patent office. If such a system was introduced, the start-up companies would have an option to decide if it was more beneficial for them to opt for the patent protection or to opt out of the patent system altogether.

The notion that a good intellectual property protection system serves as a stimulus for innovation has been further challenged by a growing predominantly empirical literature based on the US data.[33] In their study Boldrin and Levine (2013) pointed out that a properly designed patent system may contribute to increasing innovation in a specific location and time period, but the empirical evidence failed to provide clear justification of the notion that patents encourage innovation or productivity. They proposed that the patents were abolished, or alternative legislative instruments were found, which would be less prone to lobbying and rent seeking. At the same time it was stressed that some areas are more prone to the problems related to the patent system (eg, software or business method patents).

The management of the intellectual property issues by start-ups should also take into account the activities of so-called patent trolls, which received a growing attention in the literature. Patent trolls, or nonpracticing entities, do not produce any products, but make money from licensing or asserting patents against entities that produce products. Their activities have been considered to document the deficiency of the existing patent system. Patent trolls are active in the area of litigation, or they threaten businesses with the litigation. They often buy patents from bankrupt companies and follow up with the use of the technology protected by these patents. In case of its nonlicensed use they engage in the litigation. The increased amount of patenting, which took place in recent decades, made existing situation more complex. Also, under the existing patent system, patents can also be issued on broad ideas, which are easy to be developed by other companies without them realizing that the technology is under the patent protection. A patent troll may send a company the claim of the infringement of the patent and request to pay a licensing fee. The company may decide to pay the licensing fee to avoid the threat and cost of the lawsuit.

According to Lemley and Melamed (2013), in the last few years patent trolls filed about half of the patent suits. They are especially active in the computer and telecommunication industries and do not contribute to innovation activities, but obtain large settlements from costly lawsuits. In their empirical study carried out among in-house attorneys of the companies producing products using the new technology in different industries, Feldman and Lemley (2015) looked at the impact of licensing demands on new innovation activities. Their results point out that, except for the pharmaceutical industry, in most cases firms that were targeted by the lawsuits developed the technology independently and were paying the licensing fee to the patent troll so as

[33] See eg, Jaffe, A.B., 2000. The US patent system in transition: policy innovation and the innovation process. Research Policy 29 (4), 531–557. Boldrin, M., Levine, D.K., 2013. The case against patents. The Journal of Economic Perspectives. 27 (1), 3–22.

not to be sued for the use of the technology, which in reality they did not copy. They would decide to pay for the license anyway, since patent litigation is expensive and can take a long time. Feldman and Lemley also concluded that the patent licensing activities did not lead to a substantial technology transfer. This can be explained by the fact that patent demands usually take place only after the technology proves to be successful. However, the technology transfer usually happens when the technology is new. Also, it can take place in the form of the collaboration or informal transfer of know-how and not licensing of patents.

In their study of the determinants of patent suits and settlements during 1978 to 1999 in the United States, Lanjouw and Schankerman (2004) found that the litigation risk is much higher for patents that are owned by individuals and firms with small patent portfolios. At the same time the financial impact of the patent litigation is larger for small firms due to their limited financing and the lack of internal expertise to deal with the litigation cases.[34] Start-up businesses are also more exposed to the threat of litigation, perhaps, except for the pharmaceutical and chemical patentable subject areas, since they have limited resources to investigate the situation with regard to the patent protection and the needs for licensing.

Moreover, the production of new products may require the licensing of many patents, which further complicates the matter. The complexity of current technologies can diminish the effort of start-ups to license a new technology, not only because the cost of licensing can be preclusive, but also because the inherent uncertainty about what licenses are needed brings further unclarity into the issue.

7.6 Conclusions

Even though the intellectual property and its protection have a long history, the technological change and globalization processes can be considered main factors explaining why these issues have received growing attention in recent decades. Also, with the changed conditions the intellectual property-related transactions gained new character. In case of any market transaction it is necessary to establish the means of the exchange and to preclude consumers from consuming the good without paying for it.[35]

Since innovations are largely characterized by nonexclusive character, nobody can be excluded from their consumption once they were made publicly available. Once the innovation becomes known, it is then available to everyone. Thus, entities, which have not borne the costs of their production, have an incentive to gain access to them without paying (ie, there is a free-ride problem). Thus, if owners of an innovation want to ensure the return on resources invested in its development, they have to find ways to increase the transaction cost of free rides and in this way make it less

[34] See: Lerner (1995) for the conclusions based on the analysis of new biotechnology firms.

[35] For the economic analysis of the intellectual property see, eg, Landes et al. (2009), Posner (2005), Stiglitz (2008).

beneficial. The transaction cost can be increased using private-market solutions (eg, by developing the technology, which makes it harder to get free access to the innovation), or existing enforcement mechanisms (legal protection).

In this chapter we have discussed the alternatives available to businesses with regard to protecting their intellectual property and linked them to existing regulatory framework predominantly in the international, but also in the European context. It should be emphasized that the optimal choice of the protection of innovations depends, among other factors, on the type and character of an innovation, foreseen market dynamics, and also on the geographical area in which a firm operates. As we pointed out in the previous section, in the changed environment the existing intellectual property right system faces new challenges, which also impact on the business strategies of start-up business and on the optimal ways to address the intellectual property rights issues. Also, as the preliminary research indicates, the current changes in the global economy may require deep reconsideration of the existing regulatory framework to reflect the changing business models and to promote the social welfare in current societies.

References

Besen, S.M., Raskind, L.J., 1991. An introduction to the law and economics of intellectual property. The Journal of Economic Perspectives 3–27.

Boldrin, M., Levine, D.K., 2013. The case against patents. The Journal of Economic Perspectives 27 (1), 3–22.

Feldman, R., Lemley, M.A., 2015. Do patent licencing demands mean innovation?. In: Stanford Law and Economics Olin Working Paper No. 473. Stanford Public Law. Working Paper No. 2565292, UC Hastings Research Paper No. 135, available online: http://papers.ssrn.com/sol3/papers.cfm?abstract_id=2565292#%23.

Franklin, J., 2013. International Intellectual Property Law. American Society of International Law. Available online: http://www.asil.org/sites/default/files/ERG_IP.pdf.

Hall, B.H., Helmers, C., Rogers, M., Sena, V., 2012. The choice between formal and informal intellectual property: a literature review. In: NBER Working Paper Series, Working Paper No. w17983. Cambridge, MA, 35 pp.

Helpman, E., 1992. Innovation, imitation, and intellectual property rights. In: NBER Working Paper Series, Working Paper No. 4081. Cambridge, MA, 53 pp.

Jaffe, A.B., 2000. The US patent system in transition: policy innovation and the innovation process. Research Policy 29 (4), 531–557.

Landers, A.L., 2015. The antipatent: a proposal for startup immunity. Nebraska Law Review 93 (4). Article 5, available online: http://digitalcommons.unl.edu/cgi/viewcontent.cgi?article=2808&context=nlr.

Landes, W.M., Posner, R.A., Landes, W.M., 2009. The Economic Structure of Intellectual Property Law. Harvard University Press.

Lanjouw, J.O., Schankerman, M., 2004. Protecting intellectual property rights: are small firms handicapped? Journal of Law and Economics 47 (1), 45–74.

Lemley, M.A., Melamed, A.D., 2013. Missing the forest for the trolls, 113 colum. Law Review 2117, 2118–2121.

Lemley, M.A., 2005. Property, intellectual property, and free riding. Texas Law Review 83, 1031. Available online: http://papers.ssrn.com/sol3/papers.cfm?abstract_id=582602.

Lerman, C., 2015. Patent Strategies of Technology Startups: An Empirical Study. Available at SSRN 2610433. http://papers.ssrn.com/sol3/papers.cfm?abstract_id=2610433.

Lerner, J., 1995. Patenting in the shadow of competitors. Journal of Law and Economics 463–495.

Maskus, K.E., 2000. Intellectual Property Rights in the Global Economy. Peterson Institute, Washington D.C.

Mercurio, B., 2012. Beyond the text: the significance of the anti-counterfeiting trade agreement. Journal of International Economic Law 15 (2), 361–390.

Miceli, T.J., Adelstein, R.P., 2006. An economic model of fair use. Information Economics and Policy 18 (4), 359–373.

Posner, R.A., 2005. Intellectual property: the law and economics approach. Journal of Economic Perspectives 19 (2), 57–73.

Rivera, K.G., 2000. Discovering new value in intellectual property. Harvard Business Review 55. Available online: http://secure.com.sg/courses/ICI/Grab/Reading_Articles/L07_A02_Rivette.pdf.

Sell, S., 2004. Intellectual property and public policy in historical perspective: contestation and settlement. Loy. LAL Rev. 38, 267–322. Available at: http://digitalcommons.lmu.edu/llr/vol38/iss1/6.

Sichelman, T., 2012. Startups & The Patent System: A Narrative. Available online: htpp://ssrn.com/abstract=2029098.

Stiglitz, J.E., 2008. Economic foundations of intellectual property rights. Duke Law Journal 57, 1693–1724.

Ziegler, N., Gassmann, O., Friesike, S., 2014. Why do firms give away their patents for free? World Patent Information 19–25.

http://www.epo.org.

https://oami.europa.eu.

http://www.wipo.int.

https://oami.europa.eu.

http://www.iprightsoffice.org.

https://www.wto.org.

http://www.wipo.int/treaties.

http://portal.unesco.org/en.

http://ip-science.thomsonreuters.com.

Part Two

Nano and biotechnologies for eco-efficient buildings

Nano-based thermal insulation for energy-efficient buildings

Bjørn Petter Jelle
Norwegian University of Science and Technology (NTNU), Trondheim, Norway;
SINTEF Building and Infrastructure, Trondheim, Norway

8.1 Introduction

As energy use in the building sector accounts for a significant part of the world's total energy use and greenhouse gas emissions, there is a demand to improve the energy efficiency of buildings. Hence, in this respect, concepts like passive houses and zero-emission buildings are being introduced. In order to meet the demands of an improved energy efficiency, the thermal insulation of buildings plays an important role. To achieve the highest possible thermal insulation resistance, new insulation materials and solutions with low thermal conductivity values have been and are being developed, in addition to using the current traditional insulation materials in ever-increasing thicknesses in the building envelopes. However, very thick building envelopes are not desirable due to several reasons; for example, considering space issues with respect to economy, floor area, transport volumes, architectural restrictions and other limitations, material usage and existing building techniques. It should also be noted that studies (McKinsey, 2009) point out that energy-efficiency measures are the most cost-effective ones, whereas measures like solar photovoltaics and wind energy, for example, are far less cost-effective than insulation retrofit for buildings.

The objective of this work is to investigate and compare the various properties, requirements, and possibilities for traditional, state-of-the-art, and possible future thermal building insulation materials and solutions, their weaknesses and strengths, disadvantages and advantages. Furthermore, some experimental investigations have been discussed, such as nano insulation materials attempted to be made from hollow silica nanospheres (HSNS) during a sacrificial template method. The work presented herein is based on the conceptual and experimental studies by Jelle (2011a) and Jelle et al. (2011a, 2014a,b). Companies attempting a start-up creation to make nano-based thermal insulation for energy-efficient buildings in the coming years will need to have knowledge about and address efficiently the various aspects related to thermal insulation as mentioned briefly here and which will be treated and discussed in the following.

Start-Up Creation. http://dx.doi.org/10.1016/B978-0-08-100546-0.00008-X

8.2 Thermal conductivity

For a thermal building insulation material, the main key property is thermal conductivity, where the normal strategy or goal is to achieve as low thermal conductivity as possible. A low thermal conductivity (W/(mK)) enables the application of relatively thin building envelopes with a high thermal resistance (m^2K/W) and a low thermal transmittance U-value ($W/(m^2K)$). The total overall thermal conductivity λ_{tot} (i.e., the thickness of a material divided by its thermal resistance) is in principle made up from several contributions:

$$\lambda_{tot} = \lambda_{solid} + \lambda_{gas} + \lambda_{rad} + \lambda_{conv} + \lambda_{coupling} + \lambda_{leak} \qquad [8.1]$$

where λ_{tot} is total overall thermal conductivity, λ_{solid} is solid state thermal conductivity, λ_{gas} is gas thermal conductivity, λ_{rad} is radiation thermal conductivity, λ_{conv} is convection thermal conductivity, $\lambda_{coupling}$ is thermal conductivity term accounting for second-order effects between the various thermal conductivities in Eq. [8.1], and λ_{leak} is leakage thermal conductivity.

To reach as low thermal conductivity as possible, each of these thermal contributions have to be minimized. Normally, the leakage thermal conductivity λ_{leak}, representing an air and moisture leakage driven by a pressure difference, is not considered as insulation materials and solutions are supposed to be without any holes enabling such a thermal leakage transport. The coupling term $\lambda_{coupling}$ can be included to account for second-order effects between the various thermal conductivities in Eq. [8.1]. This coupling effect can be quite complex and will be neglected in the rest of this chapter. Theoretical approaches to thermal performance of vacuum insulation panels (VIP) usually assume this coupling effect to be negligible. For further information see, for example, Heinemann (2008). In general, another coupling term might also be included in Eq. [8.1]; that is, the interaction between the gas molecules and the solid state pore walls. However, as we will see later, this last coupling term is included through a factor in the expression for the gas conductivity as given in Eq. [8.2] for the Knudsen effect.

The solid state thermal conductivity λ_{solid} is linked to thermal transport between atoms by lattice vibrations (i.e., through chemical bonds between atoms). The gas thermal conductivity λ_{gas} arises from gas molecules colliding with each other and thus transferring thermal energy from one molecule to the other. The radiation thermal conductivity λ_{rad} is connected to the emittance of electromagnetic radiation in the infrared wavelength region from a material surface. The convection thermal conductivity λ_{conv} comes from thermal mass transport or movement of air and moisture. All these thermal conductivity contributions are driven by or dependent upon the temperature or temperature difference. The various thermal insulation materials and solutions utilize various strategies to keep these specific thermal conductivities as low as possible.

In addition, the thermal building insulation materials and solutions also have to fulfill a series of requirements with respect to other properties than the thermal conductivity. These other requirements may put restrictions on or provide challenges to how low thermal conductivities it will be possible to obtain with the selected materials and solutions.

8.3 Traditional thermal building insulation

The most common traditional thermal building insulation materials of today are given a short description in the following. These materials have a relatively low thermal conductivity. Moreover, note the overviews of traditional thermal insulation materials by Al-Homoud (2005) and Papadopoulos (2005). We may also envision lowering the thermal conductivity of the traditional insulation materials by applying nano-based technologies; for example, by manufacturing porous materials with the pores or voids being in the nano range. As of today, most of the pore or void volume in traditional insulation materials is made up of pores or voids from the millimeter to the nano range.

8.3.1 Mineral wool

Mineral wool covers glass wool (fiberglass) and rock wool, which normally is produced as mats and boards, but occasionally also as filling material. Light and soft mineral wool products are applied in frame houses and other structures with cavities. Heavier and harder mineral wool boards with high mass densities are used when the thermal insulation is intended for carrying loads, like on floors or roofs. Mineral wool may also be used as a filler material to fill various cavities and spaces. Glass wool is produced from borosilicate glass at a temperature around 1400°C, where the heated mass is pulled through rotating nozzles, thus creating fibers. Rock wool is produced from melting stone (diabase, dolerite) at about 1500°C, where the heated mass is hurled out from a wheel or disk, thus creating fibers. In both glass wool and rock wool dust abatement oil and phenolic resin is added to bind the fibers together and improve the product properties. Typical thermal conductivity values for mineral wool are between 30 and 40 mW/(mK). The thermal conductivity of mineral wool varies with temperature, moisture content, and mass density. As an example, the thermal conductivity of mineral wool may increase from 37 to 55 mW/(mK) with increasing moisture content from 0 to 10 vol%, respectively. Mineral wool products may be perforated, and also cut and adjusted at the building site, without any loss of thermal resistance.

8.3.2 Expanded polystyrene

Expanded polystyrene (EPS) is made from small spheres of polystyrene (from crude oil) containing an expansion agent (e.g., pentane C_6H_{12}) that expand by heating with water vapor. The expanding spheres are bound together at their contact areas. The insulation material is casted as boards or continuously on a production line. EPS has a partly open pore structure. Typical thermal conductivity values for EPS are between 30 and 40 mW/(mK). The thermal conductivity of EPS varies with temperature, moisture content, and mass density. As an example, the thermal conductivity of EPS may increase from 36 to 54 mW/(mK) with increasing moisture content from 0 to 10 vol%, respectively. EPS products may be perforated, and also cut and adjusted at the building site, without any loss of thermal resistance.

8.3.3 Extruded polystyrene

Extruded polystyrene (XPS) is produced from melted polystyrene (from crude oil) by adding an expansion gas (e.g., HFC, CO_2, or C_6H_{12}), where the polystyrene mass is extruded through a nozzle with pressure release, causing the mass to expand. The insulation material is produced in continuous lengths, which are cut after cooling. XPS has a closed pore structure. Typical thermal conductivity values for XPS are between 30 and 40 mW/(mK). The thermal conductivity of XPS varies with temperature, moisture content, and mass density. As an example, the thermal conductivity of XPS may increase from 34 to 44 mW/(mK) with increasing moisture content from 0 to 10 vol%, respectively. XPS products may be perforated, and also cut and adjusted at the building site, without any loss of thermal resistance.

8.3.4 Cellulose

Cellulose (polysaccharide $(C_6H_{10}O_5)_n$) comprises thermal insulation made from recycled paper or wood fiber mass. The production process gives the insulation material a consistency somewhat similar to that of wool. Boric acid (H_3BO_3) and borax (sodium borates, $Na_2B_4O_7 \cdot 10H_2O$ or $Na_2[B_4O_5(OH)_4] \cdot 8H_2O$) are added to improve the product properties. Cellulose insulation is used as a filler material to fill various cavities and spaces, but cellulose insulation boards and mats are also produced. Typical thermal conductivity values for cellulose insulation are between 40 and 50 mW/(mK). The thermal conductivity of cellulose insulation varies with temperature, moisture content, and mass density. As an example, the thermal conductivity of cellulose insulation may increase from 40 to 66 mW/(mK) with increasing moisture content from 0 to 5 vol%, respectively. Cellulose insulation products may be perforated, and also cut and adjusted at the building site, without any loss of thermal resistance.

8.3.5 Cork

Cork thermal insulation is primarily made from the cork oak, and can be produced as both a filler material or as boards. Typical thermal conductivity values for cork are between 40 and 50 mW/(mK). Cork insulation products may be perforated, and also cut and adjusted at the building site, without any loss of thermal resistance.

8.3.6 Polyurethane

Polyurethane (PUR) is formed by a reaction between isocyanates and polyols (alcohols containing multiple hydroxyl groups). During the expansion process the closed pores are filled with an expansion gas such as HFC, CO_2, or C_6H_{12}. The insulation material is produced as boards or continuously on a production line. PUR may also be used as an expanding foam at the building site, for example, to seal around windows and doors and to fill various cavities. Typical thermal conductivity values for PUR are between 20 and 30 mW/(mK), i.e. considerably lower than mineral wool, polystyrene, and cellulose products. However, loss of the pore gases and subsequent air permeation into the pores due to diffusion or degradation during time may increase the thermal conductivity above these values. The thermal conductivity of PUR varies with temperature, moisture content, and mass density. As an example, the thermal conductivity of PUR may increase from 25 to 46 mW/(mK) with increasing moisture content from 0 to 10 vol%, respectively. PUR products may be perforated, and also cut and adjusted at the building site, without any loss of thermal resistance. It should be noted that even if PUR is safe in its intended use it raises serious health concerns and hazards in case of a fire. During a fire PUR will when burning release hydrogen cyanide (HCN) and isocyanates, which are very poisonous. The HCN toxicity stems from the cyanide anion (CN^-), which prevents cellular respiration. Generally, HCN may be found in the smoke from nitrogen (N) containing plastics.

8.3.7 Other building materials

Thermal conductivities of other building materials, including the load-bearing ones, are normally considerably higher than the thermal conductivity values of the thermal building insulation materials, the very reason for the need and application of thermal building insulation materials. As a comparison, typical examples are wood (100–200), carbon steel (55 000), stainless steel (17 000), aluminum (220 000), concrete (150–2500), lightweight aggregate (100–700), brick (400–800), stone (1000–2000), and glass (800), where all values in parentheses are given in mW/(mK).

8.4 State-of-the-art thermal building insulation

The following is a short description of the state-of-the-art thermal building insulation materials and solutions of today. That is, the materials and solutions that are, or are considered to be, the thermal building insulation materials with the lowest thermal conductivities today. These state-of-the-art thermal insulation materials utilize to a large extent porous materials with pores in the nano range.

8.4.1 Vacuum insulation panels

Vacuum insulation panels (VIP) consist of an open porous core of fumed silica enveloped by several metallized polymer laminate layers (see Figs. 8.1 and 8.2). The VIPs represent today's state-of-the-art thermal insulation with center-of-panel thermal conductivities ranging from as low as 2 to 4 mW/(mK) in pristine non-aged condition to typically 8 mW/(mK) after 25 years aging due to water vapor and air diffusion through the VIP envelope and into the VIP core material, which has an open pore structure. Depending on the type of VIP envelope, the aged thermal conductivity after 50 and 100 years will be somewhat or substantially higher than this value (see eg, Fig. 8.3). This inevitable increase of thermal conductivity represents a major drawback of all VIPs. Puncturing the VIP envelope, which might be caused by nails and similar, causes an increase in the center-of-panel thermal conductivity to about

Figure 8.1 (left) Typical vacuum insulation panel (VIP) structure showing the main components (Simmler et al., 2005) and (right) a comparison of equivalent thermal resistance thickness of traditional thermal insulation and VIP (Zwerger and Klein, 2005).

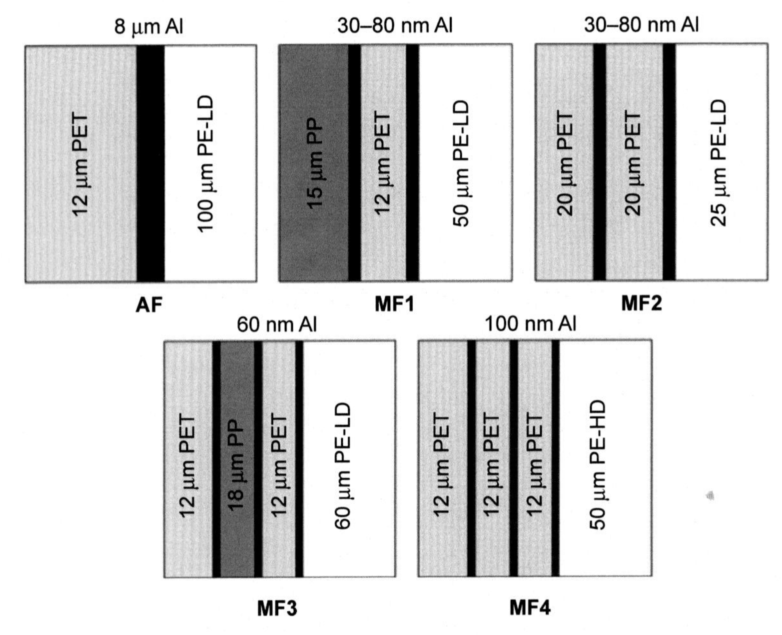

Figure 8.2 Cross-section illustration of an aluminum foil laminate (AF) and metallized polymer film multilayer laminates (MF) (Wegger et al. (2011) based on Simmler et al. (2005); see also Brunner et al. (2006) and Simmler and Brunner (2005a)).

20 mW/(mK). As a result, VIPs cannot be cut for adjustment at the building site or perforated without losing a large part of their thermal insulation performance. This represents another major disadvantage of VIPs.

Several authors have been studying various aspects of VIPs, ranging from analytical models, thermal bridges and conductivity, air and moisture penetration, aging and service life, quality control and integration of VIPs in building construction (see, e.g.,

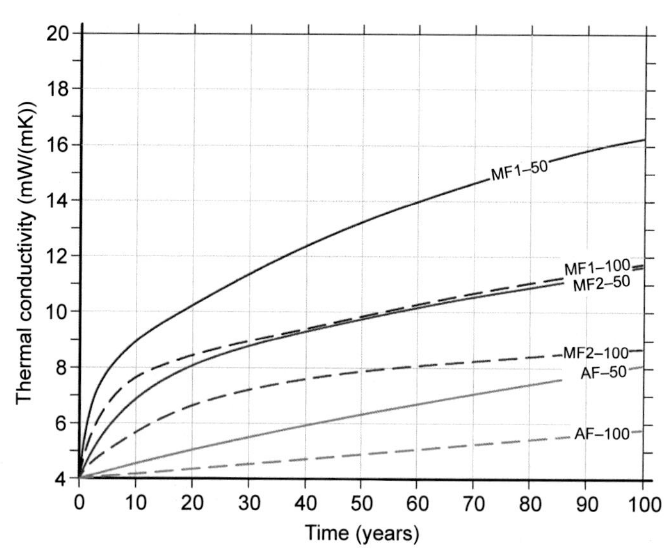

Figure 8.3 Center-of-panel thermal conductivity for vacuum insulation panels with a fumed silica core as function of elapsed time. For two different panel sizes of 50 cm × 50 cm × 1 cm and 100 cm × 100 cm × 2 cm, and for three different foil types AF, MF1, and MF2 (Baetens et al., 2010a).

Alam et al. (2011), Alotaibi and Riffat (2014), Baetens (2013), Beck et al. (2007), Boafo et al. (2014), Brunner and Simmler (2007, 2008), Brunner et al. (2014), Caps and Fricke (2000), Caps (2005), Caps et al. (2008), Fricke (2005), Fricke et al. (2006, 2008), Grynning et al. (2011), Haavi et al. (2012), Heinemann et al. (2005), Li et al. (2015), Mandilaras et al. (2014), Miesbauer et al. (2014), Mukhopadhyaya et al. (2014), Pons et al. (2014), Sallée et al. (2014), Schwab et al. (2005a,b,c,d,e), Simmler and Brunner (2005a,b), Simmler et al. (2005), Sprengard and Holm (2014), Sveipe et al. (2011), Tenpierik and Cauberg (2007, 2010), Tenpierik et al. (2007a,b, 2008), Voellinger et al. (2014), Wegger et al. (2011), Yrieix et al. (2014) and Zwerger and Klein (2005)), where among others comprehensive reviews on VIPs for building applications have been made by Tenpierik (2009), Baetens et al. (2010a) and Kalnæs and Jelle (2014).

Despite the big disadvantages of VIPs, including their relatively high costs, they do represent a large leap forward in thermal insulation for building applications. Thermal conductivities between 5 and 10 times, depending on aging time, lower than traditional thermal insulation materials like mineral wool and polystyrene products, will especially be important when trying to achieve the standard and requirements of passive houses and zero energy or zero emission buildings. Thermal insulation thicknesses up to 50 cm or more in walls and roofs are not desired (see Fig. 8.1 for a visual thickness comparison). Such thick building envelopes might require new construction techniques and skills. In addition, transport of thick building elements leads to increased costs. As an example, height restrictions may apply for passing under several bridges and through tunnels, i.e. thinner elements will bring about a more efficient transport at a reduced cost. Building restrictions during retrofitting of existing buildings (e.g., by the lawful authorities or practical restrictions concerning windows and other building parts) may also require thinner high performance thermal insulation thicknesses than traditional insulation would be able to solve. Furthermore, in areas with a high living area market value per square meter, a reduced wall thickness may involve large area savings and thus a higher value of the real estate. Simple calculations show that for such areas the application of VIPs may actually result in an economic profit (chapter: Potential cost savings by applying vacuum insulation panels).

Hence, even if the VIPs are not the ultimate solution for the future, they may be the best solution for many thermal building envelopes today and in the near future, both from a thermal energy savings and an economical point of view. VIP research and advances should be concentrated toward developing VIP envelopes capable of far better preventing air and water vapor from entering into the VIP core for longer time periods up to at least 50–100 years. Besides, the research on and application of VIPs contribute to increased knowledge and idea generation about the thermal insulation solutions of tomorrow.

It is important to note that the difference between 4 mW/(mK) (pristine condition) and 20 mW/(mK) (punctured) of 16 mW/(mK) is due entirely to gas thermal conductivity (not taking into account any changes to the solid core due to the loss of vacuum). Hence, the combined solid state and radiation thermal conductivity of fumed silica is as low as 4 mW/(mK) or in principle somewhat lower (as there is still a very small air residue inside a VIP a small part of the 4 mW/(mK) value is due to gas conduction). Thus, as it is possible to make materials with such a very low solid state and radiation conductivity, there are rather good opportunities to make a superinsulation material (SIM) functioning at atmospheric pressure by lowering the gas thermal conductivity.

8.4.2 Gas-filled panels

In principle, gas-filled panels (GFP) are close to the technology of VIPs. GFPs have been studied by Griffith et al. (1993, 1995) and Mills and Zeller (2008) among others. A review of GFPs for building applications is given by Baetens et al. (2010c). The GFPs apply a gas less thermally conductive than air (e.g., argon (Ar), krypton (Kr), and xenon (Xe)) instead of a vacuum as in the VIPs. The barrier foil and cellular structure (baffle) inside a GFP are shown in Fig. 8.4. Maintaining the low-conductive gas concentration inside the GFPs and avoiding air and moisture penetration into the GFPs are crucial to the thermal performance of these panels. A vacuum is a better thermal insulator than the various gases employed in the GFPs. On the other hand, the GFP grid structure does not have to withstand an inner vacuum as the VIPs. Low emissivity surfaces inside the GFPs decrease the radiative heat transfer. Thermal conductivities for prototype GFPs are quite high (e.g., 40 mW/(mK)), although much lower theoretical values have been calculated. Hence, the GFPs hold many of the VIPs' advantages and disadvantages. Nevertheless, the future of GFPs as thermal building insulation may be questioned or even doubtful, as compared to them the VIPs seem to be a better choice both for today and tomorrow.

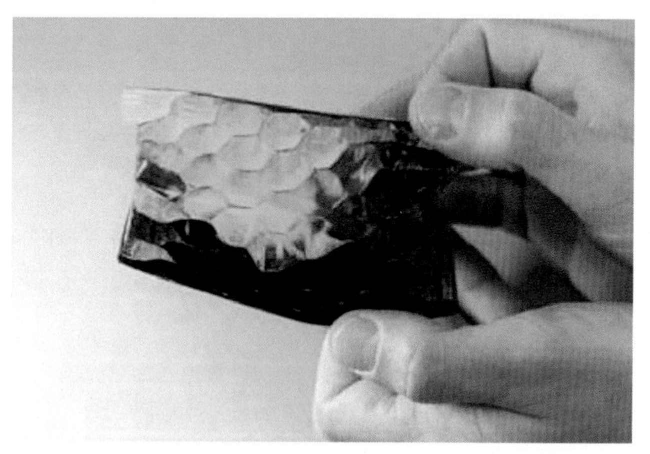

Figure 8.4 Gas-filled panel with its barrier foil and cellular structure (baffle) (LBNL, 2015).

8.4.3 Aerogels

Aerogels represent a state-of-the-art thermal insulation solution, and maybe the most promising with the highest potential of them all at the moment. Studies of aerogels have been conducted by Aegerter et al. (2011), Baetens et al. (2011), Cuce et al. (2014), Dowson et al. (2012), Gao et al. (2014c,d), Hostler et al. (2008), Ihara et al. (2015a,b,c), Jelle et al. (2015d), Koebel et al. (2012), Lee et al. (1995a,b), Levy and Zayat (2015), Schultz et al. (2005), Schultz and Jensen (2008), Smith et al. (1998), Stahl et al. (2012), Wong et al. (2014) and Zhao et al. (2012), among several others. Using carbon black to suppress the radiative transfer, thermal conductivities as low as 4 mW/(mK) may be reached at a pressure of 50 mbar. However, commercially available state-of-the-art aerogels have been reported to have thermal conductivities between 13 and 14 mW/(mK) at ambient pressure (Aspen Aerogels, 2008a,b). The production costs of aerogels are still very high. Aerogels have a relatively high compression strength, but are very fragile due to their very low tensile strength. The tensile strength may be increased by incorporation of a carbon fiber matrix. A very interesting aspect with aerogels is that they can be produced as either opaque, translucent or transparent materials, thus enabling a wide range of possible building applications. Transparent aerogels are shown in Fig. 8.5. For aerogels to become a widespread thermal insulation material for opaque applications, the costs have to be lowered substantially.

8.4.4 Phase change materials

Phase change materials (PCM) are not really thermal insulation materials, but since they may be part of the thermal building envelope and thus interesting for thermal building applications, they are mentioned within this context. PCMs change phase from solid state to liquid when heated, thus absorbing energy in the endothermic process. When the ambient temperature drops again, the liquid PCMs will turn into solid state materials again while giving off the earlier absorbed heat in the exothermic

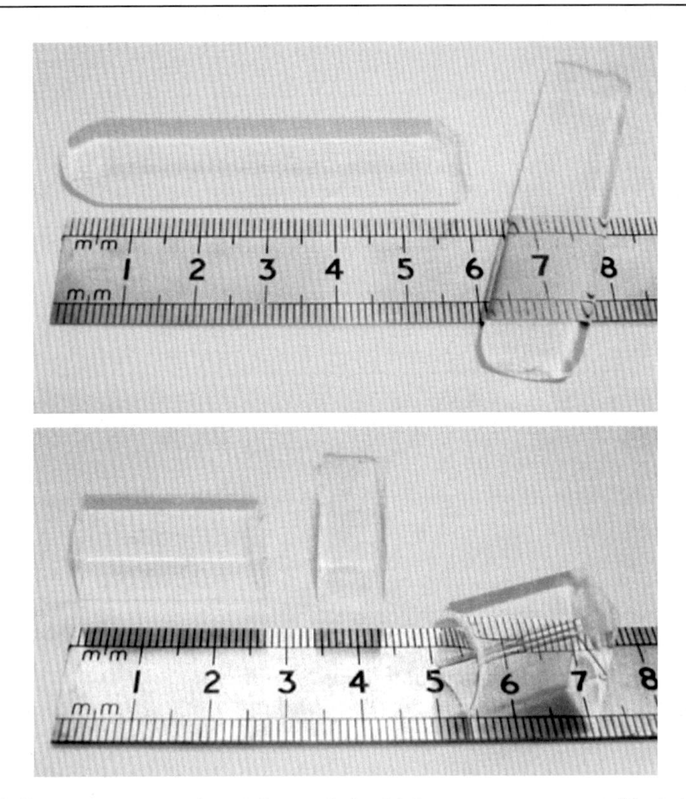

Figure 8.5 Transparent aerogel samples made by high temperature supercritical drying (top) and low temperature supercritical drying (bottom) (Dowson et al., 2012).

process. Such a phase change cycle stabilizes the indoor building temperature and decreases the heating and cooling loads. Various paraffins are typically examples of PCMs, but a low thermal conductivity (Farid et al., 2004) and a large volume change during phase transition (Hasnain, 1998) limit their building application. An overview of the main PCMs has been given by Demirbas (2006), whereas other reviews on PCMs may be found in works by Baetens et al. (2010b), Farid et al. (2004), Hasnain (1998), Kalnæs and Jelle (2015), and Khudhair and Farid (2004). A suitable phase change temperature range, depending on climate conditions and desired comfort temperatures, as well as an ability to absorb and release large amounts of heat, are important properties for the selection of a specific PCM for building applications. Corresponding melting enthalpies and melting temperatures are depicted for various groups of PCMs in the work by Dieckmann (2006).

8.5 Nanotechnology applied on thermal insulation

Shortly we will be seeing that nanotechnology may be applied as a scientific tool to make high performance thermal insulation materials. The normal focus in

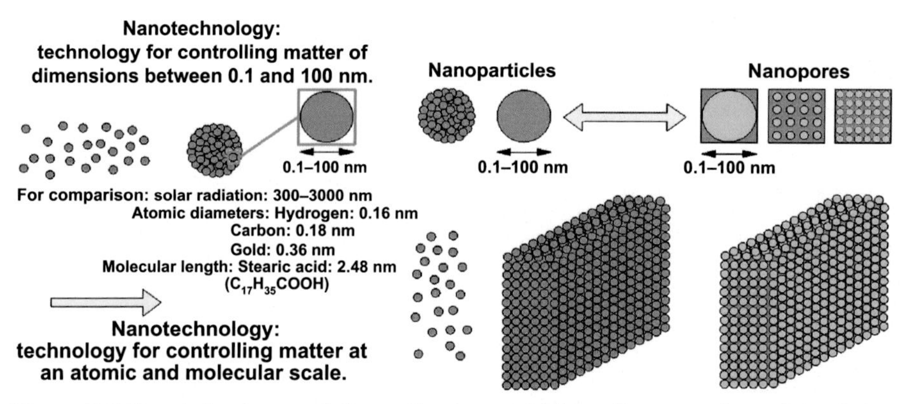

Figure 8.6 Nanotechnology and its application on high performance thermal insulation materials (Jelle, 2011a).

nanotechnology is to control matter, typically particles, of dimensions between 0.1 and 100 nm, i.e., at an atomic and molecular scale. However, for nanotechnology applied for making thermal insulation materials, the focus is shifted from particles to pores in the nano range. These aspects are visualized in Fig. 8.6.

8.6 Concepts for future thermal building insulation

The following is a short description of possible concepts for future thermal building insulation materials and solutions. That is, possible materials and solutions that are thought to become the high performance thermal building insulation materials and solutions of the future. In the heart of many of these concepts is the application of nanotechnology and especially the use of nanopores.

8.6.1 Vacuum insulation materials

A vacuum insulation material (VIM) is basically a homogeneous material with a closed small pore structure filled with vacuum with an overall thermal conductivity of less than 4 mW/(mK) in pristine condition (Fig. 8.7). The VIM can be cut and adapted at the building site with no loss of low thermal conductivity. Perforating the VIM with a nail or similar would only result in a local heat bridge, i.e., no loss of low thermal conductivity. One possible way to produce a VIM may be to envision a solid state material blowing itself up from within during the formation and subsequent expansion of an inner pore structure (Fig. 8.8). For further details on VIM refer to Jelle et al. (2010a).

8.6.2 Gas insulation materials

A gas insulation material (GIM) is basically a homogeneous material with a closed small pore structure filled with a low-conductance gas like e.g. argon (Ar), krypton (Kr) or

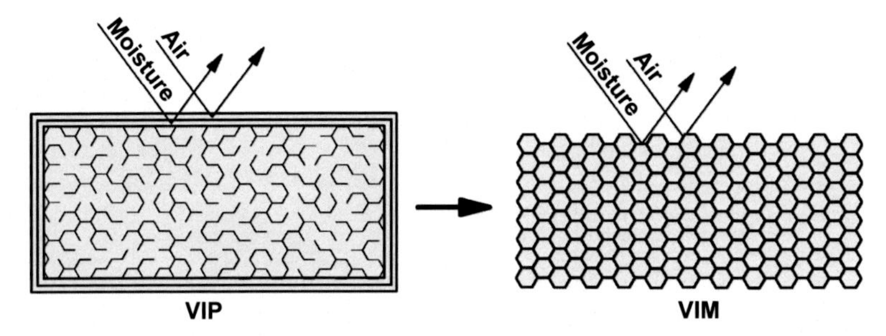

Figure 8.7 The development from vacuum insulation panel to vacuum insulation material (Jelle et al., 2010a).

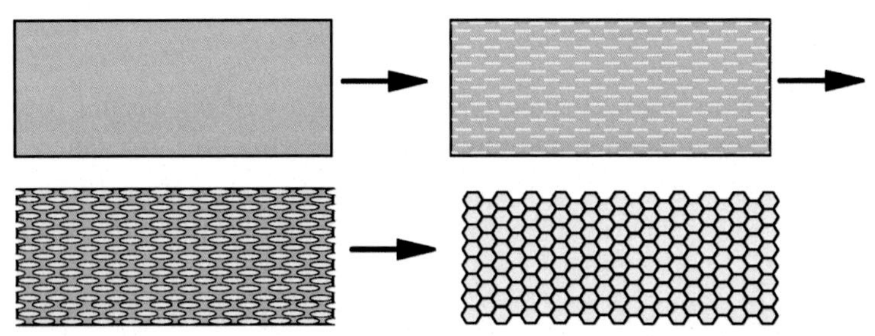

Figure 8.8 Envisioned creation of vacuum insulation material as a solid state material blowing itself up from within during the formation and subsequent expansion of an inner pore structure (Jelle et al., 2010a).

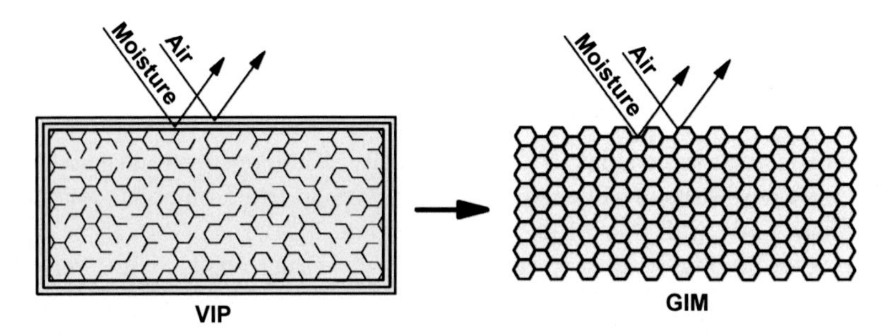

Figure 8.9 The development from vacuum insulation panel to gas insulation material (Jelle et al., 2010a).

xenon (Xe), with an overall thermal conductivity of less than 4 mW/(mK) in the pristine condition (Fig. 8.9). That is, a GIM is basically the same as a VIM, except that the vacuum inside the closed pore structure is substituted with a low-conductance gas. For further details on GIM refer to Jelle et al. (2010a).

8.6.3 Nano insulation materials

The development from VIP to nano insulation materials (NIM) is depicted in Fig. 8.10. In the NIM the pore size within the material is decreased below a certain level (i.e., 40 nm or below for air) in order to achieve an overall thermal conductivity of less than 4 mW/(mK) in the pristine condition. That is, NIM is basically a homogeneous material with a closed or open small nano pore structure with an overall thermal conductivity of less than 4 mW/(mK) in the pristine condition.

The grid structure in NIMs do not, unlike VIMs and GIMs, need to prevent air and moisture penetration into their pore structure during their service life for at least 100 years. The NIMs achieve their low thermal conductivity without applying a vacuum in the pores by utilizing the Knudsen effect. The gas thermal conductivity λ_{gas}, also including the gas and pore wall interaction, taking into account the Knudsen effect may be written in a simplified way as (Baetens et al., 2010a; Bouquerel et al., 2012; Jelle, 2011a; Jelle et al., 2010a; Kaganer, 1969):

$$\lambda_{gas} = \frac{\lambda_{gas,0}}{1 + 2\beta \mathrm{Kn}} = \frac{\lambda_{gas,0}}{1 + \frac{\sqrt{2}\beta k_B T}{\pi d^2 p \delta}} \qquad [8.2]$$

where

$$\mathrm{Kn} = \frac{\sigma_{mean}}{\delta} = \frac{k_B T}{\sqrt{2}\pi d^2 p \delta} \qquad [8.3]$$

where λ_{gas} is gas thermal conductivity in the pores, also including the gas and pore wall interaction (W/(mK)), $\lambda_{gas,0}$ is gas thermal conductivity in the pores at standard temperature and pressure (W/(mK)), β is coefficient characterizing the molecule-wall collision energy transfer (in)efficiency (between 1.5 and 2.0), k_B is Boltzmann's constant $\approx 1.38 \cdot 10^{-23}$ J/K, T is temperature (K), d is gas molecule collision diameter (m),

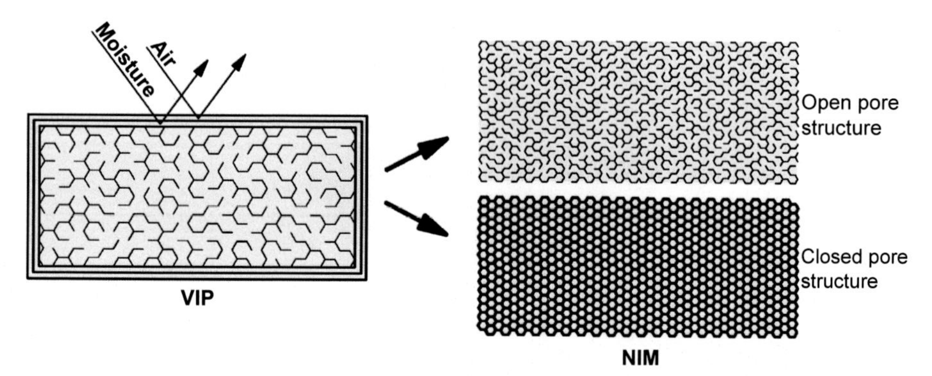

Figure 8.10 The development from vacuum insulation panel to nano insulation material (Jelle et al., 2010a).

p is gas pressure in pores (Pa), δ is characteristic pore diameter (m), and σ_{mean} is mean free path of gas molecules (m).

Decreasing the pore size within a material below a certain level (i.e., a pore diameter of the order of 40 nm or below for air), the gas thermal conductivity, and thereby also the overall thermal conductivity, becomes very low (<4 mW/(mK) with an adequate low-conductivity solid state structure) even with air-filled pores. This is caused by the Knudsen effect, where the mean free path of the gas molecules is larger than the pore diameter. That is, a gas molecule located inside a pore will hit the pore wall and not another gas molecule, where the solid state and gas interaction is taken care of by the β coefficient in Eq. [8.2]. Hence, the resulting gas thermal conductivity λ_{gas}, also including the gas and pore wall interaction, versus pore diameter and pore gas pressure, may be calculated in this simplified model and depicted as in Fig. 8.11. For further details refer to the work by Baetens et al. (2010a) and Jelle et al. (2010a).

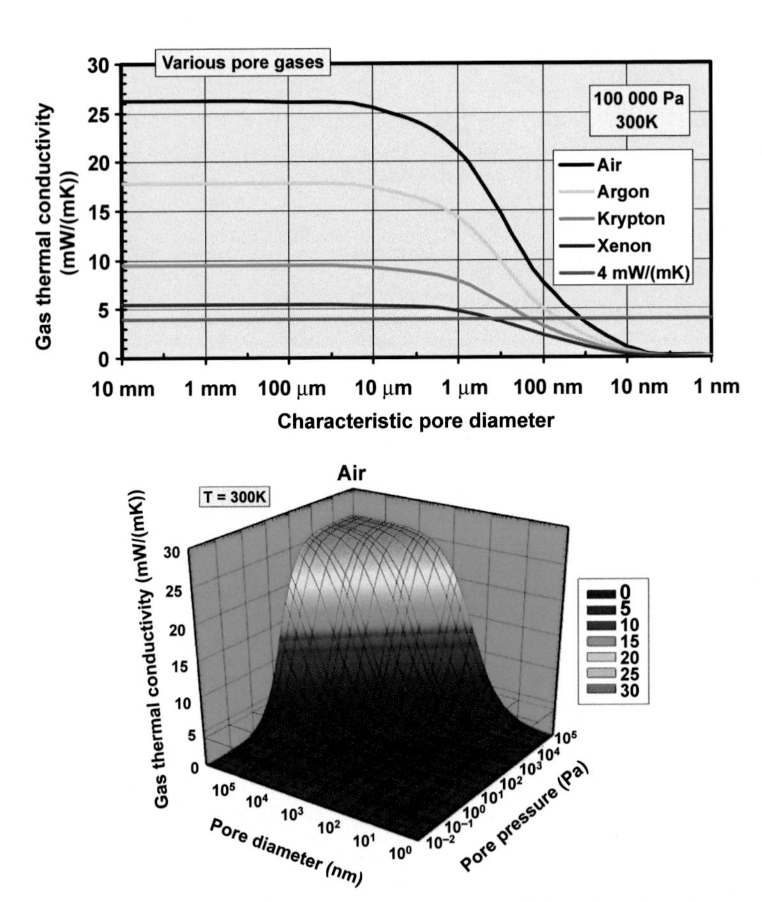

Figure 8.11 Gas thermal conductivity and a (top) 2D-plot depicting the effect of pore diameter for air, argon, krypton and xenon, and a (bottom) 3D-plot depicting the effect of pore diameter and gas pressure in pores for air (Jelle et al., 2010a).

The Stefan–Boltzmann relationship may be applied to show that the radiation thermal conductivity decreases linearly with decreasing pore diameter, where the emissivity of the inner pore walls determine the slope of the decrease. That is, the smaller the pores, and the lower the emissivity, the lower the radiation thermal conductivity will be. However, various works (e.g., Joulain et al., 2005; Mulet et al., 2002; Zhang, 2007) describe a large increase in the thermal radiation as the pore diameter decreases below the wavelength of the thermal (infrared) radiation (e.g., 10 μm), where tunneling of evanescent waves may play an important role (near field radiation effects). The work by Mulet et al. (2002) and Joulain et al. (2005) indicate that the large thermal radiation is only centered around a specific wavelength (or a few). That is, this might suggest that the total thermal radiation integrated over all wavelengths is not that large. How much this actually contributes to the total (overall) thermal conductivity is not completely known at the moment, although it is assumed to be at least rather moderate. Nevertheless, these topics are currently being addressed in ongoing research activities. The study by Jelle et al. (2010a) elaborates more on these thermal radiation issues.

The solid state lattice conductivity in the NIMs has to be kept as low as possible in order to obtain the lowest possible overall thermal conductivity. If a low-conductivity solid state lattice and a low gas thermal conductivity are achieved, and still dominate the thermal transport (i.e., larger than the thermal radiation part), then NIMs may become the high performance thermal insulation material of the future.

8.6.4 Dynamic insulation materials

A dynamic insulation material (DIM) is a material where the thermal conductivity can be controlled within a desirable range. The thermal conductivity control may be achieved by being able to change in a controlled manner:

- The inner pore gas content or concentration including the mean free path of the gas molecules and the gas–surface interaction.
- The emissivity of the inner surfaces of the pores.
- The solid state thermal conductivity of the lattice.

Two models exist for describing solid state thermal conductivity. That is, the phonon thermal conductivity (atom lattice vibrations) and the free electron thermal conductivity. We might ask if it could be possible to dynamically change the thermal conductivity from very low to very high, i.e., making a DIM. Furthermore, other fields of science and technology could inspire and give ideas about how to make a DIM, e.g., the fields of electrochromic materials, quantum mechanics, electrical superconductivity, and possible others.

For example, among electrochromic materials, i.e., materials able to change their solar radiation transmittance by an applied external voltage (Baetens et al., 2010d; Granqvist, 1995, 2005, 2008, 2012; Jelle et al., 2012a; Jelle, 2013a, 2015; Lampert, 1998, 2004), there is also a group of materials called (electrical) conducting polymers, which in addition to their ability to dynamically change the solar radiation transmittance (e.g., their absorptive and reflective properties), are also able to change their electrical conductivity by several decades from electrical insulating materials to materials that are often denoted as synthetic metals. One of these electrochromic materials and conducting polymers is polyaniline (Chiang and MacDiarmid, 1986; Huang et al., 1986; Jelle, 1993, 2013a; Jelle et al., 1992a,b, 1993a,b,c, 1998, 2007; Jelle and Hagen, 1993, 1998, 1999a,b), which by application of an external voltage may undergo color changes from a clear, transparent state to a violet color, with yellow, green, and blue as intermediate colors, the unique doping mechanisms including both redox processes and proton doping. The electrical conductivity of polyaniline may be increased by a factor of $\sim 10^{10}$ by decreasing the pH from about 4 to about 0, reaching a maximum electrical conductivity of $\sim 500\ \Omega^{-1}\text{m}^{-1}$ (Chiang and MacDiarmid, 1986). Such materials with these dynamic properties may contribute to give us ideas of how to create dynamic thermal insulation materials.

The thermal insulation-regulating abilities of DIM give these conceptual materials a great potential. Noteworthy, it remains to be demonstrated that such a robust and practical DIM can be manufactured. Refer to the work by Jelle et al. (2010a) for further details and elaborations concerning DIMs.

8.6.5 Concrete and applications of nano insulation materials

As with decreasing thermal conductivities of insulation materials, new solutions should also be sought for the load-bearing elements of the building envelope. Using concrete as an example, we might envision mixing NIMs into the concrete, thereby decreasing the thermal conductivity of the structural construction material substantially, while maintaining most or a major part of the mechanical strength and load-bearing capabilities of concrete. As concrete has a high thermal conductivity (1700–2500 mW/(mK), without and with rebars) a concrete building envelope always has to utilize various thermal insulation materials in order to achieve a satisfactory low thermal transmittance (U-value). That is, the total thickness of the building envelope will often become unnecessarily large, especially when trying to obtain passive house or zero energy building standards. Furthermore, the large CO_2 emissions connected to the production of cement imply that concrete has a large negative environmental impact with respect to global warming due to the man-made CO_2 increase in the atmosphere (McArdle and Lindstrom, 2009; World Business Council for Sustainable Development, 2002). In fact, the cement industry produces 5% of the global man-made CO_2 emissions of which (World Business Council for Sustainable Development, 2002):

- 50% is from the chemical process

  e.g., $3CaCO_3 + SiO_2 \rightarrow Ca_3SiO_5 + 3CO_2$

  $2CaCO_3 + SiO_2 \rightarrow Ca_2SiO_4 + 2CO_2$

- 40% is from burning fossil fuels
  e.g., coal and oil
- 10% is from electricity and transport uses

Different applications of NIMs as thermal insulation for concrete are given in Fig. 8.12, i.e., NIM as outdoor and/or indoor retrofitting of concrete, NIM applied in the midst of concrete, and NIM mixed together with concrete. As noted earlier, when mixing NIMs with concrete, it will be important to maintain a major part of the structural construction capabilities of concrete.

8.6.6 NanoCon

In principle, it is not the building material itself (i.e., whether it is steel, glass, wood, mineral wool, concrete, or another material) that is important. On the contrary, it is

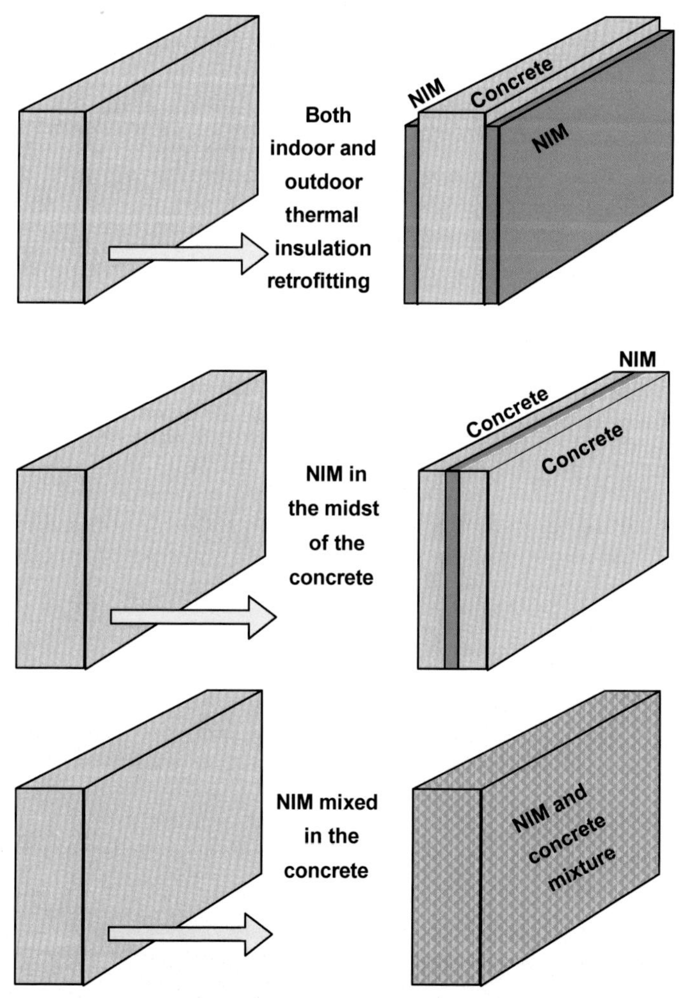

Figure 8.12 NIMs as thermal insulation for concrete: (top) NIM as outdoor and/or indoor retrofitting of concrete; (middle) NIM applied in the midst of concrete; and (bottom) NIM mixed together with concrete (Jelle, 2011a).

the property requirements or functional requirements that are crucial to the performance and possibilities of a material, component, assembly or building. Thus, we might ask if it is possible to invent and manufacture a material with the essential structural or construction properties of concrete intact or better, but with substantially (i.e., up to several decades) lower thermal conductivity? Furthermore, it would be beneficial if that new material would have a much lower negative environmental impact than concrete with respect to CO_2 emissions. Such a material may be envisioned with or without reinforcement bars, depending on the mechanical properties (e.g., tensile strength) of the material.

Regarding the above discussion a new material is introduced on a conceptual basis (Jelle et al., 2010b): NanoCon is basically a homogeneous material with a closed or open small nano pore structure with an overall thermal conductivity of less than 4 mW/(mK) (or another low value to be determined) and exhibits the crucial construction properties that are as good as or better than concrete (Fig. 8.13).

The "Con" in NanoCon is meant to illustrate the construction properties and abilities of this material, with historical homage to concrete. Essentially, NanoCon is a NIM with construction properties matching or surpassing those of concrete. Depending on the mechanical or construction properties of NanoCon, it may be envisioned both with or without reinforcement or rebars. In this definition of NanoCon, a homogeneous material is stated, although the first attempts to reach such a material might be tried by piecing or mixing several different materials together, with a final material product that on a nanoscale is not homogeneous. For example, joining NIM and carbon nanotubes in one single material might enable a very low thermal conductivity due to the NIM part and a very large tensile strength due to the carbon nanotube part. In this respect it should be noted that carbon nanotubes have a very large thermal conductivity along the tube axis. Furthermore, it is noted that the extremely large tensile strength of carbon nanotubes (63 000 MPa measured and 300 000 MPa theoretical limit) surpasses that of steel rebars (500 MPa) by more than two orders. As a comparison concrete itself without rebars has a tensile strength of 3 MPa and a compressive strength of 30 MPa. Thus, the potential impact of NanoCon is tremendously huge.

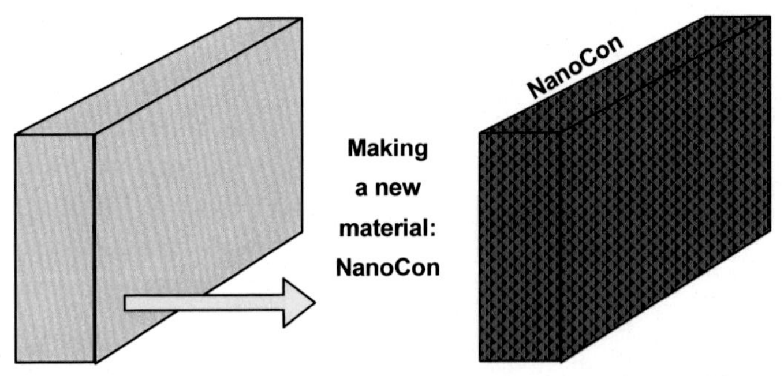

Figure 8.13 NanoCon is essentially a NIM with construction properties matching or surpassing those of concrete (Jelle et al., 2010b).

8.6.7 Other future materials and solutions?

Naturally, the ultimate thermal solution will always be subject to change as time is progressing. That is, the thermal solution we are searching for might very well be found in solutions governed by hitherto unknown principles yet to be discovered or invented. In other words, the thermal solution of tomorrow might be found in materials and solutions not yet thought of, which requires that we may have to *think thoughts not yet thought of* (Jelle et al., 2010a).

8.7 A comparison of weaknesses and strengths

Performing a comparison of the traditional, state-of-the-art, and future thermal building insulation materials and solutions from the short descriptions given earlier (referral is made to the available literature for further details and in-depth knowledge), some general trends may be seen, which will be discussed in the following.

8.7.1 Robustness of traditional thermal insulation materials

The traditional insulation materials are the robust ones with respect to perforation vulnerability and flexibility issues like, for example, whether it is possible to adapt at the building site. However, the traditional insulation materials have relatively high thermal conductivity values, which in cold climates may require all too thick building envelopes in order to reach the goals of passive houses and zero energy or zero emission buildings. In addition, the thermal conductivity increases substantially with increasing moisture content for the traditional thermal building insulation materials (i.e., a vulnerability toward moisture uptake). The traditional insulation material with the lowest thermal conductivity is polyurethane (PUR), with values between 20 and 30 mW/(mK) compared to the others' values typically ranging between 30 and 40 mW/(mK). However, loss of the pore gases and subsequent air permeation into the pores due to diffusion or degradation during time may increase the thermal conductivity of PUR. Furthermore, the toxic gas release from PUR during a fire raises serious health hazard issues. A robustness classification may be carried out (Jelle et al., 2014c), covering both different traditional and state-of-the-art thermal insulation materials.

8.7.2 Thermal conductivity of state-of-the-art thermal insulation materials

The two most promising state-of-the-art insulation materials and solutions, i.e., vacuum insulation panels (VIP) and aerogels, have considerably lower thermal conductivity values than the traditional ones. The future of gas-filled panels (GFP) is considered to be doubtful, as VIPs seem to be a better choice with respect to attaining a very low thermal conductivity and being the most robust of the two, both for now and the foreseeable future. Comparing the miscellaneous traditional and state-of-the-art thermal insulation materials and solutions, the VIP solution has definitely the lowest thermal conductivity value of them all, typically around 4 mW/(mK) in the pristine non-aged condition, whereas the typical low value for aerogel is 13 mW/(mK). However, the VIP thermal conductivity increases with time due to moisture and air penetration by diffusion, attaining a value of 20 mW/(mK) in the non-vacuum or perforated condition. The aerogel conductivity on the other hand is not considered to be increasing substantially with time, and perforations represent no problem. Both VIPs and aerogels are very expensive, but it has been demonstrated that VIPs may be cost-effective (chapter: Potential cost savings by applying vacuum insulation panels), and aerogels in their transparent or translucent state offer application areas where we may be willing to accept higher costs.

8.7.3 Thermal conductivity of future thermal insulation materials

The conceptual future insulation materials have been designed, wisely enough, to have very low thermal conductivities and at the same time be very robust with respect to aging, perforation, building site adaptations, and several other properties. If these thermal insulation materials of beyond tomorrow may be conceived within a not-too-distant future remains to be seen, but nevertheless these materials give us ambitious goals to strive at.

8.7.4 Thermal conductivity and other properties

In general, future thermal insulation materials and solutions need to have as low thermal conductivity as possible. Furthermore, the thermal conductivity should not increase substantially over a 100 year or more lifetime span. In addition, these materials and solutions should also be able to maintain their low thermal conductivity even if they are perforated by external objects like nails and similar, except the increase due to the local heat bridges. Thus, technologies based on vacuum may have problems with maintaining a low thermal conductivity over a long time span stretching over several decades, due to loss of vacuum with air and moisture uptake during the years.

A major requirement for the future thermal insulation materials is that they can be cut for adaption at the building site without losing any of their thermal insulation resistance. The VIP solution with an envelope barrier around an open pore structure supposed to maintain a vacuum does not satisfy this specific requirement, as cutting a VIP will result in a total loss of vacuum and an increase of thermal conductivity up to typically 20 mW/(mK).

Many other properties also have to be considered and addressed. These include mechanical strength (e.g., compression and tensile strength), fire protection issues either by the thermal insulation material itself or other protection means, fume emissions during fire where preferably no toxic gases should be released, climate aging durability with various climate exposures, resistance toward freezing/thawing cycles and water in general, dynamic properties (i.e., the ability to regulate the thermal insulation level), costs that should be competitive versus other thermal insulation materials, and environmental impact.

8.7.5 Requirements of future thermal insulation materials and solutions

The thermal insulation materials and solutions of tomorrow have to satisfy several crucial requirements. Table 8.1 summarizes the various properties with their proposed requirements. As it can be seen, the proposed thermal conductivity requirement in the pristine condition is a conductivity less than 4 mW/(mK), which is the typical value for non-aged VIPs. Naturally, the thermal conductivity after a certain period of time or service life is of vital importance. A conductivity less than 5 mW/(mK) after 100 years is proposed for the future thermal insulation materials and solutions to be developed. Hence, Table 8.1 represents an initial attempt to address the crucial properties and requirements of the future high performance thermal insulation materials and solutions, and will naturally be subject to change in the years to come.

Table 8.1 Selected and proposed requirements of the future high performance thermal insulation materials and solutions (see Jelle, 2011a).

Selected properties	Requirement
Thermal conductivity (pristine)	<4 mW/(mK)
Thermal conductivity (after 100 years)	<5 mW/(mK)
Thermal conductivity (after modest perforation)	<4 mW/(mK)
Perforation vulnerability	Not to be influenced significantly
Possible to cut for adaption at building site	Yes
Mechanical strength (e.g., compression and tensile)	May vary
Fire protection	May vary, depends on other protection
Fume emission during fire	Any toxic gases to be identified
Climate aging durability	Resistant
Freezing/thawing cycles	Resistant
Water resistant	Resistant
Water permeability	May vary
Dynamic thermal insulation	Desirable as an ultimate goal
Costs versus other thermal insulation materials	Competitive
Environmental impact (including energy and material use in production, emission of polluting agents and recycling issues)	Low negative impact

8.7.6 The potential of miscellaneous thermal insulation materials and solutions

Table 8.2 gives a short summary of the potential of the traditional, state-of-the-art, and possible future thermal building insulation materials and solutions with respect to becoming the high performance thermal insulation of tomorrow.

Table 8.2, in addition to being a summary, may be utilized to initiate a chain of thoughts of how to proceed beyond today's state-of-the-art thermal solutions. Also note that Table 8.2 expresses the current status for the state-of-the-art solutions of

Table 8.2 The potential of the traditional, state-of-the-art and possible future thermal building insulation materials and solutions of tomorrow (see Jelle, 2011a).

Materials	Low pristine/aged thermal conductivity	Perforation robustness	Possible building site adaption cutting	Load-bearing capabilities	A thermal insulation material and solution of tomorrow?
Traditional thermal insulation					
Mineral wool	No	Yes	Yes	No	No
EPS	No	Yes	Yes	No	No
XPS	No	Yes	Yes	No	No
Cellulose	No	Yes	Yes	No	No
Cork	No	Yes	Yes	No	No
PUR	No	Yes	Yes	No	No
State-of-the-art thermal insulation					
VIP	Yes/maybe	No	No	No	Today and near future
GFP	Maybe	No	No	No	Probably not?
Aerogel	Maybe	Yes	Yes	No	Maybe
PCM	–	–	–	–	Heat storage and release
Possible future thermal insulation					
VIM	Yes/maybe	Yes	Yes	No/maybe	Yes
GIM	Yes/maybe	Yes	Yes	No/maybe	Maybe
NIM	Yes	Yes, excellent	Yes, excellent	No/maybe	Yes, excellent
DIM	Maybe	Not known	Not known	No/maybe	Yes, excellent
NanoCon	Yes	Yes	Yes	Yes	Yes, excellent
Others	–	–	–	–	Maybe

today and the foreseen status for the beyond state-of-the-art solutions, where certain items in the table might be subject both to discussion and change. Currently, the NIM solution seems to represent the best high performance, low conductivity thermal solution for the foreseeable future. DIMs and NanoCon represent two solutions with a huge potential, with thermal insulation-regulating and load-bearing capabilities, respectively.

8.7.7 Potential cost savings by applying vacuum insulation panels

The VIPs of today are relatively costly. Nevertheless, application of VIPs may actually be directly profitable compared to traditional thermal insulation. This potential profit is due to the very low thermal conductivity values of VIPs (e.g., compared to mineral wool), which make it possible to build considerably slimmer wall constructions. The higher the market value of living area (EUR/m^2 living area), the higher the potential profit is by the application of VIPs and reduction of the wall thickness, which hence increases the living area.

Table 8.3 shows the available means for VIP purchase due to reduced wall thickness and thus increased living area, where the potential savings include the costs of traditional thermal insulation. In all the calculations shown here it is assumed that VIPs in a wall construction on average have a thermal resistance about 5–6 times higher than traditional thermal insulation during a given operation period; more specifically, that 35 cm of mineral wool corresponds thermally to 6 cm VIPs over a given period. An interior floor-to-ceiling height of 2.5 m is assumed in the calculations. For example, if the market value at the construction site is 4000 EUR/(m^2 living area), and the wall

Table 8.3 Available means for vacuum insulation panel purchase as function of living area market value and wall thickness reduction. An interior floor-to-ceiling height of 2.5 m is assumed. Mineral wool costs of 15, 18, 20, 24 and 26 EUR/m^2 are used for mineral wool thicknesses of 25, 30, 35, 40 and 45 cm, respectively, for a corresponding wall thickness reduction of 10, 15, 20, 25 and 30 cm by application of VIPs (see Jelle, 2011a).

Available means for vacuum insulation panel purchase due to living area and traditional thermal insulation savings (EUR/m^2 wall area)					
Market value living area (EUR/m^2 living area)	Reduction of wall thickness (cm)				
	10	**15**	**20**	**25**	**30**
1000	50	80	100	120	140
2000	90	140	180	220	260
3000	130	200	260	320	380
4000	170	250	330	410	490
5000	210	310	410	510	610

thickness is reduced by 20 cm (from 35 cm timber framework with mineral wool to 15 cm timber framework with VIPs), we may use up to 330 EUR/(m^2 living area) for VIP purchase before the VIP costs exceed the costs we would have had with traditional thermal insulation (Table 8.3). If we use less than 330 EUR/(m^2 living area) for the actual VIP purchase, the difference will be a clean-cut profit in favor of VIPs. The costs of traditional thermal insulation constitute only a relatively small part of the total savings. In Table 8.3 there is given mineral wool costs of 15, 18, 20, 24 and 26 EUR/m^2 for mineral wool thicknesses of 25, 30, 35, 40 and 45 cm, respectively, for a corresponding wall thickness reduction of 10, 15, 20, 25 and 30 cm by application of VIPs. These mineral wool costs may be directly subtracted from the values in Table 8.3 if we want the available means for VIP purchase due to only the living area savings.

Tables 8.4 and 8.5, with the two corresponding graphical plots in Fig. 8.14, treat a case where the wall thickness is reduced from a timber framework with 35 cm mineral wool (20 EUR/m^2) to a 15 cm timber framework with 6 cm VIPs (200 EUR/m^2), a wall thickness reduction of 20 cm with approximately the same thermal transmittance (U-value). The wall thickness reduction could have been even lower with respect to the thermal resistance of the VIPs, but a minimum wall thickness of 15 cm is chosen due to construction reasons including load-bearing properties. With a VIP thickness of 6 cm the remaining 9 cm in the timber framework if desired could be filled with mineral wool, for example, to increase the thermal resistance even further. VIPs placed in

Table 8.4 Profit in EUR/(m^2 living area) by application of vacuum insulation panels as function of living area market value where the wall thickness reduction is 20 cm for an example building of 10 m × 10 m. An interior floor-to-ceiling height of 2.5 m is assumed (see Jelle, 2011a). See Fig. 8.14.

Living area (m^2) 10 × 10	Market value living area (EUR/m^2 living area)	Increased living area gain by application of vacuum insulation panels and reduced wall thickness (m) of 0.2 (EUR/m^2 living area)	Vacuum insulation panel costs 6 cm thickness (EUR/m^2 VIP)	Traditional thermal insulation costs 35 cm thickness (EUR/m^2 insulation)	Profit due to vacuum insulation panel application (EUR/m^2 living area)
	1000	80	200	20	−100
	2000	160	200	20	−20
	3000	240	200	20	60
	4000	310	200	20	130
	5000	390	200	20	210

Table 8.5 Profit in EUR/(100 m^2 living area) by application of vacuum insulation panels as function of living area market value where the wall thickness reduction is 20 cm for an example building of 10 m × 10 m. An interior floor-to-ceiling height of 2.5 m is assumed (see Jelle, 2011a). See Fig. 8.14.

Living area (m^2) 10 × 10	Market value living area (EUR/m^2 living area)	Increased living area gain by application of vacuum insulation panels and reduced wall thickness (m) of 0.2 (EUR/100 m^2 living area)	Vacuum insulation panel costs 6 cm thickness (EUR/m^2 VIP)	Traditional thermal insulation costs 35 cm thickness (EUR/m^2 insulation)	Profit due to vacuum insulation panel application (EUR/100 m^2 living area)
	1000	8000	200	20	−10 000
	2000	16 000	200	20	−2000
	3000	24 000	200	20	6000
	4000	31 000	200	20	13 000
	5000	39 000	200	20	21 000

the middle of the wall construction with cavities on both sides will increase the robustness versus perforations. In the calculations an interior floor-to-ceiling height of 2.5 m is assumed, in addition to an example building of 10 m × 10 m, giving an interior floor area (living area) of 100 m^2 and an interior wall area of 100 m^2 after the wall thickness reduction. Tables 8.4 and 8.5, in addition to Fig. 8.14, present the profit by applying VIPs, as EUR/(m^2 living area) and EUR/(total m^2 living area), respectively. This profit represents money you have as surplus when the building has been built with VIPs (i.e., sheer profit due to VIPs). As may be seen from the tables and graphical plots, relatively large profits may be gained, especially when the market value of the living area is high.

It should be noted that the calculations are simplified and do not consider all conditions. As an example it may be mentioned that thinner timber frame walls lead to less material costs related to the timber framework. That is, there are even higher cost savings in favor of VIPs, which are not included in these calculations. Furthermore, costs and interest expenses per month related to additional loan raising by the VIP purchase compared to additional rental income due to increased living area, are not included in the calculations. All the values in Tables 8.3–8.5 are rather roughly approximated estimated values. The purchase costs of VIPs are important for the potential profit and how large it may become, and may dominate the other costs not accounted for in the above. The VIP costs of 200 EUR/m^2 for a total VIP thickness of 6 cm, which are used in the calculations, are to be regarded as an example. The VIP costs are expected to decrease considerably in the coming years.

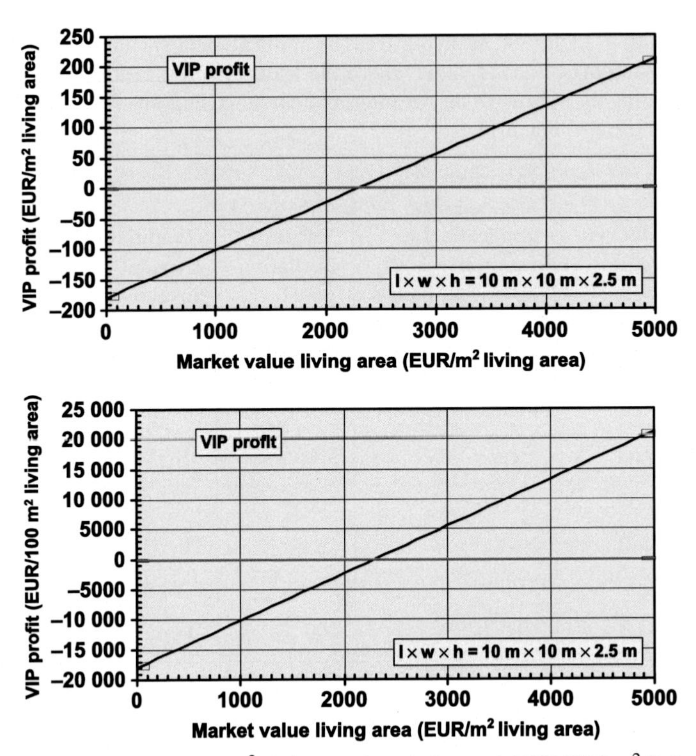

Figure 8.14 Profit in (top) EUR/(m^2 living area) and (bottom) EUR/(100 m^2 living area) by application of vacuum insulation panels as function of living area market value where the wall thickness reduction is 20 cm for an example building of 10 m × 10 m. An interior floor-to-ceiling height of 2.5 m is assumed (Jelle, 2011a). See Tables 8.4 and 8.5.

8.7.8 *Condensation risk by applying vacuum insulation panels in the building envelope*

A very low water permeability close to zero (e.g., for VIPs) may represent a condensation risk depending on the actual construction and the indoor and outdoor climate conditions, for both new buildings and thermal retrofitting of existing buildings. These issues have been addressed by Sveipe et al. (2011) and Jelle et al. (2013a) through experimental and numerical investigations of temperature and humidity conditions in building envelopes during thermal retrofitting of timber frame walls with VIPs. Hydrothermal numerical simulations and laboratory investigations for interior thermal insulation retrofit of a historical brick wall using VIPs were carried out by Johansson et al. (2014). The laboratory experiments were conducted in a large-scale vertical building envelope climate simulator as depicted in Fig. 8.15. The studies showed that adding interior VIPs could reduce the energy use substantially in brick buildings, and that various moisture considerations should be taken into account.

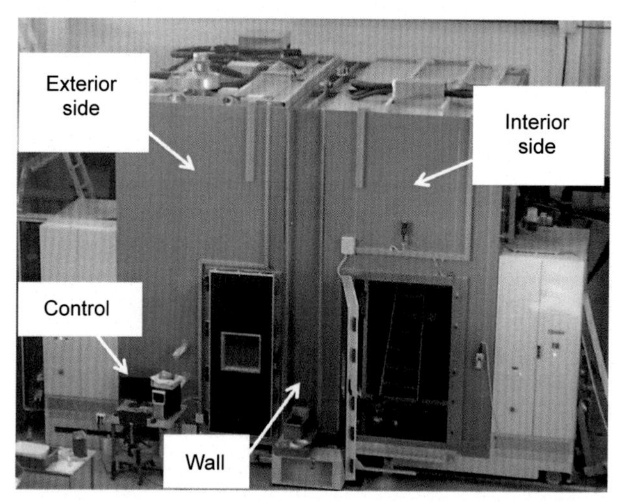

Figure 8.15 Applying a large-scale vertical building envelope climate simulator for laboratory investigations for interior thermal insulation retrofit of a historical brick wall using vacuum insulation panels (Johansson et al., 2014).

8.7.9 The cardinal weaknesses of vacuum insulation panels

Currently, VIPs have the lowest pristine thermal conductivity of all thermal insulation materials, with typical center-of-panel values around 4 mW/(mK) in the non-aged condition. Some VIPs have even been reported to have achieved center-of-panel thermal conductivities as low as 2 mW/(mK). However, as mentioned earlier the VIP thermal solution has several drawbacks, summarized into the following four cardinal weaknesses of VIPs:

- Fragility
- Perforation vulnerability
- Increasing thermal conductivity during time
- Lack of building site adaption cutting

These four VIP cardinal weaknesses all originate from the challenge of maintaining a vacuum inside a core material with a protective foil around the core, where a total loss of vacuum leads to an increase in center-of-panel thermal conductivity from typically around 4 mW/(mK) to about 20 mW/(mK) (with fumed silica core).

The fragility of VIPs represents a serious concern and a major drawback. VIPs may be handled and stored with great care, but nevertheless many panels may suddenly and unexpectedly lose all their vacuum for no evident reason. In our laboratory during our various experiments with VIPs, we have also experienced this vacuum loss for no evident reason for different types and manufacturers of VIPs. Hence, there is a risk and likely probability that several VIPs may be installed in a building envelope with the vacuum intact, and then after a relatively short time period lose their vacuum.

The perforation vulnerability or lack of perforation robustness of VIPs represents another major disadvantage. This vulnerability may imply miscellaneous restrictions to the applications of VIPs; for example, other building envelope techniques or various restrictions to perforate the building envelopes. That is, you may not put up your pictures everywhere on all the walls in the same way as you have done before. Nevertheless, despite various building techniques and restrictions, VIPs are prone to get perforated in different ways during a building's lifetime. Besides, a certain amount of VIPs are also likely to be perforated during transport and handling.

The increasing thermal conductivity during time represents yet another cardinal weakness of VIPs. Depending on the VIP type, VIP envelope foil type, and VIP size, water, oxygen, nitrogen, and other air molecules will with time diffuse through the VIP envelope and into the VIP core, thereby decreasing the vacuum inside the VIPs and hence increasing the thermal conductivity (see e.g., Fig. 8.3). That is, the thermal resistance of a building envelope based on VIPs will decrease with time, which then has to be taken into account when determining the necessary VIP thickness.

The lack of building site adaption cutting of VIPs is also a cardinal drawback. All VIPs have to be manufactured in a VIP production facility with the required dimensions, usually limited to a number of standard sizes due to cost limitations, and no adaption or cutting of the panels can be carried out at the building site as any cutting will lead to loss of vacuum. Hence, any adjustments at the building site have to be performed with other thermal insulation materials with a higher thermal conductivity than the VIPs.

Noteworthy, the heat bridge effect caused by the metallized envelope foil of VIPs, and the relatively high costs of VIPs, are not considered as cardinal weaknesses of VIPs. The heat bridge effect decreases with increasing VIP size, and is much smaller for metallized films (MF types) than for metal films (AF) (see, e.g., Fig. 8.2). Besides, the heat bridge effect will by far be outweighed by the large thermal resistance in VIPs. As commented earlier the application of VIPs may actually result in an economic profit in areas with a high living area market value per square meter, where a reduced wall thickness may involve large area savings and thus a higher value of the real estate (chapter: Potential cost savings by applying vacuum insulation panels).

8.7.10 Expanded polystyrene encapsulated vacuum insulation panels

Thermal insulation solutions with VIPs wrapped in EPS have been proposed and are also in actual use. These EPS encapsulated VIPs have an extra protection due to the EPS covering, and enables some adjustments at the building site as parts of the EPS perimeter may be cut away. Applying EPS encapsulated VIPs in two layers, ensuring there is always at least one VIP layer in the EPS joints, reduces the (EPS) heat bridges. See for example, the experimental and numerical studies by Van Den Bergh et al. (2011).

However, these EPS covered VIPs also have some drawbacks. First, the loss of vacuum in VIPs totally covered in EPS is not as easy to detect and thus increases the probability of installing VIPs with no vacuum. Contrary, in VIPs with no EPS (or other) covering it is normally very easy to detect if the vacuum has been lost, both by visual inspection of the very tight foil around the VIP core as long as a vacuum is maintained, also giving the panel a rather rigid structure, and by touching or hearing by snapping fingers at the VIP foil, where the response will be quite different for a panel with and without vacuum. Second, the EPS replaces part of the VIP area, and since the EPS thermal conductivity is considerably larger than the VIP conductivity, there is a question what configuration would have given the lowest U-values, EPS covered VIPs or only VIPs, which then have to be calculated or measured.

8.7.11 Vacuum insulation materials and gas insulation materials versus nano insulation materials

A VIM or a GIM needs to maintain a vacuum or a gas inside a closed pore structure, respectively. These materials could be cut and adapted at the building site with no loss of low thermal conductivity. In addition, perforating the VIM or GIM with a nail or something similar would only result in a local heat bridge, in other words, no loss of low thermal conductivity. The VIM grid structure has to be strong enough to withstand the vacuum inside its pores, and the air and water vapor diffusion through the grid structure and into the vacuum pores have to be small enough that the VIMs will maintain their low thermal conductivity below a certain value for at least 100 years. Keeping the vacuum inside the pores during a long service life may be the most difficult or challenging task for the VIMs. That is, the most challenging task after the VIMs have been manufactured, since making VIMs is a highly challenging task in itself. Nevertheless, when an appropriate VIM production process has been established, such a production might hopefully be found to be both efficient and competitive. With respect to GIMs, it might be easier to create a closed pore structure filled with a low-conductance gas than vacuum. Furthermore, the GIM grid structure does not have to be as strong as the VIM grid structure because a vacuum pore structure will be prone to collapse before a gas-filled pore structure. In addition, it may be easier to maintain the original low thermal conductivity within a gas-filled pore structure than in a vacuum pore structure. Comparing the VIMs and GIMs, the VIMs ultimately have the largest potential of these two as the lowest thermal conductivity is achieved in a vacuum compared to a gas-filled pore structure.

As both the VIMs and GIMs share the same disadvantage that their grid structures need to prevent air and moisture penetration into their pore structures during their service life for at least 100 years, we may ask if it is possible to envision and manufacture a high performance thermal insulation material like the VIMs and GIMs, but without their disadvantages. The answer as noted earlier is the NIM. The NIMs may obtain a very low thermal conductivity with either an open or a closed pore structure, and the NIM grid structure does not need to prevent air and water vapor from diffusing into the pores since the NIMs may attain a very low thermal conductivity with air inside the pores; see for example, the Knudsen effect depicted by Eqs. [8.2] and [8.3] and Figs. 8.10–8.11. Furthermore, perforating the NIMs do not create any local thermal bridges induced by air and water vapor leakage into the pore structure locally around the intrusion, except the thermal bridges caused by the perforating agents (e.g., nails) themselves. Hence, compared to VIMs and GIMs, the NIMs are regarded to have the largest potential. NIMs with an open pore structure will be containing air, and thus the structure has to be resistant toward various aging degradation mechanisms as air gases including different pollution will freely be admitted into the NIM pores. It is essential that water condensation in such tiny nanopores has to be prevented, or else the thermal conductivity will increase substantially and might ruin the whole concept of the NIMs.

8.7.12 The regulating potential of dynamic insulation materials

To be able to understand thermal conductance so well that we can tailor-make materials to dynamically regulate the thermal conductivity of a material from a very low to a very high conductivity has a tremendous potential. These DIMs may constitute a part of an intelligent building envelope that includes the emerging solutions phase change materials and electrochromic materials, for day/night storage/release of accumulated energy or any surplus solar energy and dynamic control of solar energy through glazing systems, respectively. Limiting ourselves to solid state thermal conduction, changes in the thermal conductivity involve changes in the chemical bonds between atoms. To change the thermal conductivity in DIMs back and forth within a desirable range may require some energy, which naturally should be as low as possible. Analogies might also be drawn toward the field of electrical superconductivity; that is, the ability to transport electrical current without resistance and hence without loss of energy. There is still much to be discovered and understood within the realms of atoms, quantum mechanics, and the matter that surrounds us. Future applications of DIMs may find large areas of use outside the building sector; for example, envision how a thermal superconductor could be utilized. Clearly, both theoretical and practical, the DIMs represent a tremendous potential for applications in buildings and several other areas.

8.7.13 The construction potential of NanoCon

The idea of making a construction material which also possesses the properties of a high performance thermal insulating material has a huge potential. The concept of NanoCon represents an ultimate solution in this respect. Even if the very ambitious goal of as low thermal conductivity as 4 mW/(mK) for NanoCon is not reached within the foreseeable future, every order of magnitude the thermal conductivity may be lowered from the high value of concrete down toward the values of the traditional and the state-of-the-art thermal building insulation materials, while still exhibiting most or a substantial part of the crucial construction properties of concrete, will be very important. Combinations of NIMs and carbon nanotubes may be utilized in order to reach the goals of a NanoCon high performance thermal insulation and construction material. Naturally, the fire resistance of NanoCon will also be important to address.

NanoCon may be envisioned both with and without the use of steel rebars. Today the corrosion of rebars in concrete constructions represents a very large worldwide problem. If concrete constructions, and ultimately NanoCon, may be used without steel reinforcement, this will have an enormous impact on the construction industry and the built environment. Thus, every effort and investment put into the research and development of NanoCon and similar materials may pay off tremendously and be worth every single penny.

8.7.14 Assessing weaknesses and strengths

During the evaluation of which thermal building insulation material or solution to choose for a specific application or building, all the important properties should be evaluated with regard to their weaknesses and strengths. As of today, no single thermal insulation material or solution exists that is superior or best in all respects. The assessment of all the pros and cons for the different properties and applications is crucial when determining which thermal insulation material or solution to choose.

8.7.15 Does the future belong to nano insulation materials, dynamic insulation materials and NanoCon?

The large challenges associated with inventing and developing thermal insulation materials like NIMs, DIMs, and NanoCon or similar concepts should neither be concealed nor understated. Considerable research efforts will be needed, both along known theoretical principles within physics, chemistry and material science, but maybe also along more unknown paths within these areas. Theoretical investigations have to be followed with concurrent experimental examinations and explorations. The efforts put into these research paths will certainly pay off sooner or later, hence leading to the development of new thermal building insulation materials and solutions.

Thus, we may conclude that NIMs, DIMs, and NanoCon may take the step or leap from the conceptual stage to being part of the high performance thermal insulation materials of beyond tomorrow. And nevertheless, whatever will be the specific thermal solutions in the near or distant future, the ideas around NIMs, DIMs, and NanoCon will undeniably contribute to the development of the future thermal building insulation materials and solutions.

8.7.16 Future research pathways

The day after and beyond tomorrow will see other thermal insulation materials and solutions than exist today. As an endnote, we may conclude that future research may beneficially be conducted along three pathways: (1) improving the existing traditional thermal insulation, (2) improving the existing state-of-the-art thermal insulation, and (3) exploring the possibilities of discovering and developing novel high performance thermal insulation materials and solutions with properties surpassing all of today's existing materials and solutions.

8.8 Experimental pathways

8.8.1 Moving from concepts to experiments

Based on the conducted conceptual studies (Baetens et al., 2010a; Jelle et al., 2009, 2010a,b; Jelle, 2011a), various experimental pathways have been attempted in the quest of making thermal superinsulation materials (SIM) and nano insulation materials (NIM) (Jelle et al., 2011a, 2014a,b), where at the moment most of these are based on fabricating hollow silica nanospheres (HSNS) by the sacrificial template method (Jelle et al., 2011a, 2013b, 2014a,b,d, 2015a,b; Gao et al., 2012, 2013, 2014b; Grandcolas et al., 2013; Sandberg et al., 2013). However, it should be noted that manufacturing a bulk material with nanopores directly, may be regarded as a more ideal and efficient way of producing a SIM, i.e., and not first making hollow nanospheres which then need to be pieced together and assembled into a bulk material. Nevertheless, currently the HSNS may represent an experimental feasible method of actually obtaining a SIM.

8.8.2 Membrane foaming method

The governing principle of membrane foaming is to produce foams with nanoscale bubbles, followed by condensation and hydrolysis within the bubble walls to obtain a silica nanofoam. In the process, gas is pressed through a membrane to obtain bubbles with controlled size as depicted in Fig. 8.16; see e.g., the studies by Bals and Kulozik (2003) and Müller-Fischer (2007). Hydrolysis and condensation of precursors at the bubble–liquid interface should result in formation of gas capsules. This method was previously used to obtain nitrogen-containing capsules with

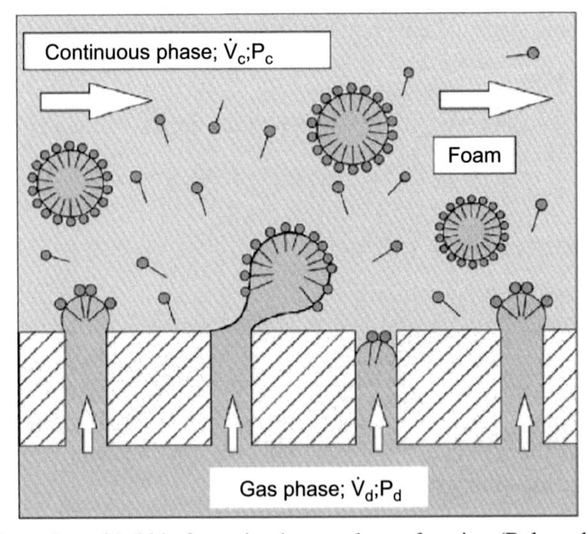

Figure 8.16 Illustration of bubble formation by membrane foaming (Bals and Kulozik, 2003).

titania-polypyrrole composite shells. Initial experiments indicated that preparation of silica nanofoams may be difficult to accomplish, which was supported by theoretical considerations. The gas pressure must be very accurately adjusted; if the pressure is too low, no bubbles will be formed and if it is too high, a continuous gas stream will be the result. The size of the bubbles may be decreased by decreasing the pore size of the membrane and adjusting its surface properties to obtain a high contact angle with the solvent (i.e., the solvent should be repelled from its surface). Furthermore, the solvent density should be rather high and its surface tension low. In principle, it should be possible to design a reaction system that fulfills these requirements, so that production of nano-sized bubbles is viable (Jelle et al., 2011a).

Production of a solid nanofoam requires that the liquid foam must be stable long enough for the reactions to proceed; for example, hydrolysis and condensation of silane precursors to form the solid network. Furthermore, if the foam is to be of interest as a thermal insulator, the foam walls must be thin in order to keep the solid contribution to the overall thermal conductivity low. Wall thicknesses of about 20 nm may be achieved if surfactant bilayers are used to stabilize the walls and the applied solvent has low viscosity and is rapidly drained from the wall interior, which is possible to achieve in water-based systems. However, the reactions are generally performed in alcohol solutions like ethanol or isopropanol. In our initial experiments no surfactant was found that could stabilize nanofoams long enough, thus work along this line has so far been abandoned (Jelle et al., 2011a).

8.8.3 Gas release method

Application of the gas release method would require simultaneous formation of nano-sized gas bubbles throughout the reaction system, followed by hydrolysis and condensation to form a solid at the bubble perimeter. Bubble formation could be achieved by either evaporation or decomposition of a component in the system. This method is similar to the process described by Grader et al. (1998), where crystals of $AlCl_3(Pr^i{}_2O)$ were heated to produce foams with closed cell structures. In this case, the crystals themselves decomposed. Upon further heating, the remaining solid dissolved in the generated solvent, whereupon a polymerization reaction occurred at the temperature of solvent evaporation. The solvent bubbles were trapped within the polymerizing gel, thus forming a stable foam with pore sizes 50–300 μm after completion of the reaction.

However, the gas release process entails several challenges. To obtain nano-sized bubbles with a sufficiently narrow size distribution, the temperature must be the same throughout the liquid phase, which would be difficult to achieve at ordinary reaction conditions. Furthermore, the reaction to form the solid shell must proceed very rapidly if the shell is to be formed before the bubbles grow too large, which would require very reactive chemicals where their application would require strict control of humidity both in the working environment and in the solvents used. Due to these practical difficulties, work along these directions has at the moment been terminated (Jelle et al., 2011a). Nevertheless, if one could succeed in applying such or similar methods, the potential for manufacturing SIMs and NIMs seems to be high.

8.8.4 Template method

When utilizing the template method, a nanoscale structure in the form of a nanoemulsion or polymer gel is prepared, followed by hydrolysis and condensation to form a solid. This procedure is applied for preparing e.g. catalysts and membrane materials. Our current approach is to prepare hollow silica nanospheres (HSNS), followed by condensation and sintering to form macroscale particles or objects. For thermal insulation applications, small pore sizes combined with small wall thicknesses are desired and required (Jelle et al., 2011a). Several methods for nanosphere fabrication are reported in the literature. Our starting point was based on the studies by Du et al. (2010) and Wan and Yu (2008), where the former ones used the template method to prepare antireflection coatings, and the latter ones described the method more in detail, where their schematic depiction of the synthesis process is shown in Fig. 8.17.

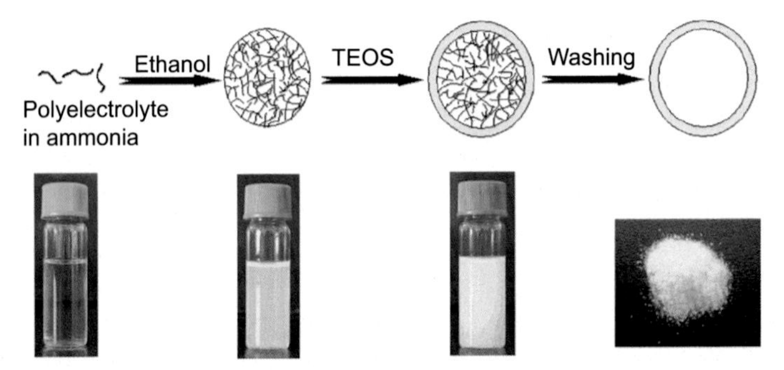

Figure 8.17 Schematic diagram showing the formation mechanism of hollow silica spheres, with corresponding real samples (Wan and Yu, 2008).

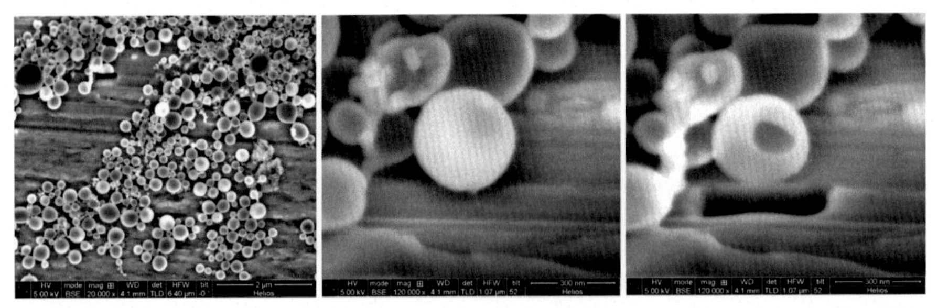

Figure 8.18 Scanning electron microscope images showing an overview of the hollow silica nanosphere sample (left, scale bar 2 μm) and an unetched sphere (middle, scale bar 300 nm), as well as the same sphere after extensive etching with a focused ion beam (right, scale bar 300 nm) (Jelle et al., 2011a).

So far our experimental work has been concentrated on making HSNS by applying polyacrylic acid (PAA) and polystyrene (PS) sacrificial templates, where the templates PAA and PS have been removed by a washing process and a heating process, respectively (the template materials diffusing and evaporating through the silica shell), where Fig. 8.18 shows scanning electron microscope (SEM) images of the first HSNS we produced.

8.9 Experimental synthesis of hollow silica nanospheres

8.9.1 Hollow silica nanosphere experimental details

Miscellaneous detailed experimental information and procedures concerning the various fabrications of hollow silica nanospheres (HSNS) are found in our earlier studies (Gao et al., 2012, 2013, 2014b; Jelle et al., 2011a; Sandberg et al., 2013). Basically, the HSNS manufacturing applies the template method as described earlier, with either polyacrylic acid (PAA) or polystyrene (PS) as sacrificial templates. The principle of the sacrificial template method for HSNS fabrication is illustrated in Fig. 8.19, whereas Fig. 8.20 depicts laboratory scale production of PS templates including an SEM image of the resulting PS template spheres.

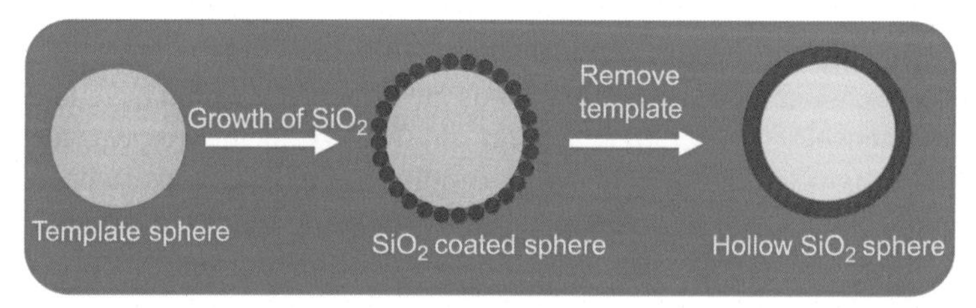

Figure 8.19 Illustration of the sacrificial template method for hollow silica nanosphere fabrication (Jelle et al., 2014a,b).

Figure 8.20 Photo from laboratory scale production (left) of polystyrene templates, and scanning electron microscope image of resulting polystyrene template spheres (Jelle et al., 2014a,b).

8.9.2 Hollow silica nanosphere results

Nano insulation materials (NIM) have been attempted to be made in the laboratory as various hollow silica nanospheres (HSNS). Fig. 8.21 shows a principle drawing of a NIM alongside a transmission electron microscope (TEM) image of actual manufactured HSNS, depicting the close resemblance from theoretical concepts to experimental synthesis attempts.

A SEM image of manufactured spherical PS templates is shown in Fig. 8.22 (left). The polystyrene (PS) templates were hence coated with small silica particles; an example is depicted in Fig. 8.22 (middle). By removal of the templates, HSNS are formed; one example is depicted in Fig. 8.22 (right). Another example of HSNS with PS templates underneath are depicted in Fig. 8.23, depicting SEM and scanning transmission electron microscope (STEM) images of HSNS before removal of the PS templates and formation of a silica shell around a void. Note the differences in the surface roughness constituted by the silica particles forming the shells of the HSNS when comparing Figs. 8.22 and 8.23. Attempting to make monodisperse PS nanospheres, the results are not always as anticipated. Sometimes only minor synthesis parameter differences may cause large and even unexpected results. Naturally, when attempting to make new materials several stepping-stones may have to be stepped onto (e.g., experimental synthesis parameter variations) before the chosen path seems to be the right one walking (actually manufacturing materials, e.g., NIMs, which are approaching the desired properties).

Thermal conductivity has been measured for various powder samples of HSNS, where the conductivity values are typically in the range 20–90 mW/(mK), though some uncertainties in the Hot Disk apparatus measurement method have to be further clarified (Grandcolas et al., 2013; Jelle et al., 2013b). In this respect, the specific powder packing of the HSNS in the bulk condition is also an issue to be addressed. The thermal conductivity is currently being attempted to be lowered by a parameter variation and optimization of the hollow silica sphere inner diameter and wall (shell)

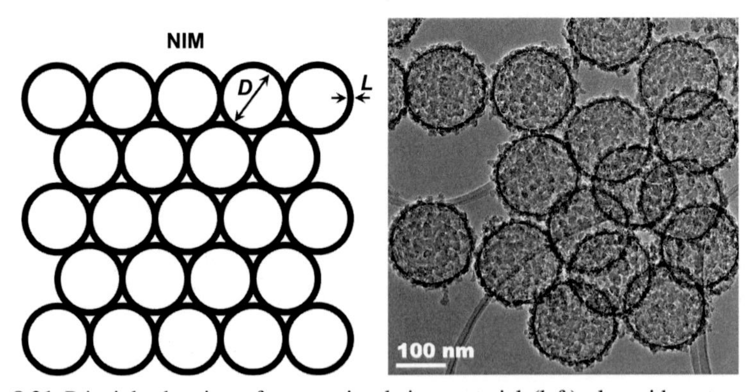

Figure 8.21 Principle drawing of a nano insulation material (left) alongside a transmission electron microscope image of actual synthesized hollow silica nanospheres (right) (Jelle et al., 2014a,b). That is, from theoretical concepts to experimental results.

Figure 8.22 Scanning electron microscope image of (left) spherical polystyrene (PS) templates, (middle) small silica particles coated around a spherical PS template, and (right) hollow silica nanospheres after removal of PS (Jelle et al., 2014a,b). See also Fig. 8.19 to compare with a principle drawing of the process.

Figure 8.23 Scanning electron microscope (left) and scanning transmission electron microscope (right) images of small silica particles coated around spherical polystyrene (PS) templates, i.e., hollow silica nanospheres before removal of PS and formation of a silica shell around a void.

thickness. Furthermore, aspects like e.g. thermal radiation, mesoporosity, powder packing at bulk scale and nanosphere packing at nanoscale should be addressed.

Life cycle analysis of NIM as HSNS has been carried out in the studies by Gao et al. (2013, 2014b) and Schlanbusch et al. (2014a,b), and follow-up investigations are currently being conducted. Initial experiments attempting to improve the thermal resistance of concrete by incorporation of aerogel have also been performed (Gao et al., 2014a; Ng et al., 2015a,b; Ng et al., 2016), where naturally any new development of NIM will be interesting for further work within this area.

For additional information on fabrication of monodisperse PS nanospheres it is referred to the work by Du and He (2008). More information on HSNS (and microspheres) may be found in the studies by Cheng et al. (2007), Du et al. (2010), Fan et al. (2011), Han et al. (2011), Kim et al. (2005), Le et al. (2004a,b), Liao et al. (2011, 2012), Liu et al. (2011), Meng et al. (2012), Peng et al. (2008), Pu et al. (2009), Teng et al. (2010), Wan and Yu (2008), Wang et al. (2010, 2012), Wu et al. (2012), Yang et al. (2008), Yuan et al. (2010), Yue et al. (2011), Zhang et al. (2004, 2009, 2012, 2013) and Zhu et al. (2005). Note also the experimental studies by Luo and Ye (2012), where they used nanocapsules as building blocks for fabricating an air-containing polymer nanofoam with low thermal conductivity (far lower than that of stagnant air of 26 mW/(mK)). Furthermore, Wicklein et al. (2015) investigated thermally insulating foams based on nanocellulose and graphene oxide, with a thermal conductivity as low as 15 mW/(mK). Examples of theoretical studies also including, for example, thermal transport and conductivity calculations may be seen in the investigations by Joulain (2008), Joulain et al. (2005), Liang and Li (2006), Mulet et al. (2002), Öchsner et al. (2009), Xu et al. (2008), Yang et al. (2012), Zhang (2007) and Zhao et al. (2012).

8.10 Start-up creation of nano-based thermal insulation

The start-up and creation of companies utilizing nanotechnologies to invent and manufacture new and novel nano-based thermal insulation materials and solutions involve both potential huge benefits and risks. The further away from commercialization the nano-based thermal insulators may seem, the larger the involved risks may seem to be. However, the ultimate benefits and payoff may then also be much larger. The risks, i.e., probability multiplied with impact, involve among other things both the possibility of not succeeding with making the thermal insulators due to e.g. miscellaneous research and technological challenges, and that some other company may come first with the same product or a better one. Moreover, the challenges are not only concerned with coming up with a very good product, but also with the launching of the product where acceptance through the whole value chain in the rather conservative construction and building sector represent a crucial hurdle to be overcome.

Any new product represents a potential failure with economical consequences, not only directly related to the specific product itself, but also related to various consequential damages to other parts of the constructions and buildings. These consequential damages may be much larger and involve much higher risks than the ones specifically to the material or product itself, e.g., also involving loss of the load-bearing structure. Thus, it will be important to test the newly developed thermal insulation materials, including durability evaluation and accelerated aging in the laboratory (Jelle, 2012b), and carry out robustness assessments (Jelle et al., 2014c).

Currently, many new thermal insulation materials and products, and other related products, are emerging. Two large pathways are seen, vacuum-based and non-vacuum-based thermal insulators. Commercial products of VIPs already exist in many varieties; a collection of these may be found in the study by Kalnæs and Jelle (2014). The low thermal conductivity in VIPs is due to a combination of utilizing vacuum, the Knudsen effect with voids in the nano range and a low solid state and radiation conductivity. A major disadvantage of VIPs as discussed earlier is the need to keep the vacuum intact in the VIP core material, with respect to the cardinal weaknesses fragility and perforation vulnerability, increasing thermal conductivity during time and lack of building site adaption cutting. In addition, heat bridge effects and high costs of VIPs are also important disadvantages. Aerogels (Levy and Zayat, 2015) are also utilizing the Knudsen effect (Jelle et al., 2015d), in this case with air-filled pores at atmospheric pressure. Thus, companies going for aerogels do not need to worry about any loss of vacuum as with VIPs. However, the fragility due to low tensile strength and high production costs of aerogels are major concerns. Furthermore, in their pristine non-aged condition VIPs attain a considerably lower thermal conductivity than aerogels.

NIMs (Jelle, 2011a; Jelle et al., 2014b) are aiming to reach as low thermal conductivities as VIPs in their pristine non-aged condition (e.g., 4 mW/(mK)) with air-filled nanopores utilizing the Knudsen effect. Hence, the NIMs will inherit the advantages of VIPs and aerogels and hopefully avoid their disadvantages. Naturally, as the NIMs are, at the moment, much further away from becoming commercial products as compared to the already existing and competitive VIP and aerogel products, there will also be several risks involved for start-up companies aimed at making NIM for the construction and building industry. Nevertheless, as already mentioned the potential benefits are, on the other hand, very large.

Various other material technologies may also be used in combination with VIPs, aerogels, NIMs and other thermal insulation materials. PCMs may be part of the thermal building envelope, absorbing and releasing heat energy according to demand (Kalnæs and Jelle, 2015). Low-emissivity materials (Jelle et al., 2015c) may be utilized to lower the thermal radiation, both macroscale products such as reflective foils in combination with thermal insulation materials and nanoscale-coated pore surfaces in porous thermal insulation materials. Electrochromic materials, where several commercial products already exist (Baetens et al., 2010d; Jelle et al., 2012a), may be used in translucent or transparent aerogels, hence enabling a dynamic and controlled regulation of the solar radiation transmittance within a high performance thermal insulation material. In order to harvest the solar radiation energy, combinations of thermal insulators with solar cell materials and building integrated photovoltaics (BIPV) (Jelle et al., 2012c), also including solar cell glazing products, may be envisioned. The development of new solar cell materials will play a crucial role in this respect (Jelle et al., 2012c; Jelle and Breivik, 2012b; Jelle, 2016). We may then start to talk about a development of multifunctional building materials and envelopes. Companies being able to exploit these material technology combinations and thus develop new and innovative products may achieve an advantage compared to others in the construction and building sector. Creative ideas and thinking are crucial for the start-up and creation of new and innovative materials and solutions, and hence also for the companies and industries wanting to exploit and develop these further.

8.11 Future perspectives for the research paths ahead

At the moment, our SIM and NIM experimental investigations are mainly focused on various attempts to tailor-make HSNS by manufacturing and applying different sacrificial templates, synthesis procedures, parameter variations and inner diameters and shell thicknesses of the nanospheres. A crucial issue will be how to assemble the HSNS together into a practical bulk material. It should be noted that future NIMs may not necessarily be based on HSNS; nevertheless the investigations on HSNS represent a possible stepping-stone toward the ultimate goal of achieving SIM and

NIM. Safety and health aspects are also important in this regard. No one wants any cases similar to the well-known asbestos case.

Although for the time being we are not pursuing fabricating NIM according to the membrane foaming and gas release methods, these methods should definitely not be forgotten as they may still represent possible pathways of achieving SIM and NIM, also directly as a bulk material without the need of piecing nanospheres together. Finally, it should be emphasized that methods and materials not included within this summary, even hitherto unknown methods and materials, may also hold the solution for future SIMs and NIMs.

Applying nanotechnology in order to make high performance thermal insulation materials like NIM and HSNS may be viewed as part of the development of new materials for the building envelope in energy-efficient buildings. A seamless and robust integration of these materials and technologies with the thermal insulation parts is important to ensure energy-efficient and durable solutions during the whole anticipated lifetime. Although not a thermal insulation material in itself, phase change materials have already been mentioned as part of the thermal building envelope (Kalnæs and Jelle, 2015). In connection to windows and fenestration in general, there is a lot of development in several fields (Jelle et al., 2012a), including smart and electrochromic windows (Granqvist, 2012; Jelle, 2013a, 2015; Lampert, 2004), window pane spacer materials (Van Den Bergh et al., 2013; Asphaug et al., 2014, 2016), low-emissivity coatings (Jelle et al., 2015c), window frame technologies (Gustavsen et al., 2008, 2011), self-cleaning surfaces (Midtdal and Jelle, 2013; Jelle, 2013c), solar cell glazing products (Jelle et al., 2012c; Jelle and Breivik, 2012a), and energy optimization (Grynning et al., 2013). Building integrated photovoltaics (BIPV) will probably become more and more important in the years to come, hence providing energy-efficient buildings with an electricity power-generating source (Breivik et al., 2013; Jelle et al., 2012c; Jelle and Breivik, 2012a,b; Jelle, 2016), where nanotechnology plays a crucial role in the development of new solar cell materials, also including the challenge of removing or avoiding snow and ice formation on the solar cell surfaces by the formation and application of superhydrophobic water-repelling and icephobic anti-icing surfaces (Dash et al., 2012; Jelle, 2013c; Kreder et al., 2016; Xiao and Chaudhuri, 2012).

To ensure energy-efficient and durable buildings many different aspects have to be considered, where satisfactory resistance toward weather exposure is one important aspect, for example, from shielding of fresh-air ventilation intakes (Jelle and Noreng, 2012) to wind-driven rain-tightness of BIPV (Breivik et al., 2013; Fasana and Nelva, 2013). Contrary to the heavy climate exposure walls and roofs are experiencing, the climate exposure for the ground may seem much less harsh; nevertheless, the ground materials and solutions must prevent penetration of air (leakage), moisture, and radon (Jelle et al., 2011b; Jelle, 2012a; Pacheco-Torgal, 2012), thus also comprising important health issues (fungi growth and radioactive radiation exposure due to moisture damages and radon ingress, respectively). Air, moisture, and radon barriers toward the ground may be combined into one barrier depending on the placement location of the barrier, and may also incorporate a low-emissivity surface coating (Jelle et al., 2015c) in order to decrease the heat loss toward the ground. Miscellaneous advantages and drawbacks of applying nanotechnology in the construction sector is reviewed by Pacheco-Torgal and Jalali (2011). Certainly, as we will experience many new and innovative materials and solutions based on nanotechnology in the coming years, we also have to address any toxicity issues to avoid any health and safety hazards. That also goes for the development of new thermal insulation materials and other technologies to be used in the building envelope like those mentioned earlier. Several of these new materials will probably be developed by research groups within the European Commission Horizon 2020 program, where the utilization of nanotechnology also will play an important role (Pacheco-Torgal, 2014).

In general, it is of major importance to investigate the durability of building materials and components, also newly developed ones, by carrying out accelerated climate aging in the laboratory (Jelle, 2011b, 2012b, 2013b; Jelle and Nilsen, 2011; Jelle et al., 2012b). Hence, performing a robustness assessment of these materials and components may also be found to be beneficial (Jelle et al., 2014c). That is, a durability and robustness evaluation of the new SIMs (when ready) should be carried out. Before then, attempts in different directions will be explored in order to synthesize the high performance thermal SIMs and NIMs of tomorrow.

8.12 Conclusions

Properties, requirements and possibilities for traditional, state-of-the-art and future thermal building insulation materials and solutions, with their advantages and disadvantages, have been investigated. Essential properties and issues raised are, among others, thermal conductivity, perforation vulnerability, building site adaptability and cuttability, mechanical strength, fire protection, fume emissions during fire, robustness, climate aging durability, resistance toward freezing/thawing cycles, water resistance, costs, and environmental impact.

Currently, no single thermal building insulation material or solution exists that satisfies all the requirements with respect to the most crucial properties. Thus, it is important to (1) choose the most suitable one from today's existing traditional and state-of-the-art thermal insulation materials and solutions, (2) conduct research and continuously improve today's existing traditional and state-of-the-art thermal insulation materials and solutions, and (3) initiate research that explores the possibilities of discovering and developing novel high performance thermal insulation materials and solutions with properties surpassing all of today's existing materials and solutions. Promising candidates for tomorrow and the future are nano insulation materials (NIM), dynamic insulation materials (DIM), and the load-bearing insulation material NanoCon, or some material or solution hitherto not yet thought of. Examples of experimental investigations have also been discussed; for example, NIMs attempted to be made from hollow silica nanospheres (HSNS) during a sacrificial template method. The thermal insulation aspects discussed within this study will be crucial to understand and address efficiently for companies attempting a start-up creation to manufacture nano-based thermal insulation for energy-efficient buildings in the years to come.

Acknowledgments

This work has been supported by the Research Council of Norway and several partners through the SINTEF and NTNU research projects, *Robust Envelope Construction Details for Buildings of the 21st Century* (ROBUST), the *Concrete Innovation Centre* (COIN), and *The Research Centre on Zero Emission Buildings* (ZEB).

References

Aegerter, M.A., Leventis, N., Koebel, M.M., 2011. Aerogels Handbook. Springer.

Alam, M., Singh, H., Limbachiya, M.C., 2011. Vacuum insulation panels (VIPs) for building construction industry – a review of the contemporary developments and future directions. Applied Energy 88, 3592–3602.

Al-Homoud, M.S., 2005. Performance characteristics and practical applications of common building thermal insulation materials. Building and Environment 40, 353–366.

Alotaibi, S.S., Riffat, S., 2014. Vacuum insulated panels for sustainable buildings: a review of research and applications. International Journal of Energy Research 38, 1–19.

Aspen Aerogels, 2008a. Spaceloft 3251, 6251, 9251. Flexible Insulation for Industrial, Commercial and Residential Applications. http://www.aerogel.com (accessed 07.10.2008).

Aspen Aerogels, 2008b. Spaceloft 6250. Extreme Protection for Extreme Environments. http://www.aerogel.com (accessed 07.10.2008).

Asphaug, S.K., Jelle, B.P., Gullbrekken, L., Uvsløkk, S., September 2–5, 2014. Accelerated ageing and durability of double-glazed sealed insulating window panes. In: Proceedings of 13th International Conference on Durability of Building Materials and Components, pp. 594–601. São Paulo, Brazil.

Asphaug, S.K., Jelle, B.P., Gullbrekken, L., Uvsløkk, S., 2016. Accelerated ageing and durability of double-glazed sealed insulating window panes and impact on heating demand in buildings. Energy and Buildings 116, 395–402.

Baetens, R., Jelle, B.P., Thue, J.V., Tenpierik, M.J., Grynning, S., Uvsløkk, S., Gustavsen, A., 2010a. Vacuum insulation panels for building applications: a review and beyond. Energy and Buildings 42, 147–172.

Baetens, R., Jelle, B.P., Gustavsen, A., 2010b. Phase change materials for building applications: a state-of-the-art review. Energy and Buildings 42, 1361–1368.

Baetens, R., Jelle, B.P., Gustavsen, A., Grynning, S., 2010c. Gas-filled panels for building applications: a state-of-the-art review. Energy and Buildings 42, 1969–1975.

Baetens, R., Jelle, B.P., Gustavsen, A., 2010d. Properties, requirements and possibilities of smart windows for dynamic daylight and solar energy control in buildings: a state-of-the-art review. Solar Energy Materials & Solar Cells 94, 87–105.

Baetens, R., Jelle, B.P., Gustavsen, A., 2011. Aerogel insulation for building applications: a state-of-the-art review. Energy and Buildings 43, 761–769.

Baetens, R., 2013. High performance thermal insulation materials for buildings. In: Pacheco-Torgal, F., Diamanti, M.V., Nazari, A., Granqvist, C.-G. (Eds.), Nanotechnology in Eco-Efficient Construction. Woodhead Publishing, pp. 188–206.

Bals, A., Kulozik, U., 2003. The influence of the pore size, the foaming temperature and the viscosity of the continuous phase on the properties of foams produced by membrane foaming. Journal of Membrane Science 220, 5–11.

Beck, A., Frank, O., Binder, M., September 18–19, 2007. Influence of water content on the thermal conductivity of vacuum panels with fumed silica kernels. In: Proceedings of the 8th International Vacuum Insulation Symposium, 2007. ZAEBayern/UniWue, Würzburg.

Boafo, F.E., Chen, Z., Li, C., Li, B., Xu, T., 2014. Structure of vacuum insulation panel in building system. Energy and Buildings 85, 644–653.

Bouquerel, M., Duforestel, T., Baillis, D., Rusaouen, G., 2012. Heat transfer modeling in vacuum insulation panels containing nanoporous silicas – a review. Energy and Buildings 54, 320–336.

Breivik, C., Jelle, B.P., Time, B., Holmberget, Ø., Nygård, J., Bergheim, E., Dalehaug, A., 2013. Large-scale experimental wind-driven rain exposure investigations of building integrated photovoltaics. Solar Energy 90, 179–187.

Brunner, S., Simmler, H., September 18–19, 2007. In situ performance assessment and service life of vacuum insulation panels (VIP) in buildings. In: Proceedings of the 8th International Vacuum Insulation Symposium, 2007. ZAEBayern/UniWue, Würzburg.

Brunner, S., Simmler, H., 2008. In situ performance assessment of vacuum insulation panels in a flat roof construction. Vacuum 82, 700–707.

Brunner, S., Gasser, Ph, Simmler, H., Ghazi, K., 2006. Investigation of multilayered aluminium-coated polymer laminates by focused ion beam (FIB) etching. Surface & Coatings Technology 200, 5908–5914.

Brunner, S., Wakili, K.G., Stahl, T., Binder, B., 2014. Vacuum insulation panels for building applications – continuous challenges and developments. Energy and Buildings 85, 592–596.

Caps, R., Fricke, J., 2000. Thermal conductivity of opacified powder filler materials for vacuum insulation. International Journal of Thermophysics 21, 445–452.

Caps, R., Beyrichen, H., Kraus, D., Weismann, S., 2008. Quality control of vacuum insulation panels: methods of measuring gas pressure. Vacuum 82, 691–699.

Caps, R., September 28–29, 2005. Monitoring gas pressure in vacuum insulation panels. In: Proceedings of the 7th International Vacuum Insulation Symposium. EMPA, Dübendorf, pp. 57–66.

Cheng, X., Liu, S., Lu, L., Sui, X., Meynen, V., Cool, P., Vansant, E.F., Jiang, J., 2007. Fast fabrication of hollow silica spheres with thermally stable nanoporous shells. Microporous and Mesoporous Materials 98, 41–46.

Chiang, J.C., MacDiarmid, A.G., 1986. 'Polyaniline': protonic acid doping of the emeraldine form to the metallic regime. Synthetic Metals 13, 193–205.

Cuce, E., Cuce, P.M., Wood, C.J., Riffat, S.B., 2014. Toward aerogel based thermal superinsulation in buildings: a comprehensive review. Renewable and Sustainable Energy Reviews 34, 273–299.

Dash, S., Alt, M.T., Garimella, S.V., 2012. Hybrid surface design for robust superhydrophobicity. Langmuir 28, 9606–9615.

Demirbas, M.F., 2006. Thermal energy storage and phase change materials: an overview. Energy Sources, Part B: Economics, Planning and Policy 1, 85–95.

Dieckmann, J., 2006. Latent Heat Storage in Concrete. Technische Universität Kaiserslautern, Kaiserslautern, Germany. http://www.eurosolar.org (accessed 09.12.2008).

Dowson, M., Grogan, M., Birks, T., Harrison, D., Craig, S., 2012. Streamlined life cycle assessment (LCA) of transparent silica aerogel made by supercritical drying. Applied Energy 97, 396–404.

Du, X., He, J., 2008. Facile size-controllable syntheses of highly monodisperse polystyrene nano- and microspheres by polyvinylpyrrolidone-mediated emulsifier-free emulsion polymerization. Journal of Applied Polymer Science 108, 1755–1760.

Du, Y., Luna, L.E., Tan, W.S., Rubner, M.F., Cohen, R.E., 2010. Hollow silica nanoparticles in UV-visible antireflection coatings for poly(methyl methacrylate) substrates. ACS Nano 4, 4308–4316.

Fan, H., Lei, Z., Pan, J.H., Zhao, X.S., 2011. Sol–gel synthesis, microstructure and adsorption properties of hollow silica spheres. Materials Letters 65, 1811–1814.

Farid, M.M., Khudhair, A.M., Razack, S.A.K., Al-Hallaj, S., 2004. A review on phase change energy storage: materials and applications. Energy Conversion and Management 45, 1597–1615.

Fasana, S., Nelva, R., 2013. Improvement of the water resistancy in the integration of photovoltaic panels on traditional roofs. Construction and Building Materials 48, 1081–1091.

Fricke, J., Schwab, H., Heinemann, U., 2006. Vacuum insulation panels – exciting thermal properties and most challenging applications. International Journal of Thermophysics 27, 1123–1139.

Fricke, J., Heinemann, U., Ebert, H.P., 2008. Vacuum insulation panels – from research to market. Vacuum 82, 680–690.

Fricke, J., September 28–29, 2005. From dewars to VIPs – one century of progress in vacuum insulation technology. In: Proceedings of the 7th International Vacuum Insulation Symposium. EMPA, Dübendorf, Switzerland, pp. 5–14.

Gao, T., Sandberg, L.I.C., Jelle, B.P., Gustavsen, A., 2012. Nano insulation materials for energy efficient buildings: a case study on hollow silica nanospheres. In: Mendez-Vilas, A. (Ed.), Fuelling the Future: Advances in Science and Technologies for Energy Generation, Transmission and Storage. BrownWalker Press, pp. 535–539.

Gao, T., Jelle, B.P., Sandberg, L.I.C., Gustavsen, A., 2013. Monodisperse hollow silica nanospheres for nano insulation materials: synthesis, characterization, and life cycle assessment. ACS Applied Materials and Interfaces 5, 761–767.

Gao, T., Jelle, B.P., Gustavsen, A., Jacobsen, S., 2014a. Aerogel-incorporated concrete: an experimental study. Construction and Building Materials 52, 130–136.

Gao, T., Sandberg, L.I.C., Jelle, B.P., 2014b. Nano insulation materials: synthesis and life cycle assessment. Procedia CIRP 15, 490–495.

Gao, T., Jelle, B.P., Ihara, T., Gustavsen, A., 2014c. Insulating glazing units with silica aerogel granules: the impact of particle size. Applied Energy 128, 27–34.

Gao, T., Jelle, B.P., Gustavsen, A., He, J., 2014d. Lightweight and thermally insulating aerogel glass materials. Applied Physics A: Materials Science & Processing 117, 799–808.

Grader, G.S., de Hazan, Y., Shter, G.E., 1998. Ultra light ceramic foams. Sol–Gel Synthesis and Processing 95, 161–172.

Grandcolas, M., Etienne, G., Tilset, B.G., Gao, T., Sandberg, L.I.C., Gustavsen, A., Jelle, B.P., September 19–20, 2013. Hollow silica nanospheres as a superinsulating material. In: Proceedings of the 11th International Vacuum Insulation Symposium (IVIS 2013), pp. 43–44. Dübendorf, Zürich, Switzerland.

Granqvist, C.G., 1995. Handbook of Inorganic Electrochromic Materials. Elsevier, Amsterdam.

Granqvist, C.G., 2005. Electrochromic devices. Journal of the European Ceramic Society 25, 2907–2912.

Granqvist, C.G., 2008. Oxide electrochromics: why, how, and whither. Solar Energy Materials and Solar Cells 92, 203–208.

Granqvist, C.G., 2012. Oxide electrochromics: an introduction to devices and materials. Solar Energy Materials and Solar Cells 99, 1–13.

Griffith, B.T., Türler, D., Arashteh, D., November 7–10, 1993. Optimizing the effective conductivity and cost of gas-filled panel thermal insulations. In: Proceedings of the 22nd International Thermal Conductivity Conference, 1993. Arizona State University.

Griffith, B.T., Arashteh, D., Türler, D., November 14, 1995. Gas-filled panels: an update on applications in the building thermal envelope. In: Proceedings of the BETEC Fall Symposium, Superinsulations and the Building Envelope, 1995, Washington, DC.

Grynning, S., Jelle, B.P., Uvsløkk, S., Gustavsen, A., Baetens, R., Caps, R., Meløysund, V., 2011. Hot box investigations and theoretical assessments of miscellaneous vacuum insulation panel configurations in building envelopes. Journal of Building Physics 34, 297–324.

Grynning, S., Gustavsen, A., Time, B., Jelle, B.P., 2013. Windows in the buildings of tomorrow: energy losers or energy gainers? Energy and Buildings 61, 185–192.

Gustavsen, A., Arasteh, D., Jelle, B.P., Curcija, C., Kohler, C., 2008. Developing low-conductance window frames: capabilities and limitations of current window heat-transfer design tools. Journal of Building Physics 32, 131–153.

Gustavsen, A., Grynning, S., Arasteh, D., Jelle, B.P., Goudey, H., 2011. Key elements of and materials performance targets for highly insulating window frames. Energy and Buildings 43, 2583–2594.

Haavi, T., Jelle, B.P., Gustavsen, A., 2012. Vacuum insulation panels in wood frame wall constructions with different stud profiles. Journal of Building Physics 36, 212–226.

Han, L., Gao, C., Wu, X., Chen, Q., Shu, P., Ding, Z., Che, S., 2011. Anionic surfactants templating route for synthesizing silica hollow spheres with different shell porosity. Solid State Sciences 13, 721–728.

Hasnain, S.M., 1998. Review on sustainable thermal energy storage technologies, Part I: heat storage materials and techniques. Energy Conversion and Management 39, 1127–1138.

Heinemann, U., Schwab, H., Simmler, H., Brunner, S., Ghazi, K., Bundi, R., Kumaran, K., Mukhopadhyaya, Ph, Quénard, D., Sallée, H., Noller, K., Kücükpinar-Niarchos, E., Stramm, C., Tenpierik, M.J., Cauberg, J.J.M., Binz, A., Steinke, G., Moosmann, A., Erb, M., September 2005. Vacuum Insulation. Panel Properties and Building Applications (Summary). HiPTI – High Performance Thermal Insulation. IEA/ECBCS Annex 39.

Heinemann, U., 2008. Influence of water on the total heat transfer in 'evacuated' insulations. International Journal of Thermophysics 29, 735–749.

Hostler, S.R., Abramson, A.R., Gawryla, M.D., Bandi, S.A., Schiraldi, D.A., 2008. Thermal conductivity of a clay-based aerogel. International Journal of Heat and Mass Transfer 52, 665–669.

Huang, W.-S., Humphrey, B.D., MacDiarmid, A.G., 1986. Polyaniline, a novel conducting polymer. Morphology and chemistry of its oxidation and reduction in aqueous electrolytes. Journal of the Chemical Society, Faraday Transactions 1: Physical Chemistry in Condensed Phases 82, 2385–2400.

Ihara, T., Grynning, S., Gao, T., Gustavsen, A., Jelle, B.P., 2015a. Impact of convection on thermal performance of aerogel granulate glazing systems. Energy and Buildings 88, 165–173.

Ihara, T., Gao, T., Grynning, S., Jelle, B.P., Gustavsen, A., 2015b. Aerogel granulate glazing facades and their application potential from an energy saving perspective. Applied Energy 142, 179–191.

Ihara, T., Jelle, B.P., Gao, T., Gustavsen, A., 2015c. Aerogel granule aging driven by moisture and solar radiation. Energy and Buildings 103, 238–248.

Jelle, B.P., Breivik, C., 2012a. State-of-the-art building integrated photovoltaics. Energy Procedia 20, 68–77.

Jelle, B.P., Breivik, C., 2012b. The path to the building integrated photovoltaics of tomorrow. Energy Procedia 20, 78–87.

Jelle, B.P., Hagen, G., 1993. Transmission spectra of an electrochromic window based on polyaniline, prussian blue and tungsten oxide. Journal of Electrochemical Society 140, 3560–3564.

Jelle, B.P., Hagen, G., 1998. Electrochemical multilayer deposition of polyaniline and prussian blue and their application in solid state electrochromic windows. Journal of Applied Electrochemistry 28, 1061–1065.

Jelle, B.P., Hagen, G., 1999a. Performance of an electrochromic window based on polyaniline, prussian blue and tungsten oxide. Solar Energy Materials and Solar Cells 58, 277–286.

Jelle, B.P., Hagen, G., 1999b. Correlation between light absorption and electric charge in solid state electrochromic windows. Journal of Applied Electrochemistry 29, 1103–1110.

Jelle, B.P., Nilsen, T.-N., 2011. Comparison of accelerated climate ageing methods of polymer building materials by attenuated total reflectance Fourier transform infrared radiation spectroscopy. Construction and Building Materials 25, 2122–2132.

Jelle, B.P., Noreng, K., April 2012. Shielding fresh air ventilation intakes. ASHRAE Journal 34–42.

Jelle, B.P., Hagen, G., Hesjevik, S.M., Ødegård, R., 1992a. Transmission through an electrochromic window based on polyaniline, tungsten oxide and a solid polymer electrolyte. Materials Science and Engineering B 13, 239–241.

Jelle, B.P., Hagen, G., Ødegård, R., 1992b. Transmission spectra of an electrochromic window based on polyaniline, tungsten oxide and a solid polymer electrolyte. Electrochimica Acta 37, 1377–1380.

Jelle, B.P., Hagen, G., Sunde, S., Ødegård, R., 1993a. Dynamic light modulation in an electrochromic window consisting of polyaniline, tungsten oxide and a solid polymer electrolyte. Synthetic Metals 54, 315–320.

Jelle, B.P., Hagen, G., Nødland, S., 1993b. Transmission spectra of an electrochromic window consisting of polyaniline, prussian blue and tungsten oxide. Electrochimica Acta 38, 1497–1500.

Jelle, B.P., Hagen, G., Hesjevik, S.M., Ødegård, R., 1993c. Reduction factor for polyaniline films on ITO from cyclic voltammetry and visible absorption spectra. Electrochimica Acta 38, 1643–1647.

Jelle, B.P., Hagen, G., Birketveit, Ø., 1998. Transmission properties for individual electrochromic layers in solid state devices based on polyaniline, prussian blue and tungsten oxide. Journal of Applied Electrochemistry 28, 483–489.

Jelle, B.P., Gustavsen, A., Nilsen, T.-N., Jacobsen, T., 2007. Solar material protection factor (SMPF) and solar skin protection factor (SSPF) for window panes and other glass structures in buildings. Solar Energy Materials and Solar Cells 91, 342–354.

Jelle, B.P., Gustavsen, A., Baetens, R., September 17–18, 2009. Beyond vacuum insulation panels – how may it be achieved?. In: Proceedings of the 9th International Vacuum Insulation Symposium (IVIS 2009) London, England.

Jelle, B.P., Gustavsen, A., Baetens, R., 2010a. The path to the high performance thermal building insulation materials and solutions of tomorrow. Journal of Building Physics 34, 99–123.

Jelle, B.P., Gustavsen, A., Baetens, R., December 5–9, 2010b. The high performance thermal building insulation materials and solutions of tomorrow. In: Proceedings of the Thermal Performance of the Exterior Envelopes of Whole Buildings XI International Conference (Buildings XI), Clearwater Beach, Florida, U.S.A.

Jelle, B.P., Tilset, B.G., Jahren, S., Gao, T., Gustavsen, A., September 15–16, 2011a. Vacuum and nanotechnologies for the thermal insulation materials of beyond tomorrow – from concept to experimental investigations. In: Proceedings of the 10th International Vacuum Insulation Symposium (IVIS 2011), pp. 171–178. Ottawa, Canada.

Jelle, B.P., Noreng, K., Erichsen, T.H., Strand, T., 2011b. Implementation of radon barriers, model development and calculation of radon concentration in indoor air. Journal of Building Physics 34, 195–222.

Jelle, B.P., Hynd, A., Gustavsen, A., Arasteh, D., Goudey, H., Hart, R., 2012a. Fenestration of today and tomorrow: a state-of-the-art review and future research opportunities. Solar Energy Materials and Solar Cells 96, 1–28.

Jelle, B.P., Nilsen, T.-N., Hovde, P.J., Gustavsen, A., 2012b. Accelerated climate aging of building materials and their characterization by Fourier transform infrared radiation analysis. Journal of Building Physics 36, 99–112.

Jelle, B.P., Breivik, C., Røkenes, H.D., 2012c. Building integrated photovoltaic products: a state-of-the-art review and future research opportunities. Solar Energy Materials and Solar Cells 100, 69–96.

Jelle, B.P., Sveipe, E., Wegger, E., Uvsløkk, S., Grynning, S., Thue, J.V., Time, B., Gustavsen, A., 2013a. Moisture robustness during retrofitting of timber frame walls with vacuum insulation panels: experimental and theoretical studies. In: de Freitas, V.P., Delgado, J.M.P.Q. (Eds.), Hygrothermal Behavior, Building Pathology and Durability, Building Pathology and Rehabilitation, vol. 1. Springer, pp. 183–210.

Jelle, B.P., Gao, T., Tilset, B.G., Sandberg, L.I.C., Grandcolas, M., Simon, C., Gustavsen, A., September 19–20, 2013b. Experimental pathways for achieving superinsulation through nano insulation materials. In: Proceedings of the 11th International Vacuum Insulation Symposium (IVIS 2013), pp. 99–100. Dübendorf, Zürich, Switzerland.

Jelle, B.P., Gao, T., Sandberg, L.I.C., Tilset, B.G., Grandcolas, M., Gustavsen, A., January 4–5, 2014a. Thermal superinsulation for building applications – from concepts to experimental investigations. In: Proceedings of the International Conference on Advances in Civil, Structural and Mechanical Engineering (ACSME 2014), pp. 97–104. Bangkok, Thailand.

Jelle, B.P., Gao, T., Sandberg, L.I.C., Tilset, B.G., Grandcolas, M., Gustavsen, A., 2014b. Thermal superinsulation for building applications – from concepts to experimental investigations. International Journal of Structural Analysis and Design 1, 43–50.

Jelle, B.P., Sveipe, E., Wegger, E., Gustavsen, A., Grynning, S., Thue, J.V., Time, B., Lisø, K.R., 2014c. Robustness classification of materials, assemblies and buildings. Journal of Building Physics 37, 213–245.

Jelle, B.P., Gao, T., Sandberg, L.I.C., Tilset, B.G., Grandcolas, M., Gustavsen, A., June 15–18, 2014d. The high performance thermal building insulation materials of beyond tomorrow – from concept to experimental investigations. In: Proceedings of TechConnect World Innovation Conference 2014 – Cleantech 2014 Energy and Efficiency Conference, pp. 296–299. Washington DC, U.S.A.

Jelle, B.P., Gao, T., Sandberg, L.I.C., Tilset, B.G., Grandcolas, M., Gustavsen, A., April 13–15, 2015a. Experimental synthesis of hollow silica nanospheres for application as superinsulation in the buildings of tomorrow. In: Proceedings of the 4th Building Enclosure Science & Technology (BEST 4 – 2015) Conference, Kansas City, Missouri, U.S.A.

Jelle, B.P., Gao, T., Sandberg, L.I.C., Ng, S., Tilset, B.G., Grandcolas, M., Gustavsen, A., May 24–26, 2015b. Development of nano insulation materials for building constructions. In: Proceedings of 5th International Symposium on Nanotechnology in Construction (NICOM5), pp. 429–434. Chicago, Illinois, U.S.A.

Jelle, B.P., Kalnæs, S.E., Gao, T., 2015c. Low-emissivity materials for building applications: a state-of-the-art review and future research perspectives. Energy and Buildings 96, 329–356.

Jelle, B.P., Baetens, R., Gustavsen, A., 2015d. Aerogel insulation for building applications. In: Levy, D., Zayat, M. (Eds.), The Sol–Gel Handbook, vol. 3. Wiley-VCH, pp. 1385–1412.

Jelle, B.P., 1993. Electrochemical and Spectroscopic Studies of Electrochromic Materials (Ph.D. thesis). Department of Applied Electrochemistry, The Norwegian Institute of Technology, Trondheim, Norway, p. 131.

Jelle, B.P., 2011a. Traditional, state-of-the-art and future thermal building insulation materials and solutions – properties, requirements and possibilities. Energy and Buildings 43, 2549–2563.

Jelle, B.P., April 12–15, 2011b. Evaluation of building products by conducting accelerated climate ageing in the laboratory. In: Proceedings of the 12th International Conference on Durability of Building Materials and Components, pp. 311–319. Porto, Portugal.

Jelle, B.P., 2012a. Development of a model for radon concentration in indoor air. Science of the Total Environment 416, 343–350.

Jelle, B.P., 2012b. Accelerated climate ageing of building materials, components and structures in the laboratory. Journal of Materials Science 47, 6475–6496.

Jelle, B.P., 2013a. Solar radiation glazing factors for window panes, glass structures and electrochromic windows in buildings – measurement and calculation. Solar Energy Materials and Solar Cells 116, 291–323.

Jelle, B.P., June 12–14, 2013b. The role of accelerated climate ageing of building materials, components and structures in the laboratory. In: Proceedings of the 7th Nordic Conference on Construction Economics and Organisation 2013, pp. 111–122. Trondheim, Norway.

Jelle, B.P., 2013c. The challenge of removing snow downfall on photovoltaic solar cell roofs in order to maximize solar energy efficiency – research opportunities for the future. Energy and Buildings 67, 334–351.

Jelle, B.P., 2015. Electrochromic smart windows for dynamic daylight and solar energy control in buildings. In: Mortimer, R.J., Rosseinsky, D.R., Monk, P.M.S. (Eds.), Electrochromic Materials and Devices. Wiley-VCH, pp. 419–502.

Jelle, B.P., 2016. Building integrated photovoltaics: a concise description of the current state of the art and possible research pathways. Energies 9, 1–30. Article no. 21.

Johansson, P., Geving, S., Hagentoft, C.-E., Jelle, B.P., Rognvik, E., Kalagasidis, A.S., Time, B., 2014. Interior insulation retrofit of a historical brick wall using vacuum insulation panels: hygrothermal numerical simulations and laboratory investigations. Building and Environment 79, 31–45.

Joulain, K., Mulet, J.-P., Marquier, F., Carminati, R., Greffet, J.-J., 2005. Surface electromagnetic waves thermally excited: radiative heat transfer, coherence properties and Casimir forces revisited in the near field. Surface Science Reports 57, 59–112.

Joulain, K., 2008. Near-field heat transfer: a radiative interpretation of thermal conduction. Journal of Quantitative Spectroscopy & Radiative Transfer 109, 294–304.

Kaganer, M.G., 1969. Thermal Insulation in Cryogenic Engineering. IPST Press (Russian version 1966).

Kalnæs, S.E., Jelle, B.P., 2014. Vacuum insulation panel products: a state-of-the-art review and future research pathways. Applied Energy 116, 355–375.

Kalnæs, S.E., Jelle, B.P., 2015. Phase change materials and products for building applications: a state-of-the-art review and future research opportunities. Energy and Buildings 94, 150–176.

Khudhair, A.M., Farid, M.M., 2004. A review on energy conservation in building applications with thermal storage by latent heat using phase change materials. Energy Conservation and Management 45, 263–275.

Kim, K.D., Choi, K.Y., Yang, J.W., 2005. Formation of spherical hollow silica particles from sodium silicate solution by ultrasonic spray pyrolysis method. Colloids and Surfaces A: Physicochemical and Engineering Aspects 254, 193–198.

Koebel, M., Rigacci, A., Achard, P., 2012. Aerogel-based thermal superinsulation: an overview. Journal of Sol–Gel Science and Technology 63, 315–339.

Kreder, M.J., Alvarenga, J., Kim, P., Aizenberg, J., 2016. Design of anti-icing surfaces: smooth, textured or slippery? Nature Reviews Materials 1, 1–15. Article no. 15003.

Lampert, C.M., 1998. Smart switchable glazing for solar energy and daylight control. Solar Energy Materials and Solar Cells 52, 207–221.

Lampert, C.M., 2004. Chromogenic smart materials. Materials Today 7, 28–35.

LBNL, 2015. Lawrence Berkeley National Laboratory. http://eetd.lbl.gov/newsletter/nl20/eetd-nl20-tt.html (accessed 15.11.2015).

Le, Y., Chen, J.-F., Wang, W.-C., 2004a. Study on the silica hollow spheres by experiment and molecular simulation. Applied Surface Science 230, 319–326.

Le, Y., Chen, J.-F., Wang, J.-X., Shao, L., Wang, W.-C., 2004b. A novel pathway for synthesis of silica hollow spheres with mesostructured walls. Materials Letters 58, 2105–2108.

Lee, D., Stevens, P.C., Zeng, S.Q., Hunt, A.J., 1995a. Thermal characterization of carbon-opacified silica aerogels. Journal of Non-Crystalline Solids 186, 285–290.

Lee, K.-H., Kim, S.-Y., Yoo, K.-P., 1995b. Low-density, hydrophobic aerogels. Journal of Non-Crystalline Solids 186, 18–22.

Levy, D., Zayat, M. (Eds.), 2015. The Sol–Gel Handbook. Wiley-VCH.

Li, H., Chen, H., Li, X., Duan, W., 2015. Degradation of VIP barrier envelopes exposed to alkaline solution at different temperatures. Energy and Buildings 93, 208–216.

Liang, L., Li, B., 2006. Size-dependent thermal conductivity of nanoscale semiconducting systems. Physical Review B 73, 153303-1–153303-4.

Liao, Y., Wu, X., Liu, H., Chen, Y., 2011. Thermal conductivity of powder silica hollow spheres. Thermochimica Acta 526, 178–184.

Liao, Y., Wu, X., Wang, Z., Yue, R., Liu, G., Chen, Y., 2012. Composite thin film of silica hollow spheres and waterborne polyurethane: excellent thermal insulation and light transmission performances. Materials Chemistry and Physics 133, 642–648.

Liu, S., Wei, M., Rao, J., Wang, H., Zhao, H., 2011. A controlled formation of cage-like nanoporous hollow silica microspheres. Materials Letters 65, 2083–2085.

Luo, Y., Ye, C., 2012. Using nanocapsules as building blocks to fabricate organic polymer nanofoam with ultra low thermal conductivity and high mechanical strength. Polymer 53, 5699–5705.

Mandilaras, I., Atsonios, I., Zannis, G., Founti, M., 2014. Thermal performance of a building envelope incorporating ETICS with vacuum insulation panels and EPS. Energy and Buildings 85, 654–665.

McArdle, P., Lindstrom, P., December 2009. Emissions of Greenhouse Gases in the United States 2008. U.S. Energy Information Administration, DOE/EIA-0573(2008).

McKinsey, 2009. Pathways to a Low-Carbon Economy. Version 2 of the Global Greenhouse Gas Abatement Cost Curve. McKinsey & Company.

Meng, Q., Xiang, S., Zhang, K., Wang, M., Bu, X., Xue, P., Liu, L., Sun, H., Yang, B., 2012. A facile two-step etching method to fabricate porous hollow silica particles. Journal of Colloid and Interface Science 384, 22–28.

Midtdal, K., Jelle, B.P., 2013. Self-cleaning glazing products: a state-of-the-art review and future research pathways. Solar Energy Materials and Solar Cells 109, 126–141.

Miesbauer, O., Kucukpinar, E., Kiese, S., Carmi, Y., Noller, K., Langowski, H.-C., 2014. Studies on the barrier performance and adhesion strength of novel barrier films for vacuum insulation panels. Energy and Buildings 85, 597–603.

Mills, G.L., Zeller, C.M., 2008. The performance of gas filled multilayer insulation. Advances of Cryogenic Engineering: Transactions of the Cryogenic Engineering Conference 53, 1475–1482.

Mukhopadhyaya, P., MacLean, D., Korn, J., van Reenen, D., Molleti, S., 2014. Building application and thermal performance of vacuum insulation panels (VIPs) in Canadian subarctic climate. Energy and Buildings 85, 672–680.

Mulet, J.-P., Joulain, K., Carminati, R., Greffet, J.-J., 2002. Enhanced radiative heat transfer at nanometric distances. Microscale Thermophysical Engineering 6, 209–222.

Müller-Fischer, N., 2007. Dynamically Enhanced Membrane Foaming (Ph.D. thesis), Dissertation ETH Number 16939. Swiss Federal Institute of Technology, Zürich, Switzerland.

Ng, S., Sandberg, L.I.C., Jelle, B.P., 2015a. Insulating and strength properties of an aerogel-incorporated mortar based on UHPC formulations. Key Engineering Materials 629–630, 43–48.

Ng, S., Jelle, B.P., Sandberg, L.I.C., Gao, T., Wallevik, Ó.H., 2015b. Experimental investigations of aerogel-incorporated ultra-high performance concrete. Construction and Building Materials 77, 307–316.

Ng, S., Jelle, B.P., Zhen, Y., Wallevik, Ó.H., 2016. Effect of storage and curing conditions at elevated temperatures on aerogel-incorporated mortar samples based on UHPC recipe. Construction and Building Materials 106, 640–649.

Öchsner, A., Hosseini, S.M.H., Merkel, M., March 18–20, 2009. Numerical simulation of sintered perforated hollow sphere structures (PHSS) to investigate thermal conductivity. In: Proceedings of the International MultiConference of Engineers and Computer Scientists 2009, vol. II (IMECS 2009), Hong Kong, China.

Pacheco-Torgal, F., Jalali, S., 2011. Nanotechnology: advantages and drawbacks in the field of construction and building materials. Construction and Building Materials 25, 582–590.

Pacheco-Torgal, F., 2012. Indoor radon: an overview on a perennial problem. Building and Environment 58, 270–277.

Pacheco-Torgal, F., 2014. Eco-efficient construction and building materials research under the EU framework programme horizon 2020. Construction and Building Materials 51, 151–162.

Papadopoulos, A.M., 2005. State of the art in thermal insulation materials and aims for future developments. Energy and Buildings 37, 77–86.

Peng, B., Chen, M., Zhou, S., Wu, L., Ma, X., 2008. Fabrication of hollow silica spheres using droplet templates derived from a miniemulsion technique. Journal of Colloid and Interface Science 321, 67–73.

Pons, E., Yrieix, B., Heymans, L., Dubelley, F., Planes, E., 2014. Permeation of water vapor through high performance laminates for VIPs and physical characterization of sorption and diffusion phenomena. Energy and Buildings 85, 604–616.

Pu, H., Zhang, X., Yuan, J., Yang, Z., 2009. A facile method for the fabrication of vinyl functionalized hollow silica spheres. Journal of Colloid and Interface Science 331, 389–393.

Sallée, H., Quenard, D., Valenti, E., Galan, M., 2014. VIP as thermal breaker for internal insulation system. Energy and Buildings 85, 631–637.

Sandberg, L.I.C., Gao, T., Jelle, B.P., Gustavsen, A., 2013. Synthesis of hollow silica nanospheres by sacrificial polystyrene templates for thermal insulation applications. Advances in Materials Science and Engineering 2013, 6, 483651.

Schlanbusch, R.D., Jelle, B.P., Sandberg, L.I.C., Fufa, S.M., Gao, T., 2014b. Integration of life cycle assessment in the design of hollow silica nanospheres for thermal insulation applications. Building and Environment 80, 115–124.

Schlanbusch, R.D., Jelle, B.P., Sandberg, L.I.C., Fufa, S.M., Gao, T., June 18–20, 2014a. Life cycle assessment integrated in the design of a new nano insulation material. In: Proceedings of the 20th Annual International Sustainable Development Research Conference (ISDRC 2014), pp. 325–335. Trondheim, Norway.

Schultz, J.M., Jensen, K.I., 2008. Evacuated aerogel glazings. Vacuum 82, 723–729.

Schultz, J.M., Jensen, K.I., Kristiansen, F.H., 2005. Super insulating aerogel glazing. Solar Energy Materials & Solar Cells 89, 275–285.

Schwab, H., Heinemann, U., Beck, A., Ebert, H.-P., Fricke, J., 2005a. Permeation of different gases through foils used as envelopes for vacuum insulation panels. Journal of Thermal Envelope & Building Science 28, 293–317.

Schwab, H., Heinemann, U., Beck, A., Ebert, H.-P., Fricke, J., 2005b. Dependence of thermal conductivity on water content in vacuum insulation panels with fumed silica kernels. Journal of Thermal Envelope & Building Science 28, 319–326.

Schwab, H., Heinemann, U., Wachtel, J., Ebert, H.-P., Fricke, J., 2005c. Predictions for the increase in pressure and water content of vacuum insulation panels (VIPs) integrated into building constructions using model calculations. Journal of Thermal Envelope & Building Science 28, 327–344.

Schwab, H., Heinemann, U., Beck, A., Ebert, H.-P., Fricke, J., 2005e. Prediction of service life for vacuum insulation panels with fumed silica kernel and foil cover. Journal of Thermal Envelope & Building Science 28, 357–374.

Schwab, H., Stark, C., Wachtel, J., Ebert, H.-P., Fricke, J., 2005d. Thermal bridges in vacuum-insulated building facades. Journal of Thermal Envelope & Building Science 28, 345–355.

Simmler, H., Brunner, S., 2005a. Vacuum insulation panels for building application – basic properties, ageing mechanisms and service life. Energy and Buildings 37, 1122–1131.

Simmler, H., Brunner, S., September 28–29, 2005b. Ageing and service life of VIP in buildings. In: Proceedings of the 7th International Vacuum Insulation Symposium. EMPA, Dübendorf, Switzerland, pp. 15–22.

Simmler, H., Brunner, S., Heinemann, U., Schwab, H., Kumaran, K., Mukhopadhyaya, P., Quènard, D., Sallèe, H., Noller, K., Kücküpinar-Niarchos, E., Stramm, C., Tenpierik, M., Cauberg, H., Erb, M., September, 2005. Vacuum Insulation Panels. Study on VIP-Components and Panels for Service Life Prediction in Building Applications (Subtask A). HiPTI – High Performance Thermal Insulation. IEA/ECBCS Annex 39.

Smith, D.M., Maskara, A., Boes, U., 1998. Aerogel-based thermal insulation. Journal of Non-Crystalline Solids 225, 254–259.

Sprengard, C., Holm, A.H., 2014. Numerical examination of thermal bridging effects at the edges of vacuum-insulation-panels (VIP) in various constructions. Energy and Buildings 85, 638–643.

Stahl, T., Brunner, S., Zimmermann, M., Wakili, K.G., 2012. Thermo-hygric properties of a newly developed aerogel based insulation rendering for both exterior and interior applications. Energy and Buildings 44, 114–117.

Sveipe, E., Jelle, B.P., Wegger, E., Uvsløkk, S., Grynning, S., Thue, J.V., Time, B., Gustavsen, A., 2011. Improving thermal insulation of timber frame walls by retrofitting with vacuum insulation panels – experimental and theoretical investigations. Journal of Building Physics 35, 168–188.

Teng, Z., Han, Y., Li, J., Yan, F., Yang, W., 2010. Preparation of hollow mesoporous silica spheres by a sol–gel/emulsion approach. Microporous and Mesoporous Materials 127, 67–72.

Tenpierik, M., Cauberg, H., 2007. Analytical models for calculating thermal bridge effects caused by thin high barrier envelopes around vacuum insulation panels. Journal of Building Physics 30, 185–215.

Tenpierik, M.J., Cauberg, J.J.M., 2010. Encapsulated vacuum insulation panels: theoretical thermal optimization. Building Research & Information 38, 660–669.

Tenpierik, M.J., Cauberg, J.J.M., Thorsell, T.I., 2007a. Integrating vacuum insulation panels in building constructions: an integral perspective. Construction Innovation 7, 38–53.

Tenpierik, M., van der Spoel, W., Cauberg, H., September 18–19, 2007b. Simplified analytical models for service life prediction of a vacuum insulation panel. In: Proceedings of the 8th International Vacuum Insulation Symposium, 2007. ZAEBayern/UniWue, Würzburg.

Tenpierik, M., van der Spoel, W., Cauberg, H., 2008. An analytical model for calculating thermal bridge effects in high performance building enclosure. Journal of Building Physics 31, 361–387.

Tenpierik, M.J., 2009. Vacuum Insulation Panels Applied in Building Constructions (VIP ABC) (Ph.D. thesis). Delft University of Technology, Delft, The Netherlands.

Van Den Bergh, S., Uvsløkk, S., Jelle, B.P., Roels, S., Gustavsen, A., September 15–16, 2011. Experimental and numerical investigations of polystyrene encapsulated vacuum insulation panels – heat transfer results. In: Proceedings of the 10th International Vacuum Insulation Symposium (IVIS-X), pp. 29–34. Ottawa, Canada.

Van Den Bergh, S., Hart, R., Jelle, B.P., Gustavsen, A., 2013. Window spacers and edge seals in insulating glass units: a state-of-the-art review and future perspectives. Energy and Buildings 58, 263–280.

Voellinger, T., Bassi, A., Heitel, M., 2014. Facilitating the incorporation of VIP into precast concrete sandwich panels. Energy and Buildings 85, 666–671.

Wan, Y., Yu, S.-H., 2008. Polyelectrolyte controlled large-scale synthesis of hollow silica spheres with tunable sizes and wall thicknesses. Journal of Physical Chemistry C 112, 3641–3647.

Wang, X., Miao, X., Li, Z., Deng, W., 2010. Fabrication of mesoporous silica hollow spheres using triblock copolymer PEG–PPG–PEG as template. Journal of Non-Crystalline Solids 356, 898–905.

Wang, F., Tang, Y., Zhang, B., Chen, B., Wang, Y., 2012. Preparation of novel magnetic hollow mesoporous silica microspheres and their efficient adsorption. Journal of Colloid and Interface Science 386, 129–134.

World Business Council for Sustainable Development, July 2002. The Cement Sustainability Initiative – Our Agenda for Action.

Wegger, E., Jelle, B.P., Sveipe, E., Grynning, S., Gustavsen, A., Baetens, R., Thue, J.V., 2011. Aging effects on thermal properties and service life of vacuum insulation panels. Journal of Building Physics 35, 128–167.

Wicklein, B., Kocjan, A., Salazar-Alvarez, G., Carosio, F., Camino, G., Antonietti, M., Bergström, L., 2015. Thermally insulating and fire-retardant lightweight anisotropic foams based on nanocellulose and graphene oxide. Nature Nanotechnology 10, 277–283.

Wong, J.C.H., Kaymak, H., Brunner, S., Koebel, M.M., 2014. Mechanical properties of monolithic silica aerogels made from polyethoxydisiloxanes. Microporous and Mesoporous Materials 183, 23–29.

Wu, W., Cao, S., Yuan, X., Zhao, Z., Fang, L., 2012. Sodium silicate route: fabricating high monodisperse hollow silica spheres by a facile method. Journal of Porous Materials 19, 913–919.

Xiao, J., Chaudhuri, S., 2012. Design of anti-icing coatings using supercooled droplets as nano-to-microscale probes. Langmuir 28, 4434–4446.

Xu, Y., Wang, J.-S., Duan, W., Gu, B.-L., Li, B., 2008. Nonequilibrium Green's function method for phonon–phonon interactions and ballistic-diffusive thermal transport. Physical Review B 78, 224303-1–224303-9.

Yang, M., Wang, G., Yang, Z., 2008. Synthesis of hollow spheres with mesoporous silica nanoparticles shell. Materials Chemistry and Physics 111, 5–8.

Yang, N., Xu, X., Zhang, G., Li, B., 2012. Thermal transport in nanostructures. AIP Advances 2, 041410-1—041410-24.

Yrieix, B., Morel, B., Pons, E., 2014. VIP service life assessment: interactions between barrier laminates and core material, and significance of silica core ageing. Energy and Buildings 85, 617—630.

Yuan, J., Zhou, T., Pu, H., 2010. Nano-sized silica hollow spheres: preparation, mechanism analysis and its water retention property. Journal of Physics and Chemistry of Solids 71, 1013—1019.

Yue, Q., Li, Y., Kong, M., Huang, J., Zhao, X., Liu, J., Williford, R.E., 2011. Ultralow density, hollow silica foams produced through interfacial reaction and their exceptional properties for environmental and energy applications. Journal of Materials Chemistry 21, 12041—12046.

Zhang, Y.-B., Qian, X.-F., Li, Z.-K., Yin, J., Zhu, Z.-K., 2004. Synthesis of novel mesoporous silica spheres with starburst pore canal structure. Journal of Solid State Chemistry 177, 844—848.

Zhang, S., Xu, L., Liu, H., Zhao, Y., Zhang, Y., Wang, Q., Yu, Z., Liu, Z., 2009. A dual template method for synthesizing hollow silica spheres with mesoporous shells. Materials Letters 63, 258—259.

Zhang, H., Zhao, Y., Akins, D.L., 2012. Synthesis and new structure shaping mechanism of silica particles formed at high pH. Journal of Solid State Chemistry 194, 277—281.

Zhang, C., Yan, H., Lv, K., Yuan, S., 2013. Facile synthesis of hierarchically porous silica nanocapsules and nanospheres via vesicle templating. Colloids and Surfaces A: Physicochemical and Engineering Aspects 424, 59—65.

Zhang, Z.M., 2007. Nano/Microscale Heat Transfer. McGraw-Hill.

Zhao, J.-J., Duan, Y.-Y., Wang, X.-D., Wang, B.-X., 2012. Effects of solid—gas coupling and pore and particle microstructures on the effective gaseous thermal conductivity in aerogels. Journal of Nanoparticle Research 14 (1024), 1—15.

Zhu, Y., Shi, J., Chen, H., Shen, W., Dong, X., 2005. A facile method to synthesize novel hollow mesoporous silica spheres and advanced storage property. Microporous and Mesoporous Materials 84, 218—222.

Zwerger, M., Klein, H., 28—29 September, 2005. Integration of VIP's into external wall insulation system. In: Proceedings of the 7th International Vacuum Insulation Symposium. EMPA, Dübendorf, Switzerland, pp. 173—179.

Nano-based phase change materials for building energy efficiency*

Dr. K. Biswas
Oak Ridge National Laboratory, Oak Ridge, TN, United States

9.1 Introduction

Phase change materials (PCMs) have been widely investigated for thermal storage in a range of applications, including integrated collector storage solar water heater (Chaabane et al., 2014), spacecraft thermal control in extreme environments (Wu et al., 2013), phase change slurries for active cooling (Lu and Tassou, 2012), thermal management of building integrated photovoltaic panels (Huang et al., 2006), and so on. Application of PCMs to buildings to take advantage of their latent heat capacities in reducing the envelope-generated heating and cooling loads has received a lot of attention in the first two decades of the 2000s (Zhou et al., 2012).

PCMs in building envelopes operate by changing phase from solid to liquid while absorbing heat from the outside and thus reducing the heat flow into the building, and releasing the absorbed heat when it gets cold outside to reduce the heat loss through the building envelope. Different approaches to PCM applications in buildings have been investigated: PCM wallboards (Zhou et al., 2007; Darkwa and Su, 2012), PCM mixed in concrete and brick (Hawes and Feldman, 1992; Cabeza et al., 2007), micro- or macroencapsulated PCM mixed with loose-fill insulation in wall cavities (Shrestha et al., 2011; Kosny et al., 2012a; Biswas and Abhari, 2014), rigid polyurethane foam incorporating fatty acid ester-based PCM (Aydin and Okutan, 2013), and macro-packaged PCM in plastic pouches (Kosny et al., 2012b). Recent experimental and numerical studies have shown the potential of PCMs in reducing indoor temperature fluctuations under different weather conditions (Meng et al., 2013; Shi et al., 2014; Ascione et al., 2014), reducing energy consumption and providing peak-load shifting (Zwanzig et al., 2013), and also providing internal humidity control (Shi et al., 2014).

In a lot of PCM applications, the low thermal conductivity of traditional PCMs is cited as a critical shortcoming. A lot of research has been focused on enhancing the thermal conductivity of PCMs by adding nanoparticles (NPs) or via nanoencapsulation

* **Notice:** This manuscript has been authored by UT-Battelle, LLC, under Contract No. DE-AC05-00OR22725 with the US Department of Energy. The US government retains and the publisher, by accepting the article for publication, acknowledges that the US government retains a nonexclusive, paid-up, irrevocable, worldwide license to publish or reproduce the published form of this manuscript, or allow others to do so, for US government purposes.

Start-Up Creation. http://dx.doi.org/10.1016/B978-0-08-100546-0.00009-1

methods. Khodadadi and Hosseinizadeh (2007) performed one of the earliest studies pertaining to the potential of nanoparticle-enhanced PCM in improving thermal storage. They predicted a higher heat release rate of nanoPCMs compared to conventional PCMs and contended that there is a clear potential for diverse thermal storage applications of nanoPCMs. Since then, there have been numerous studies related to synthesis of nano-enhanced PCMs (nanoPCMs) and the resultant improvement in thermal properties (primarily thermal conductivity). The present article is focused on nanoPCMs and their suitability for building applications. Here, a brief review of the different kinds of PCMs is provided, followed by descriptions of synthesis methods, thermophysical characteristics, and applications of nanoPCMs. Finally, recommendations for future research are provided.

9.2 Classification of phase change materials

9.2.1 Based on material

Zalba et al. (2009) and Sharma et al. (2009) reviewed PCMs for thermal storage applications and described in detail the different PCMs based on material type. The three main PCM types defined were organic, inorganic, and eutectic PCMs. Organic PCMs primarily consist of paraffins (straight chain n-alkanes), esters, fatty acids, and alcohols (Sharma et al., 2009). Under inorganic PCMs, salt hydrates are primarily used. Phase transition of salt hydrates is actually a dehydration or hydration of the salt resembling a melting or freezing process thermodynamically. Salt hydrates usually melt to other salt hydrates with fewer moles of water Eq. [9.1] or to their anhydrous forms Eq. [9.2], as follows:

$$AB \cdot nH_2O \rightarrow AB \cdot mH_2O + (n-m)H_2O \quad [9.1]$$

$$AB \cdot nH_2O \rightarrow AB + nH_2O \quad [9.2]$$

Metallics are another kind of inorganic PCMs but are seldom used due to weight-related issues (Sharma et al., 2009). Table 9.1 lists the major advantages and disadvantages of organic and inorganic PCMs. One of the drawbacks of PCMs (especially salt hydrates) is that some of them start to solidify at temperatures discernibly lower than the melting temperatures. This phenomenon is known as subcooling or supercooling, and results in inefficient utilization of PCMs.

Zalba et al. (2009) and Sharma et al. (2009) provide detailed lists of various organic and inorganic PCMs, along with their melting temperatures and latent heats of fusion. Inorganic salt hydrates have melting temperatures that are relatively uniformly distributed between 8 and 130°C, and then in the 307–380°C and 700–900°C ranges. The latent heats are in the 125–280 J/g range at low and mid-range melting temperatures and 250–450 J/g at the highest melting points. Organic PCMs have melting points in the 5–150°C range, with latent heats primarily in the 44–85 J/g range.

Finally, eutectics are minimum-melting compositions of two or more components, each of which melt and freeze congruently, forming a mixture of the component

Table 9.1 Advantages and disadvantages of the different phase change material types

Phase change material type		Advantages	Disadvantages
Organic	Paraffinic	• High latent heat of fusion • Noncorrosive • Chemical and thermal stability • Low or no subcooling	• Low thermal conductivity • Moderate flammability
	Nonparaffinic	• High latent heat of fusion • Low subcooling (fatty acids)	• Low thermal conductivity • Flammability • Varying levels of toxicity • Instability at high temperatures
Inorganic	Salt hydrates	• Higher volumetric latent heat • Higher thermal conductivity	• High degree of subcooling • Phase separation and segregation • Corrosiveness • Lack of thermal stability
	Metallics	• Higher volumetric latent heat • High thermal conductivity	• Weight issues

crystals during crystallization. Eutectics nearly always melt and freeze without segregation since they freeze to an intimate mixture of crystals, leaving little opportunity for the components to separate. Upon melting, both components liquefy simultaneously, again with separation unlikely (Sharma et al., 2009).

In addition to the solid−liquid transition PCM, there are solid−solid, solid−gas, and liquid−gas PCM. Due to the large volume changes in the gas phase, solid−gas and liquid−gas PCMs are impractical. However, solid−solid PCMs have shown some potential. Yanshan et al. (2014) synthesized solid−solid PCMs with cross-linking structures composed of polyethylene glycols at different molecular weight as energy storage ingredients and melamine as a cross-linking functional reactant. The PCMs were melamine/formaldehyde/polyethylene glycol (MFPEG) cross-linking copolymers. The authors investigated the composition and chemical structure, and thermophysical and crystallographic properties of the synthesized PCMs. The MFPEG cross-linking copolymers had high latent heats (maximum of 104−109 J/g for cooling/heating cycles) and good stability (Yanshan et al., 2014).

9.2.2 Based on packaging

Liu et al. (2015) proposed classification of encapsulated PCMs into the following categories, based on size: (1) nano-encapsulated PCM (NEPCM) (particle size ranges between 1 and 1000 nm), (2) microencapsulated PCM (microPCM) (particle size ranges between 1 and 1000 μm), and (3) macroencapsulated PCM (macroPCM) (particle size exceeds 1 mm).

Examples of nanoPCMs, both NEPCMs and PCMs with dispersed NPs, are provided in later sections. There are several studies of application of microPCM in buildings, viz. microPCMs infused in wallboards (Darkwa and Su, 2012), concrete (Hawes and Feldman, 1992; Cabeza et al., 2007), and fibrous insulation (Shrestha et al., 2011; Kosny et al., 2012a). Biswas and Abhari (2014) experimentally and numerically investigated a macroPCM consisting of a paraffin trapped by capillary action within the pores of mm-sized pellets of high-density polyethylene (HDPE). These pellets were mixed with fibrous cellulose insulation and added to an exterior test wall. Kosny et al. (2012b) evaluated fatty acid PCMs packaged in plastic pouches in a roof application. The individual pouches were about 5 × 5 cm and 2 cm high and were created by using plastic sheets of overall dimensions 0.6 × 2.4 m. Fig. 9.1 shows sample micro- and macroPCMs.

9.3 Synthesis of nano phase change materials

9.3.1 Nano-encapsulated phase change materials

The first category of nanoPCMs can be defined as NEPCMs, in which the PCM forms the core and is surrounded by a rigid or flexible membrane or shell. Micro- or NEPCMs have certain inherent advantages like less chance of the PCM reacting with the surrounding materials (more critical with inorganic PCMs) and elimination of problems related with volume change during phase transition. Liu et al. (2015) reviewed the literature related to preparation and characterization of NEPCMs and the resultant heat transfer enhancement. They identified and described in detail the following main methods of preparing nanoPCM capsules: (1) interfacial polymerization, (2) emulsion polymerization, (3) miniemulsion polymerization, (4) in situ polymerization, and (5) sol–gel method. They also discussed the suitability and relative merits of the different preparation methods for different types of PCMs and shell materials.

Sari et al. (2014) developed polystyrene(PS)/n-heptadecane micro-/nanocapsules via the emulsion polymerization method with different weight ratios of PS and heptadecane. PS is an inexpensive aromatic polymer and is also used as an insulation material in building applications and has reasonably good mechanical properties, which makes it a good candidate for shell material on encapsulated PCMs. The particle sizes ranged from 10 nm to 40 μm for a 1:2 ratio of PS and heptadecane (Sari et al., 2014). Hong et al. (2011) described the nanoencapsulation of indium NPs in silica shells using the sol–gel method. They used two different silica precursors, tetraethylorthosilicate (TEOS) and sodium silicate. The objective was to demonstrate reduced subcooling with sodium silicate-derived silica shells of thickness 50 nm, compared to TEOS-derived silica shells of thickness 100 nm. The core diameters in both cases were 200 nm. The nanoPCM with sodium silicate-derived silica shells demonstrated 22–28°C reduction in subcooling based on different cooling rates.

Park et al. (2014) described the synthesis of magnetic iron oxide (Fe_3O_4) NP-embedded PCM nanocapsules with paraffin core and polyurea shells. The synthesis

Figure 9.1 Photographs of selected micro- and macroPCMs. Top: Phase change material (PCM) microcapsules mixed with fibrous insulation (Kosny et al., 2012a) *(Kosny, J., Kossecka, E., Brzezinski, A., Tleoubaev, A., Yarbrough, D., 2012a. Dynamic thermal performance analysis of fiber insulations containing bio-based phase change materials (PCMs). Energy and Buildings 52, 122–131.)*; Middle: millimeter-sized PCM-containing HDPE pellets mixed with fibrous insulation, studied by Biswas and Abhari (2014); Bottom: macropackaged bio-based PCM in a roof application (Kosny et al., 2012b).

was done via interfacial polycondensation between tolylene diisocyanate (TDI) and ethylene diamine with various concentrations of the iron oxide NPs, resulting in particle sizes of 400–600 nm. The purpose of creating hybrid organic–inorganic capsules incorporating iron oxide particles was to overcome the relatively low thermal conductivity of the polyurea shell. Micrographs of selected NEPCMs are shown in Fig. 9.2.

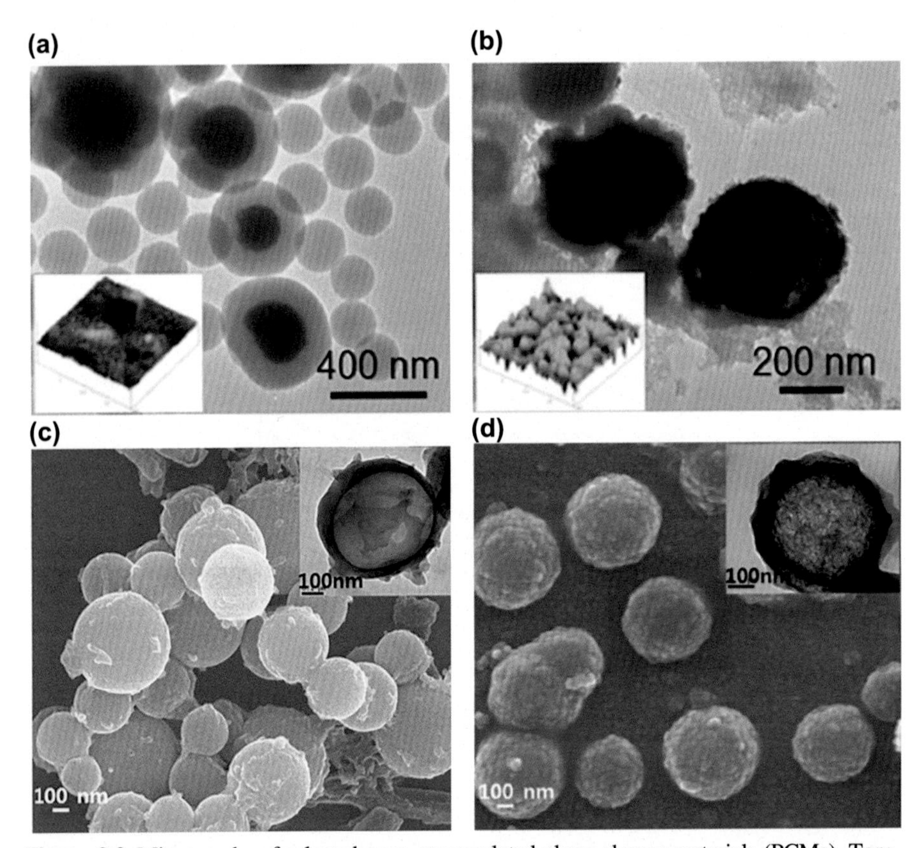

Figure 9.2 Micrographs of selected nano-encapsulated phase change materials (PCMs). Top: Indium nanoparticles encapsulated in silica derived from TEOS (a) *(Hong, Y., Wu, W., Hu, J., Zhang, M., Voevodin, A.A., Chow, L., et al., 2011. Controlling super-cooling of encapsulated phase change nanoparticles for enhanced heat transfer. Chemical Physics Letters 504, 180–184)*, and sodium silicate (b) *(Hong, Y., Wu, W., Hu, J., Zhang, M., Voevodin, A.A., Chow, L., et al., 2011. Controlling super-cooling of encapsulated phase change nanoparticles for enhanced heat transfer. Chemical Physics Letters 504, 180–184)*; Bottom: PCM nanocapsules (c) with *(Park, S., Lee, Y., Kim, Y.S., Lee, H.M., Kim, J.H., Cheong, I.W., et al., 2014. Magnetic nanoparticle-embedded PCM nanocapsules based on paraffin core and polyurea shell. Colloids and Surfaces A: Physicochemical and Engineering Aspects 450, 46–51)*, and (d) without Fe_3O_4 nanoparticles *(Park, S., Lee, Y., Kim, Y.S., Lee, H.M., Kim, J.H., Cheong, I.W., et al., 2014. Magnetic nanoparticle-embedded PCM nanocapsules based on paraffin core and polyurea shell. Colloids and Surfaces A: Physicochemical and Engineering Aspects 450, 46–51)*.

Wang et al. (2015) used a two-step Pickering emulsification process to create NEPCM with nonadecane as the core PCM and PS as the shell material. The nanoPCM capsules were observed to have fairly uniform size distribution of 734 ± 110 nm. Liu et al. (2015), in their review article, noted that the production efficiency of nanoPCM is quite low. Wang et al. (2015) claimed their method to be simple and of low-energy intensity since the nonadecane-in-water emulsion was generated by manual shaking without need for high-energy input (ie, sonication), so that it has promise for scale-up and mass production. The low-energy emulsification process was facilitated by the addition of zirconium phosphate platelets that reduced the surface tension in the water phase. Wang et al. (2015) also calculated the encapsulation efficiency of their NEPCM as 55.9%, using the following equation (Sánchez-Silva et al., 2010):

$$\text{Encapsulation efficiency}(\%) = \frac{\Delta H_{\text{NEPCM}}}{\Delta H_{\text{PCM}} \cdot \frac{m_{\text{core}}/m_{\text{shell}}}{m_{\text{core}}/m_{\text{shell}}+1}} \times 100\% \qquad [9.3]$$

In Eq. [9.3], ΔH is the latent heat (of the NEPCM and the regular PCM) and $m_{\text{core}}/m_{\text{shell}}$ is the core-to-shell mass ratio. The authors noted that the 55.9% encapsulation efficiency is lower than microPCMs but comparable to other NEPCMS.

Latibari et al. (2013) synthesized nanoPCM with palmitic acid (PA) encapsulated in silica shells using the sol–gel method. The authors studied the thermal characterization and influence of different pH values on particle size. The sol solution was created by adding TEOS and ethanol (CH_3CH_2OH) to distilled water, along with ammonium hydroxide (NH_4OH) to control the pH. The nanocapsules were created by adding the sol solution into a PA emulsion and stirring while the emulsion was kept at 70°C. On cooling to room temperature, washing with distilled water and centrifuging, a white powder was formed and it was dried at 50°C. Latibari et al. (2013) found that the encapsulation ratio, defined by Eq. [9.4], increased from 83.25% to 89.55% as the pH increased from 11 to 12. The mean diameters of the NPs were 183.7, 466.4, and 722.5 nm for pH values of 11, 11.5, and 12.

$$\text{Encapsulation ratio}(\%) = \frac{\Delta H_{\text{NEPCM}}}{\Delta H_{\text{PCM}}} \times 100\% \qquad [9.4]$$

9.3.2 Nanoparticle phase change material composites

Another category of nanoPCMs consists of PCM doped with NPs or PCM absorbed within a nanoporous matrix of another higher-conductivity material. Karunamurthy et al. (2012) studied the effect on thermal conductivity by dispersing copper-oxide (CuO) NPs in a paraffinic PCM. Two-percent-by-weight NPs were dispersed in the paraffin using an ultrasonic stirrer and were found to increase the conductivity by 76%. No discussion was presented about the long-term stability of the PCM-CuO composite. Parameshwaran et al. (2013a) embedded spherical-shaped surface-functionalized crystalline silver nanoparticles (AgNP) into organic ester PCM in varying mass ratios. They found that the AgNP did not chemically react with the PCM, enabling the nanoPCM to be chemically stable. The AgNP PCMs maintained their stability for more than 3 months with no observed precipitation.

Jeong and coworkers utilized a vacuum impregnation process to incorporate bio-based PCM into high-conductivity, porous structures like exfoliated graphite nanoplatelets (xGnP) (Jeong et al., 2013) and boron nitride (Jeong et al., 2014). In both studies, the chemical stability was investigated using Fourier transform infrared spectroscopy and it was shown that the chemical properties of the individual components did not change. The PCM molecules were retained in the pores by capillary action and surface tension and leakage of the melted PCM from the composite was prevented by these forces. Fig. 9.3 shows scanning electron microscopy images of bio-based PCM incorporated within nanoporous structures.

Sayyar et al. (2014) and Biswas et al. (2014) reported organic PCMs (fatty acids and paraffin, respectively) supported by interconnected graphite nanosheets. Graphite nanosheets have several desirable features like high specific surface area, high thermal conductivity (1250 W/m/K), and good mechanical properties. In both cases, shape-stable blends of 92%-by-weight PCM and 8%-by-weight expanded graphite nanosheets were created and incorporated into gypsum wallboards for building envelopes. Fig. 9.4 shows micrographs of the graphite nanosheets and nanoPCM containing n-heptadecane studied by Biswas et al. (2014).

Choi et al. (2014) manufactured PCM composites with different carbon additives, that is, multiwalled carbon nanotube (MWCNT), graphite, and graphene. Poly vinyl pyrrolidone (PVP) was used as a dispersion stabilizer of carbon additives. The composites were created by adding the PVP and carbon additives to the molten PCM (stearic acid) and using an ultrasonic disruptor. It was observed that without the PVP, the MWCNTs coagulated in the liquid PCM and sediments were formed in 7 days, but with the PVP the dispersion was maintained for 2 days. With graphene and graphite, sedimentation without PVP occurred in 5 and 2 days, but with PVP the dispersions were maintained for 8 and 3 days, respectively.

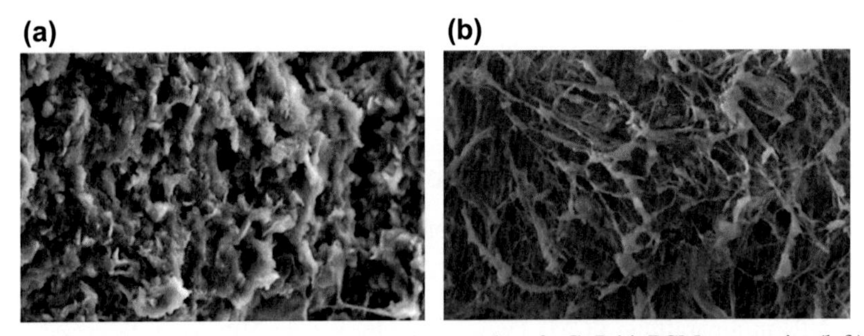

Figure 9.3 Scanning electron microscopy micrographs of xGnP-bioPCM composite (left) *(Jeong, S., Chung, O., Yu, S., Kim, S., 2013. Improvement of the thermal properties of Bio-based PCM using exfoliated graphite nanoplatelets. Solar Energy Materials & Solar Cells 117, 87–92)* and boron nitride-bioPCM composite (right) *(Jeong, S., Lee, J., Seo, J., Kim, S., 2014. Thermal performance evaluation of Bio-based shape stabilized PCM with boron nitride for energy saving. International Journal of Heat and Mass Transfer 71, 245–250).*

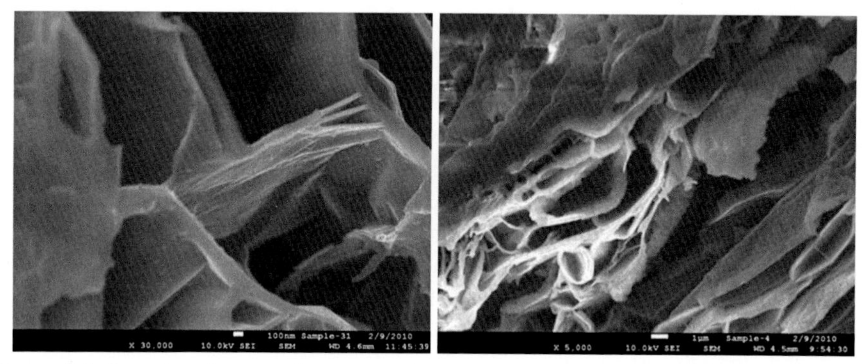

Figure 9.4 Scanning electron microscopy micrographs of expanded graphite nanosheets (left) and nanoPCM (right) studied by Biswas et al. (2014).

Alshaer et al. (2013) developed paraffin wax-based nanocomposites with MWCNT loadings of 0%, 0.25%, 0.50% and 1% by weight. Both single- and multiwalled carbon nanotubes were candidates, but the authors chose MWCNTs since they can be produced on an industrial scale. The MWCNTs had lengths ranging from 0.1 to 10 mm, outer mean diameter in the range of 10–15 nm, and number of walls between 10 and 15. The MWCNT containing granules were immersed in the molten paraffin and dispersed using a high-speed rotor, followed by sonication to homogenize the dispersion. Finally, the samples were agitated under vacuum to reduce the trapped air bubbles. Fig. 9.5 shows micrographs of MWCNTs and MWCNTs dispersed in a paraffin matrix.

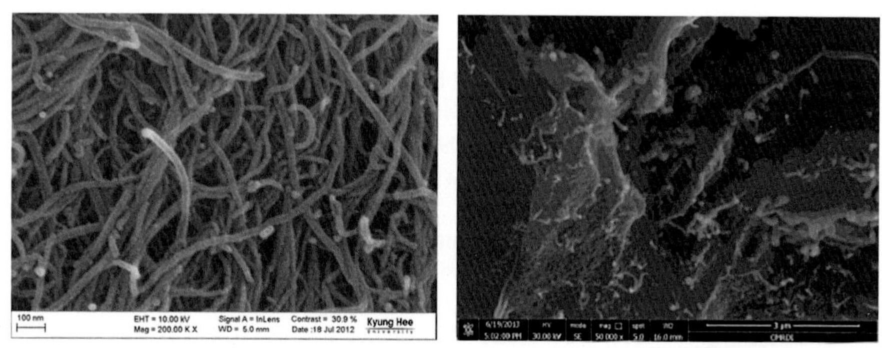

Figure 9.5 Scanning electron microscopy micrographs MWCNTs (left) *(Choi, D.H., Lee, J., Hong, H., Kang, Y.T., 2014. Thermal conductivity and heat transfer performance enhancement of phase change materials (PCM) containing carbon additives for heat storage application. International Journal of Refrigeration 42, 112–120)* and 1%-by-weight MWCNTs dispersed in a paraffin matrix *(Alshaer, W.G., Barrio E.P.D., Rady, M.A., Abdellatif, O.E., Nada, S.A., 2013. Analysis of the anomalous thermal properties of phase change materials based on paraffin wax and multi walls carbon nanotubes. International Journal of Heat and Mass Transfer – Theory and Applications (IREHEAT) 1, 297–307).*

9.4 Characterization of nano phase change materials

9.4.1 Thermophysical properties

Liu et al. (2015) reviewed the characteristics and heat transfer enhancement of NEPCMs synthesized via different polymerization and sol–gel methods. In another review article, Kibria et al. (2015) discussed the impact of NPs dispersed in PCMs on the following: (1) thermal conductivity, (2) latent heat capacity, (3) subcooling, and (4) viscosity. Parameshwaran and Kalaiselvam (2015) wrote a survey article on nanomaterial-embedded PCMs, in which they discussed the thermal energy storage properties including conductivity, phase change temperature and latent heat, nucleation, and energetic aspects of nanoPCMs.

Park et al. (2014) studied magnetic iron oxide (Fe_3O_4) NP-embedded PCM nanocapsules with paraffin core and polyurea shells and found that the thermal conductivity increased with higher concentration of Fe_3O_4 NPs, from 0.23 W/m/K to about 0.33 W/m/K. Latibari et al. (2013) reported an increase in thermal conductivity of palmitic acid by nanoencapsulation in silica shells compared to the base PCM; from 0.21 to 0.26 W/m/K to 0.47–0.77 W/m/K based on the nanocapsule sizes. Melting and freezing temperature changes were within 1.3 and 1.9°C and the maximum degradation in latent heat was 16.7% during melting and 18.2% during freezing. Karunamurthy et al. (2012) found a 75% increase in thermal conductivity by addition of 2%-by-weight CuO NPs to a paraffin. Jeong et al. (2013) prepared stable bio-based PCMs with exfoliated graphite nanoplatelets (xGnP), via vacuum-impregnation method with the goal of improving the thermal conductivity and fire retardant properties. At 75% incorporation rate the latent heat reduced to 110.6 and 115 J/g for melting and freezing compared to 149.2 and 133.5 J/g of the base PCM. Both melting and freezing points were lowered by 1.9 and 1.2°C, with a 375% increase in thermal conductivity. Biswas et al. (2014) measured the thermal conductivity of gypsum wallboard containing 20%-by-weight nanoPCM to be 0.41–0.43 W/m/K compared to 0.15 W/m/K of regular gypsum board. Yavari et al. (2011) created a nanoPCM with graphene and 1-octadecanol, and found an increase of 140% in thermal conductivity without significant loss of latent heat. Zeng et al. (2009, 2010) created nanoPCM with dispersed carbon nanotubes (CNTs) and silver nanowires and found greater thermal conductivity enhancement with silver nanowires.

Meng et al. (2012) studied various fatty acids and their mixtures with different weight percentages of CNTs (0–50%) and studied the effect on melting and crystallization temperatures and latent heats. For 10–20% CNTs the latent heats dropped by 13–27%, and the drop was higher with higher fraction of CNTs. On the other hand, the melting and freezing temperatures dropped irregularly with increasing CNT fraction. Ho and Gao (2009) looked at 5 and 10%-by-weight of nano alumina (Al_2O_3)-in-octadecane emulsions and observed minimal changes in melting and freezing points while the latent heat degraded by 7% and 13%. Alshaer et al. (2013) made an interesting observation of a significant rise in latent heat with addition of MWCNTs compared to the base paraffin (up to 8.5%).

Hong et al. (2011) found that nanoPCM with sodium silicate-derived silica shells demonstrated 22–28°C reduction in subcooling based on different cooling rates. Zhang et al. (2013) studied the role of CNTs as nucleating agents in hexadecane. The freezing temperature of the PCM slurry changed based on the concentration of the CNTs. The degree of subcooling varied from 6.5°C at 0% CNTs to a minimum of 3.5°C at 0.2% of CNTs by weight, with irregularly varying reductions at other CNT concentrations. Kibria et al. (2015) noted that "to act as a nucleating agent" the dispersed particle should have a similar structure to the base fluid. He et al. (2012) experimented with titanium oxide (TiO_2) particles dispersed in barium chloride hydrate solution ($BaCl_2-H_2O$). The energy barrier that needs to be overcome for crystal growth is a function of the specific surface free energy between the crystal nuclei and the NPs; lower the surface free energy, lower the energy barrier. In the study by He et al. (2012), the TiO_2 particles were sized about 20 nm, close to the size of the crystal nuclei of the $BaCl_2-H_2O$ solution, resulting in small surface free energy, and the degree of subcooling was near zero at the volume fraction of TiO_2 of 1.13%. A similar observation of near-zero subcooling was made by Hu et al. (2011) by mixing 5%-by-weight of aluminum nitride NPs in sodium acetate trihydrate ($CH_3COO-Na \cdot 3H_2O$). Parameshwaran et al. (2013b) investigated the nucleation and freezing in a hybrid nanocomposite-dibasic ester PCM. Fig. 9.6 shows the nucleation sites, solid–liquid interface formation, and the progress of the freezing process observed by Parameshwaran et al. (2013b).

Sayyar et al. (2014) and Biswas et al. (2014) described PCM supported within expanded graphite nanosheets. The meso- and macropores between the nanosheets provided high liquid sorption capacity and the molten PCM was immobilized within the pores by capillary action and surface tension. This provided shape-stability even when the nanoPCM was maintained beyond the melting temperature, as seen in Fig. 9.7. Jeong et al. (2013, 2014) reported similar form-stable PCMs where bio-based PCM was absorbed into porous structures of boron nitride and xGnP.

9.4.2 Test methods for thermal characterization

Characterization of the thermal storage properties of PCMs (including nanoPCMs) is usually done via differential scanning calorimetry (DSC), a standard method for thermal analysis (Gmelin and Sarge, 2000). However, several studies have shown that the measurement conditions, specifically the heating and cooling rates, may result in artifacts in the measured data. Castellon et al. (2008) (and other articles by the same group of researchers) have shown that in the continuous cooling mode and at high cooling rates, the DSC traces show significant subcooling. The degree of subcooling is sensitive to the cooling rate and is smaller at slower cooling rates. To overcome the artificial subcooling issue, a step DSC method has been proposed. Castellon et al. (2008) presented data based on both the continuous heating/cooling and the step methods. In the step method, rather than constant heating and cooling, small heating/cooling ramps are followed by periods in which the temperature is kept constant to allow the PCM

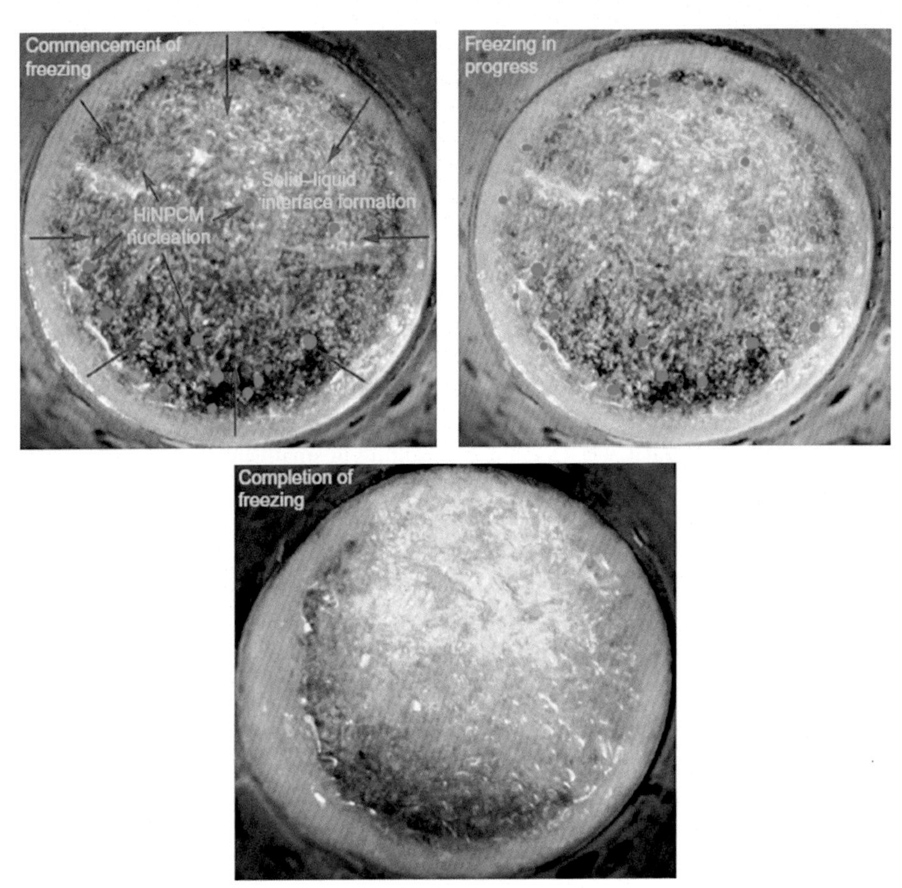

Figure 9.6 Pictorial representation of the freezing process in a hybrid nanoPCM. *(Parameshwaran, R., Dhamodharan, P., Kalaiselvam, S., 2013b. Study on thermal storage properties of hybrid nanocomposite-dibasic ester as phase change material. Thermochimica Acta 573, 106–120.)*

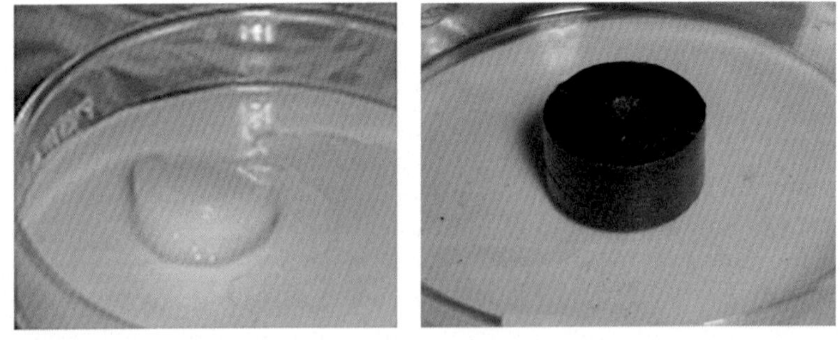

Figure 9.7 Shape stability of nanoPCM studied by Biswas et al. (2014); both heptadecane (left) and nanoPCM (right) were held at 10°C above the melting temperature.

sample to equilibrate. Unlike the constant heating and cooling mode, where heat absorption/release peaks are only seen during melting/freeing, in the step mode heat flux signal has a series of small peaks corresponding to the temperature steps. The enthalpy as a function of temperature ($h(T)$) is determined by integration of the heat flux peaks.

Another potential issue with DSC is that measurements are taken on small, mg-sized samples of the PCMs. For the DSC data to be representative of the real PCMs used in building applications, there is the inherent assumption of the homogeneity of larger-scale PCM samples, which may not be accurate. Further, heating/cooling rates of 1, 5, 10, and even 20°C per minute are used in DSC measurements. In building envelope applications, the PCMs could be subjected to much lower temperature change rates ($\sim$1°C or less per hour). Thus, the DSC data could be severely misrepresenting the PCM behavior. To overcome these issues, a recent ASTM International standard, C1784 (2013), has been developed. It is a test method analogous to the step DSC method and is performed using a heat flow meter apparatus (HFMA). This test method is based on a modification of the HFMA technology (hardware and software) developed for steady state thermal transmission property measurements of planar, homogeneous materials (ASTM C518, 2010).

The HFMA consists of two independently temperature-controlled plates that sandwich the test specimen. The plates are also equipped with surface heat flux transducers to measure the heat absorbed or released by them. ASTM C1784 is a dynamic test method in which a series of measurements are made to determine the enthalpy storage of a test specimen over a temperature range. First, both HFMA plates are held at the same constant temperature until steady state is achieved. Steady state is defined by the reduction in the amount of energy entering the specimen from both plates to a very small and nearly constant value. Next, both plate temperatures are changed by identical amounts and held at the new temperature until steady state is again achieved. The enthalpy absorbed or released by the specimen from the time of the temperature change until steady state is again achieved is recorded. Using a series of temperature step changes, the cumulative energy stored or released over a certain temperature range is determined. The storage capacity of a PCM is well defined via four parameters: specific heats of both solid and liquid phases (via slope of the enthalpy functions), phase change temperature range, and phase change enthalpy (latent heat).

Fig. 9.8 shows data from heating and cooling tests of an organic PCM. The heating tests were performed with different temperature steps. For this test method, it is recommended that the measurements should start at temperatures well below the melting initiation point of the PCM and continue until well after the melting end point is reached (or vice versa for cooling tests). This test is applicable to macroscale PCMs or products and composites containing PCMs. The rationale for developing this test method was that the mg-sized samples tested via DSC may not be representative of the relationship between temperature and enthalpy of full-scale PCM products.

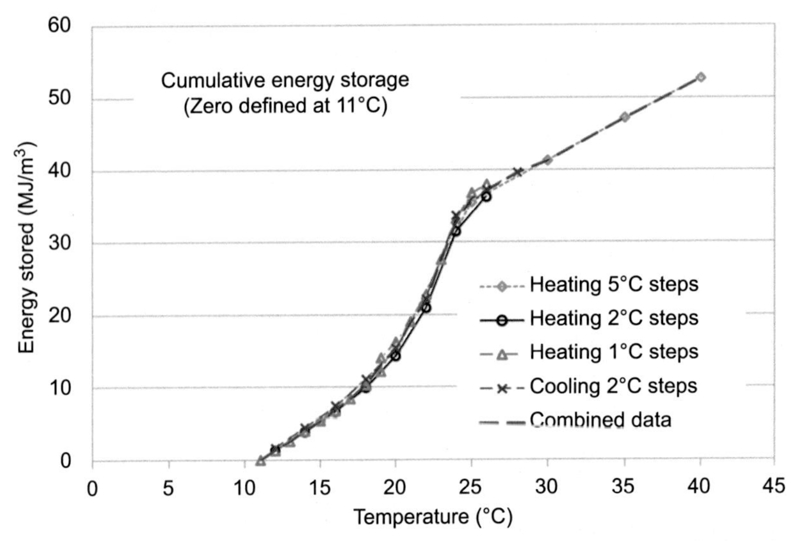

Figure 9.8 Measured volumetric energy storage of a sample phase change material from heating and cooling tests using *ASTM C1784 (2013)*[1]*. Standard test method for using a heat flow meter apparatus for measuring thermal storage properties of phase change materials and products. ASTM International, West Conshohocken, PA, USA.*

The RAL Quality PCM Association[2] was founded in 2004 by several international PCM-related enterprises with the goal of developing proper quality assurance procedures and guaranteeing the quality of thermal storage materials. The fundamental quality criteria are the stored heat as a function of temperature, the number of possible repetitions without any adverse effects, and the thermal conductivity of the storage materials, which is important for the charge and discharge time. The RAL Quality PCM Association's publication, RAL-GZ (2009), is another standard that provides guidelines for measurement and reporting of data related to thermal storage characteristics of PCMs.

Several researchers have evaluated the stability of thermal properties, chemical structure, and volume change of PCMs by subjecting them to accelerated aging via rapid melting and freezing cycles. Latibari et al. (2013) investigated properties of nano-encapsulated palmitic acid-in-silica shell PCM after 2500 thermal cycles. The latent heat of the tested PCM changed from 180.9 kJ/kg to 177.3 kJ/kg during melting and from 181.2 kJ/kg to 178.6 kJ/kg during freezing. Sari (2003) evaluated the reliability of various fatty-acid PCMs after 120, 560, 850, and 1200 thermal cycles. It was observed that the melting temperature tended to decrease after 560 and then

[1] Reprinted, with permission, from ASTM C1784-14 Standard Test Method for Using a Heat Flow Meter Apparatus for Measuring Thermal Storage Properties of Phase Change Materials and Products, copyright ASTM International, 100 Barr Harbor Drive, West Conshohocken, PA 19,428. A copy of the complete standard may be obtained from ASTM, www.astm.org.

[2] http://www.pcm-ral.de/en/quality-association.html.

1200 cycles. The latent heat also decreased with an increasing number of cycles but in an irregular manner. Sharma et al. (1999) subjected a fatty-acid PCM to limited numbers of thermal cycles (20–300) and found no degradation in melting temperature and latent heat of fusion. Karaipekli et al. (2009) and Fauzi et al. (2014) performed accelerated aging tests on eutectic mixtures of fatty acid. Karaipekli et al. (2009) did up to 5000 thermal cycles and observed irregular changes in melting temperature and latent heat; the authors contended, however, that the changes were within acceptable limits for thermal storage applications. Fauzi et al. (2014) did 200, 500, 1000, and 1500 thermal cycles of two mixtures of fatty acids: myristic acid/palmitic acid (MA/PA) (70/30, wt.%) and 5%-by-weight sodium stearate (SS) added to 70/30 (wt.%) MA/PA mixture. They also reported irregular but acceptable changes in the melting temperatures. The latent heat of MA/PA/SS decreased by 10.1% after 1500 cycles, while that of MA/PA increased by up to 6%. Some structural and volume changes were also noticed by the authors. The 1500 cycles represented 4 years of utilization, presumably by assuming one thermal cycle per day.

Similar aging tests are needed for nanoPCMs. Further, it needs to be determined if the rate of heating and cooling used in the aging tests create some artifacts in the measured data, similar to that observed in the DSC traces with respect to subcooling of PCMs (Castellon et al., 2008). Also, whether or not the changes or degradation in thermal properties are acceptable needs to be determined by coupling such aging tests with energy-savings evaluation studies. For example, if the latent heat degrades, what impact does it have on the energy-savings potential of PCM over the life of the application, and how will the cost-to-savings ratio change?

9.5 Building applications

Several researchers have presented experimental and numerical studies of melting/freezing of nanoPCMs within containers and heat exchangers. Hosseinizadeh et al. (2012) numerically investigated the unconstrained melting of a nanoPCM consisting of RT27 and copper particles in a spherical container. RT27 is a commercially available PCM[3] with a melting temperature of 28/30°C and latent heat of 179 kJ/kg. Initially the PCM was assumed to be at 6°C below its melting point and then was heated by maintaining the container wall temperatures at 5, 10, and 15°C above its melting temperature. The higher conductivity and lower latent heat resulted in higher melting rate of the nanoPCM (Hosseinizadeh et al., 2012). Dhaidan et al. (2013) experimentally and numerically studied the melting of nanoPCM under constant heat flux boundary conditions. Addition of NPs enhanced the thermal conductivity and melting rate and expedited the melting time of the PCM. The rate of increase in the melting process was higher at lower values of NP loading, which has additional benefits of higher energy storage capacity and lower cost compared to higher loadings.

Abolghasemi et al. (2012) performed a second law analysis of a thermal storage unit consisting of two concentric cylinders. The outer cylinder was filled with the PCM

[3] Rubitherm GmbH, http://www.rubitherm.eu/.

(calcium chloride hydrate, $CaCl_2 \cdot 6H_2O$) and the working fluid (water) flowed through the inner cylinder. Addition of NPs to the PCM reduced the melting period, entropy generation, and stored energy. However, the reductions were of different magnitudes; for example, in one case an 11% reduction in stored energy was accompanied by 44% reductions in both melting period and entropy generation, leading to more efficient heat transfer.

Parameshwaran and Kalaiselvam (2014) did a study aiming to improve the thermal performance and energy efficiency of chilled water-based variable air volume (VAV) air conditioning (A/C) systems integrated with the AgNP-embedded latent thermal energy storage system (NTES). The NTES-VAV A/C system was investigated with demand control ventilation (DCV) and combined DCV-economizer cycle ventilation (ECV). For comparison, a conventional constant air volume (CAV) and a VAV A/C system were also considered. These A/C systems were designed for operation in a hot and arid climate zone. The proposed NTES-VAV A/C systems achieved 36–58% and 24–51% of on-peak and per-day average energy savings, respectively, compared to a conventional CAV A/C system. For the same operating conditions, the proposed A/C system, compared with a basically similar VAV A/C system, yielded 7.5–18.6% of energy savings. However, the impact of the nanoPCM versus DCV/ECV was not isolated.

Groulx (2015) asked an interesting and relevant question: Are nanoPCMs worth it? In his numerical study, Groulx looked at the melting and freezing of a nanoPCM and regular PCM in a container. Metallic fins were assumed within the container with the base PCM. The base PCM with the fins outperformed the nanoPCM in both storing thermal energy and achieving higher heat transfer rates. The study did not consider any subcooling in the base PCM nor the reduction in subcooling in the nanoPCM, and their impact on PCM melting and freezing.

Sayyar et al. (2014) evaluated the energy impacts of a gypsum wallboard containing fatty-acid PCM incorporated with graphite nanosheets. Testing of the nanoPCM wallboard was performed in one of two side-by-side test cells of 0.28 m sides, with the other test cell containing regular gypsum board. The test cells were placed in a climatic chamber in which the temperature was varied to simulate outside temperature time-histories with slow and rapid temperature changes. The resulting temperatures within the test cells were monitored. The experimental work was coupled with simulations to estimate the energy savings. Models of a room with 1×1 m walls with and without the nanoPCM were created. Measured exterior and interior test cell wall temperatures were used as inputs to the numerical models to calculate the wall heat fluxes. The models were also used to estimate the energy demands to maintain the two cells at comfortable conditions. For the assumed conditions, a reduction in energy demand of 79% was estimated when the walls contained the nanoPCM.

Biswas et al. (2014) also studied the performance of nanoPCM-gypsum boards via experimental and numerical means. This is one of the few studies combining experimental evaluation of nanoPCM in a real building with validated numerical simulations to estimate annual energy benefits. It should be noted, however, that neither study compared the performance of the nanoPCM to regular PCM-enhanced gypsum boards.

In the study by Biswas et al. (2014), a gypsum board containing 20%-by-weight dispersed nanoPCM was installed in an exterior wood-framed wall in a conditioned building in a warm climate; the same wall contained another section covered with

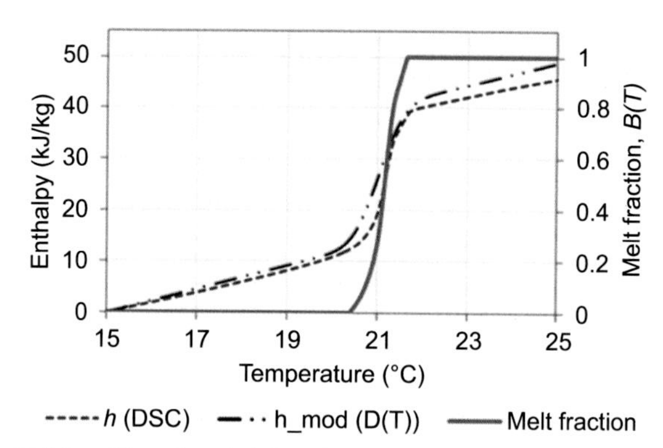

Figure 9.9 NanoPCM wallboard enthalpies (h) and melt fraction as functions of temperature. (*Biswas, K., Lu, J., Soroushian, P., Shrestha, S., 2014. Combined experimental and numerical evaluation of a prototype nano-PCM enhanced wallboard. Applied Energy 131, 517–529*).

regular gypsum board that served as the baseline. The test wall was instrumented with temperature and heat flux sensors, and was monitored over 9 months covering the summer, autumn, and winter seasons. The measured data were used to validate two-dimensional numerical models that were created to match the actual test wall. Once validated, the numerical models were used to estimate the energy savings from the nanoPCM wallboard on an annual basis. An enthalpy function, shown in Fig. 9.9, was calculated using DSC heat flow data and used as input to the model for the PCM thermal storage properties (melt fraction, specific heat, and latent heat).

For the annual simulations, appropriate exterior and interior boundary conditions were required. For the exterior side, data from typical meteorological year (TMY3) weather files were used; this included solar radiation, convection heat transfer based on exterior temperatures and wind conditions, and radiation exchange with the surroundings. On the interior side, a constant surface heat transfer coefficient was assumed to calculate the heat transfer between the wall surface and the interior conditioned space (room). Heat gains and losses at the interior wall surface were calculated and used for comparing the performance of the nanoPCM and gypsum wallboards. Finally, the time-delay in heat transfer caused by the melting/freezing of the PCM has the potential to further reduce cooling A/C electricity consumption. This is because a lot of A/C equipment is often placed outdoors and operates at higher efficiency when the outside temperatures are lower. This effect was investigated by converting the wall heat gains to electricity consumption, using assumed temperature-dependent energy efficiency ratios for heat pumps.

The thermal conductivities of the nanoPCM containing wallboard and regular gypsum board were measured using ASTM C518 (2010). The nanoPCM wallboard was tested at temperatures fully above and below the melting range of the PCM and results are listed in Table 9.2. The effective conductivity (k) was calculated as a function of the solid (s) and liquid (l) state conductivities and the melt fraction (B), as follows:

$$k(T) = k_s + (k_l - k_s)B(T) \quad [9.5]$$

Table 9.2 **Measured thermal conductivities of the gypsum and nanoPCM wallboard (Biswas et al., 2014)**

	Density (kg/m^3)	Thermal conductivity (W/m/K)	Specific heat (kJ/kg/K)	Latent heat (kJ/kg)
Gypsum	549.5	0.153	1.089	—
NanoPCM wallboard	658.5	0.410 (*s*) 0.427 (*l*)	2.312 (*s*) 2.236 (*l*)	26.2

An interesting finding of the study was that the energy savings due to the nanoPCM were sensitive to the interior heating—cooling set points. Fig. 9.10 shows the heat flows at the interior surface of a south-oriented wall or different set points (20—23.3, 20—22 and 19—21°C). Heat flows through the modeled wall with nanoPCM wallboard (NP-Gyp) and regular gypsum board (Gypsum) are shown. Fig. 9.10 also shows the calculated monthly cooling electricity consumption for the different set points.

Table 9.3 shows the percentage of annual reductions in heat gain and cooling electricity use for the different set points. During peak summer, peak daytime heat flow reductions are seen at the 19°C cooling set point (Fig. 9.10), which is in the middle of the phase change temperature range of the nanoPCM. However, the greatest reductions in annual heat gain and electricity use were observed at the 22°C cooling set point, which is slightly higher than the upper temperature limit of the phase change.

The analysis by Biswas et al. (2014) can be extended to evaluate the impact of the enhanced thermal conductivity of the nanoPCM. The thermal conductivity of nanoPCM gypsum was 0.41—0.43 W/m/K compared to 0.15 W/m/K of regular gypsum board. When the nanoPCM is fully molten/frozen, the thermal conductivity of the nanoPCM wallboard is higher than that of regular gypsum board. The higher conductivity may become a liability unless the PCM is always in the transition state. This could explain the higher calculated cooling electricity use with the nanoPCM wallboard shown in Fig. 9.10 during the two peak summer months (July and August) for the 23.3 and 22°C cooling set points. To further investigate this effect, an additional simulation was run by setting the thermal conductivity of the nanoPCM wallboard to that of regular gypsum board (0.15 W/m/K).

Fig. 9.11 compares the monthly heat gains through a south-oriented wall for the different wallboards. It is clear that with the nanoPCM wallboard with the reduced conductivity, the total monthly heat gains were minimum during the peak summer months. On an annual basis, the nanoPCM wallboard with higher conductivity reduced the heat gain by 23.5% compared to the regular gypsum board, while the nanoPCM wallboard with lower conductivity reduced the heat gain by 24.6%. Thus, the higher conductivity of the nanoPCM wasn't beneficial for this particular application. Similar energy-related investigations are needed for other potential

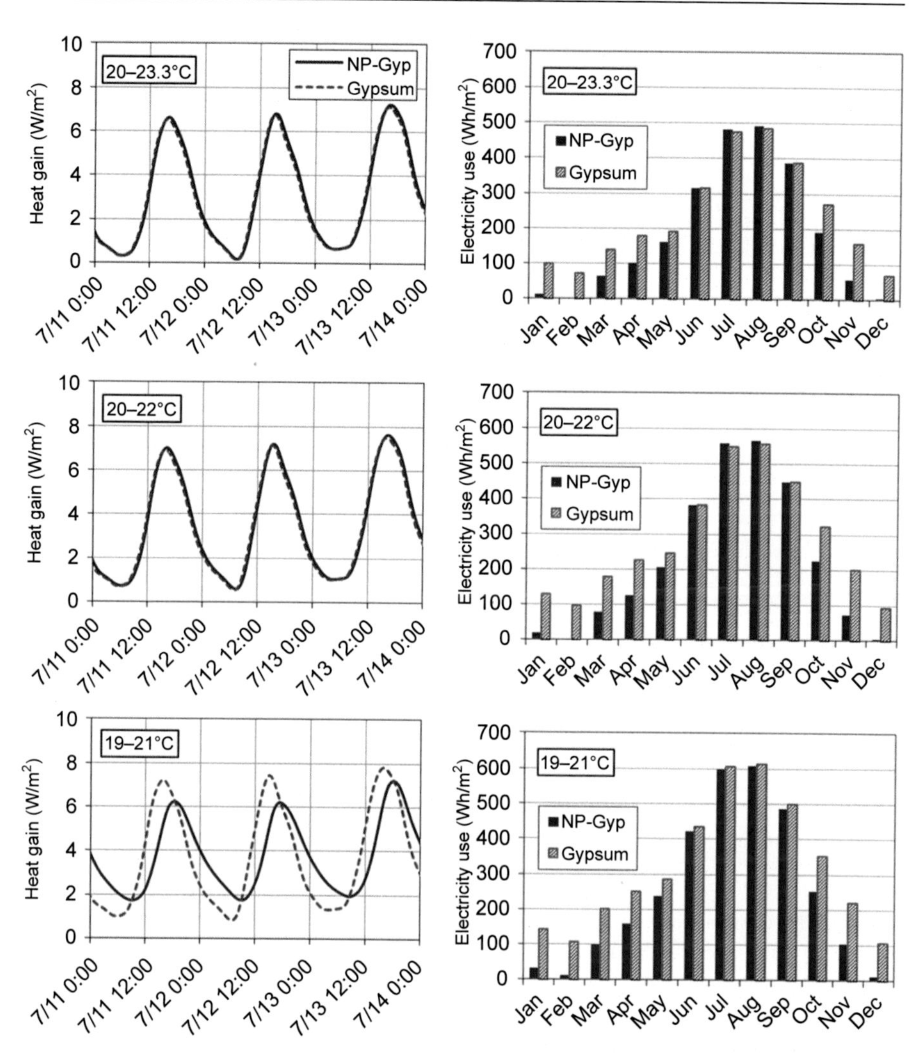

Figure 9.10 Left: calculated heat gains through the nanoPCM (NP-Gyp) and regular gypsum wallboards on a south-oriented wall; right: calculated monthly cooling electricity use (*Biswas, K., Lu, J., Soroushian, P., Shrestha, S., 2014. Combined experimental and numerical evaluation of a prototype nano-PCM enhanced wallboard. Applied Energy 131, 517-529.*)

applications to justify the development of nanoPCMs with enhanced thermal conductivity.

While experiments and modeling of the heat transfer processes within the PCM are important in characterizing PCMs, whole-building or building systems-level simulations are necessary for determining the technoeconomic feasibility of PCMs. Al-Saadi and Zhai (2013) reviewed the modeling of PCMs embedded in building enclosures. The authors discussed the different numerical formulations (enthalpy method, heat capacity method, temperature transforming method, and heat source method), PCM models

Table 9.3 Calculated annual energy savings with the nanoPCM wallboard compared to gypsum wallboard (Biswas et al., 2014)

Set points	Total heat gain (Wh/m²)		% Difference	Cooling electricity use (Wh/m²)		% Difference
	Gypsum	NP-gyp		Gypsum	NP-gyp	
19–21°C (66–70°F)	15,379.2	12,191.2	−20.73	3815.8	3010.1	−21.12
20–22°C (68–72°F)	13,768.6	10,375.2	−24.65	3427.3	2676.6	−21.90
20–23.3°C (68–74°F)	11,292.2	8639.5	−23.49	2836.5	2246.9	−20.78

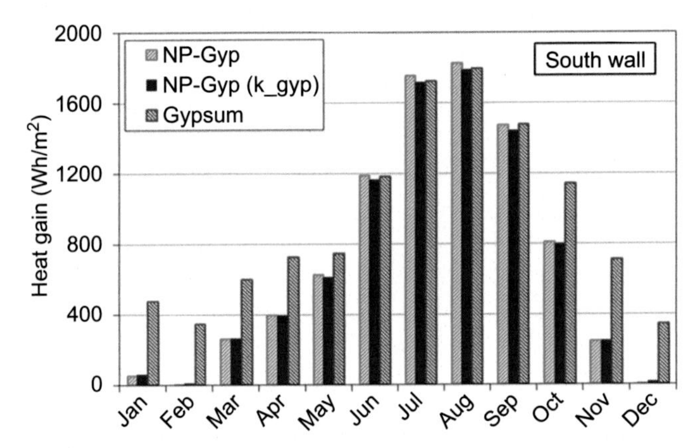

Figure 9.11 Comparison of the calculated heat gains due to nanoPCM wallboard, nanoPCM wallboard with lower conductivity (k_gyp), and regular gypsum board.

(simplified, intermediate, and sophisticated), and integration of PCM models in major building simulation tools (EnergyPlus,[4] TRNSYS,[5] ESP-r,[6] etc.), as well as their advantages and limitations. The intermediate models are most commonly used; these models are trade-offs between the rough approximations of the physical processes occurring in PCMs of simplified models and the sophisticated models created using well-validated numerical packages with optimized numerical methods. Al-Saadi and Zhai (2013) also reviewed several whole-building simulation models that perform computations on an hourly or subhourly basis while considering the dynamic interactions between all thermal-based elements related to energy consumption, including the building envelope, heating and cooling systems, lighting, and so on. The authors noted that most

[4] http://apps1.eere.energy.gov/buildings/energyplus/.
[5] http://sel.me.wisc.edu/trnsys/.
[6] http://www.esru.strath.ac.uk/Programs/ESP-r.htm.

Table 9.4 **Manufacturers and suppliers of selected PCMs**

Paraffin (55.3°C, 151.9 J/g) (Park et al., 2014)	Sigma–Aldrich, Milwaukee, WI, USA (www.sigmaaldrich.com/)
Palmitic acid (Latibari et al., 2013)	Fisher Scientific Inc. (www.fishersci.com/us/en/home.html)
Bio-based phase change material (Jeong et al., 2013; Jeong et al., 2014)	Korea C&S Corporation (www.koreacns.com/)
Capric acid and palmitic acid (Sayyar et al., 2014)	Sigma Aldrich (www.sigmaaldrich.com/)
Stearic acid (Choi et al., 2014)	Dae-Jung Chemical & Metal (daejungchem.lookchem.com/)
Paraffin wax RT65 ($C_{30}H_{62}$) (Alshaer et al., 2013)	Rubitherm Technologies Gmbh (www.rubitherm.eu/en/)
Sodium acetate trihydrate (Hu et al., 2011)	Sinopharm Chemical Reagent Co., Ltd., Shanghai, China (shreagent.lookchem.com/)
Myristic acid and palmitic acid (Fauzi et al., 2014)	Acros Organic (www.acros.com/)
N-octadecane (Dhaidan et al., 2013)	Sigma–Aldrich, St. Louis, MO, USA (www.sigmaaldrich.com/)

whole-building simulation programs incorporate PCM models based on the heat capacity method, which necessitates use of time steps of the order of minutes (rather than hourly) and makes the simulation programs computationally inefficient. Further, the most commonly used models do not have the capability to model the hysteresis or subcooling that is inherent in most PCMs. Finally, Al-Saadi and Zhai (2013) concluded by emphasizing the need for more research to investigate the performance of different PCM models under different climatic and operating conditions.

9.6 Phase change material manufacturers

The Quality PCM Association lists several PCM manufacturers from different countries as its members.[7] These members produce a wide range of PCM products, including organic and inorganic materials, as macro- and microPCMs, and PCMs for low-, ambient-, and high-temperature applications. Several of these members, like BASF[8] and Rubitherm,[9] have a global presence with worldwide partners and suppliers. Based on the articles reviewed in this chapter, Tables 9.4 and 9.5 provide brief

[7] http://www.pcm-ral.de/en/members.html.

[8] https://www.basf.com/en.html?

[9] http://www.rubitherm.eu/en//index.html.

Table 9.5 Manufacturers and suppliers of materials for preparation of nanoparticles, nanoshells, and such

Tolylene diisocyanate and ethylene diamine (for polyurea shells); Ferric chloride ($FeCl_3 \cdot 6H_2O$), ferrous chloride ($FeCl_2 \cdot 4H_2O$) (for Fe_3O_4 nanoparticles) (Park et al., 2014)	Sigma−Aldrich, Milwaukee, WI, USA (www.sigmaaldrich.com/)
Tetraethoxysilane ($SiC_8H_{20}O_4$) (98%) (precursor for silica shell) (Latibari et al., 2013)	Fisher Scientific Inc. (www.fishersci.com/us/en/home.html)
Silver nitrate (AgNP precursor) (Parameshwaran et al., 2013a)	Ranbaxy (part of Sun Pharma, www.sunpharma.com/)
Sulfuric acid-intercalated expandable graphite (to produce xGnP) (Jeong et al., 2013)	Asbury Graphite Mills, NJ, USA (asbury.com/)
Boron nitride (Jeong et al., 2014)	Momentive Performance Materials Inc. (www.momentive.com/)
Exfoliated graphite nanoplatelets (Sayyar et al., 2014)	XG Sciences (xgsciences.com/)
Interconnected graphite nanosheets (Sayyar et al., 2014)	Drzal Group, Michigan State University (www.msu.edu/)
Multiwalled carbon nanotubes (MWCNT) (Choi et al., 2014)	CNT Co., Ltd.
Graphene nanopowder (Choi et al., 2014)	Enanotec (www.enanotec.co.kr/)
Graphistrength® (with 20%-by-weight MWCNT) (Alshaer et al., 2013)	Arkema, France (www.arkema.com/en/)
Aluminium nitride (Hu et al., 2011)	Kaier Nanometer Energy & Technology Co., Ltd., Hefei, China (www.nano-powders.com/)
Sodium stearate (Fauzi et al., 2014)	Sigma Aldrich (www.sigmaaldrich.com/)
CuO nanoparticles (Dhaidan et al., 2013)	Chemistry Department, Auburn University, Auburn, AL, USA (www.auburn.edu/)
Graphite nanosheets	Xiamen Knano Graphite Technology Co. Ltd (www.knano.com.cn/)

lists of manufacturers and suppliers of PCMs and chemicals, precursors, and such, used to produce nanoshells for NEPCMs, NPs, nanosheets, and so forth, for the nanoPCM composites.

For start-up companies, especially those investing in manufacturing nanoPCMs, cost-versus-savings consideration is paramount. As discussed in the previous sections, the higher thermal conductivity of nanoPCMs may not be useful, and in some cases may be detrimental, from an energy-savings perspective. The thermophysical properties of the PCMs may need to be tailored for specific applications. As noted by Liu et al. (2015), the production efficiencies of nanoPCMs are not high enough to meet industrial demands, and it is a reasonable assumption that significant costs will be incurred in scaling up the production of nanoPCMs from laboratories to manufacturing plants. Therefore, manufacturing of PCMs, nano or otherwise, needs to consider the end use, so practical energy savings and suitable returns on investments can be realized.

9.7 Summary and conclusions

The nanoPCM research to date has primarily focused on development of nanoPCMs and characterization of the thermophysical properties or evaluation of melting rate, charging times, and so on. In comparison, there are few studies that have evaluated the actual energy-related benefits of nanoPCMs in building applications.

From a material perspective, several successful approaches to creating nanoPCMs (both encapsulated PCM and PCM-containing NPs) have been investigated. The nanoPCMs that have been developed have usually shown enhanced thermal conductivity and some degradation of latent heat. Several studies have shown that the higher conductivity and lower latent heat resulted in increased melting rate and reduced charging time in thermal storage systems or heat exchangers. However, for 10% or lower concentrations of NPs, the reductions in latent heat were within acceptable limits. The melting and freezing temperatures of nanoPCM were observed to be lowered with different concentrations of NPs, but the reductions were not monotonous with increasing concentrations. The adding of NPs also resulted in reductions in the subcooling of PCMs to various degrees based on concentration.

9.8 Future research

While PCMs have been around for several decades, nanoPCMs are still in their infancy. There are several areas where further research is needed. Whole-building simulations with PCMs are needed to estimate the potential for energy savings and for calculating the savings-to-cost ratio, but the current modeling tools are still works-in-progress. Well-validated numerical simulation tools for evaluation of nanoPCM applications are needed. Compared to macropackaged PCMs, there are few empirical studies evaluating benefits of nanoPCM in actual building applications. Such studies are needed to both evaluate the in situ performance of nanoPCMs and to generate data that can be used for validating existing and new numerical simulation programs. Further, subcooling effects (ie, different enthalpy functions for heating and cooling) are not captured by current simulation tools. The degree of subcooling itself is dependent both on the material and the rate of cooling of the PCM; the latter is a function of the application. Understanding and capturing the subcooling phenomenon is important for reliable estimates of PCM performance.

Another issue with PCMs is the limited long-term (10, 20, 30 years) data from building applications. Therefore, appropriate aging tests are needed to estimate the long-term performance of PCMs. The aging tests currently being done are primarily based on temperature cycling only. It will be interesting to see if other environmental factors, moisture for example, have any impacts on the PCM packaging, chemical stability of the materials, and so on, and cause any performance degradation.

Finally, cost-versus-savings analyses of nanoPCMs are critical for start-up companies so that they can make sound business-related decisions. The viability and economics of nanoPCMs are also related to issues like production efficiency, mass production and scale-up, development of cost-effective and low energy-intensity processes, scoping of newer materials that are well-suited to building applications, among others.

Nomenclature

B	Melt fraction
h	Enthalpy (J/g)
k	Thermal conductivity (W/m/K)
m, n	Number of water molecules
T	Temperature (K)

Subscripts

l	Fully molten state of PCM
s	Fully frozen state of PCM

Abbreviations

A/C	Air conditioning
CAV	Constant air volume
CNT	Carbon nanotube
DCV	Demand control ventilation
DSC	Differential scanning calorimetry
ECV	Economizer cycle ventilation
EDA	Ethylene diamine
EER	Energy-efficiency ratio
HDPE	High-density polyethylene
HFMA	Heat flow meter apparatus
MA	Myristic acid
MWCNT	Multiwalled carbon nanotube
NEPCM	Nano-encapsulated phase change material
NP	Nanoparticle
NTES	Nanoparticle-embedded latent thermal energy storage system

PA	Palmitic acid
PCM	Phase change material
PS	Polystyrene
PVP	Poly vinyl pyrrolidone
SAT	Sodium acetate trihydrate
SEM	Scanning electron microscopy
SS	Sodium stearate
TEOS	Tetraethylorthosilicate
TDI	Tolylene diisocyanate
VAV	Variable air volume
xGnP	Exfoliated graphite nanoplatelets

Acknowledgments

This work was supported by the Building Technologies Office (BTO) of the US Department of Energy under Contract No. DE-AC05-00OR22725 with UT-Battelle, LLC.

References

Abolghasemi, M., Keshavarz, A., Mehrabian, M.A., 2012. Heat transfer enhancement of a thermal storage unit consisting of a phase change material and nano-particles. J. Renewable Sustainable Energy 4, 043124.

Al-Saadi, S.N., Zhai, Z., 2013. Modeling phase change materials embedded in building enclosure: a review. Renewable Sustainable Energy Reviews 21, 659–673.

Alshaer, W.G., Barrio, E.P.D., Rady, M.A., Abdellatif, O.E., Nada, S.A., 2013. Analysis of the anomalous thermal properties of phase change materials based on paraffin wax and multi walls carbon nanotubes. International Journal of Heat and Mass Transfer – Theory and Applications (IREHEAT) 1, 297–307.

Ascione, F., Bianco, N., De Masi, R.F., de'Rossi, F., Vanoli, G.P., 2014. Energy refurbishment of existing buildings through the use of phase change materials: energy savings and indoor comfort in the cooling season. Applied Energy 113, 990–1007.

ASTM C1784-13, 2013. Standard test method for using a heat flow meter apparatus for measuring thermal storage properties of phase change materials and products. In: ASTM International, West Conshohocken, PA, USA.

ASTM C518-10, 2010. Standard test method for steady-state Thermal properties by means of the heat flow meter apparatus. In: ASTM International, West Conshohocken, PA, USA.

Aydin, A.A., Okutan, H., 2013. Polyurethane rigid foam composites incorporated with fatty acid ester-based phase change material. Energy Conversion and Management 68, 74–81.

Biswas, K., Abhari, R., 2014. Low-cost phase change material as an energy storage medium in building envelopes: experimental and numerical analyses. Energy Conversion and Management 88, 1020–1031.

Biswas, K., Lu, J., Soroushian, P., Shrestha, S., 2014. Combined experimental and numerical evaluation of a prototype nano-PCM enhanced wallboard. Applied Energy 131, 517–529.

Cabeza, L.F., Castellon, C., Nogues, M., Medrano, M., Leppers, R., Zubillaga, O., 2007. Use of microencapsulated PCM in concrete walls for energy savings. Energy and Buildings 39, 113–119.

Castellon, C., Gunther, E., Mehling, H., Hiebler, S., Cabeza, L.F., 2008. Determination of the enthalpy of PCM as a function of temperature using a heat-flux DSC: a study of the different measurement procedures and their accuracy. International Journal of Energy Research 32, 1258–1265.

Chaabane, M., Mhiri, H., Bournot, P., 2014. Thermal performance of an integrated collector storage solar water heater (ICSSWH) with phase change materials (PCM). Energy Conversion and Management 78, 897–903.

Choi, D.H., Lee, J., Hong, H., Kang, Y.T., 2014. Thermal conductivity and heat transfer performance enhancement of phase change materials (PCM) containing carbon additives for heat storage application. International Journal of Refrigeration 42, 112–120.

Darkwa, J., Su, O., 2012. Thermal simulation of composite high conductivity laminated microencapsulated phase change material (MEPCM) board. Applied Energy 95, 246–252.

Dhaidan, N.S., Khodadadi, J.M., Al-Hattab, T.A., Al-Mashat, S.M., 2013. Experimental and numerical investigation of melting of NePCM inside an annular container under a constant heat flux including the effect of eccentricity. International Journal of Heat and Mass Transfer 67, 455–468.

Fauzi, H., Metselaar, H.S.C., Mahlia, T.M.I., Silakhori, M., 2014. Thermo-physical stability of fatty acid eutectic mixtures subjected to accelerated aging for thermal energy storage (TES) application. Applied Thermal Engineering 66, 328–334.

Gmelin, E., Sarge, S.M., 2000. Temperature, heat and heat flow rate calibration of differential scanning calorimeters. Thermochimica Acta 347, 9–13.

Groulx, D., August 2015. Numerical study of nano-enhanced PCMs: are they worth it? In: Proc. First Thermal and Fluid Engineering Summer Conference New York, USA.

Hawes, D.W., Feldman, D., 1992. Absorption of phase change materials in concrete. Solar Energy Materials & Solar Cells 27, 91–101.

He, Q., Wang, S., Tong, M., Liu, Y., 2012. Experimental study on thermophysical properties of nanofluids as phase-change material (PCM) in low temperature cool storage. Energy Conversion and Management 64, 199–205.

Ho, C.J., Gao, J.Y., 2009. Preparation and thermophysical properties of nanoparticle-in-paraffin emulsion as phase change material. International Communications in Heat and Mass Transfer 36, 467–470.

Hong, Y., Wu, W., Hu, J., Zhang, M., Voevodin, A.A., Chow, L., Su, M., 2011. Controlling super-cooling of encapsulated phase change nanoparticles for enhanced heat transfer. Chemical Physics Letters 504, 180–184.

Hosseinizadeh, S.F., Darzi, A.A.R., Tan, F.L., 2012. Numerical investigations of unconstrained melting of nano-enhanced phase change material (NEPCM) inside a spherical container. International Journal of Thermal Sciences 51, 77–83.

Hu, P., Lu, D.-J., Fan, X.-Y., Zhou, X., Chen, Z.-S., 2011. Phase change performance of sodium acetate trihydrate with AlN nanoparticles and CMC. Solar Energy Materials & Solar Cells 95, 2645–2649.

Huang, M.J., Eames, P.C., Norton, B., 2006. Phase change materials for limiting temperature rise in building integrated photovoltaics. Solar Energy 80, 1121–1130.

Jeong, S., Chung, O., Yu, S., Kim, S., 2013. Improvement of the thermal properties of bio-based PCM using exfoliated graphite nanoplatelets. Solar Energy Materials & Solar Cells 117, 87–92.

Jeong, S., Lee, J., Seo, J., Kim, S., 2014. Thermal performance evaluation of bio-based shape stabilized PCM with boron nitride for energy saving. International Journal of Heat and Mass Transfer 71, 245–250.

Karaipekli, A., Sari, A., Kaygusuz, K., 2009. Thermal properties and thermal reliability of capric acid/stearic acid mixture for latent heat thermal energy storage. Energy Sources, Part A 31 (3), 199–207.

Karunamurthy, K., Kumar, K.M.M., Suresh, S., 2012. PCM based thermal energy storage system containing CuO nano-particles. International Journal of Nanomaterials and Technology 3, 13–16.

Khodadadi, J.M., Hosseinizadeh, S.F., 2007. Nanoparticle-enhanced phase change materials (NEPCM) with great potential for improved thermal energy storage. International Communications in Heat and Mass Transfer 34, 534–543.

Kibria, M.A., Anisur, M.R., Mahfuz, M.H., Saidur, R., Metselaar, I.H.S.C., 2015. A review on thermophysical properties of nanoparticle dispersed phase change materials. Energy Conversion and Management 95, 69–89.

Kosny, J., Kossecka, E., Brzezinski, A., Tleoubaev, A., Yarbrough, D., 2012a. Dynamic thermal performance analysis of fiber insulations containing bio-based phase change materials (PCMs). Energy and Buildings 52, 122–131.

Kosny, J., Biswas, K., Miller, W., Kriner, S., 2012b. Field thermal performance of naturally ventilated solar roof with PCM heat sink. Solar Energy 86, 2504–2514.

Latibari, S.T., Mehrali, M., Mehrali, M., Mahlia, T.M.I., Metselaar, H.S.C., 2013. Synthesis, characterization and thermal properties of nanoencapsulated phase change materials via sol-gel method. Energy 61, 664–672.

Liu, C., Rao, Z., Zhao, J., Huo, Y., Li, Y., 2015. Review on nanoencapsulated phase change materials: preparation, characterization and heat transfer enhancement. Nano Energy 13, 814–826.

Lu, W., Tassou, S.A., 2012. Experimental study of the thermal characteristics of phase change slurries for active cooling. Applied Energy 91, 366–374.

Meng, X., Zhang, H., Sun, L., Xu, F., Jiao, Q., Zhao, Z., et al., 2012. Preparation and thermal properties of fatty acids/CNTs composite as shape-stabilized phase change materials. Journal of Thermal Analysis and Calorimetry 1–8.

Meng, E., Yu, H., Zhan, G., He, Y., 2013. Experimental and numerical study of the thermal performance of a new type of phase change material room. Energy Conversion and Management 74, 386–394.

Parameshwaran, R., Kalaiselvam, S., 2014. Energy conservative air conditioning system using silver nano-based PCM thermal storage for modern buildings. Energy and Buildings 69, 202–212.

Parameshwaran, R., Kalaiselvam, S., 2015. Nanomaterial-embedded phase-change materials (PCMs) for reducing building cooling needs. In: Pacheco-Torgal, F., et al. (Eds.), Eco-Efficient Materials for Mitigating Building Cooling Needs. Woodhead, Cambridge, pp. 401–439.

Parameshwaran, R., Jayavel, R., Kalaiselvam, S., 2013a. Study on thermal properties of organic ester phase-change material embedded with silver nanoparticles. Journal of Thermal Analysis and Calorimetry 114, 845–858.

Parameshwaran, R., Dhamodharan, P., Kalaiselvam, S., 2013b. Study on thermal storage properties of hybrid nanocomposite-dibasic ester as phase change material. Thermochimica Acta 573, 106–120.

Park, S., Lee, Y., Kim, Y.S., Lee, H.M., Kim, J.H., Cheong, I.W., Koh, W.-G., 2014. Magnetic nanoparticle-embedded PCM nanocapsules based on paraffin core and polyurea shell. Colloids Surfaces A: Physicochemical Engineering Aspects 450, 46–51.

RAL-GZ 896, September 2009. Phase Change Material, Quality Assurance. German Institute for Quality Assurance and Certification E.V.

Sánchez-Silva, L., Tsavalas, J., Sundberg, D., Sánchez, P., Rodriguez, J.F., 2010. Synthesis and characterization of paraffin wax microcapsules with acrylic-based polymer shells. Industrial & Engineering Chemistry Research 49, 12204–12211.

Sari, A., Alkan, C., Döğüşcü, D.K., Biçer, A., 2014. Micro/nano-encapsulated n-heptadecane with polystyrene shell for latent heat thermal energy storage. Solar Energy Mater. Solar Cells 126, 42–50.

Sari, A., 2003. Thermal reliability test of some fatty acids as PCMs used for solar thermal latent heat storage applications. Energy Conversion and Management 44 (14), 2277–2287.

Sayyar, M., Weerasiri, R.R., Soroushian, P., Lu, J., 2014. Experimental and numerical study of shape-stable phase-changenanocomposite toward energy-efficient building constructions. Energy and Buildings 75, 249–255.

Sharma, S.D., Buddhi, D., Sawhney, R.L., 1999. Accelerated thermal cycle test of latent heat-storage materials. Solar Energy 66 (6), 483–490.

Sharma, A., Tyagi, V.V., Chen, C.R., Buddhi, D., 2009. Review on thermal energy storage with phase change materials and applications. Renewable Sustainable Energy Reviews 13, 318–345.

Shi, X., Memon, S.A., Tang, W., Cui, H., Xing, F., 2014. Experimental assessment of position of macro encapsulated phase change material in concrete walls on indoor temperatures and humidity levels. Energy and Buildings 71, 80–87.

Shrestha, S., Miller, W., Stovall, T., Desjarlais, A., Childs, K., Porter, W., et al., 2011. Modeling PCM-enhanced insulation system and benchmarking EnergyPlus against controlled field data. In: Proc. Building Simulation 2011: 12th Conf. International Building Performance Simulation Association, pp. 800–807. Available online at: http://www.ibpsa.org/proceedings/BS2011/P_1328.pdf.

Wang, X., Zhang, L., Yu, Y.-H., Jia, L., Mannan, M.S., Chen, Y., Cheng, Z., 2015. Nano-encapsulated PCM via pickering emulsification. Scientific Reports (5). http://dx.doi.org/10.1038/srep13357. http://www.nature.com/articles/srep13357.

Wu, W.-F., Liu, N., Cheng, W.-L., Liu, Y., 2013. Study on the effect of shape-stabilized phase change materials on spacecraft thermal control in extreme thermal environment. Energy Conversion and Management 69, 174–180.

Yanshan, L., Shujun, W., Hongyan, L., Fanbin, M., Huanqing, M., Wangang, Z., 2014. Preparation and characterization of melamine/formaldehyde/polyethylene glycol crosslinking copolymers as solid–solid phase change materials. Solar Energy Materials & Solar Cells 127, 92–97.

Yavari, F., Fard, H.R., Pashayi, K., Rafiee, M.A., Zamiri, A., Yu, Z., et al., 2011. Enhanced thermal conductivity in a nanostructured phase change composite due to low concentration graphene additives. The Journal of Physical Chemistry C 115, 8753–8758.

Zalba, B., Marin, J.M., Cabeza, L.F., Mehling, H., 2009. Review on thermal energy storage with phase change: materials, heat transfer analysis and applications. Applied Thermal Engineering 23, 251–283.

Zeng, J., Cao, Z., Yang, D., Xu, F., Sun, L., Zhang, X., et al., 2009. Effects of MWNTs on phase change enthalpy and thermal conductivity of a solid-liquid organic PCM. Journal of Thermal Analysis and Calorimetry 95, 507–512.

Zeng, J., Cao, Z., Yang, D., Sun, L., Zhang, L., 2010. Thermal conductivity enhancement of Ag nanowires on an organic phase change material. Journal of Thermal Analysis and Calorimetry 101, 385–389.

Zhang, X., Niu, J., Wu, J.-Y., Zhang, S., 2013. The application of modified multi-wall carbon nano-tube particles in PCM as the nucleating agent. Applied Mechanics and Materials 328, 753–757.

Zhou, G., Zhang, Y., Wang, X., Lin, K., Xiao, W., 2007. An assessment of mixed type PCM-gypsum and shape-stabilized PCM plates in a building for passive solar heating. Solar Energy 81, 1351–1360.

Zhou, D., Zhao, C.Y., Tian, Y., 2012. Review on thermal energy storage with phase change materials (PCMs) in building applications. Applied Energy 92, 593–605.

Zwanzig, S.D., Lian, Y., Brehob, E.G., 2013. Numerical simulation of phase change material composite wallboard in a multi-layered building envelope. Energy Conversion and Management 69, 27–40.

Nano-based chromogenic technologies for building energy efficiency

L. Long, H. Ye
University of Science and Technology of China, Hefei, PR China

10.1 Introduction

The building sector is responsible for 30–40% of the primary energy used in the developed countries. It is also one of the main sources of carbon dioxide emissions. As a result, improving the energy efficiency performance of a building is necessary for a sustainable and environmentally friendly society. Such improvement can be achieved through two types of technologies, the active one and the passive one. The active technologies relate to the efforts that directly reduce the energy consumption; for example, developing more efficient heating, ventilation, and air conditioning units and illuminating systems. Regarding the passive technologies, they usually work through a better-performed building envelope, although themselves do not reduce the energy consumption actively.

As a significant part of a building envelope, windows contain great potential for energy-efficient development because up to 60% of the total energy losses from buildings may be attributable to them (Gustavsen, 2008). A window impacts the energy performance in a building mainly from two aspects: radiation and thermal transmittance. It transmits solar radiation and emits long-wave thermal radiation as well as conducts heat between the indoor and outdoor environments. These features imply two approaches to improve the window: optimizing the capacity for the regulation of radiation and improving the thermal insulation performance (Jelle et al., 2012; Van Den Bergh et al., 2012).

Due to the fact that the solar radiation varies over time, the response properties of the window to the solar radiation should be changeable to realize dynamic daylight and solar energy control in buildings. In addition to liquid-crystal (Cupelli et al., 2009) and suspended-particle devices (Vergaz et al., 2008), chromogenic technologies can help the windows adjust their radiation properties in response to the demands of inhabitants (Granqvist, 2007), making them focused by both the research community and the market. In this chapter, we will first give a brief review of four typical chromogenic technologies, and then introduce performance demonstrations as well as theoretical discussions on the properties of the windows. The chapter will conclude with a section of future trends.

Start-Up Creation. http://dx.doi.org/10.1016/B978-0-08-100546-0.00010-8

10.2 Chromogenic technologies

Basically, windows with chromogenic technologies, namely, chromogenic windows, change their radiation properties by external stimuli. The most widely studied technologies involve thermochromic, electrochromic, photochromic, and gasochromic windows, whose property regulations are triggered by temperature, electrons, light, and gas, respectively. Their dynamically changing properties usually lead to a better energy performance and a higher comfort degree, so they are also called smart windows. They can be further grouped into active (electrochromic and gasochromic windows) and passive (thermochromic and photochromic windows) smart windows according to whether the property transition can be triggered artificially. For the active ones, the electrons or gas that cause the transition generally can be controlled by the residents or devices; the transition of the passive windows is conducted spontaneously by the temperature or light of the surroundings. In this section, these four types of chromogenic technologies will be briefly introduced one by one.

10.2.1 Thermochromic technology

The thermochromic window (Kamalisarvestani et al., 2013; Hoffmann et al., 2014; Lee et al., 2013; Warwick and Binions, 2014), whose transition depends on the temperature, is a widely investigated type of passive chromogenic windows. The phenomenon of thermochromism (ie, reversibly changing color with temperature) has been observed in many different compounds; for example, inorganic oxides, liquid crystals, and conjugated oligomers (Nitz and Hartwig, 2005; Seeboth et al., 2010). Vanadium dioxide (VO_2) is one of the most promising thermochromic inorganic oxides. First reported by Morin (1959), VO_2 is able to undergo a reversible transition at a phase transition temperature (T_τ): when the temperature of the material is lower than T_τ, it is monoclinic, semiconducting, and rather infrared transparent, and when the temperature is higher than T_τ, it is tetragonal, metallic, and near-infrared reflecting.

Although aforementioned features of VO_2 meet the theoretical demand of dynamic regulation of solar radiation, three inherent shortcomings block its practical application to building sectors (Li et al., 2012): (1) the solar modulation ability, which is defined as the difference in solar transmittance between its two states, is modest; (2) its luminous transmittance is relatively low compared with that of a float-glass window; and (3) the transition temperature of bulk VO_2 (approximately 68°C) is too high for building applications. For the first two issues, the regulation of the microstructure of the VO_2 film (Yao et al., 2013; Madida et al., 2014) may be a solution. For the third issue, doping with transition metal ions like tungsten (Tan et al., 2012) can significantly decrease T_τ. A VO_2 film with a T_τ as low as 30°C has been reported (Huang et al., 2011). In addition to transmittance and T_τ, the transition hysteresis width and gradient of VO_2 also play crucial roles in the energy performance (Warwick et al., 2013; Saeli et al., 2013).

10.2.2 Electrochromic technology

Compared with the thermochromic windows, electrochromic windows can change their properties in a shorter time and have a more flexible transition that is based on the user's demand rather than the uncontrollable temperature (Runnerstrom et al., 2014; Granqvist, 2014; Granqvist et al., 2014), making electrochromic technology the most available in the current market among the chromogenic windows.

A typical electrochromic window contains a functional electrochromic layer sandwiched by two protective transparent substrates. The substrates may be glass or flexible polyester foil with a high transmittance, while the functional layer consists of several films, which generally are two transparent conductors, an electrochromic film, an ion storage film, and a transparent ion conductor (electrolyte) (Granqvist et al., 2009). The most widely used transparent conductor is indium tin oxide (In_2O_3: Sn or ITO), and the electrochromic materials may be WO_3 (Huang et al., 2015; Yang et al., 2014; Vuk et al., 2014), NiO (Wen et al., 2014; Granqvist et al., 2010), and many other metal oxides (eg, Bi_2O_3, CeO_2, CoO, CuO, Fe_3O_4, Fe_2O_3, FeO, MnO_2, MoO_3, etc.) (Granqvist, 2012).

10.2.3 Gasochromic technology

Gasochromic technology can be seen as an upgrade from the electrochromic one. The gasochromic windows exhibit a better solar modulation ability than the electrochromic one, and they do not require an applied current to trigger the property transition. When a transition is needed in a gasochromic window, a certain amount of hydrogen gas will be provided in a cavity surrounded by a Pt or Pd catalyst layer and WO_3 films (Gao et al., 2014; Qu et al., 2009; Xu et al., 2001).

10.2.4 Photochromic technology

The property regulation of a photochromic window occurs when irradiated by light (in particular, ultraviolet ray), and the process is reversible when the light is removed. Typical photochromic materials may be TiO_2- and MoO_3-based films (Yao et al., 1998; Domenici et al., 2011; Ohko et al., 2003). Because of its high cost, the photochromic windows receive less attention than the other chromogenic technologies.

10.2.5 Creation of start-ups

The increasing number of high energy-consuming buildings raises the immediate need of practical chromogenic technologies. Weighing efficiency and cost, the thermochromic and electrochromic technologies may be, at present, the most promising products that merit the creation of a start-up.

Several synthesis methods of thermochromic films are available at laboratories (Warwick and Binions, 2014), such as magnetron sputtering and deposition (Saeli et al., 2009; Mlyuka et al., 2009). However, the cost of these two methods is relatively high due to the expensive equipment, making them inappropriate for mass production. Synthesis through a solution-based process can lower the cost considerably (Gao et al., 2012b). Within this method, the VO_2 particles are pretreated in a polyvinylpyrrolidone aqueous solution and then transferred to an ethanol solution to form SiO_2 shells; next, the VO_2–SiO_2 particles are further treated in an aqueous solution containing a trace amount of silane coupler; finally, a VO_2-based composite film is prepared by casting the suspension on a PET substrate. Another method through solvent–thermal and pyrolysis process was also reported to fabricate large-scale and cost-effective thermochromic films (Zhang et al., 2015).

With a better performance than the thermochromic one, the electrochromic technology takes a step further from the laboratory toward the market. ChromoGenics, a successful example that converts scientific research into practical productions, has already produced excellent electrochromic films named ConverLight Foil (seen via http://www.chromogenics.com).

10.3 Performance demonstrations

According to the previous introductions, chromogenic windows are promising components to reduce the building energy consumption. There are two types of methods to demonstrate their energy performance: one is through experiments, which involves actual material exhibitions in model rooms; and the other is through simulations, which means using computational tools to simulate the application performance. The former is direct and easy to understand, while the latter is flexible and may save time. Both of them are significant for the research of chromogenic windows, and they will be introduced successively in this section.

10.3.1 Experiments

Applying chromogenic windows to model rooms is a direct way to exhibit the windows' energy performance. Gao et al. have tested the application performances of different types of VO_2 films in two model houses of $34 \times 27 \times 29\ cm^3$ (Gao et al., 2012a). However, the experiment was relatively simple, needing an outdoor test to give a more comprehensive result. Instead of these small model houses, a full-scale testbed office has been employed by Lee et al. (2013) to evaluate a large polymer thermochromic window, which showed an energy-saving potential of this window.

Ye's group has completed a series of outdoor experiments to research the energy performance of thermochromic windows (Ye et al., 2013a; Long et al., 2015). These experiments were conducted in a Testing and Demonstration Platform for Building Energy Research, whose schematic diagram is shown in Fig. 10.1(a). The platform contains two identical testing rooms, Room A and Room B, with a dimension of $2.9 \times 1.8 \times 1.8\ m^3$ (length $\times$ width $\times$ height). The south walls of the rooms are glass curtains, and the other parts of the envelopes are made of polyurethane wrapped with metal boards, in which the polyurethane is 37 kg/m^3 in density, 1385 J/(kg K) in specific heat, and 0.0228 W/(m K) in thermal conductivity. The thicknesses of the walls

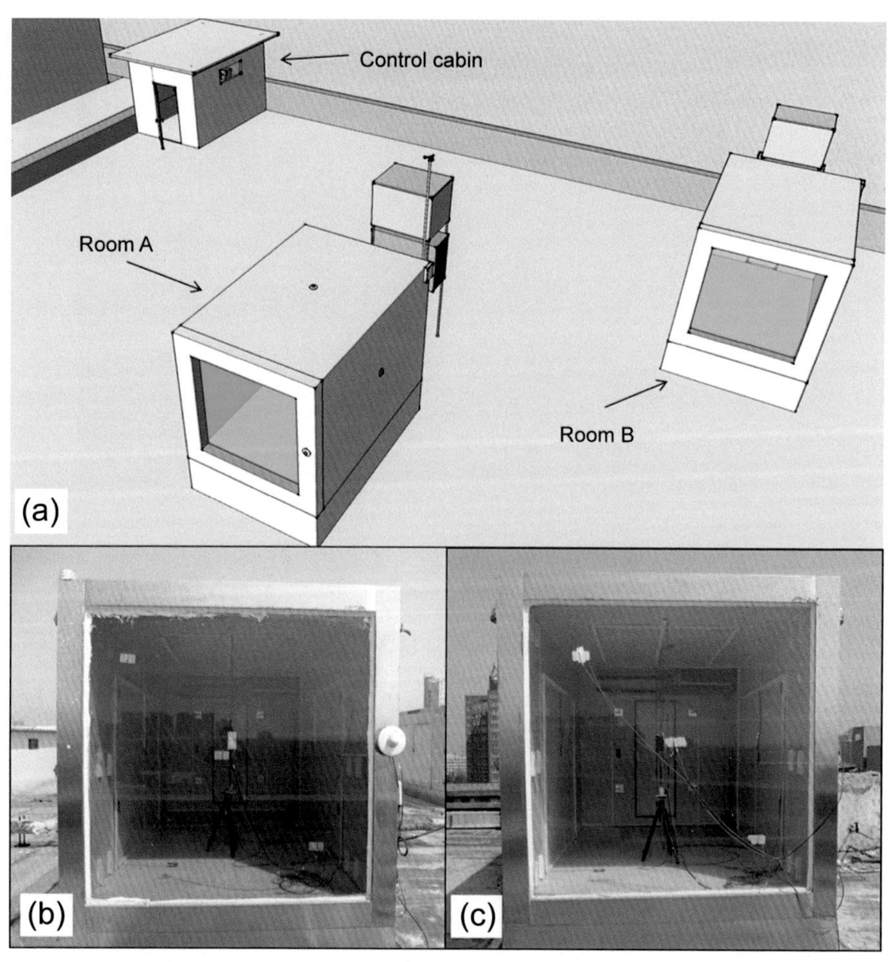

Figure 10.1 Testing and Demonstration Platform for Building Energy Research. (a) Schematic diagram. Photos of (b) Room A with VO_2 glazing and (c) Room B with ordinary glazing. Adapted from Ye, H., Long, L.S., Zhang, H.T., Xu, B., Gao, Y.F., Kang, L.T., Chen, Z., 2013a. The demonstration and simulation of the application performance of the vanadium dioxide single glazing. Solar Energy Materials and Solar Cells 117, 168–173.

and the roofs are 10 cm. The indoor temperature of the rooms can be maintained at a particular temperature through the input of cool/heat wind from the fan coil units.

The material that has been demonstrated, as shown in Fig. 10.2(a), is a type of thermal-stable PET film covered with VO_2, which was prepared through an all-solution process (Gao et al., 2012b). The phase transition temperature of the VO_2 particles on the film is 41.3°C, and the film's spectral transmittance at low temperature (in its semiconductor state or transparent state) and high temperature (in its metallic state or opaque state) are shown in Fig. 10.2(b). By sticking the film on an ordinary glass substrate, a VO_2 glazing can be formed, making it convenient for refurbishing an existing window system.

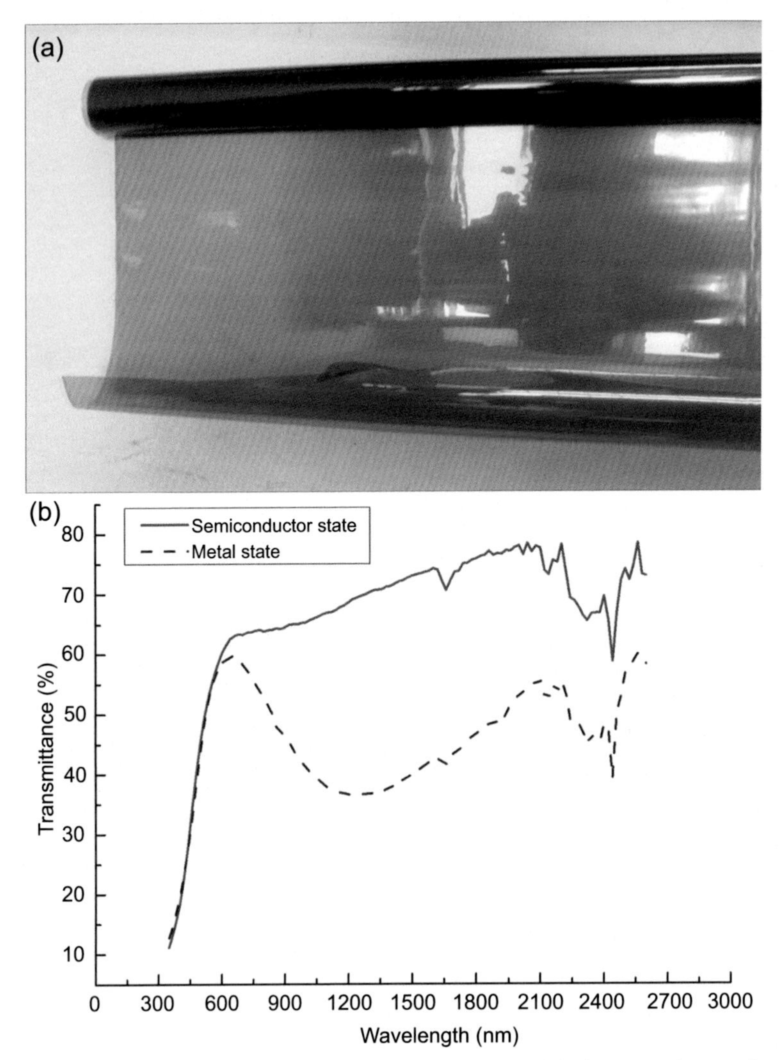

Figure 10.2 (a) Thermal-stable PET film covered with VO_2 and (b) its spectral transmittance in the semiconductor state and metallic state.
Adapted from Ye, H., Long, L.S., Zhang, H.T., Xu, B., Gao, Y.F., Kang, L.T., Chen, Z., 2013a. The demonstration and simulation of the application performance of the vanadium dioxide single glazing. Solar Energy Materials and Solar Cells 117, 168–173.

10.3.1.1 Single thermochromic glazing

During the first and second demonstrations, the VO_2 film was pasted on the single-glazed window of Room A, while Room B was set as the control without the film. The photos of these two testing rooms are shown in Fig. 10.1(b) and (c). The radiation properties of the ordinary single glazing and the VO_2 single glazing are listed in Table 10.1.

Table 10.1 Radiation properties of single glazing and double glazing

	Properties	With VO_2 film Semiconductor state	 Metallic state	Ordinary
Single glazing	Solar absorptivity	0.46	0.58	0.16
	Solar reflectance	0.10	0.07	0.07
	Solar transmittance	0.44	0.35	0.77
	Visible transmittance	0.43	0.42	0.84
Double glazing	Total solar absorptivity	0.54	0.65	0.29
	Total solar reflectance	0.12	0.08	0.11
	Total solar transmittance	0.34	0.27	0.60
	Total visible transmittance	0.38	0.37	0.76

Adapted from Long, L.S., Ye, H., Zhang, H.T., Gao, Y.F., 2015. Performance demonstration and simulation of thermochromic double glazing in building applications. Solar Energy 120, 55–64.

The first demonstration began on July 7, 2012 and lasted for 7 days. The indoor temperatures of the testing rooms were set to 20°C. The variations in the cooling load are shown in Fig. 10.3(a), which indicates that the cooling load of Room B was greater than that of Room A. This is the result of the VO_2 glazing's lower solar transmittance in the both metallic and semiconductor states compared with that of the ordinary glazing. The lower solar transmittance led to less solar radiation transmitted through the glazing, which was the main heat gain of the indoor air. Therefore, less cooling load was needed to cool down the air in Room A. The cumulative cooling load was obtained through time integration of the cooling loads. Considering the effect of the errors of measurements, cumulative cooling loads for Rooms A and B were 317.4–338.2 and 376.5–396.4 MJ, respectively. It means that the cumulative load of the room with the VO_2 glazing was at least 10.2%, up to 19.9% less than that of the ordinary glazing. However, with a decrease in solar radiation transmitted through the VO_2 glazing, the visible light transmitted into Room A was also decreased. The measured data indicated that the visible light illuminancy of Room A was approximately 80% lower than that of Room B. As a result, additional lighting load may be needed in the room with VO_2 glazing.

The second demonstration began on July 30, 2012 and lasted for 5 days. Different from the first one, the rooms here were on a passive mode under which no cooling load was input. Fig. 10.3(b) shows that the measured indoor temperature of Room A was up to 4°C lower than that of Room B in the daytime due to a lower solar transmittance of the VO_2 glazing compared with that of the ordinary glazing. At night, when there was

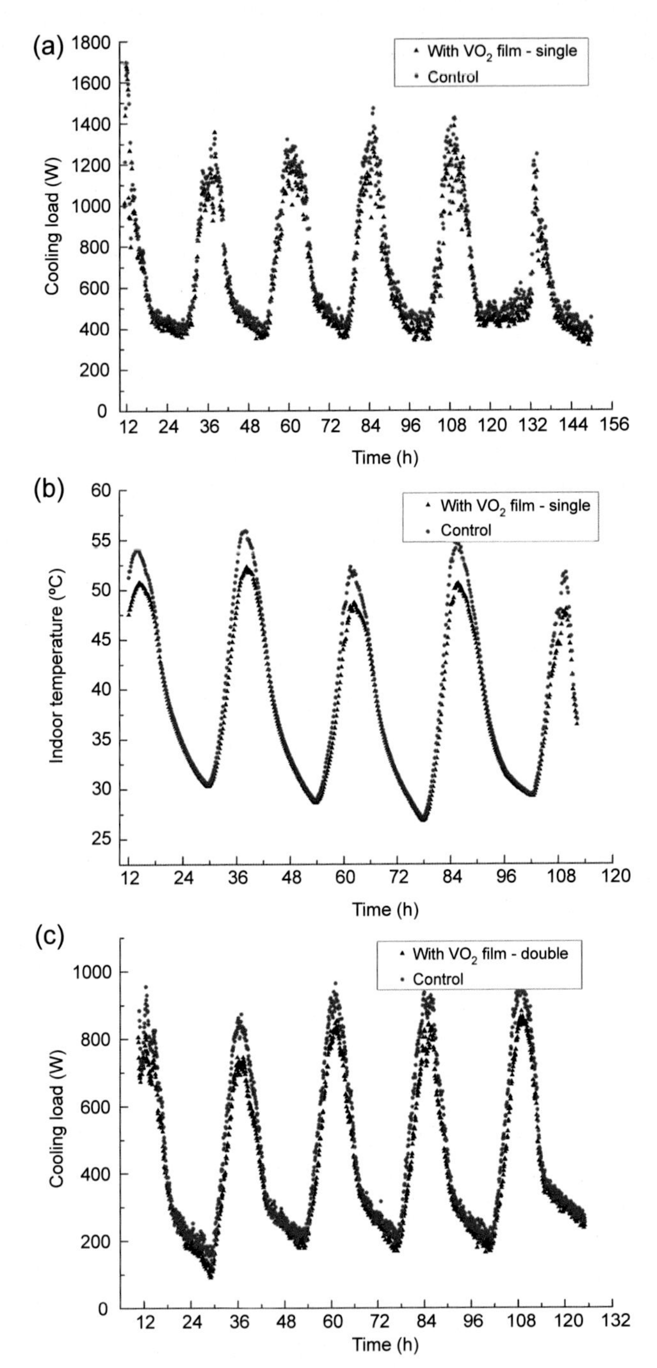

Figure 10.3 Performance demonstrations of a thermochromic window: (a) Single glazing in the active mode. *(Adapted from Ye, H., Long, L.S., Zhang, H.T., Xu, B., Gao, Y.F., Kang, L.T., Chen, Z., 2013a. The demonstration and simulation of the application performance of the vanadium dioxide single glazing. Solar Energy Materials and Solar Cells 117, 168–173.)*; (b) Single glazing in the passive mode. *(Adapted from Ye, H., Long, L.S., Zhang, H.T., Gao, Y.F., 2014. The energy saving index and the performance evaluation of thermochromic windows in passive buildings. Renewable Energy 66, 215–221.)*; (c) Double glazing in the active mode. *(Adapted from Long, L.S., Ye, H., Zhang, H.T., Gao, Y.F., 2015. Performance demonstration and simulation of thermochromic double glazing in building applications. Solar Energy 120, 55–64.)*

no solar radiation, the two temperature curves were almost identical because of the building envelope's low heat-storage capacity. It was inferred that the use of VO_2 glazing in a passive room could lower the indoor temperature in the summer and showed better performance in improving the indoor thermal comfort degree than the ordinary glazing.

10.3.1.2 Double thermochromic glazing

Double-window systems are commonly equipped in current buildings to enhance the thermal insulation performance of the transparent building envelope, making it necessary to evaluate the performance of the double windows with chromogenic materials. To do that, before the third demonstration, the testing rooms were re-installed with a double-window system containing two 4.6 mm thick glass panes of ordinary float glass. A foam spacer provided an air cavity with a thickness of 7.0 mm between the interior and exterior panes. The VO_2 film was pasted, for convenience, onto the outside surface of the exterior pane of Room A to form a VO_2 double-glazed window, and Room B was still set as the control with an ordinary double window. The solar spectral properties of the double-window systems are also listed in Table 10.1.

The third demonstration started on May 20, 2013 and lasted for 5 days. The cooling load that maintains the indoor temperature at 20°C was measured and is plotted in Fig. 10.3(c). The figure shows that during the daytime, the cooling load in the room that contained the VO_2 double glazing, Room A, is lower than that in the room with the ordinary glazing, Room B; and the daily peak load of the former is approximately 100 W less than that of the latter. At night, the presence of the VO_2 film on the glazing makes little difference due to the absence of solar radiation and the low mass envelope, which makes the cooling load in the rooms close to each other. The cumulative energy consumptions for cooling in Rooms A and B are 168.6 ± 6.0 and 190.3 ± 7.0 MJ, respectively. The results indicate that the application of the VO_2 double glazing in the testing room reduced the cooling energy consumption by $(11.1 \pm 6.4)\%$.

10.3.2 Simulations

Besides time-consuming and costly outdoor demonstrations, the performance of the chromogenic windows can be further simulated via computer programs, such as Energy Plus and TRNSYS. With the help of these tools, many researchers have exhibited the energy efficiency of chromogenic windows (Saeli et al., 2010a,b; Hoffmann et al., 2014; Xu et al., 2012). In addition to these commercial software programs, some noncommercial programs developed by the researchers themselves were also preferred due to a better understanding of thermophysical processes. With a verified program, Dussault et al. (2012) have confirmed the energy-saving potential of electrochromic windows. BuildingEnergy program was developed by Ye's group through a non-steady-state heat transfer model. It has been validated as per ANSI/ASHRAE Standard 140—2004 (Standard Method of Test for the Evaluation of Building Energy

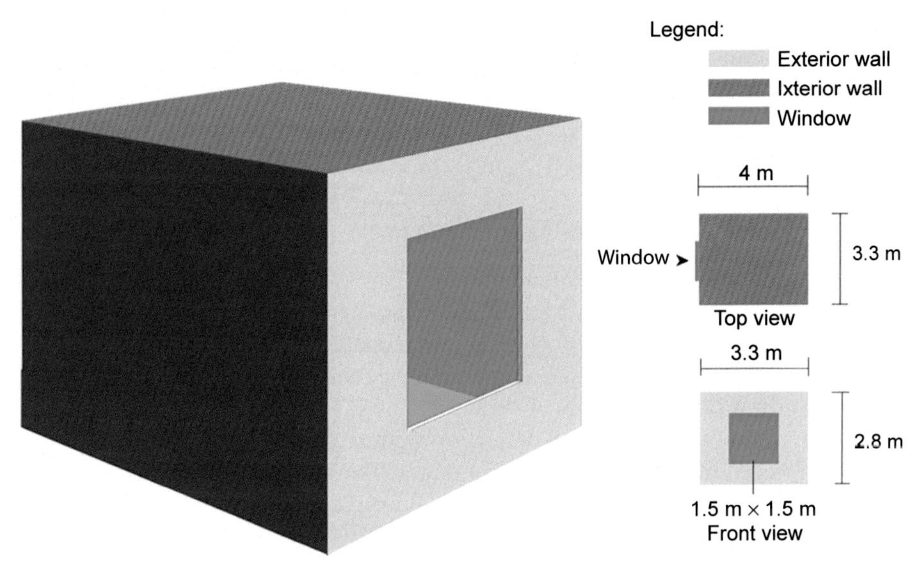

Figure 10.4 Schematic diagram of a typical room.
Reprinted from Long, L.S.,Ye, H., 2014a. Discussion of the performance improvement of thermochromic smart glazing applied in passive buildings. Solar Energy 107, 236–244.

Analysis Computer Programs) as well as through a series of experiments (Ye et al., 2012, 2013a; Long et al., 2015).

In BuildingEnergy simulations, the thermochromic windows were assumed to be applied to a mid-floor room in a multistory residential building, whose schematic diagram is shown in Fig. 10.4. The room has only one exterior wall, and contains a 1.5 × 1.5 m single window in the middle of the exterior wall.

During the simulation, the indoor temperature is maintained at 26°C through the space cooling, which is recommended by the design standard (JGJ75–2003). The application performance in Guangzhou (China) is discussed, which locates in latitude 23 N, longitude 113 E and has a cooling period from May 13 to October 17. The climate data used to simulate the performance in BuildingEnergy are the typical meteorological year data offered by the Chinese Architecture-specific Meteorological Data Sets for Thermal Environment Analysis. The simulated cooling loads of the room with a single VO_2 (the material displayed in Fig. 10.2) glazing and with an ordinary glazing are shown in Fig. 10.5(a). Although the loads during the entire cooling period were simulated, only those in one week is shown in the figure to avoid the overlapping of lines. The figure indicates that the cooling load of the room with the VO_2 glazing was lower than that with the ordinary glazing. The cooling loads can then be integrated into the cumulative cooling energy consumption. During the cooling period of a year, the cumulative cooling energy consumption with the VO_2 glazing adopted was 9.4% less than with the ordinary glazing adopted, which means that the use of VO_2 glazing could reduce cooling consumption by 22.5 kWh. Another benefit from the application of the thermochromic

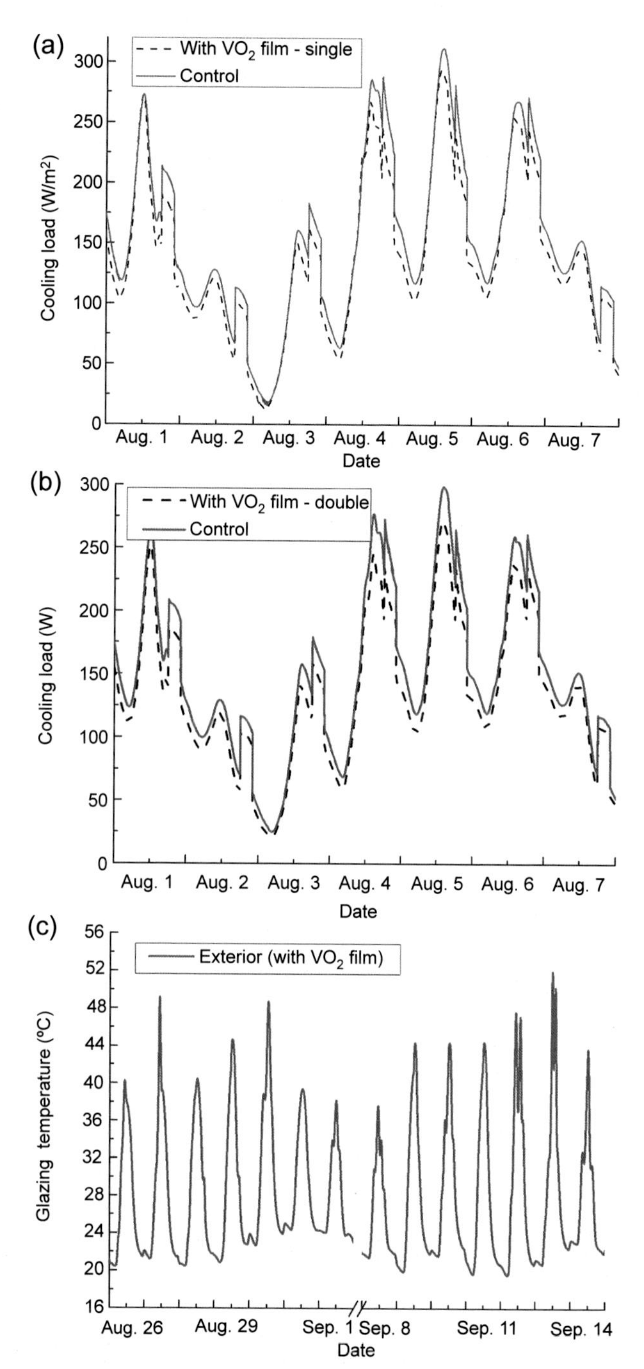

Figure 10.5 Performance simulations of a thermochromic window: (a) Rooms with single glazing. *(Adapted from Ye, H., Long, L.S., Zhang, H.T., Xu, B., Gao, Y.F., Kang, L.T., Chen, Z., 2013a. The demonstration and simulation of the application performance of the vanadium dioxide single glazing. Solar Energy Materials and Solar Cells 117, 168–173.)*; (b) Rooms with double glazing. *(Adapted from Long, L.S., Ye, H., Zhang, H.T., Gao, Y.F., 2015. Performance demonstration and simulation of thermochromic double glazing in building applications. Solar Energy 120, 55–64.)*; (c) Glazing temperature of double glazing. *(Adapted from Long, L.S., Ye, H., Zhang, H.T., Gao, Y.F., 2015. Performance demonstration and simulation of thermochromic double glazing in building applications. Solar Energy 120, 55–64.)*

glazing was to decrease the peak cooling load of the air-condictioning facility by 5.1%. This means that the air-condictioning facility with a lower load capacity could be used.

The application performance of VO_2 double glazing was also simulated via BuildingEnergy, and the cooling loads within one week are displayed in Fig. 10.5(b). The figure shows that the cooling load in the room with the VO_2 double glazing is lower than that with the ordinary glazing; and the peak cooling load of the VO_2 glazing is approximately 8.9% lower than that of the ordinary one. The results indicate that during the entire cooling period of a year, the application of the VO_2 double glazing to the standard room can reduce the energy consumption by 11.6% or 26.4 kWh without considering the lighting energy consumption.

In addition to the cooling loads, temperature of the film is another key parameter to estimate the window's performance as it reveals whether the transition of the VO_2 occurs. Although the transition temperature (41.3°C) seems too high compared with the indoor temperature (26°C), the film temperature shown in Fig. 10.5(c) can be much higher than the indoor temperature due to intense solar irradiation and high absorptivity. Regarding the specific variation, the film's temperature varies with the weather conditions on each day, leading to different status of the VO_2: it may be completely in its metallic state as its temperature can reach 50°C or it may be mostly in its semiconductor state when the highest temperature is only 36°C. Considering that the phase transition process does not occur at a particular temperature but within a range of temperatures, the seemingly high transition temperature is actually suitable for applications.

In summary, both the experiments and simulations have confirmed the energy efficiency of the chromogenic windows. However, some inherent shortcomings still exist. The next section will discuss some properties of the chromogenic windows, aiming to draw an outline of how an ideal chromogenic window is and how to improve the actual ones.

10.4 Performance improvement

As mentioned in the introduction, the radiation properties and thermal transmittance of a window are two of the dominant features that influence energy performance, implying that improvements of a window can be made from these aspects. To simplify the scenarios, we will first introduce the improvement on radiation properties on the basis of a single-glazed window because the other feature of a window, thermal insulation performance, is usually improved through a structure of multilayers. After that, double-glazed windows will be briefly discussed.

10.4.1 Radiation properties

To set a common standard for single-glazed windows, the concept of the windows with perfect radiation properties was conceived (Ye et al., 2012), and its diagrammatic

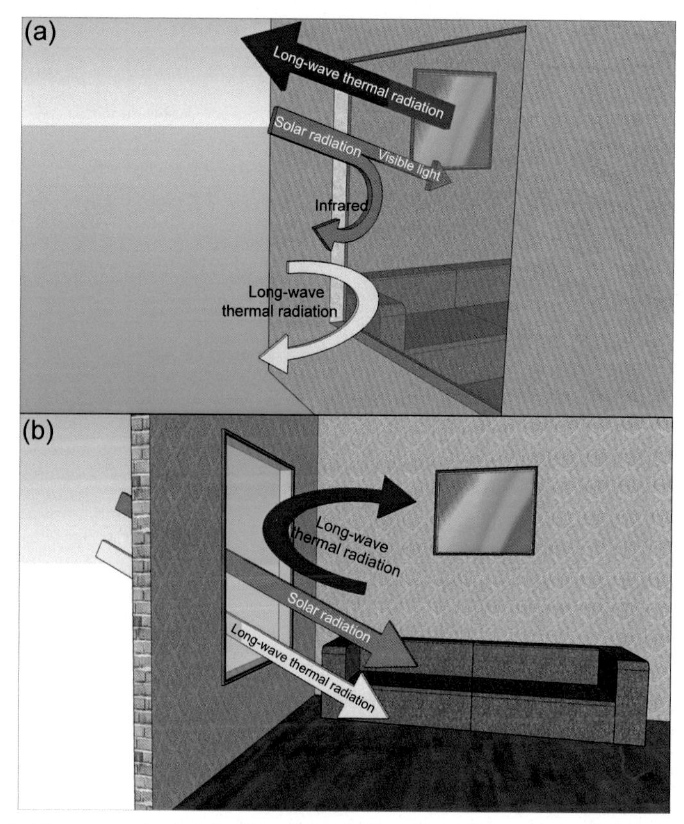

Figure 10.6 Diagrammatic sketch of perfect windows for (a) summer and (b) winter.
Reprinted from Long, L.S., Ye, H., 2014b. How to be smart and energy efficient: a general discussion on thermochromic windows. Scientific Reports 4, 6427. http://dx.doi.org/10.1038/srep06427.

sketch is shown in Fig. 10.6. In this figure, the orange arrows represent solar radiation, while the yellow arrows represent long-wave thermal radiation emitted by the outdoor environment, and the blue arrows represent the long-wave thermal radiation emitted by indoor surfaces. The perfect window for summer permits the visible light in the solar spectrum to be transmitted into the room to provide indoor illumination and enable observation, but it also reflects solar radiation in other spectra to reduce the heating of the indoor environment. Simultaneously, the long-wave thermal radiation from indoor surfaces is transmitted through the perfect window for summer, but the thermal radiation from the outdoor environment is reflected back by the window. This one-way transmission of long-wave thermal radiation allows thermal radiation to be transferred only from the indoors to the outdoors, which is advantageous for reducing the cooling load. During the heating season, the perfect window for winter transmits both the solar radiation and the long-wave thermal radiation from the outdoor environment, but it reflects the long-wave thermal radiation from the indoor surfaces. This allows the maximum possible heat compensation from the outdoors and the minimum possible

Table 10.2 Spectral properties of perfect windows

Season	Spectral properties	Solar spectrum		Long-wave thermal radiation ($\lambda > 2.5\ \mu m$)	
		Visible light (0.4–0.7 μm)	$\lambda < 0.4\ \mu m$ or $0.7 < \lambda < 2.5\ \mu m$	Outdoor	Indoor
Summer	Spectral transmittance	1	0	0	1
	Spectral reflectance	0	1	1	0
	Spectral absorptivity	0	0	0	0
Winter	Spectral transmittance	1	1	1	0
	Spectral reflectance	0	0	0	1
	Spectral absorptivity	0	0	0	0

Adapted from Long, L.S., Ye, H., 2014b. How to be smart and energy efficient: a general discussion on thermochromic windows. Scientific Reports 4, 6427. http://dx.doi.org/10.1038/srep06427.

heat loss from indoors. According to these features, the integrated radiation properties of the perfect windows are summarized in Table 10.2. With the conception of perfect windows, a dimensionless parameter, energy consumption index (ECI), is used to evaluate energy performance. The ECI of a particular window is defined as the ratio of the energy consumption of the room containing the window to the corresponding value of the same room with a perfect window.

From the description of the perfect windows, the radiation that needs to be regulated can be further sorted into two types according to the source of radiation: the long-wave thermal radiation emitted by objects at normal temperatures and the solar radiation emitted by the sun at a surface temperature of ~6000K.

10.4.1.1 Long-wave thermal radiation

The long-wave thermal emissivity (ε) of the glazing may affect the radiation heat transfer between the glazing and the indoor surface or the outdoor environment. The windows with low emissivity (low-e) coatings were considered energy efficient as they can, theoretically, hinder the long-wave thermal radiation entering the room in summer as well as block the radiation leaving the room in winter. A type of low-e glass with a high solar transmission is designed for the area in need of heating, and that with low transmittance is for the area in need of cooling. Their properties, adapted from

Table 10.3 Radiation properties of low-e glass with high or low solar transmittance and ordinary glass

Radiation property	High transmittance	Low transmittance	Ordinary glass
τ_{sol}	0.59	0.38	0.83
ρ_{sol}	0.27	0.46	0.08
ε	0.035	0.066	0.84

WINDOW software (ID 919 and 920), are listed in Table 10.3. The ECIs of these windows applied in different cities were calculated via BuildingEnergy software, and are plotted in Fig. 10.7. The figure indicates that compared with the ordinary one, the corresponding low-e window is energy efficient in different seasons and cities.

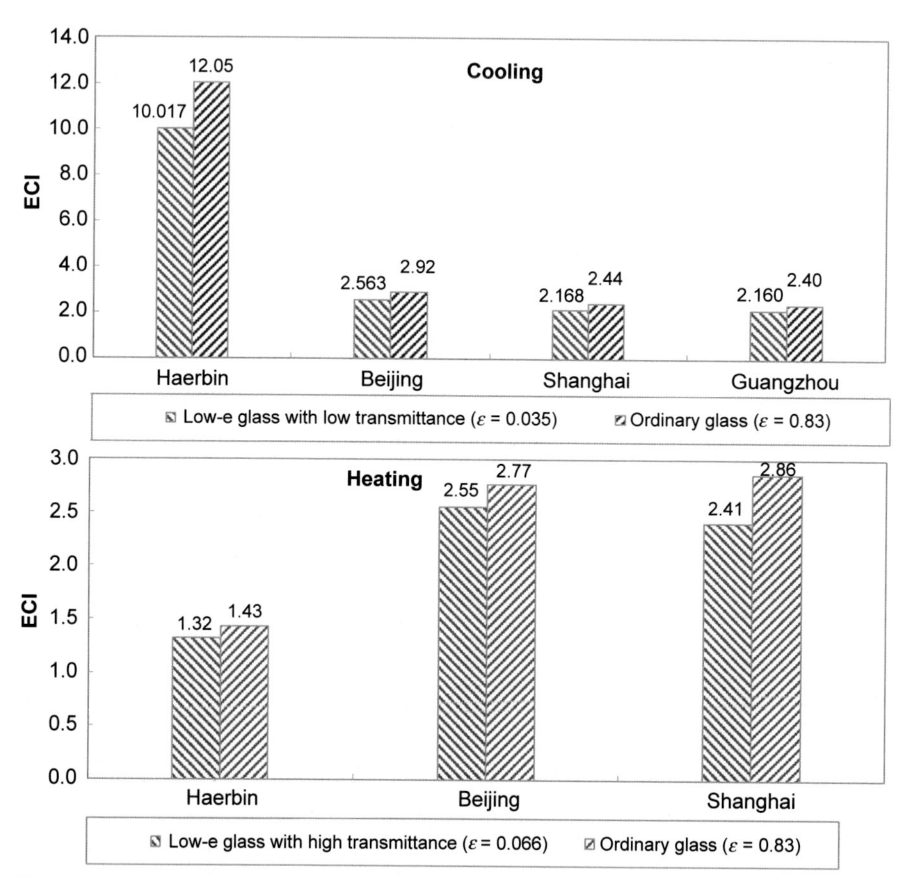

Figure 10.7 The energy consumption indexes of the low-e glass and ordinary glass, and low-e window in different seasons.
Adapted from Ye, H., Meng, X.C., Long, L.S., Xu, B., 2013b. The route to a perfect window. Renewable Energy 55, 448–455.

10.4.1.2 Solar radiation properties

Chromogenic technology is the representative one that regulates the response properties of solar radiation. As shown in Fig. 10.8(a) or (b), the functional layer of practical active chromogenic windows is an electrochromic layer consisting of five films or a gasochromic layer involving two films and a cavity. Table 10.4 shows the radiation properties of a practical electrochromic window (Nagai et al., 1999) and a practical gasochromic one (Georg et al., 1998). Bleached and colored states represent the near infrared (NIR) transmission and reflection states, respectively. The substrates of the electrochromic glazing and the gasochromic glazing are both ordinary glazing.

To observe the best performance of chromogenic windows, ideal chromogenic windows can be defined as, similar to the perfect windows, those that can not only meet the indoor lighting need but also regulate the NIR transmission in maximum

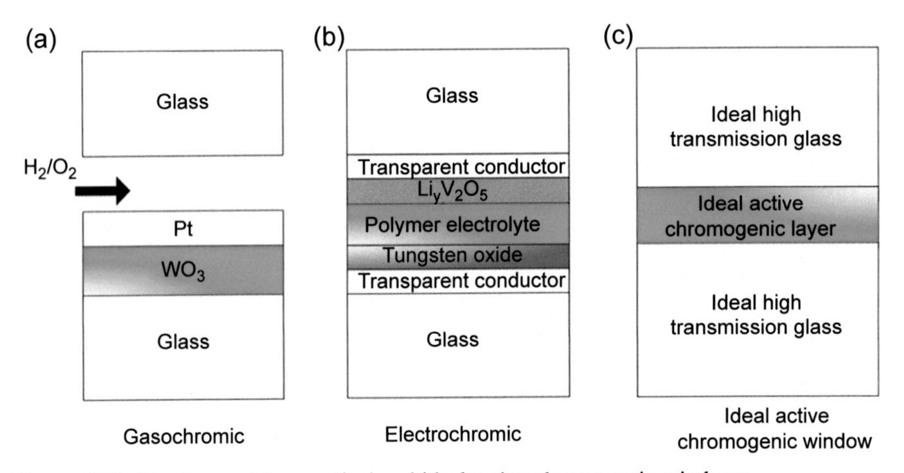

Figure 10.8 Structures of the practical and ideal active chromogenic windows.
Adapted from Ye, H., Meng, X.C., Long, L.S., Xu, B., 2013b. The route to a perfect window. Renewable Energy 55, 448–455.

Table 10.4 Radiation properties of electrochromic and gasochromic windows

Radiation property	Electrochromic windows				Gasochromic windows			
	Bleached state		Colored state		Bleached state		Colored state	
	Solar	Visible	Solar	Visible	Solar	Visible	Solar	Visible
τ	0.55	0.73	0.11	0.18	0.75	0.76	0.15	0.18
ρ	0.15	0.10	0.13	0.08	0.15	0.15	0.08	0.07
α	0.30	0.17	0.77	0.75	0.10	0.09	0.77	0.75

Adapted from Ye, H., Meng, X.C., Long, L.S., Xu, B., 2013b. The route to a perfect window. Renewable Energy 55, 448–455.

Table 10.5 Radiation characteristics of the ideal chromogenic window

Radiation property	NIR transmission state		NIR reflection state	
	Visible	NIR	Visible	NIR
τ	1	1	1	0
ρ	0	0	0	1
α	0	0	0	0

Adapted from Ye, H., Meng, X.C., Long, L.S., Xu, B., 2013b. The route to a perfect window. Renewable Energy 55, 448–455.

by regulating reflectivity. They can realize the smart shading or high transmission to solar radiation in a whole year. Fig. 10.8(c) shows the ideal chromogenic window consisting of an ideal active chromogenic layer sandwiched by two idea high transmission glasses. Table 10.5 shows the radiation characteristics of the ideal chromogenic window.

Four cities including Haerbin, Beijing, Shanghai, and Guangzhou representing four typical climates (severe cold, cold, hot summer and cold winter, and hot summer and warm winter, respectively) were selected to show the application performance. Fig. 10.9 compares the ECIs of the ideal window, electrochromic, gasochromic, and ordinary single glazing for summer and winter in each city. In the calculation, when cooling is on, the ideal and practical chromogenic windows are set in the low transmission state; when heating is on, they are set in high transmission state.

From Fig. 10.9, it is obvious that the ECI of the ideal chromogenic window is lower than those of the practical active chromogenic windows and ordinary single glazing in both summer and winter; the ECIs of the practical chromogenic windows are lower than that of ordinary single glazing in summer, but the results are contrary in winter. Thus, compared with the ordinary single glazing, the ideal chromogenic windows can reduce energy consumption in both summer and winter, but the practical chromogenic windows are efficient only in summer. With the data in Table 10.3 and Fig. 10.9, it can be seen that, when the practical chromogenic windows such as the electrochromic and gasochromic windows are in colored state, their solar radiation transmittance is lower than that of the ordinary single glazing, leading to a decrease in the heat gain and an energy-saving effect compared with the ordinary single glazing. At the same time, the visual comfort may be increased by the application of chromogenic windows because the over-illumination is also diminished. However, when the practical chromogenic windows are in bleached state in winter, their solar radiation transmittance is still lower than that of the ordinary single glazing, leading to an increase in the heating load. The solar spectrum transmittance of the gasochromic window in colored state is lower than that of the ideal chromogenic window. However, because of its high solar spectrum absorption, its ECI is still higher than that of the ideal chromogenic window. In summary, two suggestions can be made to improve the performance of the chromogenic windows: (1) to

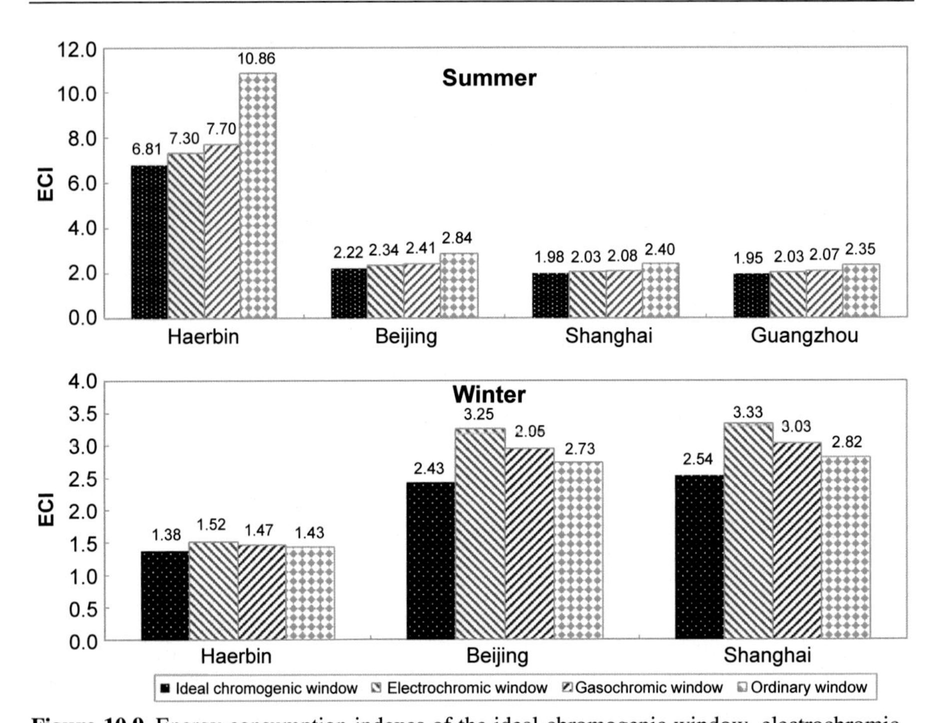

Figure 10.9 Energy consumption indexes of the ideal chromogenic window, electrochromic, gasochromic, and ordinary single glazing for summer and winter of each city.
Adapted from Ye, H., Meng, X.C., Long, L.S., Xu, B., 2013b. The route to a perfect window. Renewable Energy 55, 448–455.

increase the solar transmittance of the chromogenic windows in bleached state, and (2) to decrease the solar absorption in colored state.

Regarding the performance improvement of thermochromic windows, which are labeled as the passive chromogenic windows, the transition temperature must be equally considered because their property transition depends exclusively on temperature. The optimal transition temperature was found around 20°C. Detailed information on this issue has been elaborated in Ye's previous study (Long and Ye, 2014b).

10.4.2 Thermal transmittance

The chromogenic technologies optimize the regulation of radiation, and a double-glazed or multilayer window improves the thermal transmittance (U-value). To discuss the effect of U-value and radiation properties on energy performance, Fig. 10.10(a) depicts the energy consumptions for cooling of three types of windows. In that figure, the U-value of the ordinary single window is approximately 5.8 W/(m^2 K), whose value is obtained from the WINDOW software; the low U-value window is a double-glazing window filled with xenon with a U-value of 0.28 W/(m^2 K) (Jelle et al., 2012), and its

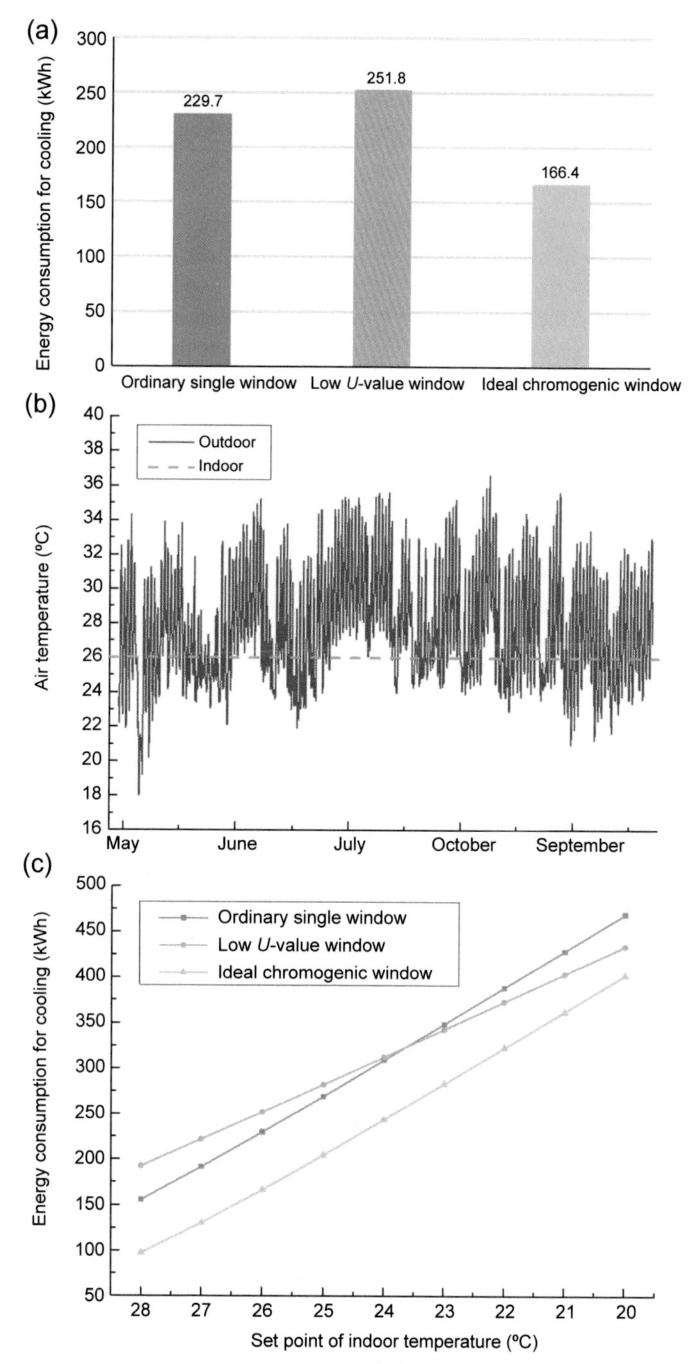

Figure 10.10 Discussions of thermal transmittance of windows. (a) Energy consumptions in rooms (with a set point of 26°C) with ordinary single window, low *U*-value window, and ideal smart window. (b) Temperature variations in Guangzhou during the cooling period. (c) Energy consumption variations along with the indoor temperature set point.
Adapted from Long, L.S., Ye, H., Zhang, H.T., Gao, Y.F., 2015. Performance demonstration and simulation of thermochromic double glazing in building applications. Solar Energy 120, 55–64.

radiation properties were assumed to be the same as those of the ordinary single window; the ideal chromogenic window is of the same *U*-value as the ordinary single window.

In Fig. 10.10(a), the room with the ordinary window requires less energy than that with the low *U*-value window, so a higher *U*-value improves the energy performance. In summer, a low *U*-value decreases the cooling load by obstructing the heat gain from the hot outdoor air to the cool indoor air. However, a low *U*-value may also increase the load by impeding the heat loss from the indoor air to the outdoor air when the outdoor air is cooler, typically at night. Fig. 10.10(b) shows that temperature of the outdoor air is generally higher than the indoor air temperature during the day but lower at night; therefore, compared with an ordinary single window, the low *U*-value window increases the cooling load at night and decreases the load during the day. If the increase is greater than the decrease, then the low *U*-value window has an overall poorer performance than the ordinary window. In other words, the set point of the indoor temperature may determine whether the heat will transfer from the outdoor air into the room, and in turn decide the performance of the low *U*-value window.

Fig. 10.10(c) shows the relationships between the energy consumptions of the windows and the set points that drop from 28 to 20°C. In the figure, the consumption increases with a decreasing set point, and the ideal smart window always displays the best performance. The room with the ordinary single window consumes less cooling energy than that of the low *U*-value window when the indoor temperature is higher than a critical point, which is between 23 and 24°C, and is nearly always lower than the outdoor air in Fig. 10.10(b). When the set point is equal to or lower than 23°C, the performance of the low *U*-value window becomes better than that of the ordinary window, signifying that a low set point provides the advantage to the low *U*-value window in summer.

In summary, a higher *U*-value will decrease the cooling energy consumption if the indoor temperature is set at a relatively low value, which is fairly common in practical applications because in a hot summer, the residents are sometimes willing to experience better thermal comfort by setting a low indoor temperature. However, the energy consumption of the room with the ideal smart window is always the lowest in both Fig. 10.10(a) and (c), which reveals that improving the regulation capacity of solar radiation is more efficient than improving the thermal insulation for a hot climate.

10.5 Conclusions and future trends

Among the four chromogenic technologies introduced in this chapter, thermochromic and electrochromic windows are currently the most promising products. The former is convenient and relatively low in cost, while the latter is more flexible and better in performance. Their efficiency was confirmed through a variety of outdoor experiments as well as numerical simulations. To approach the ideal chromogenic windows, both the optical transmittance and solar modulation ability of the current chromogenic windows should be enhanced.

In addition to the aforementioned improvements, the future chromic window can be more efficient through two aspects. The first one is to develop a selective regulation of spectral properties, for instance NIR and visible-light transmittance (Korgel, 2013; Llordes et al., 2013). The other aspect is to associate with other window technologies. The chromogenic technologies relate to only a single perspective of how a window affects the energy balance in a building (ie, thermal radiation), so a more advanced window must also contain the other perspective (ie, thermal transmittance). For example, a window with both chromogenic technologies and multilayer glazing will share the advantages of both chromogenic and multiglazing; that is, the ability to dynamically control the solar radiation into the building and block the heat from entering/leaving the building in summer/winter, resulting in a much more effective window.

References

Cupelli, D., Nicoletta, F.P., Manfredi, S., Vivacqua, M., Formoso, P., De Filpo, G., Chidichimo, G., 2009. Self-adjusting smart windows based on polymer-dispersed liquid crystals. Solar Energy Materials and Solar Cells 93, 2008–2012.

Domenici, V., Conradi, M., Remskar, M., Virsek, M., Zupancic, B., Mrzel, A., Chambers, M., Zalar, B., 2011. New composite films based on MoO_{3-x} nanowires aligned in a liquid single crystal elastomer matrix. Journal of Materials Science 46, 3639–3645.

Dussault, J.M., Gosselin, L., Galstian, T., 2012. Integration of smart windows into building design for reduction of yearly overall energy consumption and peak loads. Solar Energy 86, 3405–3416.

Gao, Y., Wang, S., Kang, L., Chen, Z., Du, J., Liu, X., Luo, H., Kanehira, M., 2012a. VO_2–Sb:SnO_2 composite thermochromic smart glass foil. Energy and Environmental Science 5, 8234.

Gao, Y., Wang, S., Luo, H., Dai, L., Cao, C., Liu, Y., Chen, Z., Kanehira, M., 2012b. Enhanced chemical stability of VO_2 nanoparticles by the formation of SiO_2/VO_2 core/shell structures and the application to transparent and flexible VO_2-based composite foils with excellent thermochromic properties for solar heat control. Energy and Environmental Science 5, 6104.

Gao, G.H., Zhang, Z.H., Wu, G.M., Jin, X.B., 2014. Engineering of coloration responses of porous WO_3 gasochromic films by ultraviolet irradiation. RSC Advances 4, 30300–30307.

Georg, A., Graf, W., Schweiger, D., Wittwer, V., Nitz, P., Wilson, H.R., 1998. Switchable glazing with a large dynamic range in total solar energy transmittance (TSET). Solar Energy 62, 215–228.

Granqvist, C.G., Lansaker, P.C., Mlyuka, N.R., Niklasson, G.A., Avendano, E., 2009. Progress in chromogenics: new results for electrochromic and thermochromic materials and devices. Solar Energy Materials and Solar Cells 93, 2032–2039.

Granqvist, C.G., Green, S., Niklasson, G.A., Mlyuka, N.R., Von Kraemer, S., Georen, P., 2010. Advances in chromogenic materials and devices. Thin Solid Films 518, 3046–3053.

Granqvist, C.G., Pehlivan, I.B., Ji, Y.X., Li, S.Y., Niklasson, G.A., 2014. Electrochromics and thermochromics for energy efficient fenestration: functionalities based on nanoparticles of In_2O_3:Sn and VO_2. Thin Solid Films 559, 2–8.

Granqvist, C.G., 2007. Transparent conductors as solar energy materials: a panoramic review. Solar Energy Materials and Solar Cells 91, 1529–1598.

Granqvist, C.G., 2012. Oxide electrochromics: an introduction to devices and materials. Solar Energy Materials and Solar Cells 99, 1–13.
Granqvist, C.G., 2014. Electrochromics for smart windows: oxide-based thin films and devices. Thin Solid Films 564, 1–38.
Gustavsen, A., 2008. State-of-the-Art Highly Insulating Window Frames—Research and Market Review. Ernest Orlando Lawrence Berkeley National Laboratory, Berkeley, CA, US.
Hoffmann, S., Lee, E.S., Clavero, C., 2014. Examination of the technical potential of near-infrared switching thermochromic windows for commercial building applications. Solar Energy Materials and Solar Cells 123, 65–80.
Huang, Z.L., Chen, S.H., Wang, B.Q., Huang, Y., Liu, N.F., Xu, J., Lai, J.J., 2011. Vanadium dioxide thin film with low phase transition temperature deposited on borosilicate glass substrate. Thin Solid Films 519, 4246–4248.
Huang, B.R., Lin, T.C., Liu, Y.M., 2015. WO_3/TiO_2 core-shell nanostructure for high performance energy-saving smart windows. Solar Energy Materials and Solar Cells 133, 32–38.
Jelle, B.P., Hynd, A., Gustavsen, A., Arasteh, D., Goudey, H., Hart, R., 2012. Fenestration of today and tomorrow: a state-of-the-art review and future research opportunities. Solar Energy Materials and Solar Cells 96, 1–28.
JGJ75–2003. Design Standard for Energy Efficiency of Residential Buildings in Hot Summer and Warm Winter Zone, Beijing.
Kamalisarvestani, M., Saidur, R., Mekhilef, S., Javadi, F., 2013. Performance, materials and coating technologies of thermochromic thin films on smart windows. Renewable and Sustainable. Energy Reviews 26, 353–364.
Korgel, B.A., 2013. Materials science: composite for smarter windows. Nature 500, 278–279.
Lee, E.S., Pang, X., Hoffmann, S., Goudey, H., Thanachareonkit, A., 2013. An empirical study of a full-scale polymer thermochromic window and its implications on material science development objectives. Solar Energy Materials and Solar Cells 116, 14–26.
Li, S.-Y., Niklasson, G.A., Granqvist, C.G., 2012. Thermochromic fenestration with VO_2-based materials: three challenges and how they can be met. Thin Solid Films 520, 3823–3828.
Llordes, A., Garcia, G., Gazquez, J., Milliron, D.J., 2013. Tunable near-infrared and visible-light transmittance in nanocrystal-in-glass composites. Nature 500, 323–326.
Long, L.S., Ye, H., 2014a. Discussion of the performance improvement of thermochromic smart glazing applied in passive buildings. Solar Energy 107, 236–244.
Long, L.S., Ye, H., 2014b. How to be smart and energy efficient: a general discussion on thermochromic windows. Scientific Reports 4, 6427. http://dx.doi.org/10.1038/srep06427.
Long, L.S., Ye, H., Zhang, H.T., Gao, Y.F., 2015. Performance demonstration and simulation of thermochromic double glazing in building applications. Solar Energy 120, 55–64.
Madida, I., Simo, A., Sone, B., Maity, A., Kana, J.K., Gibaud, A., Merad, G., Thema, F., Maaza, M., 2014. Submicronic VO_2–PVP composites coatings for smart windows applications and solar heat management. Solar Energy 107, 758–769.
Ministry of Construction of the People's Republic of China, 2003. Industry standard of the People's Republic of China.
Mlyuka, N.R., Niklasson, G.A., Granqvist, C.G., 2009. Thermochromic VO_2-based multilayer films with enhanced luminous transmittance and solar modulation. Physica Status Solidi (A) 206, 2155–2160.
Morin, F.J., 1959. Oxides which show a metal-to-insulator transition at the Neel temperature. Physical Review Letters 3, 34–36.

Nagai, J., Mcmeeking, G.D., Saitoh, Y., 1999. Durability of electrochromic glazing. Solar Energy Materials and Solar Cells 56, 309–319.

Nitz, P., Hartwig, H., 2005. Solar control with thermotropic layers. Solar Energy 79, 573–582.

Ohko, Y., Tatsuma, T., Fujii, T., Naoi, K., Niwa, C., Kubota, Y., Fujishima, A., 2003. Multi-colour photochromism of TiO_2 films loaded with silver nanoparticles. Nature Materials 2, 29–31.

Qu, J.L., Wang, Y.T., Xie, L., Zheng, J., Liu, Y., Li, X.G., 2009. Superior hydrogen absorption and desorption behavior of Mg thin films. Journal of Power Sources 186, 515–520.

Runnerstrom, E.L., Llordes, A., Lounis, S.D., Milliron, D.J., 2014. Nanostructured electro-chromic smart windows: traditional materials and NIR-selective plasmonic nanocrystals. Chemical Communications 50, 10555–10572.

Saeli, M., Binions, R., Piccirillo, C., Parkin, I.P., 2009. Templated growth of smart coatings: hybrid chemical vapour deposition of vanadyl acetylacetonate with tetraoctyl ammonium bromide. Applied Surface Science 255, 7291–7295.

Saeli, M., Piccirillo, C., Parkin, I.P., Binions, R., Ridley, I., 2010a. Energy modelling studies of thermochromic glazing. Energy and Buildings 42, 1666–1673.

Saeli, M., Piccirillo, C., Parkin, I.P., Ridley, I., Binions, R., 2010b. Nano-composite thermo-chromic thin films and their application in energy-efficient glazing. Solar Energy Materials and Solar Cells 94, 141–151.

Saeli, M., Piccirillo, C., Warwick, M., Binions, R., 2013. Thermochromic thin films: synthesis, properties and energy consumption modelling. In: Mendez-Vilas, A. (Ed.), Materials and Processes for Energy: Communicating Current Research and Technological Developments. Formatex Research Center.

Seeboth, A., Ruhmann, R., Muhling, O., 2010. Thermotropic and thermochromic polymer based materials for adaptive solar control. Materials 3, 5143–5168.

Tan, X., Yao, T., Long, R., Sun, Z., Feng, Y., Cheng, H., Yuan, X., Zhang, W., Liu, Q., Wu, C., Xie, Y., Wei, S., 2012. Unraveling metal-insulator transition mechanism of VO_2 triggered by tungsten doping. Scientific Reports-UK 2, 466–471.

Van Den Bergh, S., Hart, R., Jelle, B.P., Gustavsen, A., 2012. Window spacers and edge seals in insulating glass units: a state-of-the-art review and future perspectives. Energy and Buildings 58, 263–280.

Vergaz, R., Sanchez-Pena, J.M., Barrios, D., Vazquez, C., Contreras-Lallana, P., 2008. Modelling and electro-optical testing of suspended particle devices. Solar Energy Materials and Solar Cells 92, 1483–1487.

Vuk, A.S., Kozelj, M., Orel, B., 2014. Comparison of electrochromic devices with V- and Sn/Mo-oxide counter electrodes and (3-glycidoxypropyl)trimethoxysilane-based ormolytes with three different lithium salts. Solar Energy Materials and Solar Cells 128, 166–177.

Warwick, M.E.A., Binions, R., 2014. Advances in thermochromic vanadium dioxide films. Journal of Materials Chemistry A 2, 3275–3292.

Warwick, M.E., Ridley, I., Binions, R., 2013. The effect of transition hysteresis width in thermochromic glazing systems. Open Journal of Energy Efficiency 2, 75–88.

Wen, R.-T., Granqvist, C.G., Niklasson, G.A., 2014. Cyclic voltammetry on sputter-deposited films of electrochromic Ni oxide: power-law decay of the charge density exchange. Applied Physics Letters 105, 163502.

Xu, X.Q., Shen, H., Zhuang, L., Hu, Y.F., 2001. Gasochromic Effect of WO_3–SiO_2 Films Prepared by Sol-gel Process.

Xu, X.J., Wu, X., Zhao, C., Wang, J.X., Ge, X.T., 2012. Simulation and improvement of energy consumption on intelligent glasses in typical cities of China. Science China-Technological Sciences 55, 1999–2005.

Yang, C.Q., Zhu, Q., Zhang, S.P., Zou, Z.J., Tian, K., Xie, C.S., 2014. A comparative study of microstructures on the photoelectric properties of tungsten trioxide films with plate-like arrays. Applied Surface Science 297, 116–124.

Yao, J.N., Yang, Y.A., Loo, B.H., 1998. Enhancement of photochromism and electrochromism in MoO_3/Au and MoO_3/Pt thin films. Journal of Physical Chemistry B 102, 1856–1860.

Yao, T., Liu, L., Xiao, C., Zhang, X., Liu, Q., Wei, S., Xie, Y., 2013. Ultrathin nanosheets of half-metallic monoclinic vanadium dioxide with a thermally induced phase transition. Angewandte Chemie, International Edition in English 52, 7554–7558.

Ye, H., Meng, X.C., Xu, B., 2012. Theoretical discussions of perfect window, ideal near infrared solar spectrum regulating window and current thermochromic window. Energy and Buildings 49, 164–172.

Ye, H., Long, L.S., Zhang, H.T., Xu, B., Gao, Y.F., Kang, L.T., Chen, Z., 2013a. The demonstration and simulation of the application performance of the vanadium dioxide single glazing. Solar Energy Materials and Solar Cells 117, 168–173.

Ye, H., Meng, X.C., Long, L.S., Xu, B., 2013b. The route to a perfect window. Renewable Energy 55, 448–455.

Ye, H., Long, L.S., Zhang, H.T., Gao, Y.F., 2014. The energy saving index and the performance evaluation of thermochromic windows in passive buildings. Renewable Energy 66, 215–221.

Zhang, H., Xiao, X., Lu, X., Chai, G., Sun, Y., Zhan, Y., Xu, G., 2015. A cost-effective method to fabricate VO_2 (M) nanoparticles and films with excellent thermochromic properties. Journal of Alloys and Compounds 636, 106–112.

Façade integrated photobioreactors for building energy efficiency

S.Ş. Öncel[1], A. Köse[1], D.Ş. Öncel[2]
[1]Ege University, Izmir, Türkiye; [2]Dokuz Eylul University, Izmir, Türkiye

All of our exalted technological progress, civilization for that matter, is comparable to an axe in the hand of a pathological criminal.

Albert Einstein (Neffe, 2007)

11.1 Introduction

With the industrial age, cities, or from a different point of view, "the concentration camps of industry" became the power plant of economy and development. Today these power plants need a huge energy flow to work on. But because of the scarcity of energy sources some cities become inactive and even stand idle.

So, "what can be done?" is the primary question to answer, which is a matter of survival on this planet. From societies to individuals humanity is seeking this answer through its organized structures like governments, universities, or industries.

One answer, which this chapter will focus on, is the renewable alternatives with an inspiration from Einstein who said, "Look deep into nature and then you will understand everything better."

Microalgae, in that sense, are competitive candidates with different abilities that are important for energy, economy, and environment. With a background of billions of years of existence on the planet, microalgae played a vital role by their photosynthetic nature building up the balance in the atmosphere (Vincent, 2009). After the realization of their usage as food and feed source, industrial applications such as cosmetics and cosmeceuticals, nutritional formulations, and pharmaceuticals with regard to their valuable chemicals improved the attempts for commercialization through bigger production facilities (Spoalore et al., 2006; Becker, 2007; Borowitzka, 2013a,b; Kim, 2013; Olaizola, 2003).

With the depletion of fossil fuels a massive search on renewable energy alternatives to develop sustainable and green production technologies related to energy was started (Oncel, 2013; Chen et al., 2015). The potential of microalgae in producing biodiesel, biohydrogen, bioethanol, and biogas (Table 11.1) have made them an alternative source for energy in the 21st century (Melis et al., 2000; Chisti, 2007; Suali and

Start-Up Creation. http://dx.doi.org/10.1016/B978-0-08-100546-0.00011-X

Sarbatly, 2012; Oncel, 2013; Maity et al., 2014; Yue et al., 2014; Weber et al., 2014; Oncel et al., 2015). However the integration of microalgae to biofuel technologies do not seem to be that easy with challenging parts, limits, and bottlenecks on the metabolism, production, and commercialization; because of that microalgae began to be utilized and merged with the biorefinery concept.

The microalgae industry, with the vast usage of their products in different areas, reached a multimillion-dollar market in a few decades, started with the mimicking of natural ponds and lagoons in open systems like circular or raceways facilities (Borowitzka, 2013a,b; Oncel et al., 2015; San Pedro et al., 2015). With the extensive research, thousands of species were investigated and classified, leading to the development of special production systems that can provide the proper habitat, called photobioreactors (PBRs), which are closed systems having specific control units overseeing pH, dissolved gas, temperature, light intensity, and mixing, thus enabling the cultivation of a wider spectrum of microalgae. Through progress in engineering, different designs equipped with internal illumination sources, various types of mixers, aeration units, and baffles like tubular, flat panel, coil, helical, stirred tank, or integrated ones were developed and spread for industrial use (Tredici and Materassi, 1992; Molina Grima et al., 1999; Molina et al., 2001; Pulz, 2001; Li et al., 2015; Oncel, 2015; Ogbonna and Tanak, 2000).

Today, other than the traditional production approach, microalgae got the attention of researchers and entrepreneurs for their potential in the cities and the buildings as an aesthetic element with a target of a support for energy. Historically different applications considering the usage of plants and macroalgae as a part of construction is not new. For example, Chinese houses used dried seaweed (macroalgae) and Mediterranean houses used grape vines in the garden trellises to provide temperature control, which gave a clue for the utilization of foliage and algae integrating ecological buildings (Perez et al., 2011, 2014). For the modern applications the idea of using greenery is the merging of foliage with a vertical structure directly (Fig. 11.1) based on the attachment ability of the plant or indirectly (Fig. 11.2) by using special carriers like trellises, meshes, or strings (Ong, 2003; Köhler, 2008; Fauzi et al., 2013; Perini et al., 2013; Perez et al., 2014; Zuao and Zhao, 2015; Manson and Castro-Gomes, 2015). Keeping in mind the different terminology used in this area like vertical greenery, green façades, living walls, or biowalls mentioned in literature (Stec et al., 2005; Otteléa et al., 2011; Coma et al., 2014; Manson and Castro-Gomes, 2015; Jim, 2015), the primary aim of these systems is to gain extra space for greenery in concentrated concrete habitats and to benefit temperature control (Ip et al., 2010; Franco et al., 2012; Chen et al., 2013; Perini and Rosasco, 2013; Lianga et al., 2014; Liu et al., 2014). With the experience from these applications microalgae and PBRs are becoming new approaches for building integration.

The designs of PBR façades look fancy with a futuristic environmental friendly consideration; however, the technical background is one of the essentials and where the challenges are rooted from. A well-integrated building and PBR couple will be beneficial for energy savings with a special regard to the thermal efficiency. Acting

Table 11.1 Advantages and disadvantages of microalgal biofuels

Biofuel	Advantages	Disadvantages
Biodiesel	• High lipid production capacity (20–50% of dry weight) • Microalgal fatty acid can be utilized as aviation fuel • Noncompeting nature of microalgae with food sources • Developments in genetic engineering and metabolic engineering tools • Genetic engineering approaches enables direct synthesis of lipids to extracellular area • Short-chain lipids are promising for biodiesel conversion, which are abundant in algae • Outdoor facilities can be built-in biodiesel production purposes • Thick cellulosic wall, which provides mechanical resistance, is advantageous in outdoor conditions • Deserted areas can be utilized as outdoor microalgae production facilities • New production facilities in commercial scale offers new job opportunities especially in developing countries • Diversity of microalgae species promising for biodiesel production • Enzyme catalyst systems for conversion of fatty acids to biodiesel are promising and yield efficient • Green diesel production using hydrocracking system looks more feasible due to complex nature of algal oil • Biorefinery concept looks promising (integrating with anaerobic digestion, pyrolysis, gasification)	• Strain selection is challenging in terms of productivity (doubling time, specific growth rate), biomass and fatty acid yield and profile • Dietary fatty acids as EPA, DHA, GLA, and AA are more value-added products than biodiesel, thus nonfood fatty acid producing strain selection is required • The existing number of microalgae species for biotechnological purposes is limited • Thick cell wall is a barrier on extraction • Extraction of fatty acids requires harsh mechanic and solvent extraction techniques • Scale up of downstream processing is challenging • Harvesting, drying, and extraction are costly required steps • Change in the distribution and saturation levels of the lipid from batch to batch • Purification is a challenging step in terms of rapid saponification • Higher unsaturation levels of fatty acids and mixture of microalgal pigments • Bulk accumulation of fatty acids can be observed during stress conditions (increase in the cultivation time, reducing overall productivity) • Culture conditions should be optimized according to the microalgae strain • No known efficient photobioreactor in commercial scale • Lipid productivity and biomass productivity should be optimized • A suitable catalyst should be selected for transesterification • Glycolipids, phospholipids, and sterols are not suitable for transesterification processes

Continued

Table 11.1 **Continued**

Biofuel	Advantages	Disadvantages
	• Microalgae grow faster than land crops (harvesting can be carried out every 2–4 weeks) • Nutrient requirements are simple • Wastewater streams can be used to grow algae for biodiesel purposes	• Enzyme catalyst systems are challenging in large scales • Selection of a proper catalyst system and design is essential for hydrocracking system • Algal biodiesel is low value product compared to other valuable chemicals, thus a cost-effective understanding is vital
Biomethane	• An alternative fuel for transportation and electricity generation • Almost 50% of biogas is biomethane • Anaerobic digestion of microalgal biomass for biomethane production is promising • Waste biomass from biohydrogen, bioethanol, biodiesel industries can be utilized as substrate • Higher lipid, protein, and starch with low cellulose and no lignin is an advantage for anaerobic digestion • Digested biomass can be used as fertilizer • Biorefinery concept can be adapted with bioethanol, biohydrogen, and biodiesel production • Biomethane can directly be used for heating • Harmful algae blooms can be harvested for biomethane production • Offers the recycle of large quantity of organic compounds • Reduces the medium cost for anaerobic digestion • Once the facility is started the biogas producing reactors can be utilized for years	• Optimization of the rate of anaerobic digesters • Algal cell walls decrease the yield of biomethane generation • Microalgal productivity and downstream processing cost • Storage of biomethane is challenging • Large scale application is limited • Thus biogas is a mixture of CO_2 and CH_4, methane should be purified from CO_2; purification step requires additional equipment as scavengers • Anaerobic digestion system in large scales are required • Thus microalgal biomass quantity is low (currently) a mixture of other compounds can be added to increase the yield of the digestion; in which case optimization of digesta portions are essential • Lipid-free residues are more promising than intact cell due to thick cellulosic wall • Yield and cost-efficient technology is required • Methane portion is changeable due the degradability and composition of algal biomass • Long hydraulic retention time is required with low digestibility properties (time and energy consuming) • The economic value of biomethane is low due to low calorific value • Optimization of the bioreactor performance along with downstreaming of biogas production is time consuming

Bioethanol	• Carbon-neutral renewable liquid fuel • High amount of sugar convertible to bioethanol • Ethanol is safer gasoline additive than methyl tertiary • Butyl ether (MTBE) • Microalgae as noncompeting agricultural source prevents the conflict of food vs fuel as in agricultural compounds • Less in lignocellulosic material (lack of lignin) • Thus microalgal biomass is utilized as substrate conventional production facilities can easily be adapted • Waste algal blooms can be utilized • Remaining biomass after direct conversion can be utilized to produce biogas	• Existing of cellulose and hemicellulose fibers • Saccharification is required to increase the fermentation yield • Complex downstream processing • Direct secretion of bioethanol is still in progress • Genetic engineering techniques are required for direct secretion of ethanol via microalgae • Strict anaerobiosis is required • Direct ethanol conversion is an anaerobic bioprocess that requires starch deposits within the microalgae cells that accumulated at aerobic conditions • Low carbohydrate content of algal biomass (20% of dry weight) • Bioethanol itself is a low-value product that requires extensive processing and cost
Biohydrogen	• Biohydrogen is the cleanest energy source for transportation purposes • The combustion waste is water, which has no harm for environment • Biohydrogen can be used for household purposes as well • The energy density of biohydrogen is higher • Microalgal biohydrogen is actually a metabolic waste, which is promising for future's energy • Microalgae and cyanobacteria are unique organisms responsible from solar hydrogen conversion. • Biohydrogen production is a model for genetic engineering techniques • The hydrogen production pathway is well described • Cyanobacteria can also use nitrogenase enzymes to generate biohydrogen	• Current techniques (electrolysis, steam reforming) suffers from energy efficiency • Conventional techniques are cost required and thermodynamic losses are observed • No commercial scale is available • Biohydrogen production yield by microalgae is low (low light to fuel conversion) • Efficient hydrogen storage techniques are required • The rate of hydrogen production is changeable due to dynamic biological systems • Hydrogenase enzymes as hydrogen producing machineries within microalgae cells is highly sensitive, even to a trace amount of oxygen, or metabolic oxygen (oxygen acts as a reversible transcription inhibitors) • Strict anaerobiosis is required • Leakage of hydrogen gas in PBR system

Continued

Table 11.1 **Continued**

Biofuel	Advantages	Disadvantages
	• With optimization waste organics can also be used to produce hydrogen • Merging biohydrogen production technologies with fuel cells is a promising field for further developments	• Gas impermeable materials are required • Efficient large-scale PBR systems suited for hydrogen production is emerging • Limiting number of wild-type microalgae species capable of biohydrogen production • Two-step production (sulfur starvation) is the only known sustainable way of producing biohydrogen • Harvesting and reinoculation for anaerobiosis is required • Photosynthetic conversion efficiency is low (1–3%) • The sulfur content of the media complex is rate limiting for bio-hydrogen production in anaerobiosis • Outdoor photobioreactors are not suitable for biohydrogen production (anaerobiosis)

Figure 11.1 Traditional green wall on a façade of an Aegean summer house.
Courtesy of author.

Figure 11.2 Greenwall as an urban element near a road depicting the famous clock tower at the city of Izmir-Turkiye.
Courtesy of author.

as a shade provider, the PBRs will help to reduce the temperature of the building, supporting the air conditioning systems.

With the experiences and ideas from the construction and microalgae industry, start-up creation with regard to a focused market needs to be clarified by the entrepreneurs. Similar to the other sectors an objective investigation should be done to find the boundaries and needs of this novel market. Also the specific focus of an ecofriendly environment should be the driving force of the business plan.

Even if few examples were actualized with an experimental curiosity, merging microalgae and PBRs to buildings will have a great potential with regard to energy-sufficient buildings of the future.

At this point the aim of this chapter is to focus on the microalgae as a green volunteer in the building sector with a special emphasis on the experiences with green walls and conventional microalgae production, projecting to the future of novel solutions considering efficient buildings and start-ups.

11.2 What are microalgae?

About 3.5 billion years ago, prokaryotic microalgae started to build the atmosphere (Vincent, 2009), and about 1.5 billion years ago, endosymbiosis theory is postulated, which claims the integration of a eukaryote with a prokaryote to form another eukaryote (Gould, 2008; Parker et al., 2008; Finazzi et al., 2010; Sasso et al., 2012; Cooper and Smith, 2015). With the evolution of photosynthesis; biogeochemistry also altered and Earth evolved as a suitable place for the formation of multicellular organisms within the oxygenic atmosphere (Finazzi et al., 2010). During this period photosynthesis has been the key machinery to build a breathing habitat for all the organisms where the role of algae evolution could not be underestimated (Blakenship, 1992; Xiong et al., 2009; Giordona et al., 2005; Hohmann-Marriot and Blakenship, 2011). In this breathing atmosphere microalgae gained attention due to their growth characteristics and importance in biotechnological approaches as well as novel and futuristic design inputs.

Microalgae cells lack complex organs like roots, stems, shoots, or leaves as plants (Andersen, 2013) are not organized with a multicellular frame (colonial living forms are exceptional) such as seaweeds (macroalgae) (Parker et al., 2008). With a generally accepted definition, microalgae are microscopic, colonial, or free living organisms found in water (marine, fresh, or brackish) and soil (Becker, 2007; Sasso et al., 2012; Safi et al., 2014). Microalgae are also claimed to be responsible for the formation of reef and rock-like structures in warm and tidal areas (Andersen, 2013). However, the ecophysiological existence of microalgae is not limited to only warm areas. Microalgae can live in acidic or alkaline conditions, saline environments, deserted areas, tropical or cold waters, and also in snow cover of glaciers and sea ice channels (Zittelli et al., 2013; Guarnieri and Pienkos, 2014; Cooper and Smith, 2015). The ecosystem of algae is not even limited to water or soil environments; microalgae are also found in aeroterrestrial zones (Reissen and Houben, 2001).

Most of the eukaryotic species show coccoid-like structure of suspended cells in water where prokaryotic species are mostly organized as filaments, but there exist species as coccoid form (Belay, 2013). Microalgae, one of the largest groups of organisms, show immense morphological and functional diversity. The cellular architecture of microalgae also diversifies within the same crown, where relatively simple structures can be observed (Fig. 11.3); aesthetic architecture of individual cell morphology (Fig. 11.4) can also be clearly seen with the light microscope images.

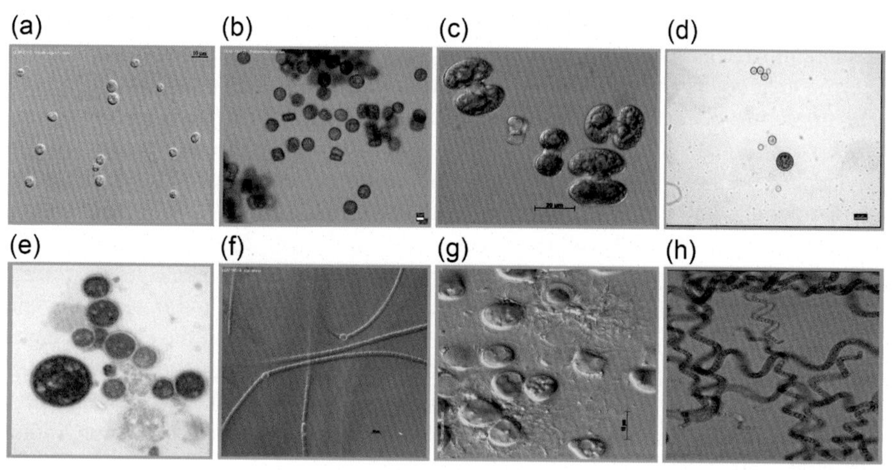

Figure 11.3 Microalgae species with simpler cellular morphology in microscopic images. (a) *Chlorella vulgaris* fo. viridis; (b) *Thalassiosira antarctica*; (c) *Cosmarium reniforme*; (d) *Golenkinia radiata* EGE-MACC; (e) *Haematococcus pluvialis* EGE-MACC in vegetative (green) and cystic form (red); (f) *Anabaena* sp.; (g) *Botrycoccus braunii*; (h) *Spirulina* sp. EGE-MACC.
Courtesy of EGE-MACC and CCAP.

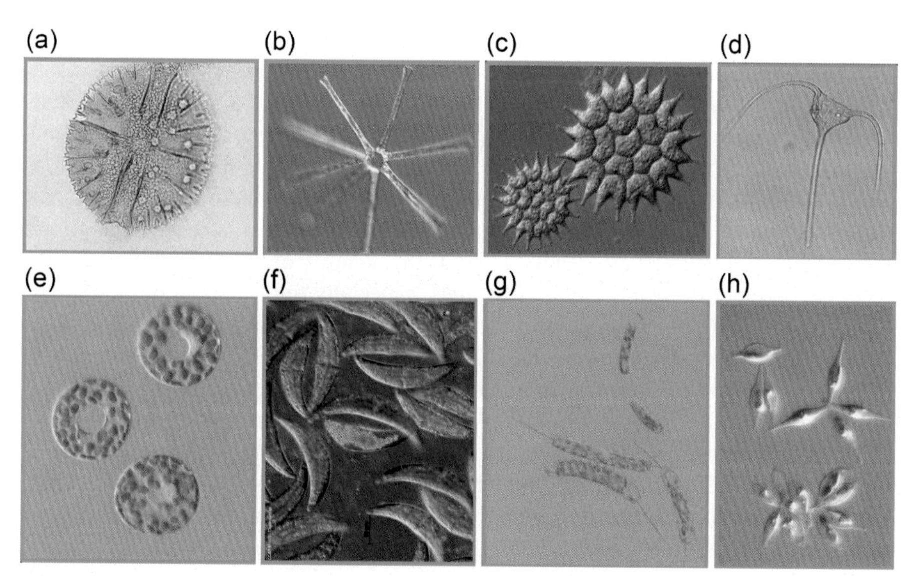

Figure 11.4 Microalgae species with complex morphology. (a) *Micrasterias* sp.; (b) *Asterionella formosa*; (c) *Pediastrum boryanum*; (d) *Ceratium horridum*; (e) *Porosira glacialis*; (f) *Closterium* sp.; (g) *Ditylum brightwellii*; (h) *Phaeodactylum tricornutum* Bohlin.
Courtesy of CCAP.

Depending on their habitats eukaryotic microalgae can develop some evolutionary adaptations such as additional cellular accessories to regulate the cell mobility to move freely in their environment or to reach to light sources (fototaxis), or adaptations in cell wall structures. Some species (eg, *Haematococcus pluvialis*, *Chlamydomonas reinhardtii*) are motile with the aid of flagella structure where the existence of flagella may depend on the culture growth stage. In some cases, under stress conditions, cells lose the flagella and become immobile (Boussiba, 2000). The eukaryotic microalgae are surrounded by a thick cellulosic cell wall (Andersen, 2013). The cell wall is composed of proteoglycan, cellulose and hemicellulose fibers, proteins, and extracellular polysaccharides (EPSs) (Morris et al., 2008). Development and thickness of the cell wall is dependent on the growth stages. Just after cell division the cell wall has a fragile structure composed of thin layers. However as cells continue to grow, the cell wall becomes thicker, which gives extra rigidity and resistance to the cells and also acts like a protecting layer against osmotic shocks and environmental stress conditions (Safi et al., 2014).

Prokaryotic microalgae on the other hand do not contain a cellulosic membrane but are surrounded by a peptidoglycan layer as in gram-negative Eubacteria (Tomaselli, 1997; Belay, 2013). In addition to the peptidoglycan layer, prokaryotic microalgae may also form EPSs. EPSs have a role on attachment of surfaces in aquatic environments and are also known to be a protection mechanism in stress conditions. The sticky structure of EPS even results in the biofilm formation on the wet rocky surfaces in natural habitats. The thickness of the EPS is changeable with the fluid movement; however, it gives a rigidity and stability to cells in biofilms (Zittelli et al., 2013).

Photosynthesis is the main energy and macromolecule source in microalgae for cell survival (Fig. 11.5). Understanding photosynthesis is essential to covering cellular mechanisms, and the overall energy balance with regard to efficiency of the photosynthesis is an important key feature (Fig. 11.6). With the aid of photosynthesis, the environmental characteristics determine the destiny of the microalgae (Barra et al., 2014). The survival and existence of microalgae are highly affected by the illumination conditions. In nature, light is the most important parameter as a growth-rate limiting factor (Molina Grima et al., 1999). The survival of a microalgae cell is dependent on the photosynthetic conversion efficiency, which is the quantity of utilized solar irradiance within the cell metabolism. In natural habitats, especially tidal areas of oceans, reefs, and deep zones the photoperiod, light distribution, spectral characteristics, daily light dose, exposure, intensity, and climate changes are major factors affecting the photosynthetic conversion efficiency (Khron-Molt et al., 2013). Most of the microalgae species are known to be obligate phototrophs. However, there are also species that are known to utilize organic carbon sources accompanied by inorganic carbon sources. When inorganic carbon is absent, microalgae cells can also metabolize organic molecules as major carbon sources; however, the molecule size determines the transport and utilization rate of the certain molecules (Perez-Garcia et al., 2011). With adaptation, microalgae can also live in heterotrophic conditions (dark) but the number of known heterotrophic species are limited (Xu et al., 2006; Perez-Garcia et al., 2011).

The eukaryotic microalgae have been evolved with organelles enveloped in a membrane system compartmentalized within the cytoplasm (Andersen, 2013). Cytoplasm

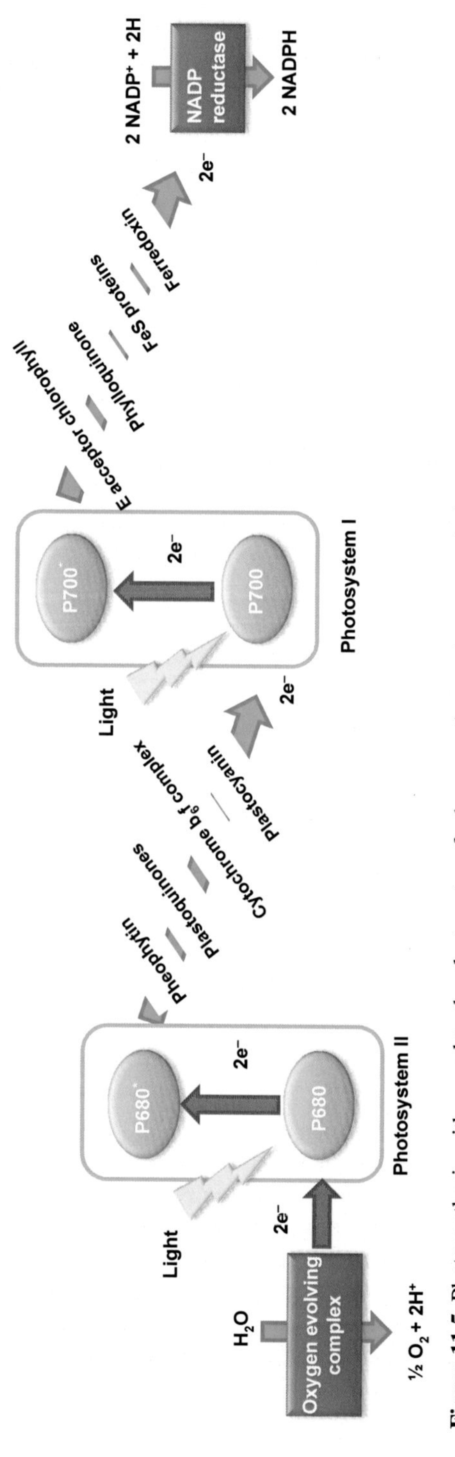

Figure 11.5 Photosynthesis with regard to the electron transfer between the two incorporated photosystems in chloroplasts.
Modified from Nelson, D.L., Cox, M.M. Lehninger, A.L., 2013. Lehninger Principles of Biochemistry, sixth ed. W.H. Freeman and Company, New York.

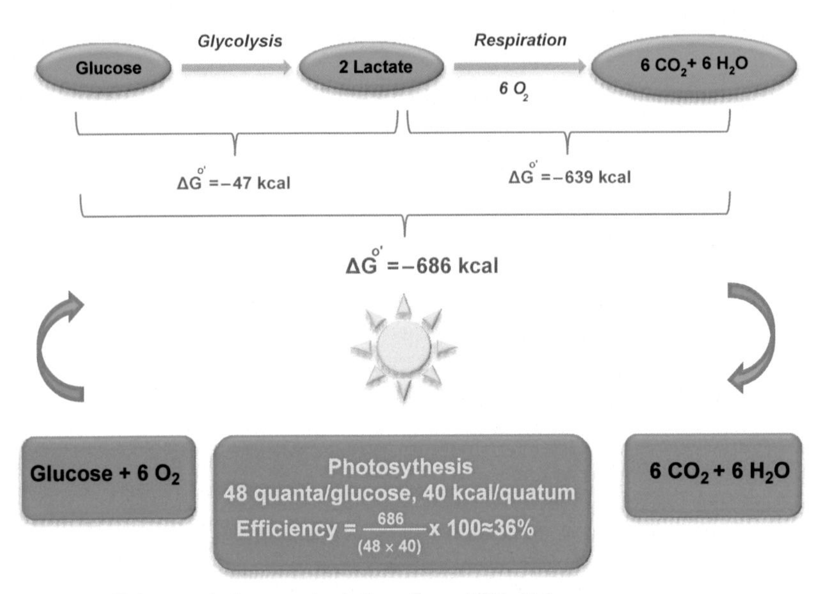

Figure 11.6 Efficiency of photosynthesis based on visible light spectrum.
Modified from Atkinson, B., Mavituna F., 1983. Biochemical Engineering and Biotechnology Handbook, The Nature Press, New York, USA, 1118 pp.

has a nucleic acid dense area, a membrane layer responsible for photosynthesis, and also macromolecular storage deposits are found in the cytoplasmic space (Tomaselli, 1997). The nucleus, mitochondria, and chloroplasts are the basic units in reproduction, regulation, energetics, and metabolic flow within individual microalgae cells (Mayfield and Franklin, 2005). Chloroplasts are the main organelles for the regulation of oxygenic photosynthesis that have specialized membranes of bilayer lipids consisting of embedded accessory pigments, proteins, and some other molecules responsible for photosynthesis called thylakoids. There are three major classes of photosynthetic pigments—chlorophylls, carotenoids, and phycobilins—each specialized on the regulation of photosynthesis (Masojidek et al., 2013).

The light for photosynthesis, which is also referred to as photosynthetically active radiation, is in the visible region (380−750 nm range). The light is absorbed by light-harvesting antennas of microalgae. After biophotolysis of water, electrons are shuttled through the thylakoid membranes. Several systematic redox reactions are accomplished along the thylakoid membrane and electrons are transported to the final electron acceptor, oxygen. The oxygen is released as a metabolic by-product and light energy is converted to macromolecules later to be utilized in cellular machinery. The regulation of photosynthesis has a similar pathway in prokaryotic microalgae. However due to the lack of organelles they do not have chloroplasts. The photosynthesis took place in specialized parallel sheets close to the cell surface named chromoplasts, and contain bacteria-like chlorophylls, chlorophyll-a (Masojidek et al., 2013). Apart from photosynthesis, there are some species with the ability of N_2 fixation.

Under stress conditions cells start to change the cell membrane composition to a more rigid and resistant form in filamentous structure called heterocysts, which preserve an oxygen-impermeable zone (Vargas et al., 1998; Han et al., 2013). Nonheterocyst-forming species can also fix N_2. Especially in nitrogen-depleted conditions, N_2 fixing prokaryotic microalgae activate specialized enzymes, nitrogenases, to convert N_2 to NH_4^+ for cells to use. These organisms have an on–off control of photosynthesis with special storage granules as cyanophycin (Sakurai, 2013). During nitrogen fixation cells can also generate hydrogen gas as a metabolic waste to balance the redox potential (Antal and Lindblad, 2005; Mathews and Wang, 2009; Maness et al., 2009; Faraloni and Torzillo, 2010), which is a promising energy source.

Under anaerobiosis the photosynthetic efficiency in eukaryotic microalgae decrease and a rapid increase in respiration rate occurs (Melis et al., 2000). Under certain photosynthetic activity (0.100) H^+ becomes the final electron acceptor because of oxygen scarcity (Melis et al., 2000; Scoma et al., 2012). Hydrogen in gas form is also released to the outer environment by eukaryotic microalgae via hydrogenase catalyzed H_2 formation process. In eukaryotic microalgae, H_2 gas is also a metabolic waste to regulated redox potential of individual cells. After introducing the aerobic conditions, hydrogenase enzymes inactivated by the oxygen and oxygenic photosynthesis is dominated within the cell (Melis and Happe, 2001; Lindblad et al., 2002; Torzillo and Seibert, 2013; Faraloni and Torzillo, 2010; Oncel and Sukan, 2011; Srirangan et al., 2011).

After photosynthesis the inorganic carbon sources are converted into organic macromolecules to be utilized for growth, reproduction, and cell maintenance. For this purpose, some of the converted molecules are stored in the cytoplasm for regulation and machinery. The main energy deposits of eukaryotic microalgae cells are starch; meanwhile prokaryotic species store glucose in glycogen form (Kruse and Hankamer, 2010; Ball et al., 2011). Under anaerobiosis, starch and glycogen sources are destructed in the respiration mechanism to provide the required energy to maintain cellular functions (Yang et al., 2000; Perez-Garcia et al., 2011).

Besides carbon storage eukaryotic microalgae cells also store lipids in granules, which are important as biofuel, pharmaceutical, and nutraceutical compounds. The lipid composition of the cell is changeable with the environmental conditions such as irradiation, pH, nutrient composition, and temperature. However, the main lipids are found in the membranes of organelles, cell walls, and cell membranes (Andersen, 2013). The lipid percentage and distribution of the cells changes with the growth stage, culture age, and species (Hu, 2013). Prokaryotic microalgae are known to accumulate up to 15–18% of lipids, but eukaryotic microalgae has a broad range of 8–50% according to the species, strain, and culture conditions (Chisti, 2007). Thus rather than prokaryotic lipids, eukaryotic microalgal lipids are of importance for biotechnological applications. The chain length of the lipid is also affected from the temperature and nutrient composition. Some species like *Chlorella*, *Nannochloropsis* sp., *Scenedesmus* sp., and *Chlorella vulgaris* have short-chain lengths fatty acids (C4–C18) (Hu et al., 2008; Kim and Hur, 2013). Eukaryotic microalgae are also an important source of omega-3 fatty acids (C22–C24) in the aquatic food web. The lipids of eukaryotic microalgae are intracellular inclusions but the long-chain hydrocarbons (up to C34) are synthesized as extracellular by certain microalgae species like *Botryococcus braunii* (Largeau et al., 1980; Banerjee et al., 2002; An et al., 2003).

Eukaryotic microalgae also have the ability to produce various types of carotenoids in primary and secondary metabolism. The primary pigments are responsible from the regulation of photosynthesis but secondary pigments are synthesized under certain environmental conditions (Yýldýrým et al., 2014). Environmental stresses, mostly irradiation and temperature and in some cases pH and salinity, trigger carotenosis, and a rapid accumulation of certain pigments is observed (Goss and Jakob, 2010; Takaichi, 2011). The regulatory role of these pigments is protecting cell and DNA damage from reactive oxygen species by antioxidant activity (Mulders et al., 2014). During pigmentation the dark green color of microalgae start to change from yellowish green to orange and reddish colors. The prokaryotic microalgae pigmentation differs from the eukaryotic microalgae. They can produce a limited number of pigments in the carotenosis chain due to the lack of responsible enzymes (Wilson et al., 2006). The main phycobiliproteins, which are covalently linked proteins with phycobilins, are abundant only in prokaryotic species. Due to the protein-containing structure, phycobiliproteins are water-soluble compounds (Patil et al., 2008). The characteristic blue color comes from the phycocyanin. They can also synthesize phycoerythrin, which gives the blackish reddish color (Becker, 2013).

Keeping the technical classification of microscopic algae as the prokaryotic (cyanobacteria) and eukaryotic (red, green, chromophyte, etc.) algae in mind (Oncel, 2013; Oncel et al., 2015), traditionally, in industry, the term microalgae covers both taxonomic groups without complicating with regard to their specific differences in subcellular structure and composition (Pulz and Gross, 2004). The roles of microalgae in a façade application designed with PBRs are crucial. Microalgae with photosynthetic properties give certain advantages for the production and continuous operation of a PBR system. Another point is that the microalgal biomass consists of value-added chemical compounds that can be evaluated as biochemicals for certain industries even at the regulation of PBR-assisted ecological buildings. Microalgae species, considering the key features like the photosynthesis, cellular structure, and valuable products, will need specific production systems technically named PBRs.

11.3 What is a photobioreactor?

Microalgae cultivation has a historical background of over 60 years of expertise starting with a single cell to a facility (Fig. 11.7). Microalgae cultivation accelerated in the early 1950s and accepted a great amount of investment later (Oncel et al., 2015). The concept of cultivating microalgae gained knowledge from core phycology studies and experiences from ancient tribes, revealing the cellular structure, morphological diversities, metabolic essentials, and divergence on the cellular composition important for biotechnology. The search on basic science of microalgal metabolism accelerated the artificial cultivation systems, which provide a reliable technology to cultivate microalgae as new model organisms in biotechnological studies. Starting with the strain isolation and understanding the natural physiology, manmade cultivation systems have been introduced to the biotechnology market (Pulz, 2001).

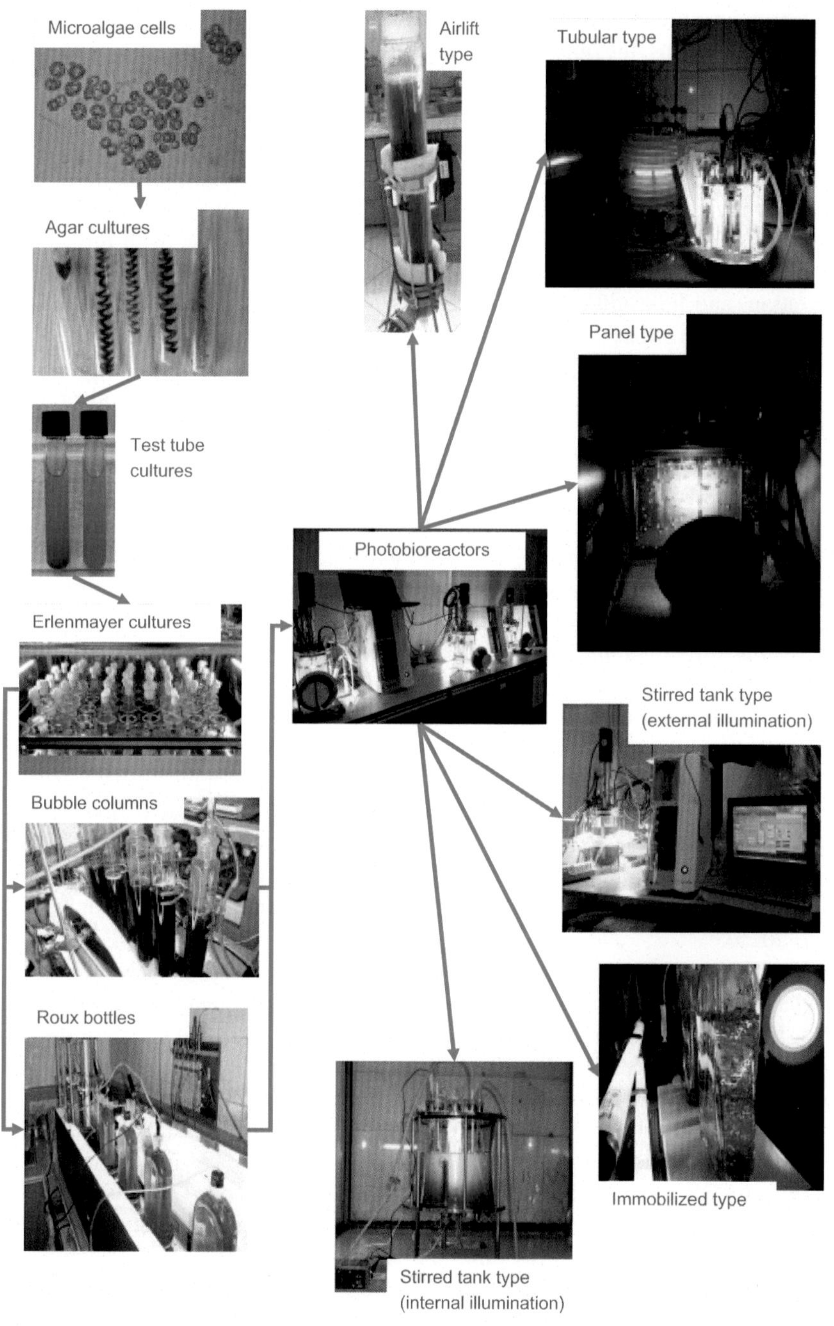

Figure 11.7 Basic steps for microalgae cultivation from single cell to a production system. Courtesy of author.

Microalgae cultivation started in natural lakes, lagoons, and artificial pond systems. As primitive designs, the open systems are conventionally used in facilities producing biomass, feed, fertilizers, and value-added chemicals (Hu and Richmond, 1996; Janssen et al., 1999; Olaizola, 2000; Borowitzka, 2013a,b). These traditional cultivation systems are beneficial with the ease in construction using cost-effective materials and other simple technical requirements (harvesting system, design materials, control units, mixing and aeration modules, etc.). Unfortunately, even if start-up is cost-effective, there are some bottlenecks on using open systems with regard to the evaporation losses, contamination risks, low product yield and quality, product standardization, and light utilization (Olivieri et al., 2014; Oncel, 2015). Thus technical challenges and drawbacks in open-system cultivation techniques ended with a clear conclusion: a technically improved microalgae cultivation system is required for promising and sustainable microalgae processing regarding commercial applications, to accept a universal production strategy that also catches the attention of new start-up creations by several aspects. Observations and evaluations on open systems highlighted new routes to cultivate different species of microalgae with a better control of the environment. With this target, closed production systems, PBRs, are emerging with sophisticated design and the introduction of technology.

PBRs are basic illuminated cultivation chambers that are partially or fully isolated from the atmosphere (Oncel, 2015). The main advantages of PBRs can be listed as (1) reducing the contamination risk of the culture, (2) reaching higher amounts of biomass, (3) enabling the cultivation of nonresistant microalgae species, (4) preventing evaporation loss of the culture volume, (5) effective sterilization for cultivation and maintenance of monoalgal and axenic cultures, (6) introducing organic nutrients to work with various nutritional regimes such as photoheterotrophic or photomixotrophic, (7) providing effective light distribution and utilization, and (8) resistance to changes in environmental conditions and better control of temperature and illumination, which are also crucial for façade applications (Molina Grima et al., 1999; Pulz and Gross, 2004; Brennan and Owende, 2010; Kunjapur and Bruce Eldigre, 2010; Oncel, 2015).

PBRs can be constructed outdoors to benefit from natural light, meanwhile protecting microalgae culture from contamination and other environmental risk factors with the specialized control units. The outdoor facilities are practical to decrease illumination cost with the usage of solar energy; however, in facilities where climatic conditions are not suitable for outdoor operations, PBRs can also be constructed for indoor operations with artificial illumination systems (Ogbonna and Tanaka, 1999; Zittelli et al., 2000a,b).

A typical PBR is constructed from (1) illumination equipment, (2) culture reservoir, and (3) control units. The integration of the three components is required to operate a functional PBR system. Providing the opportunity of batch, semicontinuous, and continuous operation modes with phototrophic, photomixotrophic, and photoheterotrophic adaptations, PBRs enable a broad spectrum of bioprocess advances (Garcia-Perez et al., 2011; Oncel et al., 2015).

The PBRs in literature for lab scale, pilot scale, and also in commercial scale show a great variety focusing on the special design elements with a target output of performance, productivity, product quality, and downstream processes (Fig. 11.8). However,

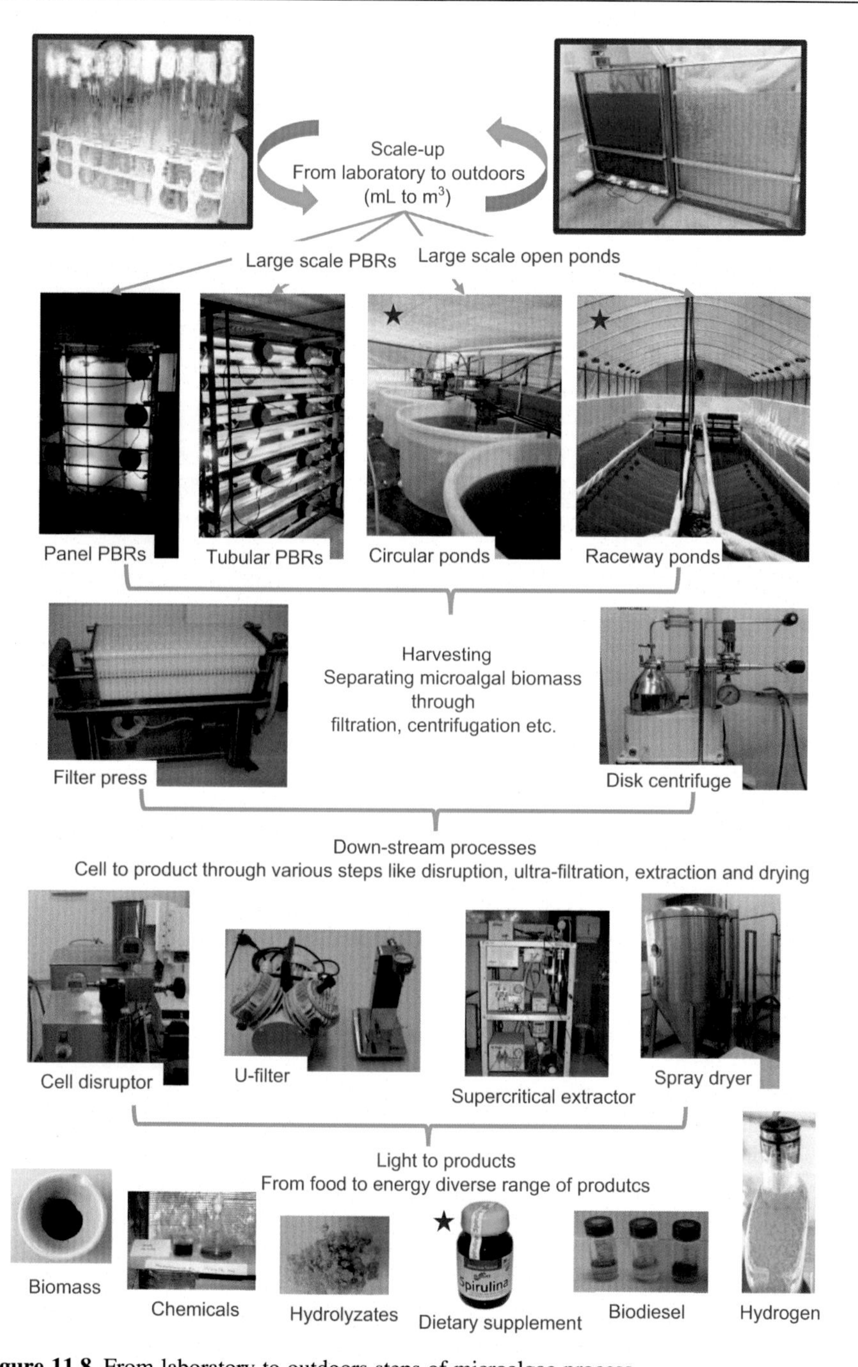

Figure 11.8 From laboratory to outdoors steps of microalgae process.
Courtesy of author. Courtesy of EGERT company, first spin-off company of Ege University and Bioengineering Department to produce microalgal products in Turkey.

existing designs are rooted from two basic PBRs known as tubular-type and panel-type PBRs. Also, traditional continuous stirred tank bioreactors (CSTRs) can also be modified as PBRs to cultivate microalgae (Borowitzka, 1999; Chen et al., 2011; Hall et al., 2003; Saeid and Chojnacka, 2015).

11.3.1 Panel-type photobioreactors

Panel-type PBRs (Fig. 11.9) are also named as flat-plate PBRs because of the wide flat surfaces derived from the same idea of solar collectors (Pulz, 2001). The compact vertical design of panel PBRs can be constructed either in indoor laboratories or outdoor production facilities. Vertical construction of the system is also a benefit for outdoor facilities because of the effective land usage.

The light pathway of the panels does not exceed 5−6 cm in laboratory conditions; however, in a larger scale the light pathway can be increased up to 10 cm (Pulz, 2001). This limit in the light path is because of the shading risk of the dense culture, which lowers the photosynthetic activity at the late stages of production. The length of the PBR is variable with the desired volume and surface area with a constant size of light penetration depth. The optimal height of the reactor does not exceed 1−2 m due to challenges in safe construction, mixing, turbulence, nutrient and gas diffusion, degassing, heat transfer, and aeration limits (Janssen et al., 2003; Kunjapur and Bruce Eldigre, 2010; Vasumathi et al., 2012).

The key design feature in panel PBR is the light utilization (photosynthetic conversion efficiency), thus high surface area-to-volume ratio is advantageous in panel-type PBR systems in terms of effective light penetration, distribution, and utilization (Pulz, 2001; Zittelli et al., 2000a,b). The overexposure of the culture to light is risky in terms of the cell loss due to higher irradiances. At higher levels of irradiation due to the light saturation limits of cellular metabolism, cell death is triggered, which is called photoinhibition (Arsalane et al., 1994; Behrenfeld et al., 1998; Masojidek et al., 2013). In diluted cultures, at the early stages of the growth, high surface area is a risk for culture loss due to photoinhibition. This is one of the operational challenges in outdoor cultures where the illumination ratio is dependent on the sun.

Naturally illuminated panel-type PBRs can be tilted by simple orienting units due to their compact design, with respect to the diurnal changes of the sun position. This tilt angle is an important design parameter in naturally illuminated cultures, where seasonal changes and climate conditions also have an impact, and helps to provide optimum conditions for light utilization and prevents the residual illumination and excessive heat (Pulz, 2001; Katsuda et al., 2006; Melis, 2007; Delevari et al., 2015). The light is absorbed from the outer surface area to the inside of the cell. Thus the outer surface is also known to be light saturated area, where the risk of photoinhibition is higher. In dense cultures, photoinhibition risk leaves in its place another risk of shading effect from the cell mass. To lower both of these risks, effective mixing and circulation of the cell within the PBR is required (Vonsak and Guy, 1992; Rodolfi et al., 2009; Belay, 2013; Giordano et al., 2005).

The mixing in traditional panel PBRs is done via aeration. The aeration unit in which multipoint spargers are preferred is placed on the bottom of the reactor. The

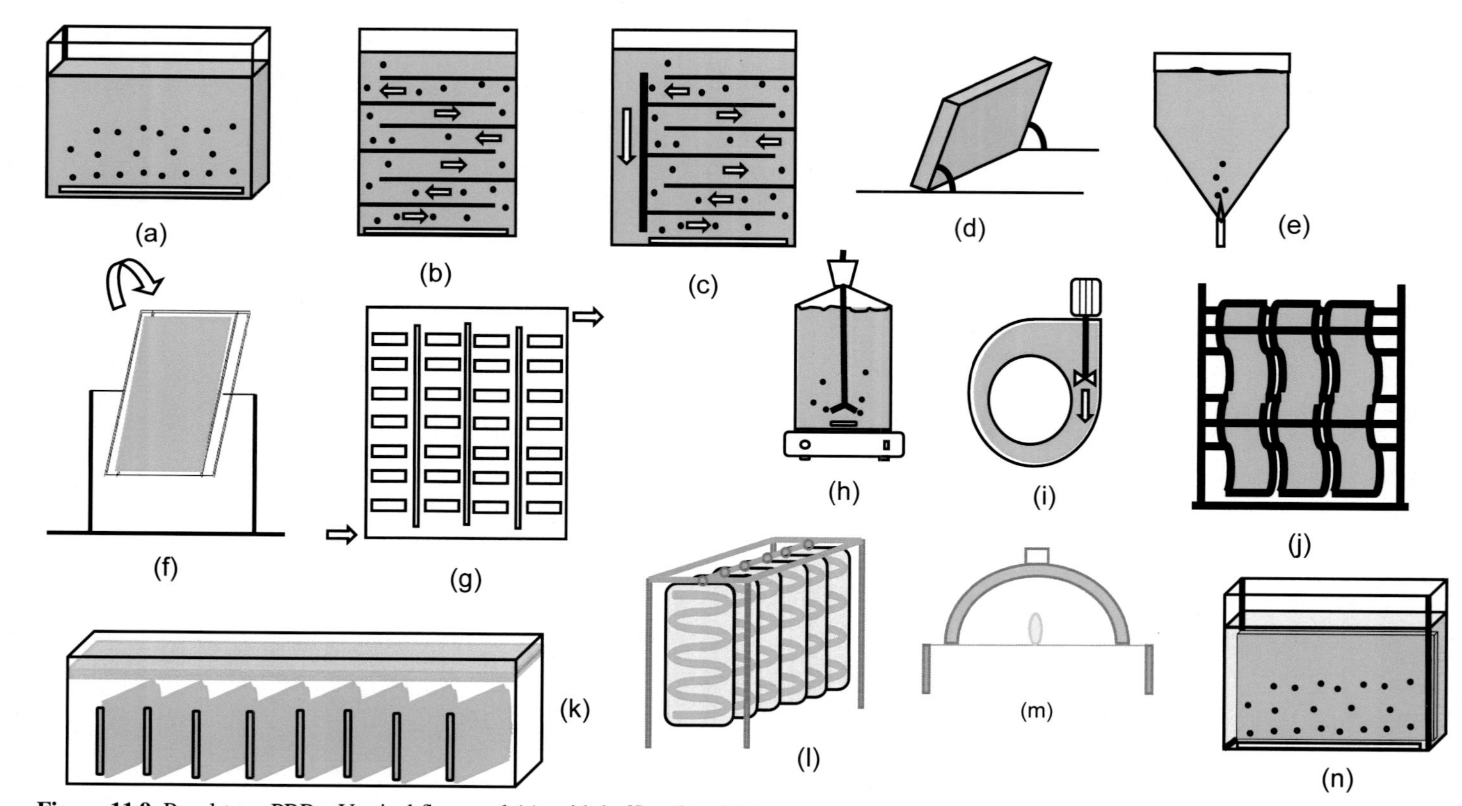

Figure 11.9 Panel-type PBRs: Vertical flat panel (a), with baffles (b), airlift type (c), inclined (d), V-shaped (e), rocking (f), integrated compartment (g), Roux-type (h), Torus type (i), accordion type (j), submerged bag type (k), with pressed tubular grid (l), dome type (m), immobilized (n) (Oncel, 2015).

uprising movement of the gas bubbles helps the mixing, gas transfer, nutrient transfer, heat transfer, as well as homogenous distribution of cells and light (Janssen et al., 2000, 2003). When pneumatically mixing conditions face aeration and transfer challenges, PBR can also be equipped with mechanical impellers. Impellers help the effective distribution of the light and also prevent the settling of the cells at high densities ($>10^8$ cells/mL) (Pulz, 2001). The gas velocity and aeration rate determine the mass transfer, which also affects gas distribution and bubble sizes.

Scale up of an individual panel PBR can be done with the increase in the surface area and culture volume with a special emphasis on the light (increase in the length and height keeping the design limits in mind). However, to increase the overall capacity, increase in the number of the PBRs with modular construction in multistacks seems to be more feasible and a preferred strategy in industrial applications.

11.3.2 Tubular-type photobioreactors

Tubular PBRs (Fig. 11.10) are special designs evolved from the combination of multiple tube systems acting as an illumination stage fed from a central reservoir of microalgae cultures. The microalgae-containing culture is pumped from the main reservoir to the individual tubes connected with U-bends, which enable the circulation of culture all along the tubes. The whole system can be drained to one or multiple reservoirs according to the scale of individual PBR (Posten, 2009). The subtypes of the tubular PBRs can be constructed as vertical, horizontal, or tilted orientation. The tubes can be lined as fences, helical, or conical according to effective light utilization (Oncel, 2015).

Tube diameter is a limiting factor on design and construction. Due to the challenges on light penetration length, tube diameters cannot exceed 10 cm, where the length of the tubes is differential according to the scale (Fernandez et al., 2013). Increase in the tube length other than construction limits also creates problems related to gas removal, culture circulation, mixing, aeration, photosynthetic activity, PBR performance, and efficiency (Oncel et al., 2015). The tubular PBRs have a smaller surface area-to-volume ratio due to the geometry of the individual tubes; however, the surface-to-volume ratio of the whole tube system can be increased with the increasement of the length (according to the limits) and tube number (Pulz, 2001; Grima et al., 2003; Chisti, 2007).

The mixing of tubular PBRs is achieved via pumps or airlift systems or in combination of both. As it is in panel PBRs, mixing determines the mass transfer ratios, gas diffusion, O_2 removal, and light utilization, and prevents settling of the cells to tube bottom (Molina et al., 2001). The design essential to the tubular PBR is management of pumping systems adequate for mixing and circulation of the culture. The identical configuration of the tubular PBRs seems to be an advantage when vertical orientation is preferred rather than horizontal by the means of O_2 removal and land utilization; however, regardless of the PBR orientation, the pumping system and required energy for fluid circulation determines the performance of the PBR (Li et al., 2014). Thus not only light utilization but also fluid dynamics is another design parameter for effective utilization of a tubular PBR.

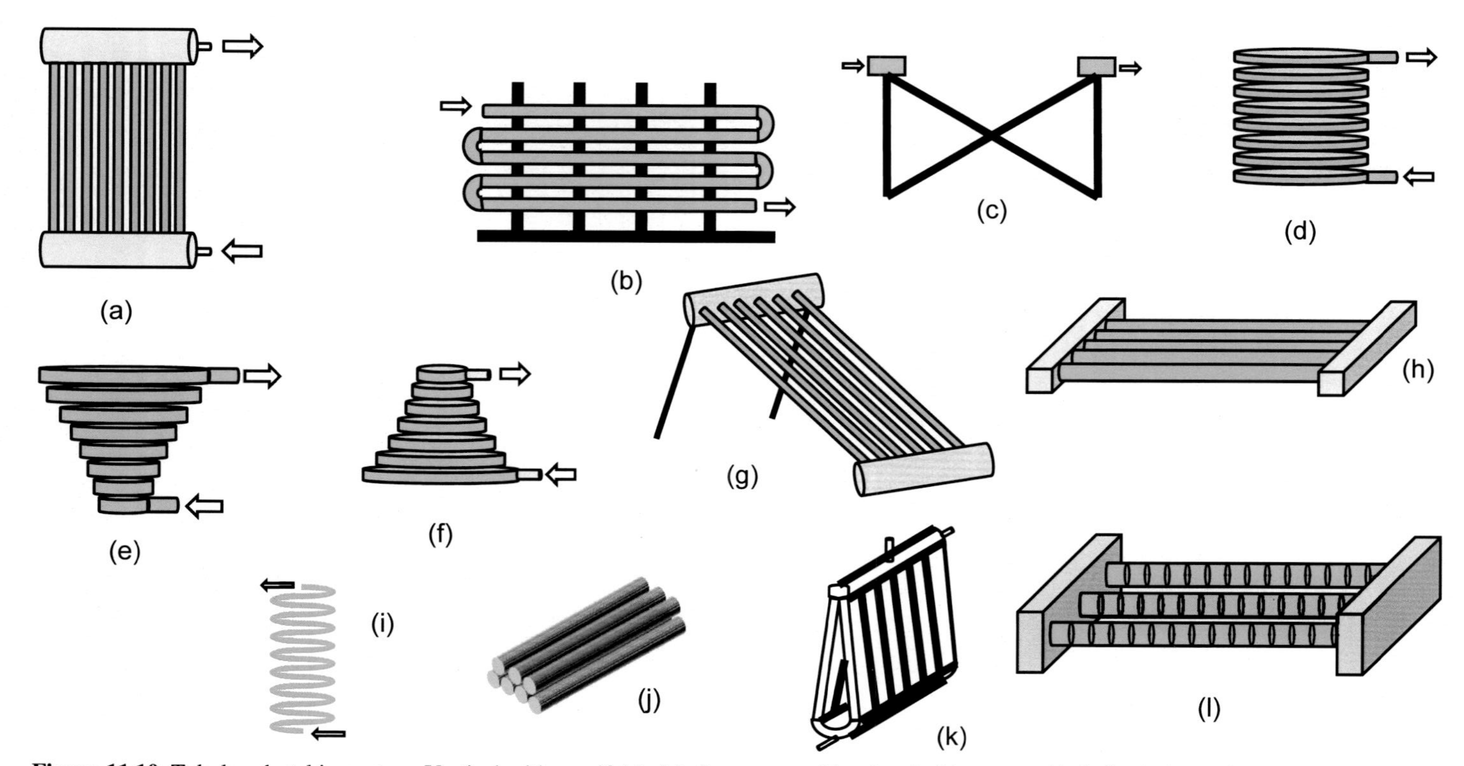

Figure 11.10 Tubular photobioreactors: Vertical with manifolds (a), fence type with u-bends (b), α-type (c), helical (d), conical (e), pyramid (f), inclined (g), horizontal (h), strongly curved (i), multi stack (j), Vertical loop like (k), with static mixers (l) (Oncel, 2015).

11.3.3 Fermenter tank photobioreactors

Tank-type PBRs, also known as fermenter types (Fig. 11.11), can be classified as stirred tanks and aerated columns. The design concept actually depends on standard bioreactors used in conventional bioprocesses usually modified with mixing and illumination systems for microalgae cultivation.

Airlift and bubble columns are vertical PBRs with no moving mechanical parts (unless modified with internal mixers) having mass transfer coefficients about 0.006 s^{-1} and low-power inputs for mixing by supplying aeration rates of 0.25 vvm (Xu et al., 2009; Posten, 2009; Dasgupta et al., 2010). To enhance the mixing efficiency, airlifts are modified with internal draft tubes (designs are called internal loop airlifts) by which the flow pattern of the culture will be steadier inside well-defined borders relative to the chaotic flow inside the bubble columns. Also by the help of the draft tube, aerated culture will rise inside the draft tube until the disengagement from air at the top of the column (Oncel, 2015). The culture will then descend in the annulus under the driving force of gravity because the degassed culture is denser than the gassed. This route is advantageous as it can harvest light more effectively in the narrow annulus between draft tube and the reactor wall (in the case of external illumination of course), it serves as an advantage (Oncel and Sukan, 2008; Oncel and Vardar-Sukan, 2009; Oncel, 2014). Also it was reported that increasing bubble size by using different spargers increases the light penetration by up to 20% while causing lower shear stress relative to the smaller bubbles (Xu et al., 2009). Column PBRs can be constructed as simple polyethylene bags usually used in aquaculture industry or in more sophisticated designs like external loop columns, divided columns, internal loop, jet loop columns with internal illumination, or mechanical mixers up to 500 L working capacity with special emphasis on light, cleaning, and construction similar to other designs (Pulz, 2001; Xu et al., 2009; Scoma et al., 2012; Tredici et al., 1992; Oncel et al., 2015).

Stirred tank bioreactors, also known as CSTRs, are conventional cultivation chambers of bioprocess applications to produce valuable chemicals from biological sources (Doran, 1995). The scale of the stirred tank reactors varies from lab to industrial. The bioreactor is fully enclosed and isolated from outer environments with sterile operation lines. The CSTRs are mechanically agitated with impellers of various designs according to the liquid characteristics of the culture broth. The agitators are responsible for heat and mass transfer, efficient aeration, and mixing, as well as sustaining the homogeneous and steady state nature of the culture. The aeration is done by multipoint circular nozzle-type spargers at the bottom of the reactor. The construction materials in classic CSTRs are glass or stainless steel.

Being the optimum reactor design in deep culture techniques with a strong experience, CSTRs also adapted to microalgae cultivation. The fundamentals of PBR design can be evaluated for microalgae cultivation in CSTR systems. CSTR-type PBRs can be used in phototrophic and also in photomixotrophic and photoheterotrophic cultivations. Because of the regulations on pharmaceutical and cosmetics, microalgal productions relevant to these industries prefer CSTRs (Tredici et al., 2009). Technically, the cylindrical shape of CSTRs have a low surface/volume ratio, which decreases the overall light harvesting capacity of the culture. However, with the efficient aeration

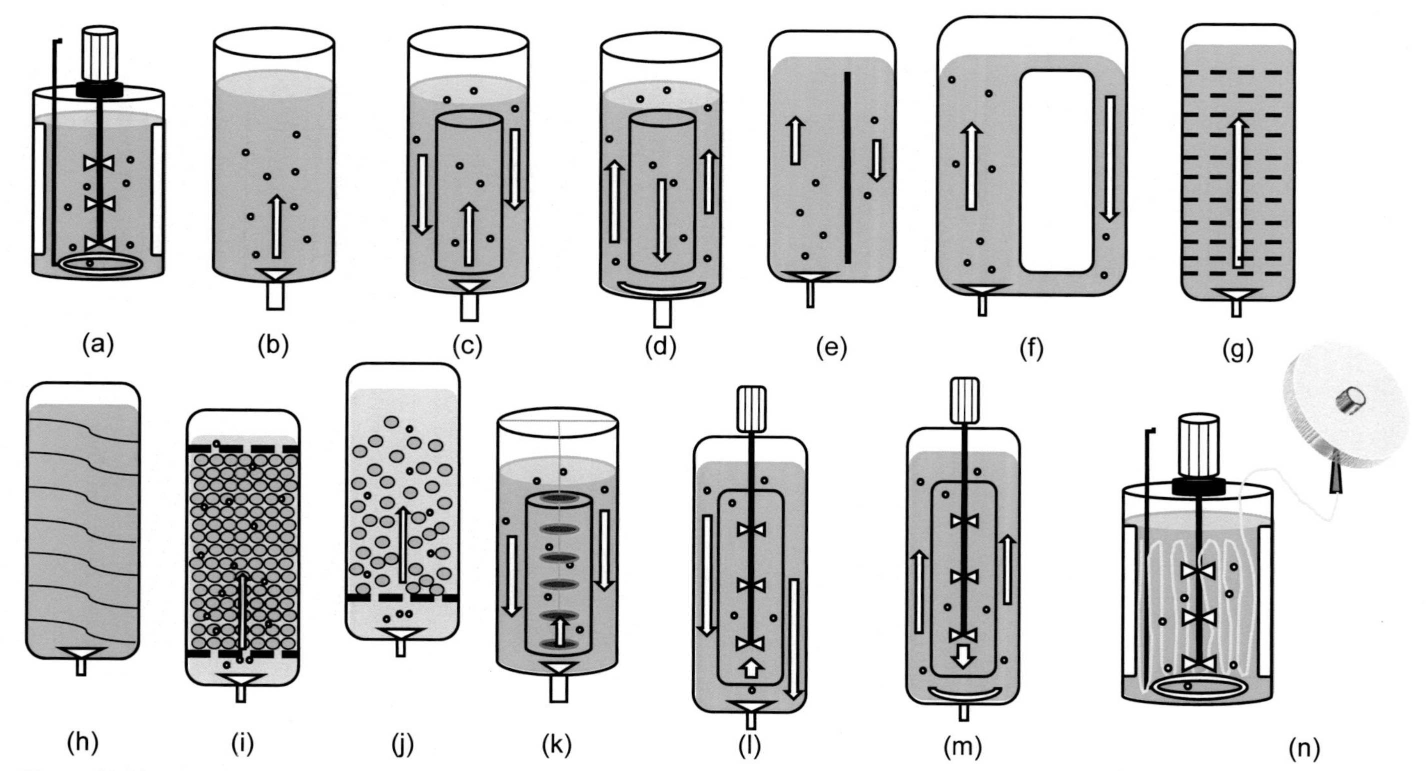

Figure 11.11 Fermentor-type photobioreactors: Stirred tank (a), Bubble column (b), internal loop draft tube sparged airlift (c), anulus sparged (d), divided column (split cylinder airlift) (e), external loop (f), perforated plate column (g), static mixer (h), packed bed (i), fluidized bed (j), draft tube-baffled airlift (k), mechanical mixer adapted draft tube sparged (l), annulus sparged (m), collector adapted stirred tank (internal illumination by optic fibers) (n) (Oncel, 2015).

dynamics derived from both gas bubbles and impeller systems, the biomass productivity of microalgal culture is high (Pulz, 2001). Light utilization is considered to be key in PBR design because CSTRs can be illuminated with different approaches like internal light sources, especially in big scales where stainless steel is the choice rather than the glass-like transparent materials that will not be preferred with regard to durability for tank construction (Oncel, 2015).

The mechanical agitation requires high energy input through an impeller system. To increase the efficiency of energy transfer to the culture baffles are preferred. Baffles protect inner circulation and fluid movements from a vortex effect; however, the settling dead zones and nonilluminating areas are potential risks to be considered during the mixing system design for microalgae cultures. The aeration is provided from the bottom of the reactor and in microalgal cultures the aeration is mostly enriched with CO_2 to provide additional inorganic carbon input (Pulz, 2001; Zittelli et al., 2013; Oncel ct al., 2015).

The main disadvantage of the fermenter-type PBR is the high capital and operational costs, thus it is not suitable for raw biomass-targeted industries like biofuel production; however, utilization of CSTR-type PBRs are advantageous if microalgal biomass and value-added chemicals are to be utilized in pharmaceutical or cosmetics as well as the food industry. As an example, astaxanthin as a natural colorant for fish and chicken has a low market price in the feed industry; however, astaxanthin used in pharmaceuticals has a price value of $30,000/kg. Considering the investment (design, construction) and operational costs (media, sterilization, downstream), tank-type PBRs are mostly preferred in the high-value chemical industry where the sterile conditions play a vital role (Bumbak et al., 2011; Stengel et al., 2011).

11.3.4 Integrated photobioreactor designs

Integrated PBRs, which are also defined as hybrid PBR systems, are a combined cultivation method of open and closed systems or closed and closed systems together. The main target of using the integrated PBRs are the utilization of the beneficial design aspects of two main cultivation strategies, thus increasing the productivity and efficiency, and also enabling the effective utilization of existing sources as nutrients, water, and light (Brennan and Owende, 2010). Another target is the reduction of cost of the overall production systems by construction of open ponds on a commercial scale (Zittelli et al., 2013). There are two strategies in the integration of the system; one is the PBRs and open ponds and the other is the PBRs with PBRs (Oncel, 2015).

In integrated systems of PBRs and open ponds, PBRs are used as the control (addition chemicals like acid, base, or air) and feeding (nutrients and especially CO_2) unit, and the open ponds are used as the illumination unit, especially to harvest solar energy in a more efficient way at outdoors. Thus attempts to preserve at least the monoalgal nature of the cultures are made with this kind of precautions. Also the chance for transferring cultures from the ponds to PBRs during night for artificial illumination will prevent biomass night loss due to respiration (Kose and Oncel, 2015). In a more technical approach the PBR and open system integration can work in phototrophic, mixotrophic, or even heterotrophic modes of culture due to its elasticity (Xu et al., 2006; Chen et al., 2013).

In a PBR–PBR integration, the strategy is to merge the advantages of different PBR types. This approach is similar to the idea in the open pond integration where each unit has its own target. A common design is the stirred tank type with a panel type or tubular type where the stirred tank acts as the control unit and the larger area panel type or tubular type as the illumination stage. The main difference with the open pond integration is because both units are PBRs the axenic culture can be protected and different illumination possibilities indoors or outdoors can be used without any contamination or product losing risks (Pulz, 2001).

Also another strategy that can be named as an integration to modify open ponds for better control is the covering or the greenhouse approach (Borowitzka, 2013a,b; Oncel et al., 2015). A covered pond targets the prevention of evaporation, dust, contamination, and animals from entering the cultures resulting in higher productivities.

One of the typical examples where the integrated systems are used is the *H. pluvialis* production at two stages. The green vegetative cells are produced at PBRs or covered open pond systems to enhance cell growth, and when cells reach the desired density the culture is transferred to the open ponds, or the cover of the open pond is removed and cells are triggered to produce carotenoids under outdoor illumination. This production strategy is one of the very basic hybrid systems but it is cost-effective and also cell growth is regulated with external coverage systems (Zittelli et al., 2013).

11.3.5 Design and scale-up parameters

Design elements, effecting parameters, and scale-up criteria of a certain PBR type is interrelated with the metabolism of a certain microalgae species for the production of a desired product (Chen et al., 2011). Apart from heterotrophic microalgae cultivation, light is the driving force for biomass productivity and product yield. There are a couple of questions to answer for an efficient PBR system that is in use. The product quality, target industry, metabolic properties of microalgae species (doubling time, specific growth rate, environmental requirements (light, salinity, pH, and temperature), photosynthetic activity, growth kinetics, nutrient requirements, etc.), desired volume, and capacity of the facility is of importance (Pulz, 2001; Rodolfi et al., 2009; Oncel et al., 2015). From this point of view there are some parameters that should be evaluated for an efficient PBR design.

11.3.5.1 Light supply and illumination strategy

Light supply and illumination strategy have a vital role in PBR design. Microalgae have a saturation point on light, thus unused light is released as excessive heat or fluorescence to the culture environment (Richmond, 2004). The released heat triggers the temperature rise and if the mixing is not efficient residual light and temperature differences may occur within the PBR. Thus the homogeneous light distribution in the PBR is of importance. The increase in the surface area-to-volume ratio positively affects the light utilization rates because light is absorbed by a wider surface rather than a narrow area. This spatial light dilution enhances growth rates and photosynthetic activity. Also the stress on the cells directed from radiation is decreased, which results in the potential of stable cultures.

An optimal PBR design enhances the penetration of required light for microalgal metabolism. The wavelength, frequency, and cellular utilization of the light are directly related to the light supply (Pulz and Scheibenbogen, 1998). Most of the commercial applications prefer to use the sun (solar radiation) as a light source due to the cost-effective production requirements. Thus control of light exposure rates in outdoor cultures will not be easy. The quantity of the light coming from the sun differs diurnally. According to the sun movement, the angle of the light beams also varies during the day. With the difference of daily radiation levels, temperature fluctuation may occur. Rising and decreasing temperature affects the solubility of CO_2 and O_2 within the PBR; an imbalance of gas distribution can be experienced (Pulz, 2001).

The exposure and utilization of the light determines the biomass productivity, CO_2 consumption, and O_2 release. In naturally illuminating cultures the microalgae shift to respiration metabolism during the night. Endogenous deposits are utilized as energy sources to continue cell survival. In this case microalgal biomass will be decreased; this is called night loss. During the night PBRs can be supported with artificial light sources to maintain photosynthesis, which requires additional input of energy.

At indoor cultures, PBRs are illuminated via artificial light sources as fluorescence lamp, LEDs, specialized light sources, solar collectors, and even spotlights are used. The artificial light can be supplied externally or internally (CSTR types). On the other hand, panel-type PBRs seem to be an advantageous design with a greater surface-to-volume ratio at outdoor cultivation.

11.3.5.2 Aeration and mixing

In PBRs, aeration and mixing cannot be separated from each other, thus the role of aeration on mixing and the effect of mixing to culture should be highlighted. Design of an optimal aeration and mixing system has two different points in microalgae cultivation, defining the efficiency of the production. One is the effects on a macroscale regarding the mass transfer in the PBR system and the other is on a microscale level regarding the maintenance of the cellular metabolism (Richmond, 1996, 2004; Oncel and Sabankay, 2012). When environmental conditions are optimum for growth, mixing will define the health status of a microalgae culture. The thick cellulosic wall gives extra rigidity and strength to microalgae cells; however, the shear stress on microalgae cells because of the air bubbles and mixing dynamics (eddies, turbulence flow) has influence on the survival rate of the microalgae cultures. Understanding the resistance limits of the selected strain may help to develop an efficient aeration and mixing system. Mixing affects the light distribution, exposure levels to radiation, and light utilization of the culture. Also mixing rate affects the transfer time and frequency between light and dark zones, which is another key factor for productivity (Oncel and Sabankay, 2012; Oncel and Kose, 2014).

On a macroscale, the aeration and mixing strategy differs with the different reactor configurations. In panel-type PBRs the length of the sparger, the number of nozzles, the nozzle diameter, and the height of the reactor affects the circulation time, kLa, gas hold-up, and bubble coalescence in the reactor (Doran, 1995). Because mixing is accomplished with aeration, mixing rate also determines the heterogeneity of the reactor. When aeration becomes inefficient PBR can also be supplemented with additional mixers. Mixers help the regulation of gas flow, heat and mass transfers, light distribution, and aeration kinetics (Pulz, 2001; Ugwu et al., 2005; Li et al., 2015). However, additional mechanical mixers also create an extra shear stress on the cells that should be kept in mind. In tubular systems the power of the pump will define the mixing and homogeneity. Static baffles can also be used to regulate the fluid flow in the PBRs in either conventional fermenter types, panel types, or even tubular types. Thus the diameter, length, and number of the baffles are effective design parameters. Air can be fed with or without CO_2. The optimized level of CO_2 in the air fed is crucial for microalgal metabolism. The CO_2 supply and the dissolvation determine the partial pressure of CO_2 and O_2 and pH dynamics of the culture broth as well as kLa.

11.3.5.3 Construction materials and reactor geometry

Construction materials of a PBR should be nontoxic and inert to microalgal metabolism. The key design parameter in PBRs is the illumination and light utilization by microalgae. Unless heterotrophic cultivation is not preferred the PBR should be made by a transparent material enabling light penetration to the culture. The absorption coefficient of the material is very important for the calculation of light absorption and light reflection. This parameter directly affects the photosynthetic efficiency and photosynthetic active radiation (Pulz, 2001; Scoma et al., 2012; Masojidek et al., 2013). At early stages of the cultivation the dilute cultures do not show shading affect; however, with the increase in the culture density shading that blocks the light utilization of the central culture will occur. With increasing density an increase in the thickness and density of the biofilm will negatively affect the cellular machinery and productivity. On the other hand, surface roughness of the material has an impact on formation of biofilms on the inner surface of the reactor. Even if effective cleaning and sterilization is done the cells stacked in the pores of the PBR material is a risk for contamination (Barclay et al., 2013). Construction material should be cheap, easy to find, easily replaceable, and suitable on larger scales. The PBRs are made of glass, PVC, and polyethylene materials.

One other design parameter related to the construction is the reactor geometry, which also varies according to the production target. The very basic reactor geometries are tubes, flat panels, and cylindrical tanks; however, with respect to light utilization, aeration, and fluid circulation, the reactor geometry can also vary under the same class of the PBRs. With design reform in PBR utilization more aesthetic designs are also introduced to the technology (Trent, 2012; http://www.marins.co.uk/).

11.3.5.4 Gas exchange and degassing

CO_2 is the sole inorganic carbon source in gas form for microalgal metabolism. In photosynthesis CO_2 is converted into organic macromolecules and O_2. As the final electron acceptor in photosynthesis, oxygen is released from the outer surface of the cell within the culture and starts to accumulate in culture liquid as dissolved form (Burlew, 1953; Vonshak et al., 1996; Pulz, 2001; Oncel et al., 2015). The gas balance of oxygen and carbon dioxide within the PBR has a vital role in the regulation of cell mechanism because photorespiration machinery of the Calvin cycle uses CO_2 but not O_2 (Perez-Garcia et al., 2011). When the ratio of carbon dioxide as a driving gas for photosynthesis to oxygen starts to decrease, the cellular machinery will face the toxic effects of the oxygen within the PBR. The accumulation of O_2 under irradiation triggers the conversion of O_2 to reactive oxygen species, which acts as a metabolic inhibitors, and results in the cell damage at the DNA level. Cell survival can be accomplished via synthesis of specified enzymes and pigments; however, under elevated levels of O_2 cells cannot survive more than 2–4 h (Pulz, 2001). Excess O_2 released from the cell is diffused to liquid and transported into the gas bubbles. With the vertical movement of the bubbles, excess O_2 is transported to the outer atmosphere. However, in tubular systems excess O_2 circulates along with the individual tubes with an increasing density. Increase in the turbulence within the reactor can be an efficient method to rapidly remove excess O_2; however, the more sustainable way is designing special O_2 purging units named degassers. Rapid accumulation of O_2 in tubes is highly toxic to cells, thus unless efficient degassing is provided, cell loss due to O_2 toxicity is the main problem in tubular PBRs (Wongluang et al., 2013). In panel PBRs degassing can be done simultaneously with aeration; also mechanical impellers may be beneficial for the rapid removal of excess gas. In short tube lengths continuous feeding of the culture to the main tank can be effective for rapid gas removal. In longer tubes, a gas–liquid separator on the top of the reactor prevents the circulation of gas bubbles within the horizontal tubes. The circulation time of the fluid determines the residence time of gas bubbles rich with O_2; thus the pumping capacity, liquid velocity, and length of the degasser are main design parameters in tubular PBRs (Pulz, 2001; Tredici, 2004; Kunjapur and Bruce Eldigre, 2010).

11.3.5.5 Control elements

Controlling the environmental conditions in a PBR system is vital for high performance. In outdoor environments microalgae culture is affected by the daily and seasonal fluctuations in light intensities and temperature. Changes in pH, dissolved O_2 and CO_2 ratio, and light distribution are internal factors responsible for the survival of cultures (Tredici, 2004, 2010; Hindersin et al., 2014). Thus minimizing the negative effects of environmental conditions and balancing the cultivation conditions is important for a better productivity and reliability of a certain PBR system. For this purpose introducing control systems to PBR design will fit the benefits of the survival rate and also productivity and quality of the culture. Keeping in mind the other parameters of process, light and temperature will need extra attention, especially for outdoors. Because the irradiance level from sun cannot be manipulated, the negative effect of the sun is controlled via controlling other parameters. Solar trackers with auto control units can be utilized for adjustment of the tilt angle (Masojidek et al., 2009; Hindersin et al., 2014). In artificially illuminated cultures the light–dark cycles can be controlled with an on/off switch determined according to illumination time. A turbidity meter can be coupled with an artificial illumination system where increasing the levels of light supply with the increasing cell density can be achieved simultaneously. Thus shading effect and inadequate illumination can be prevented. With a temperature control system the culture can be maintained at a constant temperature, which enables the optimum conditions (Borowitzka, 2013a,b; Zittelli et al., 2013; Ming et al., 2014). The temperature can be kept constant via cooling fingers and jackets, and at outdoor conditions sometimes PBRs can be utilized as submerged systems in pools. Also sprinklers are utilized to cool down the PBR volume (Oncel et al., 2015).

11.4 Potential role of photobioreactor systems in building

Thinking of the green wall experience in the construction sector, other than aesthetic progress of the buildings, this can be an environment- and energy-friendly approach with a direct influence on the urban area and population. With regard to urban wild life, green walls increase bioactivities related to animal and insect populations by providing an opportunity for denser greenery, which fades away in the cities or locked in between concentrated areas of parks and gardens (Tzoulas et al., 2007; Madrea et al., 2015). Also they act like photosynthetic biofilters over the buildings, decreasing particle accumulation, which has a direct effect on the local air quality and in a similar way acts like a bioinsulator to decrease noise levels (Feng and Hewage, 2014; Susorova et al., 2014; Azkorra et al., 2015; Guillaume et al., 2015). On the other hand, the green wall applications have a local effect on the urban climate and act as a heat island preventer with the ability of shading decreasing wall temperatures, which directly affects the energy consumption for acclimatization of the buildings (Alexandria and Jones, 2008; Cheng et al., 2010; Sunakorna and Yimprayoona, 2011; Hunter et al., 2014; Safikhani et al., 2014; Dijedjiga et al., 2015; Kontoleon and Eumorfopoulou, 2010; Shewaka and Mohamed, 2012).

These are some critical facts about green walls and can be a starting point for the PBRs. PBRs are a key issue to discuss and clarify their potential in building applications. The main advantage of the microalgae in PBRs, which may be the main difference from the green walls, is that microalgae in the PBRs can be cultivated in a very short period of time and the density of the culture will be homogeneous inside the PBR, whereas the plants need a longer lag phase to spread over the wall and need extra attention and management to be homogeneously distributed over the surface. Also the microalgae species can be cultivated all year, resisting seasonal effects due to the controlled environment inside the PBR, which is superior to season-dependent green walls and maintenance. Technically the success of merging microalgae and the PBR system with the building can be evaluated with the impact on the environment, thermal comfort, air quality, light, and acoustics.

Starting with the environment, PBR systems and microalgae may be a part of urban life similar to the plants of the green walls and novel solutions and designs show the potential of PBR usage as a part of urban life (Figs. 11.12–11.14). But

Figure 11.12 Open photo bioreactor: urban fountain.
Courtesy of R. Crevera, Cervera and Pioz Architects.

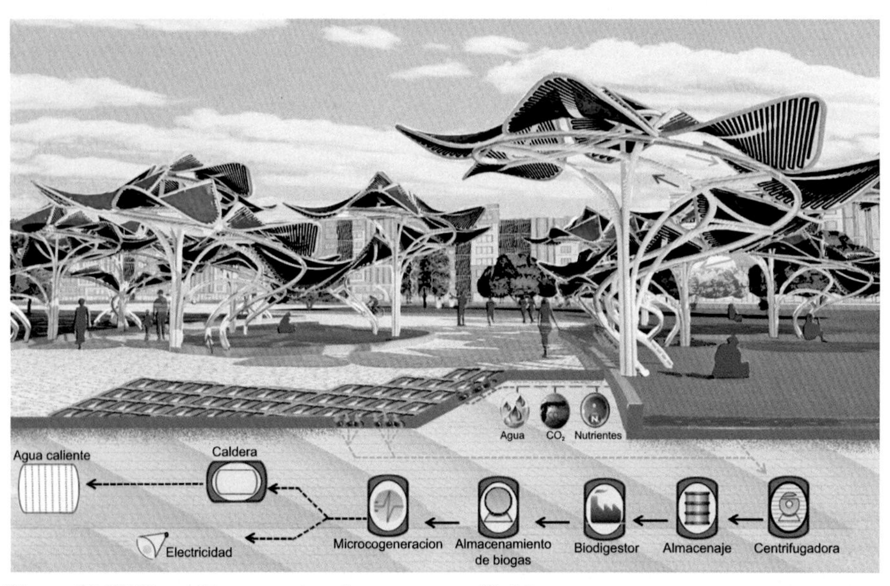

Figure 11.13 Photo bioreactor in urban spaces: artificial trees.
Courtesy of R. Cervera, Cervera and Pioz Architects.

because microalgae will be isolated in PBRs they may not have a direct influence on the environment with regard to diversity of animals and insects like plants. On the other hand this may be an advantage according to some residents, who can be reluctant to have unwanted neighbors like insects on their buildings.

Similar to the green walls, microalgae PBRs also act as a filtration system in a more indirect way. Microalgae cultured in PBRs do not have a direct contact with

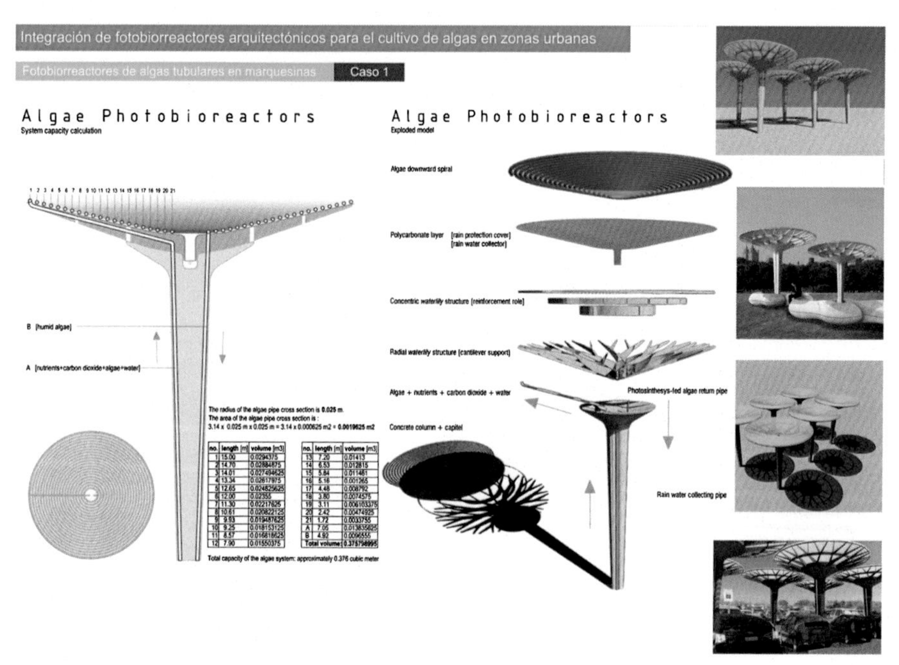

Figure 11.14 Parking canopies generating a closed circle photobioreactor. Courtesy R. Cervera, Cervera and Pioz Architects.

the outdoor environment like plants of green walls. They will act as a filter aid when integrated by an air supply system. Air is an important tool for microalgae, which is used as a mixing agent as well as for transfer of nutrients with its CO_2 content. In industrial scale microalgae productions, CO_2 is enriched in the air feed to enhance the productivity, but in a building this can be applied by indoor air, which is naturally CO_2-enriched from respiration of the human population inside. This way air supplied inside the PBRs will flow in a biofilter and with the help of photosynthetic microalgae, air quality will be enhanced. A simulation presented by Ryong and Hoon (2014) concluded a decrease of 13% in CO_2 level when they compared a normal office building with 200 employees to an algae façade; one annually is a good sign of focus.

The thermal comfort by PBRs can be a strong point of these systems. Similar to the ease of control that helps to target the air quality, thermal comfort can be accomplished by the compact design of PBRs. Indoor and outdoor heat differences can be exchanged, which will act as a bioheat exchanger (Figs. 11.14 and 11.15). With this aspect, the cooling of a PBR system can be accomplished via transporting the warmer culture to a cooler one and vice versa. This ability of culture transfer will also serve to prevent problems related to high temperatures or light intensities, which limit the usage of green walls all over the building. In high-temperature or light-intensity periods of time, like during summer or noon, microalgae culture facing the high temperatures

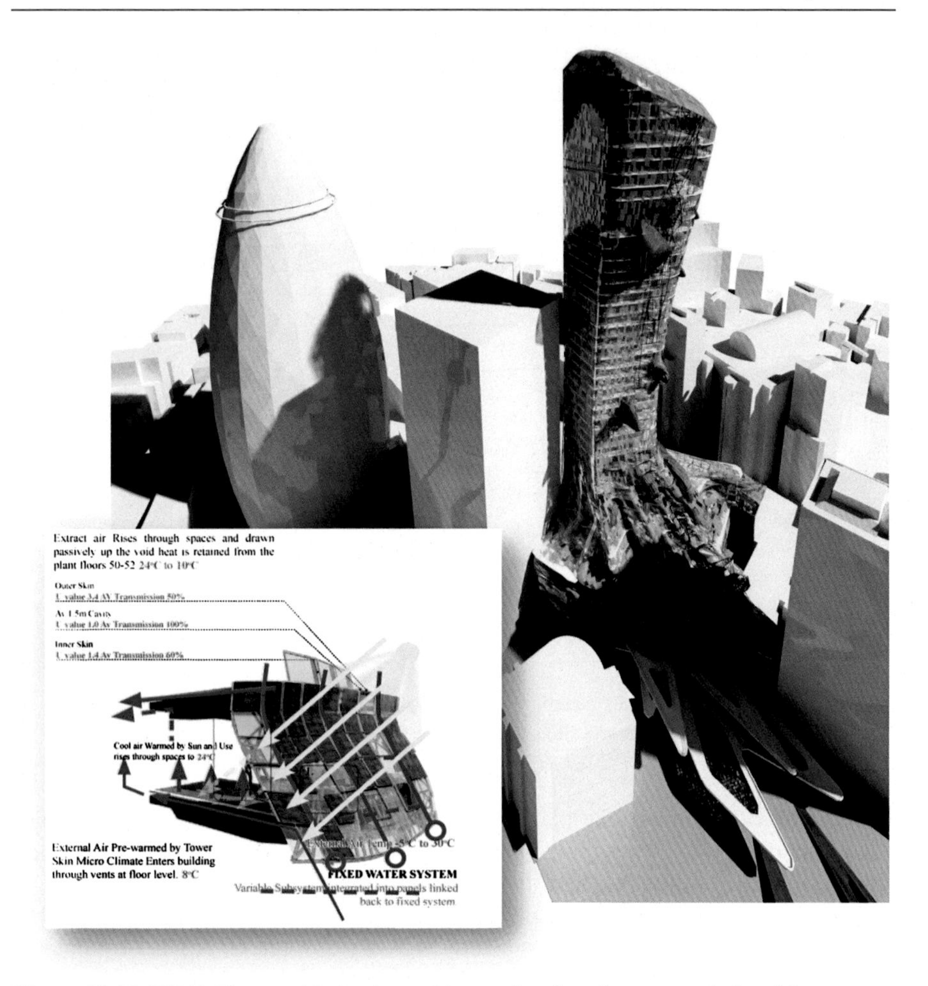

Figure 11.15 FSMA Tower with the thermal interaction focusing on panel photobioreactors. Courtesy of D. Edwards, Dave Edwards Design Ltd.

or light intensities can be transferred or pumped around the building façades to prevent the risk of productivity loss. In a similar way this ability can be used like a security valve during low-temperature or light-intensity times, like during winter or afternoon, by transferring the culture through illuminated façades. On the other hand, like the Trombe wall, PBRs will act as a heat control unit where absorbed thermal energy can be manipulated by changing the PBR's position on the façade (Fig. 11.16) depending on the outdoor temperature and irradiation (Cheng et al., 2010; Susorova et al., 2013; Kim et al., 2014). The air flow between the façade and PBR will play a key role like in the indirect green façade applications, but can be adjusted more easily because of the compact PBR design compared to the high weight plant carrying frames (Fig. 11.17). Again according to the simulation of Ryong and Hoon (2014), PBR

Figure 11.16 Photobioreactors "SolarLeaf" mounted on the façade of the Algae House building with an adjustable orientation to track the sun movement.
Courtesy of M. Kerner, SSC-Germany.

façades with regard to temperature effect can save more than 33% with regard to fuel usage and 10% with regard to electricity usage in a building (Susurova et al., 2013).

Light as a major factor on the microalgal growth should have a special emphasis on façade PBR systems. Similar to the green wall experience where differences of 80,000 lux was reported compared to outside illuminance during mid-summer (Perez et al., 2011), PBR systems will also act as a shading element, which prevents heat accumulation on the building. Especially with dense cultures PBR systems will lower light penetration, which will have a direct effect on the building heat absorption if mounted on the façade. But different from the green walls, PBR systems can be directly integrated as a window or mounted easily over the windows like a bio-curtain by which with the changing density of the culture the light penetration can be adjusted (Figs. 11.18 and 11.19). Also the compact design of the PBR systems can provide the change of orientation on the façade according to the light direction, which will improve the light interaction between the inner and outer space of the building.

With regard to acoustics, the sound insulation may be effective similar to an extra window over the façade. According to some studies using green walls as acoustic insulators showed the beneficial effects depending on parameters like the sound frequency, thickness of the foliage, and construction materials. The sound absorption by the green surface around 60 dB proved better results; the dominated range in the cities (Wong et al., 2010; Guillaume et al., 2015) will also encourage similar experiments for a newer concept of PBR façades.

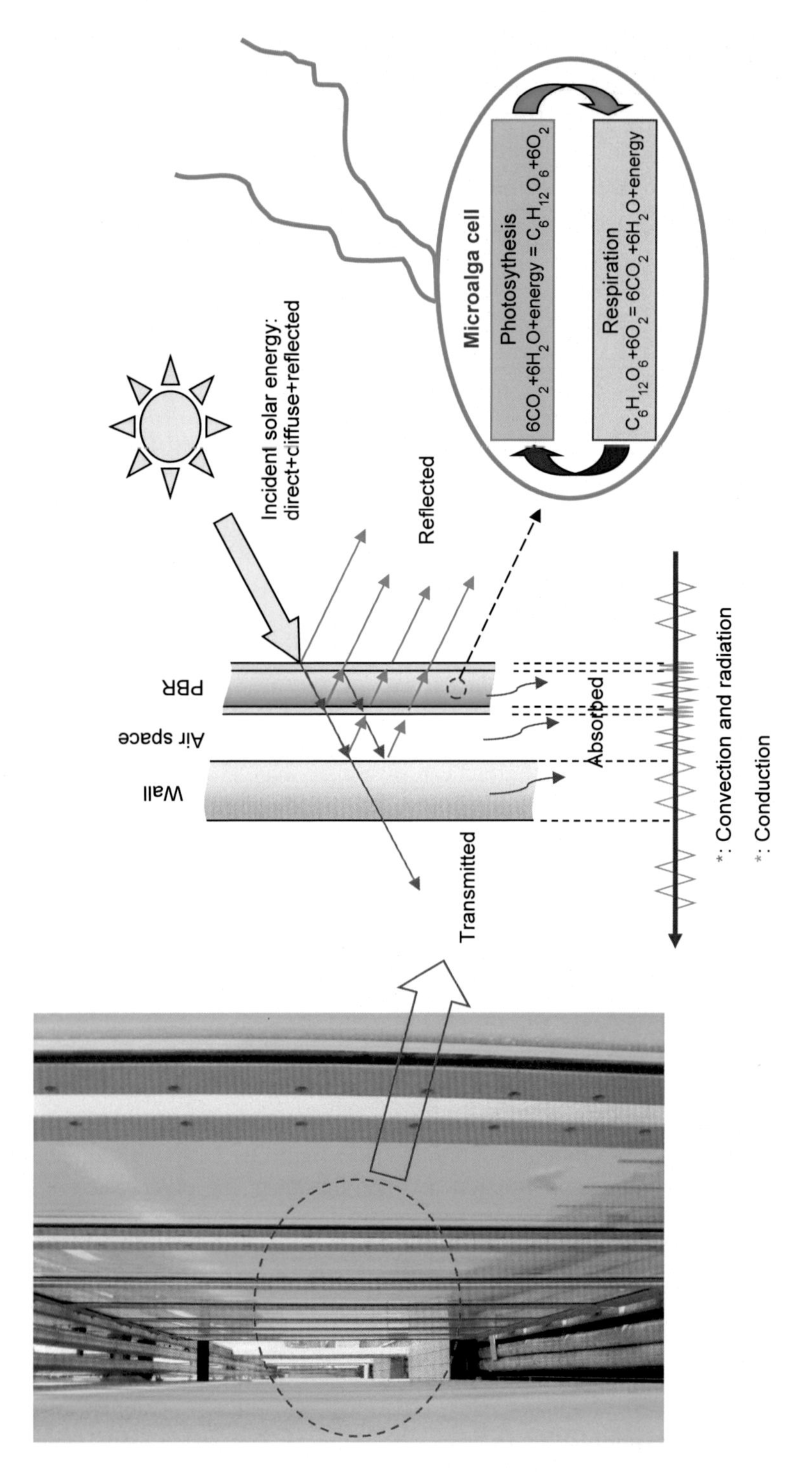

Figure 11.17 The energy transfer from the sun to the cell with a special emphasis on PBR façade, explained on the photograph from the Algae House. Courtesy of M. Kerner, SSC-Germany.

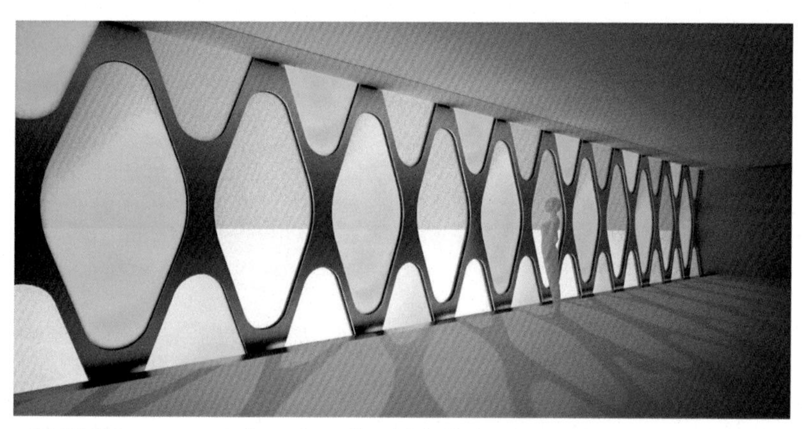

Figure 11.18 First concepts of curtain walls with built-in diamond-shaped photobioreactors for a feasibility study conducted by XTU in 2009–2010 for ICADE.
Courtesy of Olivier Scheffer, XTU Architects.

Figure 11.19 Effect of dense culture on light penetration. World's first prototypes of curtain-wall integrated PBRs, installed on a dedicated test bench.
Courtesy of Olivier Scheffer, XTU Architects.

11.5 The realization of a façade photobioreactor-integrated building for the future

With the limited experience on PBR façades some critical points for design should be considered to have a clear vision toward real-life applications. Even if the PBR experience in conventional microalgal industry is strong enough to have some clues about façade potential, a different sector like construction will open different views to PBR design (Table 11.2). These will lead to alternative approaches and solutions compared to the traditional applications.

Starting with the materials, façade PBR systems should focus on lightweight materials with feasible costs because the main cost, which can rise up to 80% of the total investment, in the microalgal industry is the PBR systems (Molina-Grima et al., 2013; Stephans et al., 2010; Davis et al., 2011; Silva et al., 2015). This will be a key issue in the diffusion to sector-like building applications where the costs are critical. This will also be an important factor when considering the existing buildings where the attraction should be needed. On the other hand the PBRs should be durable enough to resist the outdoor environment, keeping in mind the average building service life. According to an economic simulation on a green wall, where 50 years of building service life was considered, it showed that the system may not pay off its cost depending on various parameters like construction material, maintenance, lag time of plants to cover the wall surface, and plant life (Perini et al., 2011, 2013; Liua et al., 2014). Projection to a PBR system, cost and pay-off balance will be a very important issue to consider in attracting potential customers (Norsker et al., 2011; Singh and Sharma, 2012).

PBR design should also consider the easy maintenance in addition to the lowered costs, keeping in mind the cleaning of the PBR façade. One of the problems in conventional PBRs is that the settlement or sticking of the microalgae cells on the walls should be considered as a key issue when designing. Standard procedures like telescopic booms and hanging window cleaners can work for outdoor surfaces of the buildings, but PBRs will need novel solutions like automatic systems to clean the reactor interior unless a loss of control in contamination can be experienced. Solutions should also focus on the construction of mountable and movable parts different than the stable conventional applications that will be important for building applications keeping in mind the construction quality with leakage prevention. Speaking of cleaning, PBRs will also need more sophisticated solutions with regard to the possible contamination risks, or biotechnologically speaking, sterilization systems where each unit should be sterilized according to specific procedures like high temperatures or chemical concentrations with defined duration. Other than vapor sterilization, which is a high energy-consuming process, chemicals or UV can be energy-efficient alternatives, but they should all be selected to have a compatible integration with the building. Even if all methods are familiar for conventional systems, applications need to be modified according to the buildings used.

Another point for brainstorming will be the need of milking outdoor light for microalgae culture in the most effective way. Thinking of the interaction of light with culture productivity PBRs should have orientation ability according to outdoor illumination with a special emphasis on temperature change. The building sector has great experience in window construction and mounting on buildings but it should

Table 11.2 **Key features of a photobioreactor design for a building façade application**

Key feature	Target	Comment for progress
Materials	Durable, cheap, lightweight, comply with the building legislations	Developing technologies for glass manufacture like special glasses for high wind speed (or other physical shocks) resistance.
Illumination	Using sunlight as the main light source with the highest efficiency, keeping in mind the low and high oscillations that may cause a stress in the culture	With the developing technologies photochromic glasses or electronically tinting glass films will catalyze the light-adjusting applications in the building photobioreactor (PBR) façades. Also orientation capability and modular units, with energy-efficient pumping system to transfer between high and low intensity façades will increase the adaptation capability and productivity of the cultures. Using special lenses like Fresnel or light scattering nano-/microparticles for the PBRs or even using solar collectors and optical fiber systems like in the Himawari system for illumination will increase the efficiency. For a better building integration light tubes to catch more sunlight will also help to have a high-quality culture in an energy-efficient way.
Temperature	Adaptation to dynamic outdoor conditions keeping in mind the oscillations between low and high values that may cause a stress in the culture	Special glass materials like light-adjustable photochromic clever glasses, electronic glass films, or integrated thermal shields will serve as a control system that will prevent high thermal radiation absorption that will yield temperature increase. Also orientation capability and modular units, with an energy-efficient pumping system to transfer between high and low temperature façades will increase the adaptation capability and productivity of the cultures.
Water	Utilization of water in the most efficient way	Utilization of wastewater or recycle streams like rain water for the microalgae production.

Continued

Table 11.2 Continued

Key feature	Target	Comment for progress
Air	Utilization of indoor and outdoor air to feed the microalgae culture focusing on CO_2 accumulation or particle load; in other words, air quality inside the building	Using energy-efficient compressing systems to feed the air to the PBRs integrated with the building interior. The key will be the air quality that should be monitored to give clues to the compressing system about the air circulation inside the PBRs. During high CO_2 accumulation air can be recycled inside the PBRs more than the moderate levels (keeping in mind the shear stress risk with high air bubbling PBRs should have a proper sparging system design). Also another key feature to consider is based on the nature of photosynthesis where the microalgae culture acts as a CO_2 producer during dark hours (night). Because of that PBR exhaust systems should be directed outdoors during night.
Biomass and valuable products	Microalgae biomass as a valuable source should be evaluated to support the building expenditures	Microalgae biomass that is an important source for different sectors can be sold to the related companies to be processed for the market. Also in the case of extensive application several buildings can be connected with a sewage-like piping system where the produced biomass can be further processed in a central plant. This will decrease the facility expenses with regard to several topics like downstream, transportation, and labor. This approach may become a design feature of urban areas that will catalyze the application interest and serve as a strong support for a stable microalgae market.
Culture maintenance	Sustaining healthy cultures with easy adaptation and high productivities considering the inoculation, harvesting, contamination prevention, and control	Considering the type of building it may not be feasible to have all the detailed process units like a laboratory or a pilot plant for inoculation cultures. A practical solution can be a central unit having qualified units that will be responsible from various buildings. So in case of any problems this center will respond in a feasible way having a standardization.

be transferred or modified according to PBRs and microalgae. Different glazing technologies and materials for glass or transparent similars will serve as a catalyst for more efficient PBRs.

PBRs, being an interface-like unit for microalgae culture, will also need specific design elements like control systems with regard to pH, temperature, dissolved oxygen, degas systems, down-stream systems, pumping units, mixing and aeration systems, which should be considered in a façade application. Some solutions may cover the central processing system for online controls but keeping in mind the need of fast response to prevent the negative effects like sudden pH changes, which will not be easy especially on a large façade comprised of many PBRs (in other words, high volumes). Also maintenance during the operation should consider the intervention of a PBR where on a façade it will not be possible to shut down the process for a single unit. So the design needs to focus on the operation of a series of PBR systems without underestimating the problem potential of a single one.

Other than the basic considerations of a PBR system some support topics can help to have an acceptable investment on these systems. Some of these topics can be listed as the valuable products, wastewater treatment and recycle, carbon credits and subsidies, and real estate value, which will have a direct influence on the attention of these systems.

Considering valuable products from microalgal biomass that have potential in various industries will play a key role in the motivation of constructing PBR façades. The economic value of valuable chemicals like astaxanthin (up to $3000/kg) will serve as support for the feasible operation. In that case a building will act as a renewable and environment-friendly green factory that can be part of a societal supply chain for food, chemicals, and feed-like products. Also the usage of microalgae as a potential renewable energy source will be an important point to highlight for a self-sufficient sustainable building. Microalgal biomass can be used for biodiesel, bioethanol, or biogas production and some species of microalgae can even produce biohydrogen and bioethanol directly by cellular reactions (Oncel, 2013).

Thinking about the potentials of microalgae in wastewater treatment (Pittman et al., 2011), PBR façades can be integrated to use the wastewater or recycle streams like rain water. The utilization of wastewaters for nutrient support will increase the process feasibility by cutting the costs for new nutrients. On the other hand the efficient water usage will be very important for sustainability and environmental impact. The key point is that wastewaters may need some pretreatment like settlement or sterilization before usage, which may bring an economic load and risks that will not be preferred in a building application. But the microalgal biomass from the façades can be transferred to a separate wastewater treatment system to be used, or water in the PBRs can be supplied from a water treatment facility in order to lower the risks in application. Thinking about the environment and the carbon footprint, using the microalgae as an alternative approach to reach a self-sustainable building, new legislations and subsidies by the governments will strongly benefit the spreading of applications. Especially pilot plant investigations using flue gas integrated PBR systems like in the lime industry gave promising results (Pulz, 2001), which can be a good motivation for future applications (Fig. 11.20).

Figure 11.20 Concept of an industrial symbiosis between a waste incineration plant and microalgae biofaçades, using CO_2 and energy from the plant for microalgae cultivation. Courtesy of Olivier Scheffer, XTU Architects.

But to some degree the real catalyst for an ordinary applicant will be the beneficial impact on the real estate price; in other words, real-life measure will be the money in the pocket of the consumer. With the developing market the interest in green buildings and green solutions are rapidly increasing with an impact on the price tag of the property (Elaine, 2013; Yaron and Noel, 2013; World Green Building Council, 2013). With a special emphasis on higher resale values, higher rental rates, higher occupancy rates, lower operating expense, higher operating income, lower capitalization rates, and higher productivity gains compared to standard buildings, the importance of the green solutions and energy-efficient ecofriendly approaches impact on the value of the estate are apparent (Ng, 2013; World Green Building Council, 2013). Other than its visual attraction a proper design with a truly merged PBR system will increase the private investors' attention.

Even if the market is not mature, money will always be an issue and the increasing awareness about global warming, ecofriendly solutions, pollution, and other environmental topics, which are seen more dramatically in daily life, will help future applications. For a projection of the future, baby steps are taken on this topic.

One of the pioneering applications to integrate a PBR system with a building was presented by a group of researchers from the Czech Republic and Italy based on the experimental unit experience in 2000–01 in which a glass roof was constructed with solar concentrators to investigate the effect of high irradiances on cultures (Masojidek et al., 2003). The system was a 450 L roof-installed interior tubular PBR for *Arthrospira platensis* and *H. pluvialis* cultivation; these two microalgae models are very important in conventional productions. The systems used linear Fresnel lenses with movable frames for focal adjustment on the roof and façade (Fig. 11.21). The productivity with regard to better light utilization had increased, which is a good projection for façade systems' potential (Masojidek et al., 2003, 2009).

Other than the technical approaches in traditional microalgae processes a catalyst like the "Algae Competition" in 2011 organized under the guidance of frontiers in the area showed the different ideas that can evolve to reality (Henrikson and Edwards, 2012). There are only a few examples of potential real-life applications like the one presented by a collaboration between Splitterwerk, Arup, SSC, and Colt International in 2013 (Splitterwerk and Arup, 2013) during an international building exhibition in Hamburg (Fig. 11.22), with the patent pending design (Patent No: EP 2 359 682 A1); or the one presented by X-Tu, CNRS, and University of Nantes (Fig. 11.19), which also applied for a patent (Patent No: WO 2013/011240 A3), that show the increased attention to this topic and catch the conventional curiosity. With a case study point of view these two examples should be investigated in a more detailed way. The first full building application with a complete automatically controlled central system for cultivation in Hamburg by the German consortium cultivates microalgae as a renewable source. Each PBR unit, called the SolarLeaf, comprise a 2.5 × 0.7 m vertical unit with a capacity of 24 L. By using 129 panel PBRs on the façade covering 200 m^2 the building expected to produce 30 kWh/m^2a biomass and 150 kWh/m^2a heat, which supports the lowering of CO_2 emission by 2.5 tons/year (Wurm, 2013). The solar

Figure 11.21 Photobioreactor façade and roof system based on Fresnel lenses for microalgae cultivation.
Courtesy of J. Masojidek and G. Torzillo.

Figure 11.22 The Algae House, International Building Exhibition, Hamburg 2013. Courtesy of M. Kerner, SSC-Germany.

energy can be converted by 38% efficiency to heat by the PBR façades that will be used by the building via heat exchangers, heat pumps, and thermal energy store. The produced microalgal biomass will be a potential source for biogas by converting 8–10% of the solar energy to biomass and will act as a biofilter to utilize inside CO_2 emissions reaching a carbon neutral building (Splitterwerk and Arup, 2013; Wurm, 2013). Similarly the French consortium, which constructed the first curtain-wall PBR prototypes in the University of Nantes (Fig. 11.23 through Fig. 11.27), shares the PBR expenses by an efficient building integration through a symbiosis of thermal energy, light quality, and air quality. The theoretical studies also justify the potential of 300 m^2 façade system integrated to a waste processing plant that can simultaneously produce 0.7–1 tons/year microalgal biomass with a capture of 1–1.8 tons CO_2/year, which shows the reasons for interest in PBR façades (Pruvost et al., 2014).

A detailed scenario presented for the FSMA tower design (Fig. 11.23) foresees absorption of 250,000 t/y of CO_2 by the construction of a 2.1 Ha algae panel over the tower façade. The façades can produce enough biomass for the production of 450 tonnes biodiesel, which can be converted 4.6×10^6 kWh/y, enough for 120 houses. Due to the energy efficiency the tower can save 50% considering heating requirements (Henrikson and Edwards, 2012).

When thinking about the future, the potential of PBR façades should not be underestimated. Even futuristic ideas of self-sufficient green buildings presented by different designers (Figs. 11.24–11.28) may be considered interesting; due to the aesthetic value the science behind these designs with a special emphasis on microalgae will trigger the realization in the near future.

Figure 11.23 Pilot project for the first 200 m^2 biofaçade equipped with curtain-wall integrated PBRs, CSTB Headquarters, Champs-sur-Marne, France.
Courtesy of Olivier Scheffer, XTU Architects/SymBIO2.

11.6 Microalgae, a green volunteer for a better building: looking from an objective perspective for a start-up

Understanding the microalgae production and application potential blending with the knowledge of their biology and evolution is very important. Wild life observations of early civilizations lead to the usage of microalgae as a source for food and feed at the early development of microalgal biotechnology (Borowitzka, 2013a,b; Oncel et al., 2015). This historic acquaintance followed by metabolic advances make the

Figure 11.24 FSMA Tower on the panorama of London.
Original image courtesy of AVR London. FSMA Tower added by D. Edwards, courtesy of Dave Edwards Design Ltd.

Figure 11.25 Microalgae photobioreactor integrated in façades: double glass modules.
Courtesy of R. Cervera, Cervera and Pioz Architects.

Figure 11.26 Utopic concept of a depolluting floating city, using algae and other plants to clean water and transform CO_2 into O_2.
Courtesy of Olivier Scheffer, XTU Architects.

construction of microalgal biotechnology possible. The observations on biogeography, ecophysiological studies assisted and supported with genomics, proteomics and metagenomics, microalgae (Felczykowska, 2012; Reijnders et al., 2014) can be utilized as a promising raw material for several industries such as pharmaceuticals, nutraceuticals, cosmetics, food, feed, and biofuels as the leading ones (Fig. 11.29).

Today considering rents, health and well-being of occupants, indoor air quality, total 10-year costs, occupancy rates, building value, ongoing operations, and maintenance costs, energy efficiency is the most important factor for sustainability of a green building (Yaron and Noel, 2013). Within this regard, keeping the limits and bottlenecks in mind, microalgae can be counted as a promising source for the development of a sustainable building in the 21st century (Table 11.3). The key is to catch the attention of the applicants by answering the main question, why use microalgae in a building?

The idea of using microalgae in building applications is advantageous because (1) microalgae grow 20–30 times faster than land crops so are easier to manage compared to the plants; (2) the doubling time of most of the microalgal species are less than 24 h (16–18 h for *Chlorella* species but some species can be exceptional like 48–72 h for *Botryococcus barunii*), so one should not need to wait long to see the product—in the case of a building, a green façade; (3) microalgae cultivation is an environment-friendly technology so less footprint with regard to the environmental risks; (4) most of the algae species or algal substances are categorized as generally recognized as safe, so as a raw material they can be utilized for human and animal health supplement purposes, and will have a supportive effect considering the economic benefit; (5)

Figure 11.27 First feasibility studies conducted by XTU in 2009 for ICADE on algae façades for a high-rise project in La Défense, Paris.
Courtesy of Olivier Scheffer, XTU Architects.

microalgae are able to sequester CO_2 as an inorganic carbon source. CO_2 sequestration is an advantageous metabolic ability for the world, which is suffering from elevating levels of carbon emissions (http://www.globalcarbonatlas.org/?q=en/emissions) so this will be beneficial for a higher air quality inside the building and even outside air; (6) microalgae production is independent from the seasonal changes when considering the controlled environment in PBRs so the façade will have a sustainable culture all year long; (7) the nutrient requirements are cost-effective and even wastewaters can be utilized for nutrient so it will be beneficial with regard to the economy and maintenance; (8) microalgae has a flexible metabolism, which can be used at nonrelated industries like wastewater treatment so they can be useful as a treatment unit for a building; (9) microalgae will not require horizontal cultivation areas so the production can be done on the façades efficiently; and (10) microalgal production does not require only fresh water sources, thus marine, brackish, acidic, or alkaline water can be utilized

Figure 11.28 Concept of a large office building with algae biofaçades for local fresh Spirulina production/consumption, Reinventer. Paris competition, Paris Batignolles - N2, with BPD Marignan.
Courtesy Olivier Scheffer, XTU Architects/SymBIO2/BPD Marignan.

efficiently (Chisti, 2007; Oncel et al., 2015; Riberio et al., 2015). Considering all these basic advantages, the balance between the innovative approaches and the economy plays the key role for progress (Fig. 11.30). Today speaking of PBR façades the financial side should be supported by the developments and solutions to the technical challenges to increase the chance of investment.

With a special emphasis on the companies that will play the main role in the development of the market, there seems to be a gap. From an objective perspective there are actually some potential candidates rather than real players. For the classification of these candidates one branch will be related with the biotechnology and the other branch will be related with the construction technology.

Considering the biotechnology branch, one of the highest chances can be given to the microalgae producers that have a background in industrial applications. These will be the companies that have the outdoor production experience and PBR design and operation capabilities. These companies also have the economic projection of the market and are able to transfer their experience rapidly to building applications. On the other hand green façade companies that are specialized in plant applications directly on the building will have a very high chance in the market as well. These companies have background knowledge considering specific needs in the building integrations,

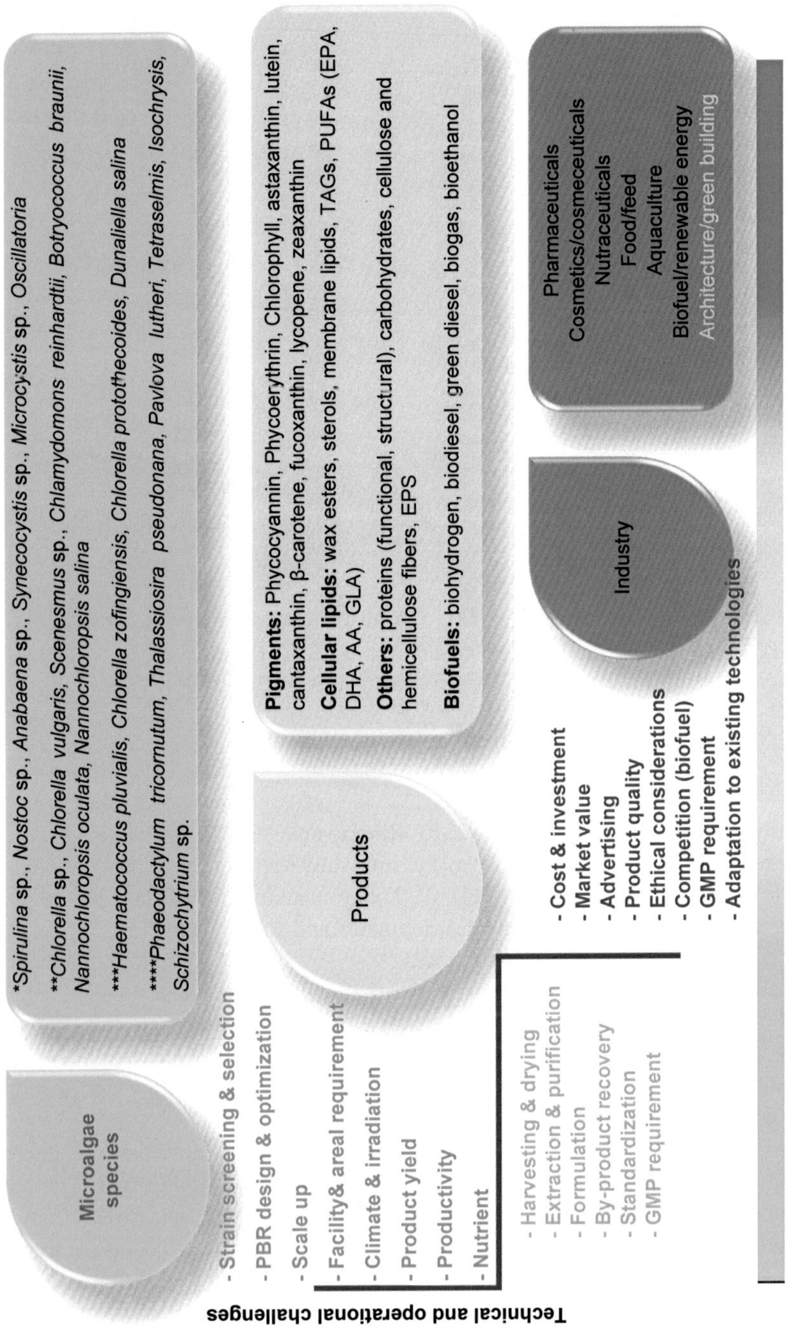

Microalgae species

**Spirulina* sp., *Nostoc* sp., *Anabaena* sp., *Synecocystis* sp., *Microcystis* sp., *Oscillatoria*

***Chlorella* sp., *Chlorella vulgaris, Scenesmus* sp., *Chlamydomons reinhardtii, Botryococcus braunii, Nannochloropsis oculata, Nannochloropsis salina*

****Haematococcus pluvialis, Chlorella zofingiensis, Chlorella protothecoides, Dunaliella salina*

*****Phaeodactylum tricornutum, Thalassiosira pseudonana, Pavlova lutheri, Tetraselmis, Isochrysis, Schizochytrium* sp.

Technical and operational challenges

- Strain screening & selection
- PBR design & optimization
- Scale up
- Facility& areal requirement
- Climate & irradiation
- Product yield
- Productivity
- Nutrient

Products

Pigments: Phycocyannin, Phycoerythrin, Chlorophyll, astaxanthin, lutein, cantaxanthin, β-carotene, fucoxanthin, lycopene, zeaxanthin

Cellular lipids: wax esters, sterols, membrane lipids, TAGs, PUFAs (EPA, DHA, AA, GLA)

Others: proteins (functional, structural), carbohydrates, cellulose and hemicellulose fibers, EPS

Biofuels: biohydrogen, biodiesel, green diesel, biogas, bioethanol

- Harvesting & drying
- Extraction & purification
- Formulation
- By-product recovery
- Standardization
- GMP requirement

Industry

- Cost & investment
- Market value
- Advertising
- Product quality
- Ethical considerations
- Competition (biofuel)
- GMP requirement
- Adaptation to existing technologies

Pharmaceuticals
Cosmetics/cosmeceuticals
Nutraceuticals
Food/feed
Aquaculture
Biofuel/renewable energy
Architecture/green building

Figure 11.29 Technical and operational challenges on production and commercialization of microalgae-based products and services.

Table 11.3 SWOT analysis with a special emphasis on the microalgae and photobioreactor systems for façades

	Strengths	Weaknesses	Opportunities	Threats	Steps for progress
Microalgae	Various strains with different specifications	Only limited experience mostly focused on conventional strains	Potential strains that are being investigated and ongoing search for new species from nature	Application problems from lab to real life	Selection of high productivity strains with resistance in outdoors
PBRs	Various designs with laboratory and conventional experience	Lack of experience on building applications	Increasing attention and awareness for environment-friendly sustainable buildings with green solutions like microalgae integration	Specific and tight regulations for real-life building applications	A realistic scaling and integration solutions by a holistic approach from PBR design to building with regard to important parameters like physical design criteria and microalgae cultivation needs
Production process	Background knowledge and experience on laboratory and conventional scales from inoculation to downstream operations	Need of different solutions and approaches for the real-life building applications keeping in mind the needs of existing or new buildings with various heights and façade forms	Known or under development of immense types of designs blending background knowledge with futuristic scenarios with regard to rooted experiences in building construction	Need of complex solutions for microalgae cultivation like control systems, contamination prevention, or downstream units that may raise a question on the applicability to the buildings	The interdisciplinary collaboration between biologists and engineers to find new solutions or modify existing ones for the realization of the building application considering all the steps

Continued

Table 11.3 **Continued**

	Strengths	Weaknesses	Opportunities	Threats	Steps for progress
Energy	Experience from green façades and some simulation studies considering the advantages related to energy efficiency of buildings will serve as a positive starting point	Lack of knowledge with regard to real life applications considering microalgae production steps	New process designs and analysis with regard to energy efficiency	Lack of knowledge and real-life applications that will attract the potential applicants, especially in a dynamic sector like construction where time and money is the key feature	Development of new approaches from the knowledge of existing with a projection on building application considering all the steps from PBR design to microalgae production keeping in mind the energy efficiency
Economy	Knowledge on the economy issues related to microalgal production other than the photobioreactor (PBR) and building construction	Lack of knowledge on feasibility and economic analysis considering building applications considering all the stages of the process integrated to a building	Economy analysis considering supporting points like biomass value, wastewater treatment potential, and renewable energy potential of microalgae, real estate value increase of building, legislation and politics for the incentives of green buildings, and issues highlighting carbon credits	Lack of knowledge on the building applications raising questions about the feasibility that stops the demand, which prevents the breaking of this vicious circle	New applications, designs, and developing technology will catalyze the awareness and will act as a base for the construction of a sustainable market leading feasibility

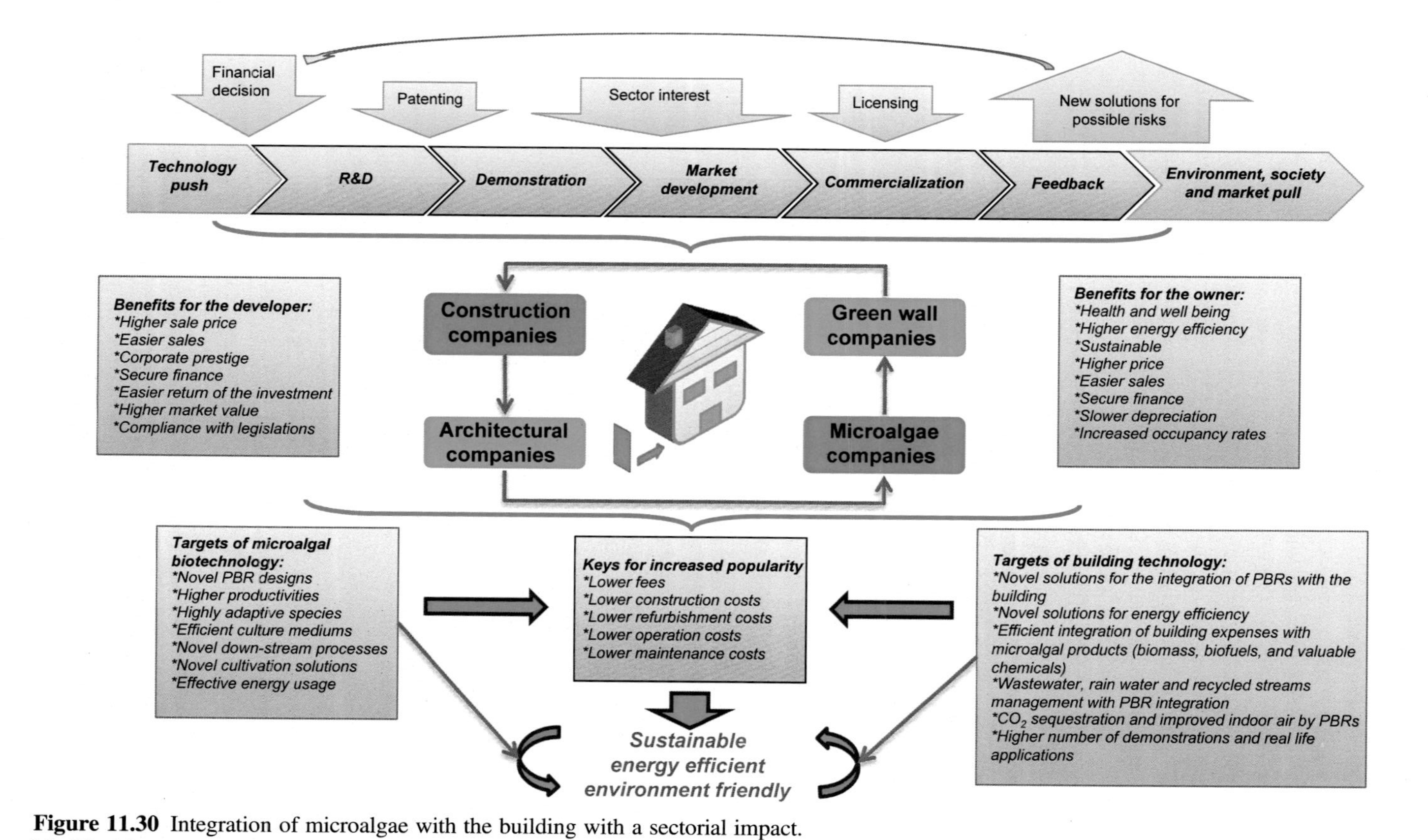

Figure 11.30 Integration of microalgae with the building with a sectorial impact.
Modified from Yaron, G., et al., 2013. Does Building Green Create Value? MBA, BSc, Light House Sustainable Building Centre Society, Vancouver, BC. www.lhsbc.com; Ng, 2013; Oncel, S.S., 2015. Biohydrogen from microalgae, uniting energy, life, and green future. In: Kim, S.K. (Ed.), Handbook of Marine Microalgae. Biotechnology Advances, Academic Press, pp. 159–196.

data about energy interactions with the buildings, outdoor experience considering dynamic environment with different plant species and connections with the building sector.

Considering the construction technology branch, potential candidates are the architectural companies. Architecture will act as a catalyst for the popularity of the microalgae PBR façades with the novel designs and will act as a bridge between the construction sector and the biotechnology sector. The aesthetic touch in the nature of the architecture will attract the market directly. On the other hand construction companies having strong experience with regard to the engineering and the real estate sector should be inside the game, otherwise the market cannot be developed strongly. Construction companies that are experienced in green buildings will get the higher chance in the market and will easily transform their needs comprising PBR applications. These companies will lead the different members in the periphery of the building industry like architects, mechanical engineers, construction engineers, environmental engineers, urban planners, finance offices, and real estate agents for the maturation of the market. These companies also have the knowledge about the consumer's desire and point of view for a building, which is a critical parameter for the acceptability of the new technology and design.

But to reach a stable market a different player that will focus directly on the sector or an interface of these two branches will be needed to blend the experiences. This ideal candidate should have the experience to route this interdisciplinary sector related with biotechnology and construction technology to reach the target of sustainable and environment-friendly buildings and in a bigger scale city with a healthy interaction of the life and urban environment. Considering today's economic approach an interface or a director company will be more realistic rather than a stand-alone company that will control everything, which actually will help to catalyze the maturation of the sector.

So to start up a creation in this area the aforementioned ideal player need to be specific on some key points for a strong business. One, ecofriendly and sustainable are the two important concepts that will need to be in the center of the business plan. As a start-up considering an efficient built environment with regard to these two concepts will increase the chance for survival. The following is the awareness of each topic in the sector like the market for microalgae, their products, potential areas of alternative applications, and in the target of merging to the building sector, the possible pros and cons. In the start-up stage these points need to be clear or at least enough analysis should be done to have a vision. The following will be the issues of economy, which should never be underestimated or risked. Financing opportunities, tax, and other payments will be specified and costs should be realistic, especially with regard to the microalgae PBRs and building application considering all the infrastructure. In the start-up creation stage the well-established strategies and B plans will help to prevent the possible risks. Also another key point is that the realistic goals that will play an important role in the business should be set. Supporting the goals with a well-designed road map with a special emphasis on the competition, technological progress, economic feedbacks, industrial links, market analysis, legal issues, public relations, management, organization structure, minimum viable product, intellectual property, and know-how will need extra attention in a novel area like microalgae building integration.

Each effort to shorten the lag time to profitable application PBR façade system will have more chance in commercial projections. This will be a good opportunity for the startup companies that want to take the lead in this fertile ground which is directly related with the new and existing buildings and urban areas.

11.7 Conclusion

Today the façade PBR systems, in other words microalgae, for integration with the building is not a mature technology and there is still a long journey to reach a stable market volume even if the experiences on conventional microalgae production, applications of green walls with plants, or green building constructions have a background on the sectoral basis. But this situation also gave a very important radius of action with a chance of higher progress for the new players. Taking the huge economic value of the construction sector into consideration with a special emphasis on the existing buildings and the new ones the PBR façade applications should be highlighted as an emerging area that should not be underestimated.

Targeting the ultimate expectations will also need realistic fundamentals, otherwise facing disappointment is inevitable. The key is the objective vision, related to the interdisciplinary nature of this technology, which should blend the needs and facts from two different branches, biotechnology and construction technology. The ideal player of this area should control and direct these two branches in one main stream to have a successful application considering all the targets for a sustainable, energy-efficient, environment-friendly, and economically feasible green building. With limited applications it is not easy to speculate about the PBR façade building but the aesthetic progress blended with the energy and side incomes like biomass may catalyze the spreading of the applications. Increased demonstrations will be important for the public awareness and interest thus feedbacks from the end users will play a vital role in the future.

A famous proverb, "All roads lead to Rome," which originates from the ancient Roman saying, "*Mille viae ducunt homines* per *seacula Romam* (A thousand roads lead man forever to Rome)," should again be considered because seeking the shortest road to the environmentally friendly, sustainable, and wealthy future is the key point for today and tomorrow. And we should never forget the quote from Thomas Edison, "Just because something doesn't do what you planned it to do doesn't mean it's useless."

Acknowledgments

The authors would like to thank to Rosa Cervera (Cervera and Pioz Architects, Madrid, Spain) for the different designs of the urban applications of microalgae and PBR façades; Olivier Scheffer (X-TU Architects, Paris, France) for the designs and pictures of the PBR façades and the detailed information; Dave Edwards (Dave Edwards Design Ltd., London, UK) for the FSMA tower design and all the valuable information; Prof. Jiri Masojidek (University of South Bohemia, Czech Republic) and Dr. Giuseppe Torzillo (CNR, Florence, Italy) for the Fresnel lens-modified PBR system and the information; Martin Kerner (SSC-Hamburg, Germany) for the pictures of "The Algae House"; the CCAP and Ege-MACC culture collection for the permission for the microscopic images of diverse microalgae cells. The authors would also like to thank Prof. Jack Legrand (University of Nantes, France) and Robert Henrikson (Ronore Enterprises, Inc., USA) for their valuable support of the chapter. And the authors would like to thank again all the contributors for their kindness and help.

References

Alexandria, E., Jones, P., 2008. Temperature decreases in an urban canyon due to green walls and green roofs in diverse climates. Building and Environment 43, 480–493.

An, J.Y., et al., 2003. Hydrocarbon production from secondarily treated piggery wastewater by the green alga *Botryococcus braunii*. Journal of Applied Phycology 15, 185–191.

Andersen, R.A., 2013. The microalgal cells. In: Richmond, A.R., Hu, Q. (Eds.), Handbook of Microalgal Cultures. Wiley, pp. 3–21.

Antal, T., Lindblad, P., 2005. Production of H_2 by sulphur-deprived cells of the unicellular cyanobacteria *Gloeocapsa alpicola* and *Synechocystis* sp. PCC 6803 during dark incubation with methane or at various extracellular pH. Journal of Applied Microbiology 98, 114–120.

Arsalane, W., Rousseau, B., Duval, J.C., 1994. Influence of the pool size of the xanthophyll cycle on the effects of light stress in a diatom: competition between photoprotection and photoinhibition. Photochemistry and Photobiology 60, 237–243.

Atkinson, B., Mavituna, F., 1983. Biochemical Engineering and Biotechnology Handbook. The Nature Press, New York, USA, 1118 pp.

Azkorra, Z., et al., 2015. Evaluation of green walls as a passive acoustic insulation system for building. Applied Acoustics 89, 46–56.

Ball, S., Colleoni, C., Cenci, U., Raj, J.N., Tirtiaux, C., 2011. The evolution of glycogen and starch metabolism in eukaryotes gives molecular clues to understand the establishment of plastid endosymbiosis. J. Exp. Bot 62, 1775–1801.

Banerjee, A., et al., 2002. *Botryococcus braunii*: a renewable source of hydrocarbons and other chemicals. Critical Reviews in Biotechnology 22, 245–279.

Barclay, W., Apt, K., Dong, D.X., 2013. Commercial Production of Microalgae Via Fermentation, pp. 134–145.

Barra, L., et al., 2014. The challenge of ecophysiological biodiversity for biotechnological applications of marine microalgae. Marine Drugs 12, 1641–1675.

Becker, E.W., 2007. Micro-algae as a source of protein. Biotechnology Advances 25, 207.

Becker, E.W., 2013. Microalgae for Human and Animal Nutrition. Handbook of Microalgal Culture: Applied Phycology and Biotechnology, second ed. Wiley, pp. 461–503.

Behrenfeld, M., et al., 1998. Compensatory changes in photosystem II electron turnover rates protect photosynthesis from photoinhibition. Photosynthesis Research 58, 259–268.

Belay, A., 2013. Biology and industrial production of *Arthrospira* (*Spirulina*). In: Richmond, A., Hu, Q. (Eds.), Handbook of Microalgal Cultures. Wiley, pp. 339–358.

Blakenship, R.E., 1992. Origin and early evolution of photosynthesis. Photosynthesis Research 1992 (33), 91–111.

Borowitzka, M.A., 1999. Commercial production of microalgae: ponds, tanks, tubes and fermenters. Journal of Biotechnology 70, 313–321.

Borowitzka, M.A., 2013a. Dunaliella: Biology, Production, and Markets, pp. 259–368.

Borowitzka, M.A., 2013b. Energy from microalgae. In: Borowitzka, M.A., Moheimani, N.R. (Eds.), Algae for Biofuels and Energy. Springer Science+Business Media Dordrecht, pp. 1–16.

Boussiba, S., 2000. Carotenogenesis in the green alga *Haeamatococcus pluvialis*: cellular physiology and stress response. Physiologia Plantorum 108, 111–117.

Brennan, L., Owende, P., 2010. Biofuels from microalgae—a review of technologies for production, processing and extractions of biofuels and co-products. Renewable and Sustainable Energy Resources 14, 557–577.

Bumbak, F., et al., 2011. Best practices in heterotrophic high-cell-density microalgae processes: achievements, potential and possible limitations. Applied Microbiology and Biotechnology 91, 31–46.

Burlew, J.S., 1953. Current status of the large-scale culture of algae. In: Burlew, J.S. (Ed.), Algal Culture: From Laboratory to Pilot Plant. Carnegie Institution of Washington Publication, Washington DC.

Chen, C.Y., et al., 2011. Cultivation photobioreactor design and harvesting of microalgae for biodiesel production: a critical review. Bioresource Technology 102, 71–81.

Chen, Q., et al., 2013. An experimental evaluation of the living wall system in hot and humid climate. Energy and Buildings 61, 298–307.

Chen, H., He, C., Wang, Q., 2015. Microalgal biofuel revisited: An informaticsbased analysis of developments to date and future prospects. Applied Energy 155, 585–598.

Cheng, C.Y., et al., 2010. Thermal performance of a vegetated cladding system on facade walls. Building and Environment 45, 1779–1787.

Chisti, Y., 2007. Biodiesel from microalgae. Biotechnology Advances 25, 294–306.

Cooper, M.B., Smith, A.G., 2015. Exploring mutualistic interactions between microalgae and bacteria in the omics age. Current Opinion in Plant Biology 26, 147–153.

Dasgupta, C.N., Gilbert, J.J., Lindblad, P., Heirdorn, T., Borgvang, S.A., Skjanes, K., Das, D., 2010. Recent trends on the development of photobiological processes and photobioreactors for the improvement of hydrogen production. International Journal of Hydrogen Energy 35, 10218–10238.

Davis, R., et al., 2011. Techno-economic analysis of autotrophic microalgae for fuel production. Applied Energy 88, 3524–3531.

Delevari, H.A., et al., 2015. Using fluorescent material for enhancing microalgae growth rate in photobioreactors. Journal of Applied Phycology 27, 67–74.

Dijedjiga, R., et al., 2015. Analysis of thermal effects of vegetated envelopes: integration of a validated model in a building energy simulation. Energy and Buildings 86, 93–103.

Doran, P.M., 1995. Bioprocess Engineering Principles, first ed. Elsevier Academic Press, San Diego.

Elaine, L.M., 2013. Impact of Green Buıldıngs on the Value of Property, Elaine Ng L. M., A Dissertation Submitted in Part Fulfillment of the Degree of Master of Science Built Environment: Facility and Environment Management Bartlett School of Graduate Studies. University College London, 73 pages.

Li, J., et al., 2015. Design and characterization of a scalable airlift flat panel photobioreactor for microalgae cultivation. Journal of Applied Phycology 27, 75–86.

Faraloni, C., Torzillo, G., 2010. Phenotypic characterization and hydrogen production in *Chlamydomonas reinhardtii* Q_B binding D1 protein mutants under sulfur starvation, changes in chlorophyll fluorescence and pigment composition. Journal of Phycology 46, 788–799.

Fauzi, A.F., Malek, N.A., 2013. Green building assessment tools: evaluating different tools for green roof system. International Journal of Education and Research 1 (11), 1–14.

Felczykowska, A., 2012. Metagenomic approach in the investigation of new bioactive compounds in the marine environment. Acta Biochimica Polonica 59, 501–505.

Feng, H., Hewage, K., 2014. Lifecycle assessment of living walls: air purification and energy performance. Journal of Cleaner Production 69, 91–99.

Fernandez, F.G.A., Sevilla, J.M.F., Molina, E., 2013. Photobioreactors for the production of microalgae. Sci. Biotechnol 12, 131–151.

Finazzi, G., Moreau, H., Bowler, C., 2010. Genomic insights into photosynthesis in eukaryotic phytoplankton. Trends in Plant Science 15, 565–572.

Franco, A., et al., 2012. Wind tunnel analysis of artificial substrates used in active living walls for indoor environment conditioning in Mediterranean buildings. Building and Environment 51, 370–378.

Giordano, M., Beardall, J., Raven, J.A., 2005. CO_2 concentrating mechanisms in algae: mechanisms, environmental modulation, and evolution. Annual Review of Plant Biology 56, 99–131.

Goss, R., Jakob, T., 2010. Regulation and function of xanthophyll cycle-dependent photoprotection in algae. Photosynthesis Research 106, 103–122.

Gould, S.B., 2008. Plastid evolution. Annual Review of Plant Biology 59, 491–517.

Grima, E.M., et al., 2003. Recovery of microalgal biomass and metabolites: process options and economics. Biotechnology Advances 20, 491–515.

Guarnieri, M.T., Pienkos, P.T., 2015. Algal omics: unlocking bioproduct diversity in algae cell factories. Photosynthesis Research 123, 255–263.

Guillaume, G., et al., 2015. Numerical study of the impact of vegetation coverings on sound levels and time decays in a canyon street. Science of the Total Environment 502, 22–30.

Hall, D.O., Fernández, F.G., Guerrero, E.C., Rao, K.K., Grima, E.M., Biotechnol Bioeng, 2003. Outdoor helical tubular photobioreactors for microalgal production: modeling of fluid-dynamics and mass transfer and assessment of biomass productivity. Apr 5 82 (1), 62–73.

Han, D., Li, Y., Hu, Q., 2013a. Biology and commercial aspects of *Haematococcus pluvialis* products. In: Richmond, A., Hu, Q. (Eds.), Handbook of Microalgal Cultures. Wiley, pp. 388–405.

Henrikson, R., Edwards, M., 2012. Imagine Our Algae Future. Visionary Algae Architecture and Landscape Designs, 163 pages.

Hindersin, S., Leupold, M., Kerner, M., Hanelt, D., 2014. Keye parameters for outdoor biomass production of *Scedenesmus obliquus* in solar tracked photobioreactors. Journal of Applied Phycology 26, 2315–2325.

Hohmann-Marriot, M.F., Blakenship, R.E., 2011. Evolution of photodynthesis. Annual Review of Plant Biology 62, 515–548.

Hu, Q., Richmond, A., 1996. Productivity and photosynthetic efficiency of Spirulina platensis as affected by light intensity, algal density and rate of mixing in a flat plate photobioreactor. J. Appl. Phycol 8, 139–145.

Hu, Q., 2013. Environmental effects on cell composition. In: Richmond, A., Hu, Q. (Eds.), Handbook of Microalgal Cultures. Wiley, pp. 114–122.

Hu, Q., et al., 2008. Microalgal triacylglycerols as feedstocks for biofuel production: perspectives and advances. The Plant Journal 54 (4), 621–639.

Huntera, M., et al., 2014. Quantifying the thermal performance of green façades: a critical review. Ecological Engineering 63, 102–113.

Ip, K., et al., 2010. Shading performance of a vertical deciduous climbing plant. Building and Environment 45, 81–88.

Jansenn, M., et al., 2000. Scale up aspects photobioreactors: effects of mixing induced light/dark cycles. Journal of Applied Phycology 12, 225–237.

Janssen, M., et al., 1999. Specific growth rate of *Chlamydomonas reinhardtii* and *Chlorella sorokiniana* under medium duration light/dark cycles: 13–87 s. Journal of Biotechnology 70, 323–333.

Janssen, M., et al., 2003. Enclosed outdoor photobioreactors: light regime, photosynthetic efficiency, scaleup and future prospects. Biotechnology and Bioengineering 81, 193–210.

Jim, C.Y., 2015. Greenwall classification and critical design-management assessments. Ecological Engineering 77, 348–362.

Katsuda, T., et al., 2006. Effect of flashing light from blue light emitting diodes on cell growth and astaxanthin production of *Haematococcus pluvialis*. Journal of Bioscience and Bioengineering 102, 442–446.

Kim, D.G., Hur, S.B., 2013. Growth and fatty acid composition of three heterotrophic *Chlorella* species. Algae 28 (1), 101–109.

Kim, S.K., 2013. Marine cosmeceuticals. Journal of Cosmetic Dermatology 13, 56–67.

Kim, Tae-Ryong, Han, Seung-Hoon, 2014. Analysis for energy efficiency of the algae façade, focused on closed bioreactor system. KIEAE Journal 14 (6), 15–21.

Köhler, M., 2008. Green facades a view back and some visions. Urban Ecosystems 11, 423–436.

Kontoleon, K.J., Eumorfopoulou, E.A., 2010. The effect of the orientation and proportion of a plant-covered wall layer on the thermal performance of a building zone. Building and Environment 45, 1287–1303.

Kose, A., Oncel, S., 2015. Properties of microalgal enzymatic protein hydrolysates: biochemical composition, protein distribution and FTIR characteristics. Biotechnology Reports 6, 137–143.

Krohn-Molt, I., et al., 2013. Metagenome survey of a multispecies and alga-associated biofilm revealed key elements of bacterial-algal interactions in photobioreactors. Applied and Environmental Microbiology 79 (20), 6196–6206.

Kruse, O., Hankamer, B., 2010. Microalgal hydrogen production. Current Opinion in Biotechnology 21, 238–243.

Kunjapur, A.M., Bruce Eldigre, R., 2010. Photobioreactor design for commercial biofuel production from microalgae. Industrial and Engineering Chemistry Research 49, 3516–3526.

Largeau, C., et al., 1980. Site of accumulation of hydrocarbon in *Botryococcus brauni*. Phytochemistry 19, 1043–1051.

Li, J., et al., 2014. Design and characterization of a scalable airlift flat panel photobioreactor for microalgae cultivation. J Appl Phycol 27, 75–86.

Lianga, T.C., Hiena, w. N., Jusuf, S.K., 2014. Effects of vertical greenery on mean radiant temperature in the tropical urban environment. Landscape and Urban Planning 127, 52–64.

Lindblad, P., et al., 2002. Photoproduction of H_2 by wild type *Anabaena* PCC7120 and a hydrogen uptake deficient mutant: from laboratory experiments to outdoor culture. International Journal of Hydrogen Energy 27, 1271–1281.

Liu, Y., et al., 2014. Cost-benefit analysis on green building energy efficiency technology application: a case in China. Energy and Buildings 82, 37–46.

Madrea, F., et al., 2015. Building biodiversity: vegetated façades as habitats for spider and beetle assemblages. Global Ecology and Conservation 3, 222–233.

Maity, J.P., et al., 2014. Microalgae for third generation biofuel production, mitigation of greenhouse gas emissions and wastewater treatment: present and future perspectives – a mini review. Energy 78, 104–113.

Maness, P.C., et al., 2009. Photobiological hydrogen production – prospects and challenges. Microbe 4, 6.

Manson, M., Gomes, J., 2015. Green wall systems: a review of their characteristics. Renewable and Sustainable Energy Reviews 41, 863–871.

Masojidek, J., et al., 2003. A closed solar photobioreactor for cultivation of microalgae under supra-high irradiance: basic design and performance.

Masojidek, J., et al., 2009. A two-stage solar photobioreactor for cultivation of microalgae based on solar concentrators. Journal of Applied Phycology 21, 55–63.

Masojidek, J., Torzillo, G., Koblizek, M., 2013. Photosynthesis in microalgae. In: Richmond, A., Hu, Q. (Eds.), Handbook of Microalgal Culture: Applied Phycology and Biotechnology Second Edition. Wiley, pp. 21–36.

Mathews, J., Wang, G., 2009. Metabolic pathway engineering for enhanced biohydrogen production. International Journal of Hydrogen Energy 34, 7404–7416.

Mayfield, S.P., Franklin, S.E., 2005. Expression of human antibodies in eukaryotic micro-algae. Vaccine 23, 1828–1832.

Melis, A., et al., 2000. Sustained photobiological hydrogen gas production upon reversible inactivation of oxygen evolution in the green alga *Chlamydomonas reinhardtii*. Plant Physiology 122, 127–135.

Melis, A., 2007. Photosynthetic H_2 metabolism in *Chlamydomonas reinhardtii* (unicellular green algae). Planta 226, 1075–1086.

Melis, A., Happe, T., 2001. Hydrogen production. Green algae as a source of energy. Plant Physiology 127, 740–748.

Ming, L., Hua, D., Liu, H., 2014. Photobioreactor with ideal light–dark cycle designed and built from mathematical modeling and CFD simulation. Ecological Engineering 73, 162–167.

Molina Grima, E., et al., 1999. Photobioreactors: light regime, mass transfer, and scale up. Journal of Biotechnology 70, 231–248.

Molina, E., et al., 2001. Tubular photobioreactor design for algal cultures. Journal of Biotechnology 92, 113–131.

Molina-Grima, et al., 2013. Downstream processing of Cell mass and products. In: Richmond, A., Hu, Q. (Eds.), Handbook of Microalgal Cultures. Wiley, pp. 267–309.

Morris, H.J., et al., 2008. Utilisation of *Chlorella vulgaris* cell biomass for the production of enzymatic protein hydrolysates. Bioresource Technology 99, 7723–7729.

Mulders, K.J.M., et al., 2014. Phototrophic pigment production with microalgae: biological constraints and opportunities. Journal of Phycology 50, 229–242.

Neffe, J., 2007. Einstein: A Biography (translated by Shelley Frisch). Farrar, Straus and Giroux, New York, p. 256.

Nelson, D.L., Cox, M.M., Lehninger, A.L., 2013. Lehninger Principles of Biochemistry, sixth ed. W.H. Freeman and Company, New York.

Norsker, N.H., et al., 2011. Microalgal production: a close look at the economics. Biotechnology Advances 29, 24–27.

Ogbonna, J.C., Tanaka, H., 2000. Light requirement and photosynthetic cell cultivation: development of processes for efficient light utilization in photobioreactors. Journal of Applied Phycology 12, 207.

Ogbonna, J.C., Soejima, T., Tanaka, H., 1999. An integrated solar and artificial light system for internal illumination of photobioreactors. Journal of Biotechnology 70, 289–297.

Olaizola, M., 2000. Commercial production of astaxanthin from *Haematococcus pluvialis* using 25,000-liter outdoor photobioreactors. Journal of Applied Phycology 12, 499–506.

Olaizola, M., 2003. Commercial development of microalgalbiotechnology: from the test tube to the marketplace. Biomolecular Engineering 20, 459–466.

Olivieri, G., Salatino, P., Marzocchella, A., 2014. Advances in photobioreactors for intensive microalgal production: configurations, operating strategies and applications. Journal of Chemical Technology and Biotechnology 89, 178–195.

Oncel, S., Sukan, F.V., 2011. Effect of light intensity and the light: dark cycles on the long term hydrogen production of Chlamydomonas reinhardtii by batch cultures. Biomass Bioenergy 35, 1066–1074.

Oncel, S.S., Vardar-Sukan, F., 2009. Photo-bioproduction of hydrogen by *Chlamydomonas reinhardtii* using a Semi-continuous process regime. International Journal of Hydrogen Energy 34, 7592–7602.

Oncel, S., 2013. Microalgae for a macro energy world. Renewable and Sustainable Energy Reviews 26, 241–264.

Oncel, S., 2014. Focusing on the optimization for scale up in airlift bioreactors and the production of *Chlamydomonas reinhardtii* as a model microorganism. Ekoloji 23 (90), 20–32.

Oncel, S., Kose, A., 2014. Comparison of tubular and panel type photobioreactors for biohydrogen production utilizing *Chlamydomonas reinhardtii* considering mixing time and light intensity. Bioresource Technology 151, 265–270.

Oncel, S., Sabankay, M., 2012. Microalgal biohydrogen production considering light energy and mixing time as the key features for scale up. Bioresource Technology 121, 228–234.

Oncel, S., Sukan, F.V., 2008. Comparison of two different pneumatically mixed column photobioreactors for the cultivation of *Arthrospira platensis* (*Spirulina platensis*). Bioresource Technology 99, 4755–4760.

Oncel, S.S., et al., 2015. From the ancient tribes to modern societies, microalgae evolution from a simple food to an alternative fuel source. In: Kim, S.K. (Ed.), Handbook of Marine Microalgae. Biotechnology Advances, Academic Press, pp. 127–144.

Oncel, S.S., 2015. Biohydrogen from microalgae, uniting energy, life, and green future. In: Kim, S.K. (Ed.), Handbook of Marine Microalgae. Biotechnology Advances, Academic Press, pp. 159–196.

Ong, B.L., 2003. Green plot ratio: an ecological measure for architecture and urban planning. Landscape and Urban Planning 63, 197–211.

Otteléa, M., et al., 2011. Comparative life cycle analysis for green façades and living wall systems. Energy and Buildings 43, 3419–3429.

Parker, M.S., Mock, T., Armbrust, E.V., 2008. Genomic insights into marine microalgae. Annual Review of Genetics 42, 619–645.

Patil, G., Chethana, S., Madhusudhan, M.C., Raghavarao, K.S., 2008. Fractionation and purification of the phycobiliproteins from *Spirulina platensis*. Bioresource Technology 99, 7393–7396.

Perez-Garcia, O., et al., 2011. Heterotrophic cultures of microalgae: metabolism and potential products. Water Research 45 (1), 11–36.

Perez, G., et al., 2014. Vertical Greenery Systems (VGS) for energy saving in buildings: a review. Renewable and Sustainable Energy Reviews 39, 139–165.

Perez, G., Rincon, L., Vila, A., Gonzales, J.M., Cabeza, L.F., 2011. Behaviour of green facades in Mediterranean Continental climate. Energy Conversion and Management 52, 1861–1867.

Perini, K., Rosasco, P., 2013. Cost-benefit analysis for green façades and living wall. Building and Environment 70, 110–121.

Perini, K., et al., 2011. Greening the building envelope, façade greening and living wall systems. Open Journal of Ecology 1 (1), 1–8.

Perini, K., et al., 2013. Vertical greening systems, a process tree for green façades and living walls. Urban Ecosystems 16, 265–277.

Pittman, J.K., Dean, A.P., Osundeko, O., 2011. The potential of sustainable algal biofuel production using waste water resources. Bioresource Technology 102, 17–25.

Posten, C., 2009. Design principles of photobioreactors for cultivation of microalgae. Engineering in Life Sciences 9 (3), 165–177.

Pruvost, J., Gouic, B.L., Legrand, J., 2014. Symbiotic integration of photobioreactors. In: A Factory Building Façade for Mutual Benefit between Buildings and Microalgae Needs, 21st International Congress of Chemical and Process Engineering CHISA 2014 Prague 17th Conference on Process Integration, Modelling and Optimisation for Energy Saving and Pollution Reduction PRES 2014, Praha, Czech Republik; 08/2014.

Pulz, O., Scheibenbogen, K., 1998. Photobioreactors: design and performance with respect to light energy input. In: Scheper, T. (Ed.), Bioprocess and Algae Reactor Technology, Apoptosis, Advances in biochemical engineering/biotechnology, vol. 59. Springer, Berlin, Heidelberg, New York, p. 123.

Pulz, O., 2001. Photobioreactors: production systems for phototrophic microorganisms. Applied Microbiology and Biotechnology 57, 287–293.

Pulz, O., Gross, W., 2004. Valuable products from biotechnology of microalgae. Applied Microbiology and Biotechnology 65, 635–648.

Reijnders, M.J.M.F., et al., 2014. Green genes: bioinformatics and systems-biology innovations drive algal biotechnology. Trends in Biotechnology 32, 617–626.

Reisser, W., Houben, P., 2001. Different strategies on aeroterrestrial algae in reacting to increased levels of UV-B and ozone. Nova Hedwigia 123, 291–296.

Riberio, L.A., et al., 2015. Prospects of using microalgae for biofuels production: results of a Delphi study. Renewable Energy 11, 799–804.

Richmond, A., 1996. Efficient utilization of high irradiance for production of photo-autotrophic cell mass: a survey. Journal of Applied Phycology 8, 381–387.

Richmond, A., 2004. Handbook of microalgal culture: biotechnology and applied phycology. Blackwell Science Ltd. J. Appl. Phycol. 16, 159–160.

Rodolfi, L., et al., 2009. Microalgae for oil: strain selection, introduction of lipid synthesis and outdoor mass cultivation in a low-cost photobioreactor. Biotechnology and Bioengineering 102 (1), 100–112.

Ryong, T., et al., 2014. Analysis for energy efficiency of the algae façade-focused on closed bioreactor system. KIEAE Journal 14 (6), 15–21.

Saeid, A., Chojnacka, K., 2015. Toward production of microalgae in photobioreactors under temperate climate. Chemical Engineering Research and Design 93, 377–391.

Safi, C., et al., 2014. Morphology, composition, production, processing and applications of *Chlorella vulgaris*: a review. Renewable and Sustainable Energy Reviews 35, 265–278.

Safikhani, T., et al., 2014. A review of energy characteristic of vertical greenery systems. Renewable and Sustainable Energy Reviews 40, 450–462.

Sakurai, H., 2013. Phtobiological hydrogen production: Bioenergetics and challenges for its practical application. Journal of Photochemistry and Photobiology C 17, 1–25.

San Pedro, A., González-López, C.V., Molina-Grima, A.E., 2015. Outdoor pilot production of *Nannochloropsis gaditana*: influence of culture parameters and lipid production rates in raceway ponds. Algal Research 8, 205–213.

Sasso, S., et al., 2012. Microalgae in the postgenomic era: a blooming reservoir for new natural products. FEMS Microbiology Reviews 36, 761–785.

Scoma, A., et al., 2012. Outdoor H2 production in a 50-L tubular photobioreactor by means of a sulfur-deprived culture of the microalga *Chlamydomonas reinhardtii*. Journal of Biotechnology 157, 620–627.

Shewaka, S.M., Mohamed, N.M., 2012. Green facades as a new sustainable approach towards climate change. Energy Procedia 18, 507–520.

Silva, A.G., et al., 2015. Life cycle assessment of biomass production in microalgae compact photobioreactors. GCB Bioenergy 7, 184–194.

Singh, R.N., Sharma, S., 2012. Development of suitable photobioreactor for algae production: a review. Renewable and Sustainable Energy Reviews 16, 2347–2353.

Splitterwerk and Arup, 2013. The Algae House, ISBN 978-3-7212-0907-5.

Spolaore, P., et al., 2006. Commercial applications of microalgae. Journal of Bioscience and Bioengineering 101, 87–96.

Srirangan, K., Pyne, M.E., Chou, C.P., 2011. Biochemical and genetic engineering strategies to enhance hydrogen production in photosynthetic algae and cyanobacteria. Bioresource Technology 102, 8559–8604.

Stec, W.J., van Paasen, A.H.C., Maziarz, A., 2005. Modelling the double skin facade with plants. Energy and Buildings 37, 419–427.

Stengel, D.B., Connan, S., Popper, Z.A., 2011. Algal chemodiversity and bioactivity: sources of natural variability and implications for commercial application. Biotechnology Advances 29, 483–501.

Stephens, E., et al., 2010. Future prospects of microalgal biofuel production systems. Trends in Plant Science 15 (10), 554–564.

Suali, E., Sarbatly, R., 2012. Conversion of microalgae to biofuel. Renewable and Sustainable Energy Reviews 16, 4316–4342.

Sunakorna, P., et al., 2011. Thermal performance of biofacade with natural ventilation in the tropical climate. Procedia Engineering 21, 34–41.

Susurova, I., et al., 2013. A model of vegetated exterior facades for evaluation of wall thermal performance. Building and Environment 67, 1–13.

Susurova, I., et al., 2014. The effects of climbing vegetation on the local microclimate, thermal performance, and air infiltration of four building facade orientations. Building and Environment 76, 113–124.

Takaichi, S., 2011. Carotenoids in algae: distributions, biosyntheses and functions. Marine Drugs 9, 1101–1118.

The Business Case for Green Building, 2013. A Review of the Costs and Benefits for Developers, Investors and Occupants. World Green Building Council, 124 pages.

Tomaselli, L., 1997. Morphology, ultrastructure and taxonomy. In: *Spirulina Platensis* (*Arthrospira*): Physiology, Cell Biology and Biotechnology.

Torzillo, G., Seibert, M., 2013. Hydrogen production by microalgae. In: Richmond, A., Hu, Q. (Eds.), Handbook of Microalgal Culture: Applied Phycology and Biotechnology Second Edition. Wiley, pp. 417–444.

Tredici, M., 2004. Mass production of microalgae: photobioreactors. In: Richmond, A. (Ed.), Handbook of Microalgal Mass Cultures. Blackwell, Oxford, pp. 178–214.

Tredici, M.R., Materassi, R., 1992. From open pond to alveolar panel: the Italian experience. Journal of Applied Phycology 4, 221.

Tredici, M.R., 2010. Photobiology of microalgae mass cultures; understanding the tools for the next green revolution. Biofuels 1, 143–162.

Tredici, M.R., Zitelli, G.C., Biagiolini, S., 1992. Influence of turbulence and areal density on the productivity of *Spirulina platensis* grown outdoor in a vertical alveolar panel. In: First European Workshop on Microalgal Biotechnology, Bergholz-rehbrucke, pp. 58–60.

Tredici, et al., 2009. Advances in microalgal culture for aquaculture feed and other uses. In: Burnell, G., Allan, G. (Eds.), New technologies in aquaculture: Improving production efficiency, quality and environmental management Woodhead Publishing. Ltd/CRC Press LCC, Boca Raton, pp. 610–676.

Trent, J., 2012. Offshore Membrane Enclosures for Growing Algae (OMEGA) – A Feasibility Study of Wastewaters to Biofuels. NASA Ames Research Center. CEC-500-2013-143.

Tzoulas, K., et al., 2007. Promoting ecosystem and human health in urban areas using green infrastructure: a literature review. Landscape and Urban Planning 81, 167–178.

Ugwu, C.U., Ogbonna, J.C., Tanaka, H., 2005. Characterization of light utilization and biomass yields of *Chlorella sorokiniana* in inclined outdoor tubular photobioreactors equipped with static mixers. Process Biochemistry 40 (11), 3406–3411.

Vargas, M.A., Rodríguez, H., Moreno, J., Olivares, H., Del Campo, J.A., Rivas, J., Guerrero, M.G., October 1998. Biochemical composition and fatty acid content of filamentous nitrogen-fixing cyanobacteria. Journal of Phycology 34 (5), 812–817.

Vasumathi, K.K., Premalatha, M., Subramanian, P., 2012. Parameters influencing the design of photobioreactor for the growth of microalgae. Renewable and Sustainable Energy Reviews 16, 5443–5450.

Vincent, W.F., 2009. Cyanobacteria. Elsevier, pp. 226–232.

Vonshak, A., Guy, R., 1992. Photoadaptation, photoinhibition, and productivity in the blue-green alga *Arthrospira platensis*, grown outdoors. Plant Cell Environment 15, 613–616.

Vonshak, A., et al., 1996. Light and oxygen stress in *Arthrospira platensis* (*Cyanobacteria*) grown outdoors in tubular reactors. Physiologia Plantarum 97, 175–179.

Weber, J., et al., 2014. Biotechnological hydrogen production by photosynthesis. Engineering in Life Sciences 1–15.

Wilson, A., et al., 2006. A soluble carotenoid protein involved in phycobilisome-related energy dissipation in cyanobacterial. Plant Cell 18, 992–1007.

Wong, N.H., et al., 2010. Thermal evaluation of vertical greenery systems for building walls. Building and Environment 45, 663–672.

Wongluang, P., Chistib, Y., Srinophakun, T., 2013. Optimal hydrodynamic design of tubular photobioreactors. Journal of Chemical Technology and Biotechnology 88, 55–61.

Wurm, J., 2013. Photobioreactors on facades for energy generation, alternative technologies in the building envelope. In: Int. Rosenheim Window and Facade Conference, pp. 83–87.

Xiong, J., et al., 2009. Molecular evidence for the early evolution of photosynthesis. Science 289, 1724–1730.

Xu, H., Miao, X.L., Wu, Q.Y., 2006. High quality biodiesel production from a microalgae *Chlorella protothecoides* by heterotrophic growth in fermenters. Journal of Biotechnology 126, 499–507.

Xu, L., et al., 2009. Microalgal bioreactors: challenges and opputinuties. Engineering Life Sciences 9 (3), 178–189.

Yang, C., Hua, Q., Shimizu, K., 2000. Energetics and carbon metabolism during growth of microalgal cells under photoautotrophic, mixotrophic and cyclic light-autotrophic/dark-heterotrophic conditions. Biochemical Engineering Journal 6, 87–102.

Yaron, G., et al., 2013. Does Building Green Create Value? MBA, BSc. Light House Sustainable Building Centre Society, Vancouver, BC. www.lhsbc.com.

Yıldırım, A., et al., 2014. Carotenoid and fatty acid compositions of an indigenous *Ettlia texensis* isolate (*Chlorophyceae*) under phototrophic and mixotrophic conditions. Applied Biochemistry and Biotechnology 172, 1307–1319.

Yue, D., You, F., Snyderba, S.W., 2014. Biomass-to-bioenergy and biofuel supply chain optimization: overview, key issues and challenges. Computers and Chemical Engineering 66, 36–56.

Zittelli, C., Rodolfi, L., Tredici, M.R., 2000a. Mass cultivation of marine microalgae under natural, mixed and artificial illumination. In: Abstracts of the 4th European Workshop on Biotechnology of Microalgae. Bergholz-Rehbrücke, Germany.

Zittelli, G.C., Pastorelli, R., Tredici, M.R., 2000b. A modular flat panel photobioreactor (MFPP) for indoor mass cultivation of *Nanochloropsis* sp. Under artificial illumination. Journal of Applied Phycology 12, 521–526.

Zittelli, G.C., et al., 2013. Photobioreactors for mass production of microalgae. In: Richmond, A., Hu, Q. (Eds.), Handbook of Microalgal Culture: Applied Phycology and Biotechnology Second Edition. Publisher: Wiley Oxford, pp. 225–266.

Zuao, J., Zhao, Z., 2015. Green building research current status and future agenda: a review. Renewable and Sustainable Energy Reviews 30, 271–281.

Biotechnologies for improving indoor air quality

12

G. Soreanu
Technical University "Gheorghe Asachi" of Iasi, Faculty of Chemical Engineering and Environmental Protection, Department of Environmental Engineering and Management, Iasi, Romania

Motto: Moving back to microorganisms and plants....

12.1 Introduction

Indoor air quality (IAQ) is directly related to well-being, comfort, and health of indoor building occupants. IAQ is affected by pollution with various airborne contaminants such as gaseous compounds, fine particles (nonbiological), and bioaerosols (Soreanu et al., 2013; van Ras et al., 2005). One of the most typical indicators of poor air quality is "sick building syndrome" (Soreanu et al., 2013). In some cases, severe health problems known as "building-related illness" and "multiple chemical sensitivity" can develop (Luengas et al., 2015; Welch, 1991; Hort et al., 2014). Indoor air pollution is considered one of the top five risks to public health (Soreanu et al., 2013; Hort et al., 2014). According to Luengas et al. (2015), about 4.3 million deaths in 2012 have been reported by the World Health Organization (WHO) as being related to indoor air pollution. Improving IAQ could result in important economical benefits at a country level. For example, improving IAQ is expected to save more than $20 billion annually in the United States for instance, by increasing worker productivity and decreasing the health costs of the subjects (Guieysse et al., 2008; Luengas et al., 2015).

Increasing energy savings in new buildings can affect the IAQ by airtight or sealed construction or insufficient air exchange (Luengas et al., 2015; Pickett and Bell, 2011). The use of the present air cleaner systems is not efficient, economic, or safe (Soreanu et al., 2013). In a smart eco-efficient built environment, which is based on some key component symbiosis (Fig. 12.1), environmentally friendly and cost-effective tools should be used for enhancing IAQ. A green candidate in this sense is the biological air filtration, which can remove almost all chemical pollutants from indoor air. Practically, plant or microbial biosystems can be used alone or in combination with other systems in order to overcome the drawbacks of the conventional air cleaning systems (Soreanu et al., 2013; Alfonsín et al., 2013; Padhi and Gokhale, 2014). Overall, the use of the biotechnologies for various building-related purposes (not only for indoor air application, but also for development of sustainable

Start-Up Creation. http://dx.doi.org/10.1016/B978-0-08-100546-0.00012-1

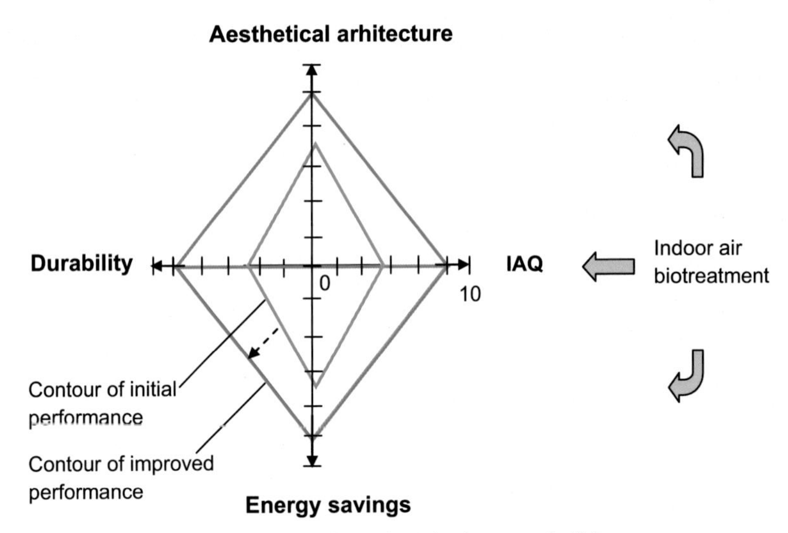

Figure 12.1 Key components (but not limited to) in the smart-building concept, which are related to sustainability. Moving forward optimized performance scenario (score 10 = maximum performance).

biomaterials and bioprocesses; Ivanov et al., 2015; Pacheco-Torgal, 2015) is a challenge, but also a significant step toward eco-efficiency in the smart-building concept.

This chapter points out the main aspects involved in applications of biotechnologies for indoor air pollution mitigation. Indoor environments types, their pollution issues and technological aspects of biological air filtration, including types of biosystems and their application as a tool for improving IAQ, are analyzed, compared, and evaluated in this study.

12.2 Issues of air pollution in indoor environments

12.2.1 Classification of indoor environments

An indoor air pollution profile depends on the particularities of the indoor environment, which can be classified as follows:

1. Residential: blocks of flats, apartments, houses, caravan, hotels
2. Nonresidential
   a. Nonindustrial settings
      i large public access: theaters, museums, cinemas, libraries, restaurants, bars, shopping malls, supermarkets, sporting or other recreation facilities (arena, swimming pools, dancing halls), event halls (conferences, shows, exhibitions, banquets), religious buildings
      ii large public transit: underground (metro) stations, railway stations, airport, underground ruttier and pedestrian passages

iii institutions: educational (university buildings and campus, research centers, schools, kindergartens, training centers), medical (hospitals, policlinics, including cabinets and laboratories), local and governmental organizations (agencies, city halls, parliament), prisons
iv well-being places: balneo-climateric treatment policlinics; hair/beauty salons and spas
v others: offices buildings, banks, service agencies, garages, etc.
vi military closed or semiclosed space of submarines, ships, underground constructions like command centers, ammunition stores
vii astronautic closed space of missiles, research station and bases

b. Industrial settings: manufacturing buildings and their associated facilities (offices building, storage buildings, laboratories, ateliers, etc.)

c. Agricultural settings: animal or plant farms and their associated facilities

The aspects of indoor air pollution for such environments should be addressed in respect to the following circumstances:

i location: rural or urban; earth or space; meteorological conditions (wind direction, temperature, precipitation regime), pollution risk (natural or anthropogenic phenomena historical data), development level
ii area/occupancy ratio: large or small
iii ventilation: natural or artificial
iv life occupancy: permanent or nonpermanent (eg, cottages, schools); variable/stable (eg, same occupants with same lifestyles)
v lifestyle: smoker or nonsmoker habitants; electrical or fuel-based heaters or cooking equipments; habitant activities (eg, home workers, painters)
vi interior components: rooms, kitchens, corridors, floors, stair space, elevators
vii functionality: uni- or multifunctional (eg, commercial complex including shopping area, restaurants, supermarket, conference rooms in a single building)
viii occupancy regulations: smoking or nonsmoking rooms, pets or pet-free buildings
ix mobility: stationary or mobile (eg, buses, airplane cabins)

These remarks highlight the complexity of indoor environments. Supporting information is provided elsewhere (eg, OSHA, 2011; Shimer et al., 2005). Many of the enumerated places are subject to a large air circulation, which can supply the substrates for the air-treatment biotechnologies.

12.2.2 Indoor air pollutants: types, sources, impacts

The main types of air pollutants that can contaminate the indoors are as follows (Soreanu et al., 2013; van Ras et al., 2005; OSHA, 2011; Shimer et al., 2005; Kelley and Gilbert, 2013):

- **Gaseous compounds**: Refer to the volatile organic compounds (VOCs) (eg, aliphatic and aromatic hydrocarbons, alcohols, aldehydes, chlorinated compounds, many other emitted indoor pollutants from various in-house materials and products) and inorganic compounds such as carbon oxides, sulphur oxides, nitrogen oxides, which are products of combustion, or radon emitted from underground rocks or construction materials; ozone can be also emitted indoors from office machines, some air purifiers, and ozone generators. Exposure to one or more of these gaseous compounds appears to be involved in headache, ocular, nasal, cutaneous and throat irritation, asthma aggravation, as well as damage to liver or brain

(OSHA, 2011; Shimer et al., 2005). In addition, VOC exposure can induce various forms of cancer (OSHA, 2011; Shimer et al., 2005).

- **Fine particles** (nonbiological): Refer to the suspended particulate matter (PM) (eg, PM10, PM2.5) often emitted indoors from combustion processes, smoking, printing, or from building materials during restoration activities; such particles can contain heavy metals, but also pesticides and asbestos. It is well known that exposure to PM is associated with respiratory diseases, including infections and aggravated asthma, and also lung cancer.
- **Bioaerosols**: Refer to the bacteria, fungi, viruses, that can contaminate the air via improper air filters, (de)humidifiers, molds, pets, insects, sick building occupants, or inadequate housekeeping practices; biocontaminants can be attached to the dust particles. Allergic reactions and infectious diseases can occur from exposure to biocontaminated air.

The indoor environment can be associated with a closed or semiclosed space that is more susceptible to pollutant accumulation than an open-like outdoor environment. As earlier indicated, pollutant accumulation in indoor air occurs via many sources, which can be classified as the following:

- **Outdoor sources**: Road transport, industry, construction activities, power stations, refineries, agriculture, wastewater plants, contaminated sites, contaminated waters, volcanic eruptions, sandstorms
- **Underground sources**: Foundation materials, underground parkings, groundwaters, leachate and gases from landfills
- **Indoor sources**: Construction materials, paints, furniture, platforms, carpets, cosmetics, perfumes, sprays, recipients, cleaning products, insect repellents, pets, human metabolism, secondary reactions, tobacco smoking, fuel combustion, cooking, restoration

The pollution sources can be characterized by different emission regimes; for example, they can be passive and permanent (eg, fugitive emissions from building materials; leaking from pipes) or active and temporary (eg, high emissions during cooking activities) (Luengas et al., 2015).

The number of contaminants in indoor air can be very large (eg, hundreds of VOCs can be detected in an indoor environment). Their individual concentrations in air are usually below the exposure limits for the workplaces recommended by the Occupational Safety and Health Administration (Soreanu et al., 2013), which in turn can be higher than the outdoor limits set in Directive 2008/50/EC for the ambient air quality. Even in low concentrations, their cumulative level and synergic effect, or the prolonged exposure, are involved in the developing of health problems. For example, “sick building syndrome” includes symptoms such as ocular, nasal, cutaneous irritations, allergies, respiratory dysfunction, headache, and fatigue, which can be associated with the simultaneous presence of many of these compounds in indoor air (Soreanu et al., 2013).

The air pollution levels indoors may be 2 to 5 times or occasionally 100 times greater than outdoors (Tanner et al., 2013; Missia et al., 2010; Gawrońska and Bakera, 2014). From the point of view of the outdoor air, the indoors act as multiple pollution sources. When emitted outdoors, the indoor air pollutants contribute to the main environmental impacts, expressed as human toxicity potential, global warming potential, aquatic toxicity, acidification potential, eutrophication potential, ozone depletion potential, and photochemical ozone creation potential (European Commission, 2006).

Other impacts refer to indoor space deterioration (eg, degradation of building materials by the combustion products, or molds development under inappropriate humidity conditions) and the costs involved for its restoration. Generally speaking, it is difficult to identify the major source that contributes to the indoor air pollution, taking into consideration the variety of the contamination possibilities.

An indoor air pollution directive has not been launched yet; however some countries already adopted specific regulations with regard to the high-risk toxic compounds for indoor air. Among the indoor air pollutants currently regulated by several organizations in different countries are the following compounds (Luengas et al., 2015): formaldehyde, carbon monoxide, naphthalene (Canada, France, WHO), benzene (France, WHO, China), toluene (Canada), trichloroethylene, tetrachloroethylene (France, WHO), nitrogen dioxide (France, WHO, China), acrolein, acetaldehyde (France), benzo(α) pyrene (WHO, China), sulphur dioxide, and TVOC (total volatile organic compounds) (China). Many of them are known as human carcinogens (formaldehyde, benzene, benzo(α) pyrene) or classified as possibly human carcinogens (naphthalene, trichloroethylene), while CO and NO_2 are often associated with asthma prevalence (WHO, 2010). Also, trichloroethylene and tetrachloroethylene appear to be related to neurological dysfunctions (WHO, 2010).

In a smart eco-efficient built environment, air treatment methods should be able to remove the indoor air pollutants for a healthy environment in a sustainable manner.

12.2.3 Treatment methods of indoor air

Some indoor air pollution issues could be eliminated by the life style and the implementation of specific measures for pollution prevention at the source, but others could not be controlled and additional measures for indoor air treatment must be considered. An accessible abatement method is air dilution by ventilation, but the effective pollutant removal involves the use of air-cleaning systems. The actual air-cleaning systems are mainly based on mechanical filtration and electrostatic precipitation for suspended particle removal, adsorption, photocatalytic oxidation, and ozonation for gaseous compound removal, or UV photolysis for biocontaminant removal (Soreanu et al., 2013; Luengas et al., 2015; Cruz et al., 2014a). Some of the drawbacks associated with the use of these systems for indoor air treatment include the limited performance in the removal of multiple pollutants, the limited life time of some filter materials or catalysts, the risk of contamination by hazardous secondary products, and the high investment or operating cost (Soreanu et al., 2013; Luengas et al., 2015; Padhi and Gokhale, 2014). Currently, the most commercialized air-cleaning systems are those based on suspended particle removal. Among these filters, the most popular is the high-efficiency particle arresting filter, which can remove more than 95% of particles of 0.3–6 μm, including spores of bacteria and fungi, dust mites, and other allergens (Luengas et al., 2015). The commercial heating, ventilating, and air conditioning (HVAC) system contains about 15–20% outdoor air and 80–85%

recirculated indoor air. Particularly, a reduction by 20% of VOCs in the supply air of HVAC will result in the reduction of outdoor ventilation rate by 50% (Sidheswaran et al., 2011).

A new trend in indoor air treatment refers to the use of biological methods (microorganisms and plants) for the removal of a wide range of chemical pollutants. Conventional (microbial) biological filtration is successfully used for the removal of air pollutants and it is expected to remove indoor air pollutants as well. Plant filtration is already used for indoor air purification (Soreanu et al., 2013). Overall, when considering biosystems for indoor application, there is a general perception related to the potential contamination by actinomycetes and fungal spores and relative humidity (RH) increase, which can be avoided by using appropriate operating conditions and reactor designs (Soreanu et al., 2013).

Overall, the system performance, the energy savings, and the environmental impact should be considered in the design of the new air treatment units that could contribute to the building eco-efficiency. According to Luengas et al. (2015), the energy consumption for various technologies varies in the following order: photolysis, ozonation, membrane separation (**highest energy consumption**) > mechanical filtration, electrostatic filtration, adsorption, photocatalysis (**moderated energy consumption**) > microbial biofiltration, plant biofiltration (**lowest energy consumption**).

As can be seen, the biotechnologies offer a good opportunity for the development of the cost-effective and environmentally friendly air treatment systems. A combination between different technologies could overcome the drawbacks of the air cleaning systems (Soreanu et al., 2013; Luengas et al., 2015).

12.3 Biotechnologies for air treatment: a brief theoretical background

12.3.1 Concept basics and principles

The particularity of the air treatment biotechnologies is the use of natural bioagents (microorganisms and/or plants) for pollutant removal. This method is considered green, robust, accessible, and cost-effective (Syed et al., 2006; Soreanu et al., 2013; Luengas et al., 2015; Alfosin et al., 2013; Shareefdeen et al., 2005). The biological air filtration is carried out in specific bioreactors, namely biofilters (BFs), biotrickling filters (BTFs), bioscrubbers (BSs), and so on (Section 12.3.2) that can remove a large spectrum of gaseous pollutants found at low concentrations in large quantities of air, and thus is well suited for end-of-pipe polishing purposes. Most popular applications refer to VOC and malodorous compound removal from air fluxes. The use of biological processes for air treatment dates back to 1950, but the interest in this application increased after 1970 (Alfosin et al., 2013; Shareefdeen et al., 2005; Devinny et al., 1999).

Biological filtration of contaminated air is an aerobic process, where the pollutants act as energy, carbon, and other nutritional sources for the involved bioagents. In this process, the bioagents carry out the conversion of the harmful compounds (such as VOCs) to less harmful compounds (such as carbon dioxide, water, and biomass) by specific enzymes. Biodegradation is an exothermic process, where temperature gradients of 2–5°C are often recorded (Padhi and Gokhale, 2014; Dastous et al., 2008). Particularly, the plant-assisted bioconversion is a complex process, where both the plant as well as its rhizosphere microorganisms living around the roots contribute to the overall process performance (Soreanu et al., 2013). Depending on the type of bioagent, the biological process can be classified as:

- **Microbial**: Refers to the bioconversion carried out by sole microorganisms
- **Plant**: Refers to the bioconversion carried out by plants and their associate rhizosphere microorganisms (phytodegradation and rhizosphere biodegradation)

The fundamental mechanism of these processes is more or less the same (Padhi and Gokhale, 2014; Soccol et al., 2003) and involves a contact between a solid media (biosupport), a nutrient solution, and the gaseous phase delivering the substrates (pollutants) for the bioagents (Fig. 12.2). Particularly, a visual biofilm is gradually developed on the solid media or suspended cells or cell aggregates in the liquid medium as an indicator of the microbial-driven biological process. Microbiological reaction takes place via the microorganisms attached in the fixed biofilm and/or suspended in the liquid phase. From this point of view, the biological systems function

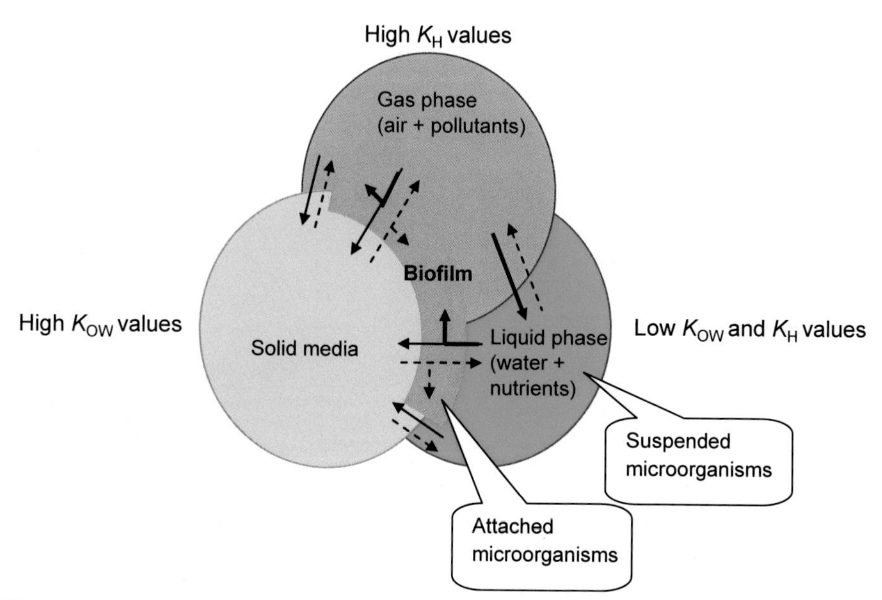

Figure 12.2 Pollutant transfer routes during the microbial biofiltration of the contaminated air. K_H = Henry's Law constant; K_{OW} = octanol-water partition coefficient; main arrows indicate the favorable mass transfer; dashed lines indicates the reverse phenomena (limited mass-transfer due volatilization, desorption etc.); the transfer of the reaction products between the compartments can be also associated with the dashed lines.

as an attached and/or suspended microorganism growth reactor (Soreanu et al., 2010a). Pollutant transfer between the different compartments of the system takes place via different mechanisms (adsorption, absorption, diffusion, convection, ionic exchange, etc.). Under nonlimited mass-transfer conditions (high substrate availability for bio-agents) and the presence of needed microbial growth nutrients, high efficiency bio-filtration performance is expected, subject to the substrate biodegradability and the cell density.

Gaseous pollutants with high solubility in water, low volatility from water (low Henry's Law constant), and high biodegradation rate such as the alcohols, aldehydes, and some aromatic compounds are easily removed from air by biofiltration. Other gases such as methane, siloxanes, and some chlorinated compounds are low-soluble and low-biodegradable and involve the use of additional tools for the enhancing of the biofiltration performance (eg, the use of surfactants, enzymes, specific operating conditions, etc.) (Soreanu et al., 2011; Shareefdeen et al., 2005).

12.3.2 Types of bioreactors

12.3.2.1 Conventional bioreactors

Conventional bioreactors for air treatment use microorganisms for pollutant removal and are BFs, BTFs, and BSs (Fig. 12.3). Particularly, the packing bed in BFs and BTFs serves as support media for the microorganisms and can be natural (compost, lava rocks, wood chips, etc.) or artificial (plastic, ceramic, or metal frames/pieces, with a random or structured arrangement). Nutrient solutions can be synthetic-defined medium prepared from water and nutrients such as ammonium

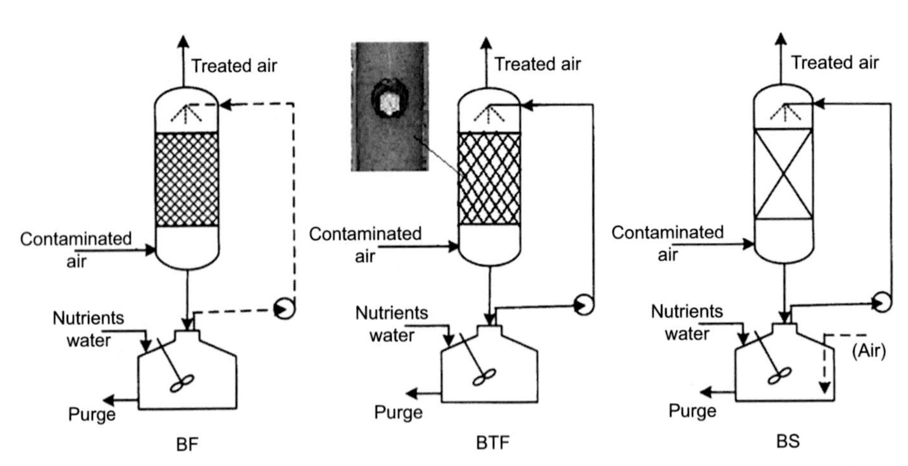

Figure 12.3 Conventional bioreactors for air treatment (BF, BTF, BS). *(Adapted from Syed, M., Soreanu, G., Falletta, P., Béland, M., 2006. Removal of hydrogen sulfide from gas streams using biological processes – a review. Canadian Biosystems Engineering 48, 2.1–2.14.)* The picture shows a packing media used in a BTF treating biogas for H_2S removal at Wastewater Technology Centre, Burlington, ON, Environment Canada.

or nitrate salt, source of organic nitrogen, phosphate, potassium, bicarbonate, sulfate, microelements, and other such, or undefined medium (eg, effluent from a biological wastewater treatment unit). The work principle of these bioreactors is almost the same, except for some particularities depicted as follows (Shareefdeen et al., 2005; Alfonsín et al., 2013; Padhi and Gokhale, 2014; Syed et al., 2006):

BF: The polluted air is continuously fed into a bioreactor, while the nutrient solution is discontinuously added; BF functions are to ensure an attached microorganism growth and all subsequent processes such as sorption and biodegradation leading to pollutant removal. A packing material with high porosity, buffering capacity, nutrient availability, and moisture retention capacity is usually recommended, in order to ensure the microorganisms growth. BFs are suitable for treating low-soluble pollutants. BF can also be used as an open-system for passive biofiltration in field application.

BTF: The polluted air and the nutrient solution are continuously fed into the bioreactor; BTF functions as an attached or combined attached/suspended microorganism growth reactor, where the pollutant removal occurs in the fixed biofilm and eventually in the moving solution; an inert packing material is usually recommended, in order to avoid the excessive biofilm development and the bed clogging. Uniformity of liquid distribution has a significant influence on the process performance. BTFs are suitable for treating high-soluble pollutants.

BS: The polluted air and the nutrient solution are continuously fed into an absorption unit packed with an inert media, from which the liquid is transferred in a subsequent unit, where the biological conversion is favored. BS functions as a suspended microorganism growth reactor. In comparison with BF and BTF, BS better supports high variations in pollutant loading rates, but is usually more expensive.

Conventional bioreactors are commonly implemented for industrial air treatment (off-gases from chemical industry, wastewater treatment plants, etc.) (Padhi and Gokhale, 2014; Syed et al., 2006). Rotating drum BFs, horizontal flow BFs, continuous fluidized bed bioreactors, two-phase partitioning bioreactors or biopolymeric coating-based bioreactors are some of the new microbial bioreactors derived from the conventional bioreactors.

The development of conventional bioreactors for indoor air treatment is in an early stage. Some concerns in this case refer to the possible substrate limitation due to the low pollutant concentration in the inlet gas or spore contamination in the outlet gas, and discharge of the nutrient-containing liquid medium.

12.3.2.2 Plant bioreactors

The concept of plant bioreactors is based on the work principle of conventional bioreactors, but includes a botanical component (plants) for pollutant removal. In plant bioreactors, the vegetarian part and the rhizosphere microorganisms exist in symbiosis and both are involved in the pollutant removal. There are two main categories of botanical configurations: potted plants (PPs) and plant-assisted (botanical) biotrickling filters (PBTFs) (Soreanu et al., 2013):

PPs: Associated with air biofiltration in a passive manner; in some test chambers, a contaminated air flux can be used instead of a passive exposure; also, if air is forced through the root zone, the process is close to a conventional biofiltration; PP irrigation is occasionally performed; when the polluted air is contacted with the PPs (plant, soil, microorganisms), biological uptake and bioconversion take place, leading to pollutant removal.

PBTFs: A more sophisticated configuration, consisting of a BTF equipped with hydroponic plants (Fig. 12.4), is continuously operated with polluted air and nutrient solution; PBTFs are engineered for high air-flow-rate processing.

Some botanical prototypes were implemented for indoor air treatment across North America, as part of the green building concept (Soreanu et al., 2013). Plant-assisted bioreactors are suitable for applications where the contaminant concentration in the processed air is small (eg, pollutants of indoor air). The limitation in terms of the substrate availability is compensated by the exudates generated by plants. Despite the recorded progress since early 2000, botanical biofiltration of indoor air is not completely investigated (Soreanu et al., 2013).

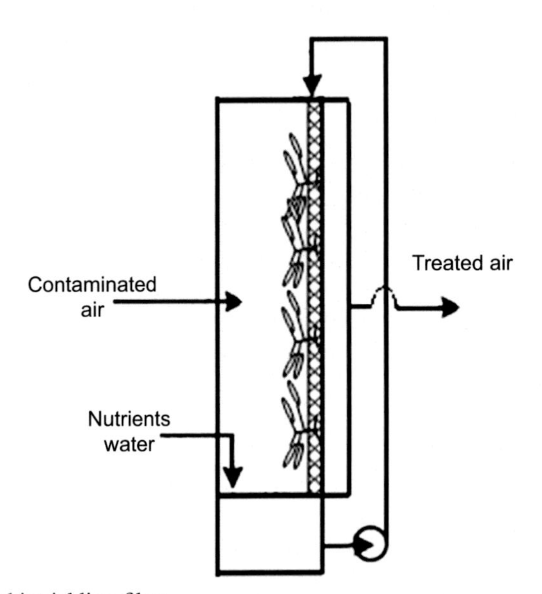

Figure 12.4 Plant biotrickling filter.
Adapted from Soreanu, G., Dixon, M., Darlington, A., 2013. Botanical biofiltration of indoor gaseous pollutants – a mini-review. Chemical Engineering Journal 229, 585–594.

12.3.2.3 Membrane bioreactors

These reactors are equipped with a gas-permeable membrane array instead of a packing bed. The volume occupied by the membrane module is smaller than for other separation systems. The membrane can be polymeric or inorganic and allows the selective transfer of the pollutant from air in a recirculated nutrient solution, via a diffusion process. The pollutant capture and conversion takes place at the interface membrane-liquid, where a microbial biofilm tends to develop (Soreanu et al., 2011; van Ras et al.,

2005; Wang et al., 2013; Zhao et al., 2011). Microbial contamination of air is avoided as it is not directly in contact with the biofilm. Membrane bioreactors can be microbiological (eg, Fig. 12.5) or botanical (membrane array is incorporated into the root zone of PBTFs) (van Ras et al., 2005; Soreanu et al., 2013).

Membrane bioreactors are supposed to overcome the mass-transfer limitations of the conventional or plant bioreactors, but are more costly (Estrada et al., 2015). Membrane bioreactors can simultaneously control the pollutant, humidity, and spore levels in air (van Ras et al., 2005; Soreanu et al., 2013). Particularly, membrane bioreactors are currently investigated for their potential to address all limitations related to application in space (the occupied volume and the level of pollutants, humidity, and spores) (van Ras et al., 2005; Soreanu et al., 2013).

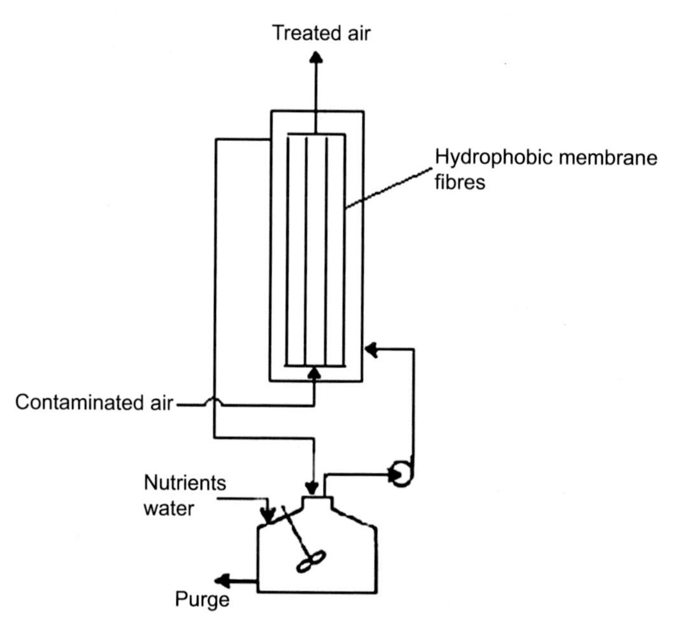

Figure 12.5 Simplified diagram of a microbial membrane bioreactor.
Adapted from van Ras, N., Krooneman, J., Ogink, N., Willers, H., D'Amico, A., di Natale, C., et al., 2005. Biological air filter for air-quality control. In: Wilson, A. coordination, Elmann-Larsen, B., ESA SP-1290 (Eds.), Microgravity Applications Programme: Successful Teaming of Science and Industry, ESTEC, Noordwijk, Netherlands. ESA Publications Division, ISBN 92-9092-971-5, 2005, pp. 270–280; Soreanu, G., Béland, M., Falletta, P., Edmonson, K., Svoboda, L., Al-Jamal, M., et al., 2011. Approaches concerning siloxane removal from biogas – a review. Canadian Biosystems Engineering 53, 8.1-8.18; Soreanu, G., Lishman, L., Dunlop, S., Behmann, H., Seto, P., 2010b. An assessment of oxygen transfer efficiency in a gas permeable hollow fibre membrane biological reactor. Water Science and Technology 61 (5), 1165–1171; and Zhao, K., Xiu, G., Xu, L., Zhang, D., Zhang, X., Deshusses, M. A., 2011. Biological treatment of mixtures of toluene and n-hexane vapours in a hollow fibre membrane bioreactor. Environmental Technology 32 (6), 617–623.

12.3.2.4 Hybrid modules

Enhanced performance hybrid modules can be configured by coupling bioreactors with nonbiological units for indoor air purification. Such configurations are in early stages of development and are not yet commercialized. So far, the following combinations appear to be suitable for indoor air treatment (Luengas et al., 2015):

1. Bioreactor – Photocatalytic oxidation reactor
2. Bioreactor – Adsorption unit

The nonbiological units can act as pre- or posttreatment stage within these hybrid modules. If located before the bioreactor, the photocatalytic reactor can destroy the recalcitrant compounds (eg, trichloroethylene), facilitating their subsequent biodegradation. If located after the bioreactor, the photocatalytic reactor can act as a disinfection unit, as it destroys the potential microorganisms released from the bioreactor. For example, spore deactivation efficiency for a fungal perlite-based BF coupled with photocatalytic oxidation reactor was 98% (Luengas et al., 2015). The photocatalytic stage could sustain the overall system performance when the biological process is inhibited or limited due to specific pollutant or load fluctuations (Luengas et al., 2015). Alternatively, the biological and adsorption processes could be achieved in the same unit, by using a mixed bed tailored for these purposes. This last configuration option supports fluctuating conditions during the start-up and steady operation, but also starvation periods (Luengas et al., 2015).

12.3.3 Evaluation of bioreactor performance

The common criteria for the evaluation of air treatment bioreactors are as follows (Soreanu et al., 2009, 2010c; Dastous et al., 2008; Devinny et al., 1999; Wang et al., 2013):

Removal efficiency (RE): Indicates the percentage of pollutant removed by the bioreactor. Under optimum conditions, an RE of 100% can be obtained.

$$\mathrm{RE} = \frac{c_0 - c_f}{c_0} \cdot 100, \% \qquad [12.1]$$

Loading rate (LR): Indicates the amount of pollutant loaded per media (eg, packing bed) and time; it can be expressed as volumetric or surface loading rate:

$$\mathrm{LR} = \frac{Q \cdot c_0}{V}, \ \mathrm{g}/\left(\mathrm{m}^3 \cdot \mathrm{d}\right) \qquad [12.2]$$

$$\mathrm{LR} = \frac{Q \cdot c_0}{A}, \ \mathrm{g}/\left(\mathrm{m}^2 \cdot \mathrm{d}\right) \qquad [12.3]$$

Elimination capacity (EC): Indicates the amount of pollutant removed per media (eg, packing bed) and time. EC increases with the increase of LR up to a critical value (corresponding to a maximum RE), after which the behavior changes. Thus, the maximum EC is different than the critical EC. EC values between 10 and 300 $g/(m^3h)$ are common for conventional bioreactors.

$$\mathrm{EC} = \frac{Q\cdot(c_0 - c_f)}{V},\ \mathrm{g}/\left(\mathrm{m}^3\cdot\mathrm{d}\right) \quad [12.4]$$

$$\mathrm{EC} = \mathrm{LR}\cdot\mathrm{RE} \quad [12.5]$$

Carbon dioxide production (P_{CO_2}): Considered an indicator of the biological process, being associated with the biomass production and oxidation of the pollutant. The formation of biomass is weak when the theoretical (calculated for oxidation of pollutant) and the actual mass ratios of CO_2 produced/organic pollutant oxidized are about the same.

$$P_{\mathrm{CO_2}} = \frac{Q\cdot\left(c_{f(\mathrm{CO_2})} - c_{0(\mathrm{CO_2})}\right)}{V},\ \mathrm{g}/\left(\mathrm{m}^3\cdot\mathrm{d}\right) \quad [12.6]$$

where c_0 and c_f = initial (inlet) and final (outlet) pollutant concentration; $c_{0(\mathrm{CO_2})}$ and $c_{f(\mathrm{CO_2})}$ = initial and final CO_2 concentration; Q = gas flow rate; V = media bulk volume; A = media area.

In addition, some indicators are also useful for bioreactor diagnosis and design (Soreanu et al., 2009, 2008; Dastous et al., 2008; Devinny et al., 1999; Iranpour et al., 2005):

The pressure drop: An indicator of the media fouling, increasing with the biomass production. From this point of view, the control of the biomass production in bioreactors treating high biodegradable compounds could be a challenge. The maximum pressure drop in a typical BF should be less than 10 cm of water, from both technical as well as economical points of view.

Empty bed residence time (EBRT), true residence time (τ), and interstitial gas velocity (v): These indicators are important for the design purposes and can reveal the influence of the media characteristics (eg, packing bed voidage or porosity) on the process performance. In practice, the bioreactors are usually operated at residence time ranging between seconds and minutes. Longer residence time (low air flow rate or high reactor height) is usually associated with higher performance.

$$\mathrm{EBRT} = \frac{V}{Q},\ \mathrm{min} \quad [12.7]$$

$$\tau = \mathrm{EBRT}\cdot\alpha,\ \mathrm{min} \quad [12.8]$$

$$v = \frac{Q}{S\cdot\alpha},\ \mathrm{cm/min} \quad [12.9]$$

where α = porosity (eg, volume of void space/volume of filter material); S = cross-section area.

12.3.4 Factors influencing bioreactor performance

As mentioned earlier, the bioprocess performance is strongly influenced by the nature of the pollutant, especially in terms of solubility, volatility, and affinity for porous materials, which are directly related to the mass-transfer rate from gas to water and other system compartments, but also biodegradability of the pollutant. Considering a favorable scenario from this point of view, the following important factors influencing bioreactor performance are as follows:

Type of microorganisms: Indigenous or specific isolated microorganisms can be used for pollutant removal from air. For example, the compost used in the biofiltration process can contain 1 billion microorganisms per gram (Shareefdeen et al., 2005). Common bacteria identified in the microbial bioreactors treating air fluxes are *Pseudomonas*, *Alcaligenes*, *Bacillus*, *Xanthomonas*, *Mycobacterium*, *Rhodococcus*, and *Xanthobacter*. Some fungi such as *Phanerochaete chrysosporium, Tramtes versicolor, Pleurotus ostreatus, and Bjerkandera adusta* are also able to carry out the elimination of air pollutants, including hydrophobic compounds such as benzene or styrene. The microbial bioreactors are typically studied at mesophilic conditions (15−40°C), but there are microorganisms that can efficiently perform air treatment under psychrophilic (<10°C) or thermophilic conditions (>45°C). Microbial activity is usually favorable for a pH range between 4 and 8, with fungi being more tolerant to acid conditions and bacteria more tolerant to alkaline conditions (Shareefdeen et al., 2005).

Plant-assisted bioreactors are subject to a similar approach in terms of their rhizosphere microorganisms (exemplified in the next paragraph).

Type of plants: Plants such as potted *Hemigraphis alternata* (Purple waffle), *Tradescantia pallida* (Purple heart), *Hedera helix* (English Ivy), *Asparagus densiflorous* (Asparagus fern), *Hoya camosa* (Variegated wax), and *Crassula portulacea* (Crassulaceae) as well as hydroponically grown plants such as *English Ivy, Dracaena godseffiana, Spathiphyllum Mauna Loa* (Peace lily), *Adiantum raddianum, Rhododendron obtusum, Vriesea splendens, Dieffenbachia picta*, and *Plagiomnium cuspidatum* have been studied with promising results for VOC removal from indoor air. Also, PPs such as *Eucalyptus viminalis, Magnolia kobus, Populus nigra, Nicotiana tabacum*, and *Leucaena leucocephala* appear to be able to remove inorganic compounds from air. Rhizosphere microorganisms such as bacteria, followed by fungi, protozoa, and algae contribute to pollutant removal in plant bioreactors. Additionally, some bacteria from genera *Pseudomonas, Enterobacter, Azotobacter, Alcaligenes, Arthrobacter, Bacillus, Serratia*, and *Rhisobium* can stimulate the plant growth (Soreanu et al., 2013).

Water content: (micro)Biological activity is negatively affected by a dry environment. For example, the water content of a BF media should be maintained at around 40−60% by weight (Hajizadeh and Rezaei, 2013; Baltrėnas et al., 2015; Padhi and Gokhale, 2014). Water is provided through the water or nutrient solution supply and the gas prehumidification. Excessive water content should be avoided as it leads to the water hold up, preferential gas channeling, pollutant mass-transfer limitation, and appearance of anaerobic microzones with the development of bad-smelling anaerobic bioprocesses (Ondarts et al., 2010; Shareefdeen et al., 2005). Decreasing water content (media drying) leads to percolation and channeling. Low water content can decrease the RE of hydrophilic compounds, but could enhance the RE of hydrophobic compounds. For example, the compost can become hydrophobic if dried (Ondarts et al., 2010).

Nutrient solution composition: Operating the microbial bioreactors under nutrient-rich conditions is not economical and is often related to the excessive biofilm development, especially when treating easily biodegradable compounds. A limitation in nutrient supply could reduce the microorganism growth (Shareefdeen et al., 2005). According to Dastous et al. (2008), the nitrogen requirement for the readily biodegradable compounds (such as ethanol) is smaller than in the case of lower biodegradable compounds (such as methanol). The same authors show that the importance of nitrogen concentration decreases with the LR. Careful attention should be paid to the selection of N-nutrient sources, as some of them are not quickly converted or can have an inhibitory effect on the microorganisms growth (Soreanu et al., 2005; Veillette et al., 2012).

Particularly when using plants for removal of pollutants from air, a competition for nutrients between plants and the associated rhizosphere microorganisms can occur, especially under nutrient- limiting conditions. On the other hand, the microorganisms support plant life in highly polluted environments (Soreanu et al., 2013).

Alternatively, the initial adding of slow-release solid fertilizers to the packing media of microbial or plant BFs and periodic water irrigation could replace the procedure based on nutrient solution supply (Son and Striebig, 2001; Torpy et al., 2014).

Gaseous phase composition: For both the microbial and the plant bioreactors, an increase in pollutant concentration can result in bioreactor saturation with pollutants, with some pollutants being subject to desorption or to simply passing without treatment, and thus the decrease of the RE (Cruz et al., 2014a; Devinny et al., 1999). Also, high pollutant concentration can inhibit the microbial growth and activity, can affect the oxygen availability, or can be toxic to the plants, which in turn affects the bioreactor performance (Padhi and Gokhale, 2014; Soreanu et al., 2013).

Uptake of the pollutants by plants appears to be independent of the functional group or the molecular weight of the treated pollutants, but for the degradation studies the functional groups could play an important role in the removal performance (Cruz et al., 2014a). For typical BFs, the following biodegradation order is mentioned by Padhi and Gokhale (2014): oxygenated hydrocarbons > linear alkanes > aromatic hydrocarbons. Chlorinated VOCs appears to be the most recalcitrant for biodegradation.

The performance of the bioreactors treating mixed pollutants is certainly different than in the case of single-pollutant removal. In this case, competition for substrate or antagonistic effects (eg, desorption of a pollutant by other pollutant, etc.) can occur (Cruz et al., 2014a; Ménard et al., 2012; Zhao et al., 2011).

Gas flow rate: When using dynamic bioreactors, an increase in air flow rate results in the decrease of the residence time and incomplete pollutant capture and bioconversion (Devinny et al., 1999; Soreanu et al., 2009). Also, a desiccation phenomena could occur in the BFs operated at high air flow rate, which in turn affects the biofiltration performance. For this reason, the air is prehumidified in special towers (this option also removes the PM) for at least 95% saturation degree (Leson and Winer, 1991). In practice, air flow rates between 1000 and 50,000 m^3/h at low pollutant concentration (up to 1000 mg/m^3) are considered suitable for conventional biofiltration (Devinny et al., 1999). Typical gas flow rates used in laboratory pilot-scale biofiltration studies are up to 1–2 m^3/h.

Temperature: For plants, an increase in temperature determines an increase in the permeability of the cuticle and thus of the pollutant diffusion rate into the plant. Moreover, microbial growth in soil and other solid media is positively influenced by the temperature increase, which in turn can enhance the overall bioreactor performance (Cruz et al., 2014a). On the other hand, an increase in temperature could diminish the dissolution of the pollutant (and oxygen) in the water phase and sorption in the solid phase, which can affect the bioreactor performance (Padhi and Gokhale, 2014; Leson and Winer, 1991). For a dynamic plant bioreactor treating indoor air, the favorable impact of the temperature decrease on the removal performance was greater than the reduction in microbial activity (Darlington and Dixon, 2000; Darlington et al., 2001).

Light: Light levels in the indoor environment are about 9–14 μmol/m^2/s, which is considered low for the photosynthetic activity of many plants (Cruz et al., 2014a). Increasing light intensity usually favors the plants' biofiltration performance, especially when the pollutant uptake occurs via stomata. However, no significant difference in terms of the performance between light or dark experiments has been observed in some studies, likely due to a microbial dominant route rather than phytodegradation (Cruz et al., 2014a; Soreanu et al., 2013). Light is an important parameter in photobioreactors (Zhang et al., 2007; Syed et al., 2006).

12.4 Application of biotechnologies for improving air quality in indoors

12.4.1 Opportunities and challenges of using bioreactors in indoor air treatment

The use of the biotechnologies for indoor air applications is subject to some criteria accomplishment, especially in terms of RH and spore contamination (van Ras et al., 2005; Luengas et al., 2015). Particularly, RH should be maintained between 40% and 60% from the point of view of human health, but also in order to avoid mold development and building deterioration (Pacheco-Torgal and Jalali, 2012; Soreanu et al., 2013). All these issues can be overcome by using membrane-based bioreactors or by using appropriate operating conditions; for example, by operating the bioreactors at cooler temperatures (Soreanu et al., 2013; Darlington et al., 2001, 2000).

There is no evidence on the pathogenic fungal spores or bacterial contamination of space where plant bioreactors were installed. Moreover, dust and mold spores are retained on the plant leaves or captured in the liquid phase of the bioreactors (Soreanu et al., 2013). Plant leaves should be cleaned periodically in order to maintain leaf–gas exchange (Soreanu et al., 2013). Conventional bioreactors can also reduce the level of microorganisms in the processed air (Leson and Winer, 1991; Chung et al., 2004).

The start-up phase could be long (eg, months), especially for the biodegradation of poorly biodegradable VOCs such as methyl-tert-butyl ether (Iranpour et al., 2005).

In addition, volatiles released (eg, such as ethylene) as a part of normal plant metabolism and their accumulation in a closed environment can occur when using botanical processes; however, this phenomena does not appear to be a major health concern. These releases remain at extremely low levels that do not affect the IAQ under normal environmental conditions (van Ras et al., 2005). However, an integration of VOC removal rates with VOC emission rates should be considered for a more realistic evaluation of bioreactor performance in practice (Cruz et al., 2014a). Also, CO_2 release during VOC mineralization or respiration of plants does not accumulate in indoor air, being fixed by plants during the photosynthetic process (Soreanu et al., 2013).

Commercially available indoor air purification plant-based biosystems are produced by companies such as (Torpy et al., 2015): Nedlaw Living Walls (Canada), Phytofilter Technologies Inc. (United States), CASE (United States), Junglefy (Australia). As mentioned earlier, such biosystems have the potential to contribute to the green building concept. For example, the Cambridge City Hall (ON, Canada), which was equipped in 2007 with a 110 m^2 PBTF (credited to Nedlaw Living Walls) in a central atrium, in 2009 received the Gold LEED Certification (Leadership in Energy and Environmental Design) (Soreanu et al., 2013).

12.4.2 Removal of specific indoor air pollutants

Many of the studies related to indoor air purification by biological processes were performed by using plant-based biosystems. Several aspects related to the process performance of these systems, pollutant conversion pathways and plant psychology, are reviewed in Soreanu et al. (2013), Cruz et al. (2014a), and Torpy et al. (2015). Additional information is presented in the following in order to outline the possibilities of using various biotechnological tools (microbial or plant-based) from the perspective of the indoor air pollutant type. According to Hort et al. (2014), the concentration of a single gaseous contaminant in indoor air can range from less than 1 $\mu g/m^3$ up to several hundred $\mu g/m^3$. On the other hand, the indoor air contains a wide range of pollutants and their cumulative or interaction effect could be significant. With some exceptions, the following studies are usually carried out close to such concentration range.

12.4.2.1 Microorganism removal

The potential of PPs in bacterial aerosol removal was investigated by Choi et al. (2013). This study confirms that the antimicrobial effects of various plant extracts can be extrapolated toward air purification field. The antimicrobial effects appear to be related to some plants' secondary metabolites with an antiseptic effect such as terpenoids and other natural VOCs emitted by plants. The experiments were performed using a plastic air-sealed clear chamber of 300 L, equipped with 20 plant pots (*Cupressus macrocarpa*) already acclimated. Soil has been covered in order to avoid the contamination of processed air by soil bacteria. The system was operated at 27°C and a fluorescence light intensity of 81 μmol/(m^2 s). Air was passed through the chamber at a flow rate of 200 mL/min for experimental periods of one or a few days. The exhaust air was sparked through different culture media inoculated by bacteria of the following species: *Bacillus subtilis*, *Staphylococcus aureus*, *Escherichia coli*, and *Pseudomonas aeruginosa*, in order to investigate the antimicrobial effects of the plant emissions. In addition, the CO_2 exchange was evaluated. The presence of light induced an increase in O_2 concentration and a decrease in CO_2 concentration in the exhaust gas. The following VOCs have been identified in the exhaust gas: α-pinene, β-pinene, and linalool. These volatiles were associated with the inhibition of the growth of gram-positive bacteria (*B. subtilis*, *S. aureus*), but not of gram-negative bacteria (*E. coli*, *P. aeruginosa*). Other plants that release volatile components with an antimicrobial effect against gram-negative bacteria should be considered in the last case.

According to Chung et al. (2004), the bioaerosol concentration in the outlet air of the BFs is similar to that encountered in indoor air or slightly higher than in outdoors. Moreover, Leson and Winer (1991) showed that the BFs were actually able to reduce the high concentrations of microorganisms in the processed air.

12.4.2.2 Formaldehyde removal

Sigawia et al. (2011) show that airborne formaldehyde, one of the most common and toxic indoor air pollutants, can be removed in a continuous fluidized bed bioreactor by using an immobilized formaldehyde-oxidizing enzyme from yeasts *Hansenula polymorpha*. The concentration of formaldehyde in inlet air ranged from 0.3 ppm to 18.5 ppm and the removal efficiency of the bioreactor was above 90%.

The potential of plants in formaldehyde removal was demonstrated for different biosystems. For example, potted *Fatsia japonica* and *Ficus benjamina* were tested by Kim et al. (2008) for formaldehyde removal by using a 1 m^3 airtight chamber equipped with three pots of the same plant species, already acclimated. The plants were exposed for 5 h (during the day or the night) to an air flow rate of 6 L/min and a formaldehyde concentration of 2 μL/L. The light intensity was 20 ± 2 μmol/m^2/s. The time interval to reach 50% removal performance was smaller for *F. japonica* than for *F. benjamina* (96 min versus 123 min, respectively). For both plants, the aerial plant part's contribution to the removal efficiency was significant higher during the day than during the night. Particularly, the root zone contribution to the removal efficiency was dominant during the night, but comparable with the contribution of the aerial parts during the day. About 90% of formaldehyde removed by the root zone was due to the microorganisms and roots, while 10% was due to the adsorption by the growing medium.

Su and Liang (2015) show that formaldehyde is readily taken up from air by the *Chlorophytum comosum* leaves within an air–plant–water system. Most of the formaldehyde accumulated in shoots, while a portion was transported downward to the rizosphere solution and a portion was volatilized. The same plant *C. comosum* performed better than aloe (*Aloe vera*) and golden pothos (*Epipremnum aureum*) in a PP-soil system used for formaldehyde removal (Xu et al., 2011).

Formaldehyde can be toxic to plants at very low concentrations. Zhou et al. (2015) proposed a specific designed transgenic geranium that was able to purify a formaldehyde-rich indoor environment, without inhibition by long-term exposure conditions.

12.4.2.3 Toluene removal

Toluene was used as an indoor VOC model in a biofiltration study carried out by Hort et al. (2014). A hybrid system, consisting of a typical BF coupled to an adsorber, was used for the treatment of air contaminated with toluene. The BF (10 cm diameter) was equipped with mature green waste compost (23 cm height). The adsorber (25 cm diameter) was equipped with activated carbon (over 2.9 cm height) and was located after the BF. Particularly, the compost is characterized by poor adsorption capacity for toluene. The experiment was carried out for about 3 months, at an air flow rate of 8.28 L/min and inlet toluene concentrations between 17 $\mu g/m^3$ and 52 $\mu g/m^3$, with the peaks close to 733 $\mu g/m^3$ for 2–3 h every day. LR ranged between 4.2×10^{-3} and 1.4×10^{-2} g/(m^3 h) under normal conditions, while the LR during the peaks was about 0.20 g/(m^3 h). The EBRT was about 13 s for the BF and 0.1–0.27s for the adsorber. The BF was able to efficiently remove the toluene under the entire loading rate range, but was sensitive to the changes in inlet concentrations, which in turn were well handled by the adsorber. Overall, the removal efficiency of the hybrid system was about 100%. Moreover, the adsorber was able to deal with the emissions of particles and microorganisms from the BF. Based on these results, the authors recommend the hybridization of biological process as the most promising option for the improvement of IAQ.

In a PPs study, Sriprapat et al. (2014a) screened several plant species in terms of their potential to remove toluene or ethylbenzene from indoor air. The plant species were cultivated in pots (0.1×0.1 m^2) containing soil and coco coir in a ratio of 1:1. The plants (covered pot) were exposed to an initial contaminant concentration of 20 ppm (toluene or ethylbenzene) in separate glass chambers of 15.6 L, at 32°C, under 12 h of dark and light cycles. A leaf area of 0.013 m^2 was considered for each experiment. The highest toluene removal efficiency was recorded for *Sansevieria trifasciata*, while the highest ethylene benzene removal efficiency was recorded for *C. comosum* (eg, about 80% and 90%, respectively, after 72 h experiment).

12.4.2.4 Particulate matter removal

Gawrońska and Bakera (2014) show that higher plants such as *C. comosum* L. (spider plant) can be used not only for gaseous pollutant removal in indoor air (eg, formaldehyde, benzene, toluene, cigarette smoke, ozone, nitrogen dioxide, carbon oxide, ammonia), but also for phytoremediation of PM, which is one of the most harmful pollutants for human health. Their study was carried out in free-air conditioning rooms of buildings with different activities (a dental clinic, a perfume-bottling room, an office, an apartment, a suburban house). Potted C. comosum L and aluminum plates (used as control surface for PM deposition) were exposed to the indoor air of the previously mentioned rooms for 2 months, under natural light and ambient temperature (18–23°C). The results indicate that these plants have the ability to accumulate on their leaves a significant amount of PM with different fraction sizes (0.2–100 μm) in all rooms. The amount of PM accumulated on aluminum plates was significantly lower than the amount accumulated on plant leaves, indicating that factors other than gravitation were involved in the PM accumulation on leaves. The dominant accumulated fractions were the large PM. On the other hand, fine PM accumulated to a greater extent as in-wax PM than surface PM, as they attached more tightly to leaves and phytostabilized more effectively.

12.4.2.5 Mixed volatile organic compound removal

A microbial membrane bioreactor investigated by van Ras et al. (2005) was able to reduce VOCs such as chlorobenzene, 1,2-dichloroethane, alcohols, methane, acetone, and BTEX (benzene, toluene, ethylbenzene, xylene), at an elimination capacity of 2–26 $g/m^3/h$ for inlet contaminant concentrations of 20–200 mg/m^3 and retention times of 8–23 s. The bioreactor showed an operational stability up to a year and its performance was not affected by long starvation or microgravity. A plant membrane bioreactor prototype was also able to significantly reduce the level of methylethylketone and toluene from an air stream, which was considered sufficient by van Ras et al. (2005) to demonstrate the proof of concept.

The role of microorganisms in plant bioreactors treating VOCs was investigated by Russell et al. (2014) at Drexel University, where a five-story vertical biowall was installed. The biowall is a PBTF equipped with plants rooted into an inorganic, porous textile material. The plant roots are continuously irrigated with recirculated nutrient solution. The indoor air is drawn through the wall using a fan, allowing the VOCs to transfer from air to the liquid, where they are subject to biodegradation by microorganisms. The treated air is then delivered to other zones in the building via the mechanical system. VOCs removal efficiency of about 25–90% can be obtained in such a biosystem. The following plants have been monitored: *Croton* "Mammy", Ficus elastica (rubber tree), *Schefflera arboricola* "Gold Capella", *S. arboricola*, unknown *Ficus* sp. and *Algerian ivy*. The biowall-grown roots exhibited enriched levels of bacteria from the genus *Hyphomicrobioum*, which are VOCs degraders that are able to break down aromatic and halogenated compounds as often found in indoor environments. According to these authors, these bacteria play an essential role in VOC removal in plant bioreactors treating indoor air. The proof of the PBTF concept has been previously reported by Darlington et al. (2000, 2001) and Darlington (2004) and summarized by Soreanu et al. (2013). Similarly, a vertical planted cylinder equipped with hydroponic plants and trickled with nutrient-rich water was recently investigated within a European project (CETIEB report). This biosystem was able to reduce the VOC level by 50% in a registrar's office at the University of Stuttgart.

A BTF with a diameter of 9 cm, height of 100 cm, and equipped with pottery pieces was tested by Lu et al. (2010) for VOC removal from the indoor air of a newly renovated room. The BTF was inoculated with sludge from a municipal sewage treatment plant and operated at a gas flow rate of 600 L/h, surface liquid velocity of 3.14 m/h, pH of 6−7 and temperature of 30°C. When the air stream contained 0−6.5 mg/m^3 formaldehyde, 2.2−46.7 mg/m^3 benzene, 0.5−28.2 mg/m^3 toluene, and 4.1−59.0 mg/m^3 xylene, the VOC removal efficiency was about 100%, 65−70%, 93%, and 85−90%, respectively. The treated air can be recirculated to the room. The following bacterial species isolated from BTF have been associated with this performance: *Pseudomonas* sp., *Kocuria sp.*, *Arthrobacter sp.* and *Bacillus sp.* Increasing the temperature up to 40°C resulted in a decrease of BTF performance by about 1.3 for all VOCs except formaldehyde, for which degradation was attributed to the thermotolerant bacteria. The better performance in formaldehyde removal was explained by its higher solubility in water in comparison with the other tested hydrophobic compounds.

12.4.2.6 Mixed volatile organic compound and inorganic gaseous compound removal

A BF with 0.1 m inner diameter, 0.40 m height, filled with compost was tested by Ondarts et al. (2010) for the removal of VOCs (butanol, butyl acetate, limonene, trichloroethylene (TCE), toluene, undecane, formaldehyde) and inorganic compounds (nitrogen dioxide) from indoor air. The concentration of these compounds in the inlet air was within the range of 32 (eg, for undecane) and 140 (eg, for NO_2) μg/m^3. The BF was operated for 75 days at a gas flow rate of 0.6 m^3/h (EBRT of 23 s) and a RH of about 80%. The compost was watered every 2−5 days, depending on the experimental design. All pollutants except TCE were removed with removal efficiency close to 100%. After 40 days, a decrease in removal efficiency down to 55.5, 77.8, and 13.9 was observed for butanol, formaldehyde, and toluene, likely due to the drying of the packing bed. NO_2 removal efficiency was 86.3−99.6%. TCE was not biodegraded, but it was subject to sorption and desorption phenomena over the course of the test period.

12.4.2.7 Reduction of carbon dioxide level

Indoor CO_2 concentration should be maintained below the maximum CO_2 concentration for comfort acceptability, which is 0.1% (Torpy et al., 2014, 2015; Zhang et al., 2007). Torpy et al. (2014) investigated the possibility of reducing the CO_2 level via photosynthetic process of plants under different light conditions. Several common indoor plants, 12 months in age, were investigated. The plants were grown in a mixture of composted hardwood, sawdust, composted bark fines, and coarse river sand, amended with a slow-release fertilizer, before acclimatization to light treatments. The highest CO_2 reduction rate (approximately 657 mg CO_2/m^2 leaf area/h) was recorded for the high light acclimatized *Dypsis lutescens* at 350 μmol PAR/m^2/s. These authors estimated that about 15 potted *D. lutescens* could reduce ventilation requirements by approximately 6%.

According to Sriprapat et al. (2014b), the plant species could be classified upon the type of photosynthetic activity (eg, C3, crassulacean acid metabolism (CAM), facultative CAM, C4), which involves a different CO_2 uptake. A change in photosynthetic activity can influence the VOC removal by changing the VOC stomatal conductance and VOC entry into the plant (Cruz et al., 2014b; Sriprapat et al., 2014b). An increase in photosynthetic activity may increase the root exudate production and the associated microbial activity involved in VOC degradation (Cruz et al., 2014b).

New trends refer to CO_2 capture in microalgae photobioreactors and enzyme membrane bioreactors (Zhang et al., 2007; Sardá and Pioz, 2015). For example, the CO_2 removal rate of a microalgae-based photobioreactor of 3 L volume and containing 7.2 g chlorella dry weight was 0.118 g/(L h), under the following operating conditions: air flow rate of 0.6 L/min air, CO_2 concentration of 0.1%, light intensity of 500 lux, temperature of 25–30°C, and pH of 9.5–8.5. An enhanced CO_2 removal efficiency (about 60%) was reported for novel membrane-microalgae photobioreactors (Zhang et al., 2007). Moreover, a laboratory-scale enzyme-based membrane system using carbonic anhydrase demonstrated a potential for CO_2 capture of about 90% (Figueroa et al., 2008).

12.4.3 Global performance relevance

In practice, the performance of air treatment biotechnologies should be evaluated in terms of not only a single pollutant removal, but also in terms of mixed pollutant removal and the environmental impact associated with the implementation of the considered biotechnology. Overall, biotechnologies are considered to have a low environmental impact compared to other technologies, but different biotechnologies can have a different impact. The environmental profile of air treatment biotechnologies can be obtained through Life Cycle Assessment by considering several environmental impacts such as (Alfonsín et al., 2013; Soreanu, 2014) global warming, eutrophication, acidification, photochemical oxidation, malodorous air, human toxicity, and aquatic toxicity. This approach is relatively new in the field of biofiltration for air pollution control. Alfonsín et al. (2013) performed the environmental assessment of four BFs equipped with different packing materials (1: spherical clay pellets covered with compost; 2: coconut fiber and sludge-based carbon; 3: peat and heather; 4: pine bark). The air flux contaminated by a complex gaseous mixture (ammonia, hydrogen sulphide, dimethyl sulphide, dimethyl disulphide, methyl isobutyl ketone, α-pinene, hexanal, toluene, limonene, butyric acid) has been treated in BFs. The lowest impact values have been obtained for the bioreactor 2, while the bioreactor 1 obtained the worst marks for most of the considered impacts, likely due to the significant pressure drop and the considerable energy demand (Alfonsín et al., 2013). The same authors show the importance of the balance between the environmental assessment and the operational behavior in terms of removal efficiency in the selection of bioreactors.

12.4.4 Future trends

The green building concept is specifically related to indoor environmental quality and human health, while focusing on minimizing impact to the environment through reduction in energy and water usage (Allen et al., 2015). A multicriteria assessment of various biotechnologies should be performed in order to identify the best option for indoor air treatment method.

For instance, an estimation of the number of plants that could improve the IAQ in a PPs scenario is still contradictory and appears to be far beyond the reasonable practicality (Soreanu et al., 2013). In contrast, the dynamic engineered systems are more efficient and suitable for esthetical integration within the green building design (Soreanu et al., 2013; Wang and Zhang, 2011). Careful monitoring of nutrient levels and assuring their optimum concentrations could contribute to the enhanced performance of these biosystems. Transgenic plants could be considered in the design of plant bioreactors for both environmental and economical benefits (Soreanu et al., 2013). Postbioreactor antimicrobial materials could be developed for integration in a hybrid module (Porosnicu et al., 2015). Overall, enhanced performance hybrid modules and membrane bioreactors (microbiological or botanical) that can avoid contamination of air by fungal and actinomycetes spores and can handle RH and the irregular concentrations (pollution peaks) in indoor air could be further developed (Soreanu et al., 2013; Hort et al., 2014).

12.5 Conclusions

This study addresses the issues of indoor air pollution and the possibilities of improving IAQ via biotechnological tools. The available information indicates that, with an appropriate design, the biotechnologies for indoor air purification can be implemented within the green built environment. Overall, the investigation of the pollutant removal mechanism should be performed to deal with the multiple pollutants under a range of conditions. An understanding of the impact of biotechnology in terms of environmental and economical benefits, as well as protection of human health, is critical to its acceptance for implementation in indoor environments. An appropriate combination with other technologies appears to be able to assure the expected results.

References

Alfonsín, C., Hernández, J., Omil, F., Prado, O.J., Gabriel, D., Feijoo, G., et al., 2013. Environmental assessment of different biofilters for the treatment of gaseous streams. Journal of Environmental Management 129, 463–470.

Allen, J.G., MacNaughton, P., Laurent, J.G.C., Flanigan, S.S., Eitland, E.S., Spengler, J.D., 2015. Green buildings and health. Current Environmental Health Reports 2, 250–258.

Baltrėnas, P., Zagorskis, A., Misevičius, A., 2015. Research into acetone removal from air by biofiltration using a biofilter with straight structure plates. Biotechnology and Biotechnological Equipment 29 (2), 404–413.

CETIEB – Final publishable summary. http://cordis.europa.eu/docs/results/285/285623/final1-cetieb-finalreport-final-publishable-summary-text-incl-figures.pdf (accessed 30.10.15.).

Choi, Y.K., Lee, S., Jeon, H.J., Jung, H., Kim, H.S., Kwon, K.K., Kim, S.M., Song, K.G., Lee, I.S., Lee, B.U., Kim, H.J., 2013. Development and evaluation of a plant-based air filter system for bacterial growth control. African Journal of Biotechnology 12 (16), 2027–2033.

Chung, Y.C., Lin, Y.Y., Tseng, C.P., 2004. Operational characteristics of effective removal of H_2S and NH_3 waste gases by activated carbon biofilter. Journal of the Air and Waste Management Association 54, 450–458.

Cruz, M.D., Christensen, J.H., Thomsen, J.D., Müller, R., 2014a. Can ornamental potted plants remove volatile organic compounds from indoor air? A review. Environmental Science and Pollution Research International 21, 13909–13928.

Cruz, M.D., Müller, R., Svensmark, B., Pedersen, J.S., Christensen, J.H., 2014b. Assessment of volatile organic compound removal by indoor plants – a novel experimental setup. Environmental Science and Pollution Research International 21, 7838–7846.

Darlington, A., Dixon, M., 2000. The biofiltration of indoor air III: air flux and temperature and removal of VOCs. In: Proceedings of the 2000 USC-TRG Conference on Biofiltration and Air Pollution Control, Los Angeles, California, October 19–20, 2000, pp. 269–276.

Darlington, A., Chan, M., Malloch, D., Pilger, C., Dixon, M.A., 2000. The biofiltration of indoor air: implications for air quality. Indoor Air 10, 39–46.

Darlington, A.B., Dat, J.F., Dixon, M.A., 2001. The biofiltration of indoor air: air flux and temperature influences the removal of toluene, ethylbenzene and xylene. Environmental Science and Technology 35, 240–246.

Darlington, A., 2004. Room Air Cleansing Using Hydroponic Plants, United States Patent No. US 672791B2, 2004.

Dastous, P.A., Nikiema, J., Soreanu, G., Bibeau, L., Heitz, M., 2008. Elimination of methanol and ethanol by biofiltration: an experimental study. Water Air and Soil Pollution – Focus 8 (3), 275–286.

Devinny, S.J., Deshusses, A.M., Webster, S.T., 1999. Biofiltration for Air Pollution Control. CRC/Lewis Publishers, Boca Raton, FL.

Estrada, J.M., Bernal, O.I., Flickinger, M.C., Muñoz, R., Deshusses, M.A., 2015. Biocatalytic coatings for air pollution control: a proof of concept study on VOC biodegradation. Biotechnology and Bioengineering 112 (2), 263–271.

European Commission, 2006. Integrated Pollution Prevention and Control – Reference Document on Economics and Cross-Media Effects.

Figueroa, J.D., Fout, T., Plasynski, S., McIlvried, H., Srivastava, R.D., 2008. Advances in CO_2 capture technology – the U.S. Department of Energy's carbon sequestration program (Review). International Journal of Greenhouse Gas Control 2, 9–20.

Gawrońska, H., Bakera, B., 2014. Phytoremediation of particulate matter from indoor air by *Chlorophytum comosum* L. plants. Air Quality, Atmosphere and Health. http://dx.doi.org/10.1007/s11869-014-0285-4.

Guieysse, B., Hort, C., Platel, V., Munoz, R., Ondarts, M., Revah, S., 2008. Biological treatment of indoor air for VOC removal: potential and challenges. Biotechnology Advances 26 (5), 398–410.

Hajizadeh, Y., Rezaei, M., 2013. Biodegradation of formaldehyde from contaminated air using a laboratory scale static-bed bioreactor. International Journal of Environmental Health Engineering 2 (6), 18–25.

Hort, C., Platel, V., Socharda, S., Luengas Munoz, A.T., Ondarts, M., Reguer, A., et al., 2014. A hybrid biological process of indoor air treatment for toluene removal. Journal of the Air and Waste Management Association 64 (12), 1403–1409.

Iranpour, R., Cox, H.H.J., Deshusses, M.A., Schroeder, E.D., 2005. Literature review of air pollution control. Biofilters and biotrickling filters for odor and volatile organic compound removal. Environmental Progress 24 (3), 254–267.

Ivanov, V., Chu, J., Stabnikov, V., 2015. Basics of construction microbial biotechnology (Chapter 2). In: Pacheco Torgal, F., et al. (Eds.), Biotechnologies and Biomimetics for Civil Engineering. Springer International Publishing, Switzerland. http://dx.doi.org/10.1007/978-3-319-09287-4_2.

Kelley, S.T., Gilbert, J.A., 2013. Studying the microbiology of the indoor environment (Review). Genome Biology 14, 202. http://dx.doi.org/10.1186/gb-2013-14-2-202.

Kim, K.J., Kil, M.J., Song, J.S., Yoo, E.H., 2008. Efficiency of volatile formaldehyde removal by indoor plants: contribution of aerial plant parts versus the root zone. Journal of the American Society for Horticultural Science 133 (4), 521–526.

Leson, G., Winer, A.M., 1991. Biofiltration: an innovative air pollution control technology for VOC emissions. Journal of the Air and Waste Management Association 41 (8), 1045–1054.

Lu, Y., Liu, J., Lu, B., Jiang, A., Wan, C., 2010. Study on the removal of indoor VOCs using biotechnology. Journal of Hazardous Materials 182, 204–209.

Luengas, A., Barona, A., Hort, C., Gallastegui, G., Platel, V., Elias, A., 2015. A review of indoor air treatment technologies. Reviews in Environmental Science and Biotechnology. http://dx.doi.org/10.1007/s11157-015-9363-9.

Ménard, C., Ramirez, A.A., Nikiema, J., Heitz, M., 2012. Effect of trace gases, toluene and chlorobenzene, on methane biofiltration: an experimental study. Chemical Engineering Journal 204–206, 8–15.

Missia, D.A., Demetriou, E., Michael, N., Tolis, E.I., Bartzis, J.G., 2010. Indoor exposure from building materials: a field study. Atmospheric Environment 44, 4388–4395.

Ondarts, M., Hort, C., Platel, V., Sochard, S., 2010. Indoor air purification by compost packed biofilter. International Journal of Chemical Reactor Engineering 8, 14 art. A54.

Occupational Safety and Health Administration (OSHA), 2011. Indoor Air Quality in Commercial and Institutional Buildings. OSHA, pp. 3430–3504. https://www.osha.gov/Publications/3430indoor-air-quality-sm.pdf (accessed 26.08.15.).

Pacheco-Torgal, F., Jalali, S., 2012. Earth construction: lessons from the past for future eco-efficient construction. Construction and Building Materials 29, 512–519.

Pacheco-Torgal, F., 2015. Introduction to biotechnologies and biomimetics for civil engineering (Chapter 1). In: Pacheco Torgal, F., Labrincha, J.A., Diamanti, M.V., Yu, C.P., Lee, H.K. (Eds.), Biotechnologies and Biomimetics for Civil Engineering. Springer International Publishing, Switzerland.

Padhi, S.K., Gokhale, S., 2014. Biological oxidation of gaseous VOCs – rotating biological contactor a promising and eco-friendly technique. Journal of Environmental Chemical Engineering 2, 2085–2102.

Pickett, A.R., Bell, M.L., 2011. Assessment of indoor air pollution in homes with infants. International Journal of Environmental Research and Public Health 8, 4502–4520.

Porosnicu, C., Butoi, B., Dinca, P., Jepu, I., Lungu, C.P., Diaconu, M., et al., 2015. Anti-microbial testing of the wall green module developed for sustainable enhancing of wellbeing and comfort in indoor buildings. Extended abstract. In: Cement Based Materials and Environmental Approaches for a Sustainable Agriculture (Proceedings of CSA 2015 Joint International Conference), Iasi, Romania, October 22–25th, 2015, pp. 165–166.

Russell, J.A., Hu, Y., Chau, L., Pauliushchyk, M., Anastopoulos, I., Anandan, S., et al., 2014. Indoor-biofilter growth and exposure to airborne chemicals drive similar changes in plant root bacterial communities. Applied and Environmental Microbiology 80 (16), 4805–4813.

Sardá, R.S., Pioz, J.G., 2015. Architectural bio-photo reactors: harvesting microalgae on the surface of architecture (Chapter 7). In: Pacheco Torgal, F., Labrincha, J.A., Diamanti, M.V., Yu, C.P., Lee, H.K. (Eds.), Biotechnologies and Biomimetics for Civil Engineering. Springer International Publishing, Switzerland.

Shareefdeen, Z., Herner, B., Singh, A., 2005. Biotechnology for air pollution control – an overview. In: Shareefdeen, Singh (Eds.), Biotechnology for Odor and Air Pollution Control. Springer-Verlag, Berlin Heidelberg.

Shimer, D., Phillips, T.J., Jenkins, P.L., Lum, S., Webber, W., Robert, M., et al., 2005. Indoor Air Pollution in California. Report to the California Legislature. California Environmental Protection Agency – California Air Resources Board. http://www.arb.ca.gov/research/indoor/ab1173/rpt0705.pdf (accessed 28.08.15.).

Sidheswaran, M.A., Destaillats, H., Sullivan, D.P., Cohn, S., Fisk, W.J., 2011. Energy efficient indoor VOC air cleaning with activated carbon fiber (ACF) filters. Building and Environment. http://dx.doi.org/10.1016/j.buildenv.2011.07.002.

Sigawia, S., Smutok, O., Demkiv, O., Zakalska, O., Gayda, G., Nitzan, Y., Nisnevitch, M., Gonchar, M., 2011. Immobilized formaldehyde-metabolizing enzymes from Hansenula polymorpha for removal and control of airborne formaldehyde. Journal of Biotechnology 153 (3–4), 138–144.

Soccol, C.R., Woiciechowski, A.L., Vandenberghe, L.P.S., Soares, M., Neto, G.K., Thomaz-Soccol, V., 2003. Biofiltration: an emerging technology. Indian Journal of Biotechnology 2, 396–410.

Son, H.K., Striebig, B.A., 2001. Ethylbenzene removal in a multiple-stage biofilter. Journal of Air Waste Management Association 51 (12), 1689–1695.

Soreanu, G., Al-Jamal, M., Béland, M., 2005. Biogas treatment using an anaerobic biosystem. In: Proceedings of the 3rd Canadian Organic Residuals Recycling Conference, Calgary, AB, Canada, June 1–4, 2005, pp. 502–513.

Soreanu, G., Falletta, P., Béland, M., Edmonson, K., Seto, P., 2008. Study on the performance of an anoxic biotrickling filter for the removal of hydrogen sulphide from biogas. Water Quality Research Journal of Canada 43 (2/3), 211–218.

Soreanu, G., Béland, M., Falletta, P., Ventresca, B., Seto, P., 2009. Evaluation of different packing media for anoxic H_2S control in biogas. Environmental Technology 30 (12), 1249–1259.

Soreanu, G., Falletta, P., Husain, H., Seto, P., 2010a. Process optimization of H_2S removal from biogas in an anoxic biotrickling filter. In: Proceeding of IWA World Water Congress and Exhibition, Montreal, QC, Canada, 19–24 September, Paper. IWA-2653.

Soreanu, G., Lishman, L., Dunlop, S., Behmann, H., Seto, P., 2010b. An assessment of oxygen transfer efficiency in a gas permeable hollow fibre membrane biological reactor. Water Science and Technology 61 (5), 1165–1171.

Soreanu, G., Falletta, P., Béland, M., Edmonson, K., Ventresca, B., Seto, P., 2010c. Empirical modelling and dual-performance optimisation of a hydrogen sulphide removal process for biogas treatment. Bioresource Technology 101 (23), 9387–9390.

Soreanu, G., Béland, M., Falletta, P., Edmonson, K., Svoboda, L., Al-Jamal, M., et al., 2011. Approaches concerning siloxane removal from biogas – a review. Canadian Biosystems Engineering 53, 8.1–8.18.

Soreanu, G., Dixon, M., Darlington, A., 2013. Botanical biofiltration of indoor gaseous pollutants – a mini-review. Chemical Engineering Journal 229, 585–594.

Soreanu, G., 2014. Prevenirea Poluarii si Protectia Mediului. Ed. Performantica. Pollution Prevention and Environmental Protection, Iasi.

Sriprapat, W., Suksabye, P., Areephak, S., Klantup, P., Waraha, A., Sawattan, A., et al., 2014a. Uptake of toluene and ethylbenzene by plants: removal of volatile indoor air contaminants. Ecotoxicology and Environmental Safety 102, 147–151.

Sriprapat, W., Boraphech, P., Thiravetyan, P., 2014b. Factors affecting xylene-contaminated air removal by the ornamental plant *Zamioculcas zamiifolia*. Environmental Science and Pollution Research 21, 2603–2610.

Su, Y., Liang, Y., 2015. Foliar uptake and translocation of formaldehyde with Bracket plants (*Chlorophytum comosum*). Journal of Hazardous Materials 291, 120–128.

Syed, M., Soreanu, G., Falletta, P., Béland, M., 2006. Removal of hydrogen sulfide from gas streams using biological processes – a review. Canadian Biosystems Engineering 48, 2.1–2.14.

Taner, S., Pekey, B., Pekey, H., 2013. Fine particulate matter in the indoor air of barbeque restaurants: elemental compositions, sources and health risks. Science of the Total Environment 454–455, 79–87.

Torpy, F.R., Irga, P.J., Burchett, M.D., 2014. Profiling indoor plants for the amelioration of high CO_2 concentration. Urban Forestry and Urban Greening 13, 227–233.

Torpy, F.R., Irga, P.J., Burchett, M.D., 2015. Reducing indoor air pollutants through biotechnology (Chapter 8). In: Pacheco Torgal, F., Labrincha, J.A., Diamanti, M.V., Yu, C.P., Lee, H.K. (Eds.), Biotechnologies and Biomimetics for Civil Engineering. Springer International Publishing, Switzerland.

van Ras, N., Krooneman, J., Ogink, N., Willers, H., D'Amico, A., di Natale, C., Godia, F., Albiol, J., Perez, J., Martinez, N., Dixon, M., Llewellyn, D., Eckhard, Fir., Zona, G., Fachecci, L., Kraakman, B., Demey, D., Michel, N., Darlington, A., 2005. Biological air filter for air-quality control. In: Wilson, coordination, A., Elmann-Larsen, B., ESA SP-1290, ESTEC (Eds.), Microgravity Applications Programme: Successful Teaming of Science and Industry. ESA Publications Division, Noordwijk, Netherlands, ISBN 92-9092-971-5, pp. 270–280.

Veillette, M., Girard, M., Viens, P., Brzezinski, R., Heitz, M., 2012. Function and limits of biofilters for the removal of methane in exhaust gases from the pig industry (mini-review). Applied Microbiology and Biotechnology 94, 601–611.

Wang, Z., Zhang, J.S., 2011. Characterisation and performance evaluation of a full-scale activated carbon-based dynamic botanical air filtration system for improving indoor air quality. Building and Environment 46, 758–768.

Wang, Z., Xiu, G., Wu, X., Wang, L., Cai, J., Zhang, D., 2013. Biodegradation of xylene mixture from artificial simulated waste gases by capillary membrane bioreactors. Chemical Engineering Journal 229, 508–514.

Welch, L.S., 1991. Severity of health effects associated with building-related illness. Environmental Health Perspectives 95, 67–69.

World Health Organization (WHO), 2010. WHO Guidelines for Indoor Air Quality: Selected Pollutants. World Health Organization Regional Office for Europe, Copenhagen. http://www.euro.who.int/__data/assets/pdf_file/0009/128169/e94535.pdf (accessed 28.08.15.).

Xu, Z., Wang, Li, Hou, H., 2011. Formaldehyde removal by potted plant–soil systems. Journal of Hazardous Materials 192, 314–318.

Zhang, Y., Fan, L., Zhang, L., Chen, H., 2007. Research progress in removal of trace carbon dioxide from closed spaces. Frontiers of Chemical Engineering in China 1 (3), 310–316.

Zhao, K., Xiu, G., Xu, L., Zhang, D., Zhang, X., Deshusses, M.A., 2011. Biological treatment of mixtures of toluene and *n*-hexane vapours in a hollow fibre membrane bioreactor. Environmental Technology 32 (6), 617–623.

Zhou, S., Xiao, S., Xuan, X., Sun, Z., Li, K., Chen, L., 2015. Simultaneous functions of the installed DAS/DAK formaldehyde-assimilation pathway and the original formaldehyde metabolic pathways enhance the ability of transgenic geranium to purify gaseous formaldehyde polluted environment. Plant Physiology and Biochemistry 89, 53–63.

Bio-based plastics for building facades

C. Köhler-Hammer, J. Knippers, M.R. Hammer
University of Stuttgart, Stuttgart, Germany

13.1 Introduction

In the construction industry, plastics for pipes, gaskets, insulation, or vapor barriers are indispensable. Plastics are increasingly being used in interior and exterior spaces, because the use of novel digital planning and production methods allows buildings with free geometries to be designed more frequently. Free forms can be easily achieved with plastics. Additional benefits of this building material include their corrosion resistance, their creative versatility, and their low thermal conductivity or low weight compared with, for example, glass. Plastics are produced from crude oil, which is, however, a finite resource, which will lead to steadily rising oil prices. Therefore, polymers based on renewable raw materials, which are available indefinitely, are much more interesting chemical compounds. The properties of plastics or bio-based plastics, for instance their biodegradability or durability, depend on their chemical structure (Endres and Siebert-Raths, 2011). Some bio-based plastics are nonbiodegradable and, equally, some petroleum-based plastics are biodegradable (European Bioplastics, 2015).

> *According to European Bioplastics, a plastic material is defined as a bio-plastic if it is either bio-based, biodegradable or features both properties.*
>
> *European Bioplastics (2015)*

Agricultural wastes can even be considered for the production of high-tech materials. After use, they can be burned in a CO_2-neutral manner.

During the combustion of biomaterial, the same amount of carbon dioxide is released that was originally taken up by the plants during their growth. Any mineral additives must, of course, be deducted correctly. The use of bioplastics is growing in the electronics and automotive industry. Hence, the question arises as to whether and in what way plastics produced from renewable raw materials can be used in the construction industry.

13.2 Feedstock

Bio-based polymers are mainly obtained from renewable raw materials such as starch, cellulose, or vegetable oils (European Bioplastics, 2015).

Start-Up Creation. http://dx.doi.org/10.1016/B978-0-08-100546-0.00013-3

According to estimates published by the Association of European Bioplastics, the production volume in 2017 will be approximately 6 million tonnes, which requires only 0.02% of the world agricultural land (European Bioplastics & Institute for Bioplastics and Biocomposites, 2013a,b).

Hence, the necessary land use for bioplastics is low and the reproach that bio-based plastics are competing with the production of food is unwarranted. Furthermore, in the future, more cellulose-containing raw materials (eg, grass), agricultural refuse, and waste from food production and nonfood plants will be used in the manufacture of bioplastics (Natureworks, 2015).

13.3 Ecological advantages and resource efficiency

When, for example, bioplastics are used in electronic devices, a high percentage of renewable resources is important if these products are to be marketed as green products. In the case of packaging made of bio-plastics, the compostability is more relevant. Buildings are increasingly being environmentally certified. Materials or components from predominantly renewable raw materials might improve their environmental classification.

Table 13.1 Energy use and greenhouse gas emissions of plastic production

Material (granule)	Energy use (MJ/kg)	Greenhouse gas emissions (kg CO_2 eq./ kg polymer)	System boundary	Effective	References
Polylactide current technology	42.2	1.3	Cradle-to-gate	2010	Vink et al. (2010)
Polylactide biorefinery	35	0.8	Cradle-to-gate	2010	Vink et al. (2010)
Cellulose acetate	89	2.2	Cradle-to-gate	2010	Windsperger et al. (2010)
Bio-polyethylene terephthalate (polyethylene furanoate)	26	1.4	Cradle-to-grave	2012	Eehart et al. (2012)
Polyvinylchloride	59.3	1.9	Cradle-to-gate	2010	Vink et al. (2010)
Polycarbonate	113	7.6	Cradle-to-gate	2010	Vink et al. (2010)

Bio-based plastics will conserve the finite resource of crude oil, especially if less energy is required for their manufacture, but this is not generally the case. Similar considerations apply to CO_2-emissions during bioplastic production (Table 13.1).

Cradle-to-gate includes the material and energy flows up to the completion of the plastic granules. Cradle-to-grave encompasses the entire life cycle of the production of the plastic to the operating phase through to disposal.

The production of 1 kg polylactide (PLA) requires about 63% less energy than that to produce 1 kg polycarbonate (PC) (Vink et al., 2010). Over the next few years, PLA will be produced in biorefineries, enabling an even larger saving in energy. The production of cellulose acetate is currently biology-based to a maximum extent of 50% and is no longer advantageous with respect to energy consumption and the emission of greenhouse gases.

In the production of polyvinylchloride (PVC), approximately 34% less energy is required than that for cellulose acetate. Because bioplastics differ little from conventional plastics optically, ecological advantages can be made visible via labeling (eg, TÜC certified or Vinçotte seal).

13.4 Recycling and disposal

Bio-based plastics that are designed for uses with long functional duration cannot be composted because of their chemical structure, either on the compost heap or in industrial composting facilities. This concerns, for example, cellulose acetate, drop-in solutions such as bio-polyethylene, or PLA blends that contain flame retardants.

During biological degradation, bioplastics are decomposed to CO_2, water, and biomass. The energy stored in the plastic cannot be used in this method of disposal. Biodegradable plastics are especially used in the packaging industry and in agriculture.

If thermoplastic bio-based plastics are used, the recycling of production waste is possible. This has also been carried out with the Mock-Up Arboskin (see Section 13.13). It is not possible to recycle and to compost thermosetting bio-based plastics.

13.5 Technical and design aspects

In addition to functional and constructional demands, aesthetic and haptic aspects are also important for architectural uses. The optical quality includes the surface quality (eg, scratch resistance), which also contributes to the durability, colorfastness, and aesthetic appearance of the material. In particular with plastics and hence with bioplastics, an overly plastic-like appearance should be avoided. Shining, matt, rough or structured surfaces and therefore diverse effects of the plastic can be created by the surface character of the molding tools.

The color, surface, or light permeability of the bioplastic is also important. If, for example, only a deep brown material is feasible, then the application possibilities of the material are potentially low.

From their completion, some bioplastics exhibit a yellowness (eg, PLA). This yellowing is falsely associated with signs of aging, despite the bio-based material changing over time. Other important requirements that have to be taken into account include the uniqueness of every construction project, the creation of the project, and broad processing possibilities in general.

For example, with regard to molded components, the feasibility of using the various surface structures that are the result of the thermoforming of the semifinished products (eg, extruded solid sheets) or their treatment by Computer Numerical Control (CNC)-cutting can be considered. See also Section 13.12.

13.6 Requirements

Whether bio-based plastics are suitable for use in buildings depends on the application requirements. Generally, the EU Construction Products Regulation is valid for all products.

According to the basic standards of EU Construction Products Regulation for building, the stability and mechanical strength of the material must be guaranteed, in addition to its low flammability (European Union, 2011). Before new building products appear on the market, their application suitability must be established. For new or nonregulated building products, for example, and for bioplastic facade claddings, the usability must be proven. Usability proofs at the European level, so-called European Technical Approval (ETA), are given to a manufacturer of building products on request, if a harmonized European norm is not available.

Most polymers, thus also biopolymers, are easily flammable. For both interior and exterior uses in buildings, the behavior of the relevant material when burned must be improved by the addition of protection against flaming.

Facade claddings from bio-based plastic should be low flammable, so that they can be applied in all building classes. Normally, flammable materials can be used in buildings with a maximum of 7 m in height and, at most, two utilization units that are smaller than 400 m^2. According to the German standard of rear-ventilated facade claddings, the material should resist temperatures from −20 to 80°C (German standard DIN, 18516-1, 2010).

Material stability has to be provided for an economically adequate period. These are, on average, 30 years for facade claddings (Bund Technischer Experten, 2008).

13.7 Possible thermoplastic bio-based plastics and material selection

The work-to-application possibilities of bio-based plastics in buildings focus on thermoplastics, which can readily be transformed and processed. Moreover, they can be recycled. Thermosetting bio-plastics show better heat resistance and durability than thermoplastics. The high ratio of renewable raw materials (more than 90%) and the low price of about 2 €/kg argue strongly for the material PLA, a lactic-acid-based plastic. PLA has a relatively high hardness and therefore also surface hardness, which is comparable with the values of the technical plastics Acrylonitrile butadiene styrene (ABS) and polyamide (Endres and Siebert-Raths, 2011). It shows excellent mechanical qualities by having an elastic modulus with an average of 4 GPa (NatureWorks LLC, 2014, p. 2) and is therefore a stiff plastic. The technical plastic ABS has an E-modulus of 2.3 GPa (Kern GmbH, 2014). The heat resistance of the part-crystalline biopolyester PLA is approximately 58°C (HDT B). The biopolymer polyhydroxybutyrate is made by bacteria and is also suitable because of its high level of renewable resources and its high heat resistance. However, it is two to three times more expensive than pure PLA (Endres and Siebert-Raths, 2011).

According to the proportion of plasticizer that is necessary for the processing of cellulose acetate, the percentage of renewable raw materials amounts to approximately 30–50%.

Cellulose, which is a major component of plant cell walls, is esterified with an acid (eg, acetic acid or acetic anhydride) to primary acetate (Brockmann, 2000). The material costs compared with PLA are approximately twice as high (Endres and Siebert-Raths, 2011). Cellulose acetate has a heat deflection temperature (HDT-B) of 86°C (Mazzuccelli, 2010).

There are companies like SUPLA Material Technology, Purac, Tecnaro, Teijin, or NEC Corporation that are already producing long-durable bioplastics and there are commercial opportunities in this market.

13.7.1 Improving behavior to fire by compounding plastics with fire retardants

The oxygen content of the air amounts about to 21%. PLA and PC have a limiting oxygen index (LOI) of 26 (Farrington et al., 2005).

This means that, for combustion, a value of about 26% of oxygen is required, which is more than is present in air. The result indicates that PLA will probably have a lower demand for flame retardants than cellulose acetate (LOI 18) or polyethylene (Tozzi-Spadoni et al., 2009).

An LOI value of less than 21% indicates easily flammable materials. Materials with an LOI index between 21 and 28 burn slowly. PLA, therefore, has been preferentially tested for modification in order to be adapted to architecture uses. Because of its higher heat resistance, cellulose diacetate has also been considered for modifications.

For the improvement of behavior to fire, ecologically safe flame retardants have been used that are free of halogens. These are phosphate-based or sulphur-based and thus lower the content of renewable raw materials in the final compound, because they are not biologically based.

If the bioplastic has to remain clear, it must be examined to determine whether the elected flaming retardant has an effect on its light permeability. When plastics are mixed with additives, a variety of light indexes of refraction can be obtained. A clear bioplastic can thus become blurred or opaque in mixes with other substances.

Figure 13.1 Polylactide with diverse flame retardants: various diaphaneity rates.

Table 13.2 Polylactide with diverse flame retardants: various diaphaneity rates

Specimen	1	2	3	4	5	6
Composition	PLA	PLA + 10 wt% APP	PLA + 8 wt% TPP	PLA + 2 wt% KSS	See specimen 1	See specimen 3
Diameter of specimen (mm)	4	4	4	4	2	2

PLA, polylactide; *APP*, ammonium-polyphosphate; *TPP*, triphenylphosphate; *KSS*, aromatic sulphonate salt.

We have shown that only triphenylphosphate (TPP) does not change the transparency of PLA (see Fig. 13.1). We have also tested ammonium polyphosphate (APP) and an aromatic sulphonate salt used as flaming protection for PC (Table 13.2).

In addition, for opaque use in facades, we have tested expandable graphite and RP. Both flame retardants lead to an opaque black or red plastic.

13.8 Fire tests

To test the burning behavior of the flame-retardant-modified biopolymers, it is useful, initially, to test at the material level according to the test standard UL 94: 2013-03 (Underwriters Laboratories, 2013). However, subsequent tests at the component level are inevitable.

The Tests for Flammability of Plastic Materials for Parts in Devices and Applications (Underwriters Laboratories, 2013) of the Underwriters Laboratories (UL) are used in the modification of bio-based polymers in a first assessment of the formulation in the process of adaption to architectural applications. The adaptation of a biopolymer is an iterative process. Many batches and test specimens have to be produced and examined. Consequently, testing would be too costly at the component level.

In the vertical burning tests according to UL 94: 2013-03, the fire class V-0, V-1, or V-2 can be achieved. If the fire goes out on the specimen after removal of the flame within a maximum of 10 s and after the second flame treatment within no more than

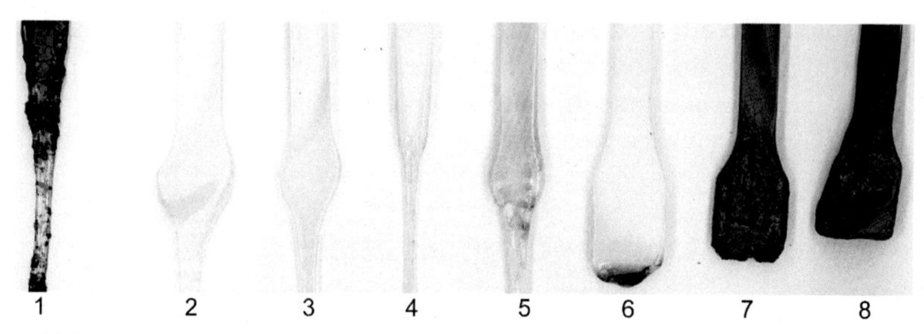

Figure 13.2 Specimen after fire tests.

30 s, this corresponds to fire classification V-0. However, no drops should emerge that can ignite cotton wool placed at a distance of 30 cm from the test object. The flame applications are performed for 10 s each. The fire class V-0 describes, at the material level, that the examined material at the tested thickness is flame retardant. For promising formulations, the flame treatment was increased to 15, 30, or more seconds, so that the test conditions could more closely attain the component level.

The fire classification of our PLA specimens was determined at a sample thickness of 4 mm according to the standard UL 94 vertical burning tests at material level.

The fire tests to UL 94 V showed that the dosage of TPP of initially 15% and 10% in PLA could be minimized because the burning times were only an average of zero to 3 s (see specimen in Fig. 13.2, no. 2, 3). Although droplets were evident in the second flame treatment, these did not burn and therefore did not ignite the cotton wool placed nearby.

At least 20 wt% of TPP was added to cellulose acetate (CA) in order to achieve fire class UL 94 V-0. Accompaniments such as long-lasting smoke (30 s) and a strong sooting of the rod (see Fig. 13.2, no.1) are present in both 10 wt% and in the double dose of 20%. For these reasons, this biopolymer was not considered further.

PLA specimens with the flame retardant potassium diphenyl sulfone sulfonate, which is conventionally used for the petrochemical-based polyester PC, burn down until the retaining clip, if not extinguished earlier. Fig. 13.2 shows test specimens that have been extinguished (no. 4, 5). The dosage was tenfold higher than recommended.

The flame-retardant Ammonium polyphosphate (APP) performs at the material level to a flame-resistant material (UL 94-V0). Immediately after removal of the flame, the fire on the specimen instantly disappears, as is the case in the second flame exposure (see Fig. 13.2, no. 6). In a further step, we could examine whether the dosage of 10 wt% can be reduced.

A flame-retardant PLA material can be produced with red phosphorus (RP). The after-burning time is approximately four times longer than for the compound of PLA and APP and is within the scope of the conditions for classification in fire class UL94-V0. It forms a charred surface layer (see Fig. 13.2, no. 7), which acts as an insulator and prevents the spread of fire (intumescent).

The dosage of 12.5 wt% expandable graphite is insufficient for achieving the fire classification V-0. The test specimens are highly flammable. The fire extinguishes after an average of 180s and burns down almost to the retaining clip. Specimen no. 7 in Fig. 13.2 has been previously extinguished. The insulating encrusted ash layer can be recognized.

The fire protection effect of expandable graphite can probably be improved with a combination of APP or zinc borate, which act synergistically here (EFRA – Europäischer Verband der Flammschutzmittelhersteller, 2004). Furthermore, a higher dose of the expandable graphite should be tried.

13.9 Heat resistance and possibilities

The heat deflection temperature of plastics is the ability of a specimen to maintain its shape under certain loading conditions up to a certain temperature (Grellmann and Seidler, 2011, p. 591).

The heat distortion temperature (HDT) is determined according to DIN EN ISO 75. Here, the test specimen is loaded by the three-point bending principle. The maximum bending stress in the specimen is 0.45 MPa (method B) (Grellmann and Seidler, 2011).

The HDT is the temperature at which the specimen has an obtained elongation of 0.2% (Grellmann and Seidler, 2011).

Because materials in outdoor applications should withstand temperatures of up to +80°C, the temperature stability of PLA (PLA standard HDT-B 52–56°C) must be improved. In order to achieve this, nucleating agents were added in our experiments with PLA; these agents increase the degree of crystallinity of the partially crystalline polymer and thus, the heat resistance (Table 13.3).

For the PLA-TPP-Compound B, the HDT drops at a dose of 7% by weight for TPP from 52°C for the PLA to an average of 46.7°C. TPP, in addition to its function as a flame retardant, acts as a plasticizer. The softening point and thus the melting range of the PLA compounds thereby decrease. The compound is transparent. By increasing the mold temperature from 25 to 100°C and by using a longer cooling time of about 240 s instead of 20 s, the heat resistance can be increased to 80°C

Table 13.3 Lower light transmittance with improved heat resistance[a]

Specimen	Composition	Manufacturing	HDT-B (°C)	Light transmittance (%)
1	PLA + 7 wt% TPP	Cooling temperature 25°C; cooling time 20 s	45.6	59–63
2	See specimen 1	Cooling temperature 100°C; cooling time 180 s	59.7	2.9–3.0
3	See specimen 1	Cooling temperature 100°C; cooling time 240 s	79.9	2.7–2.8

TPP, triphenylphosphate.
[a]The table does not contain all used additives.

Figure 13.3 Lower light transmittance with improved heat resistance.

by increasing the degree of crystallization (Köhler-Hammer, 2015). The light transmittance decreases and therefore the material remains translucent.

Similar results were obtained for PLA, which was compounded with the flame-retardant APP and a nucleating agent. In the conventional production of the specimens, an average heat distortion temperature of 55°C occurs. If these specimens are tempered afterward for 4 min at 50°C, an HDT-B can be obtained with an average of 79.17°C. The nucleating agent contained in the compound contributes to faster crystallization of the PLA-APP network. The heat resistance is sufficient for outdoor applications. However, the proportion of the nucleating agent can be increased to 1–2 wt% in order to achieve a more rapid crystallization. The addition of short fibers would have the same effect.

Instead of tempering, heat resistance can be increased during the shaping process in the production of moldings via injection molding by increasing the mold temperature and by extending the cooling time.

If full panels for facade applications are made by extrusion, they should be allowed to cool slowly. Mostly, because of poor thermal conductivity, a waterbath is used for cooling, to ensure that the heat is dissipated faster. This step in the process can be omitted but, in that case, a smaller quantity of production per machine hour results.

The reworking of conventional extruded sheets by vacuum forming is another possibility. On the one hand, through the thermal treatment of the forming process, the heat distortion resistance increases by a secondary crystallization. Without these measures, the HDT-B of the PLA-APP composite would be about 55°C (NatureWorks LLC, p. 1). On the other hand, facades for free-form building can be produced (1.8 design options, 1.9 Facade Mock-Up).

13.10 Resistance to weather

UV light in the wavelength range of 280–400 nm can have an aging effect on plastics (Eyerer et al., 2005, p.122). This is the case if the proportion of absorbed energy radiation is so high that bonds between the molecules are broken up. According to DIN 50,035, aging describes the "totality of all irreversibly occurring chemical and physical processes in a material over time" (German standard DIN 50035, 2009). This is triggered in addition to high-energy radiation, heat, temperature cycling, mechanical loads, chemical influences, or combinations thereof. It occurs irrespective of whether the polymer is bio-based or not.

To test the effects of UV radiation on the modified PLA compound, both aging over time and irradiation by outdoor exposure in real time have been tested. For architectural applications in visible interior and exterior spaces, the way that materials convert optically over time is relevant, in addition to any changes in mechanical performance.

The mechanical properties before and after aging by artificial or natural irradiation, temperature, humidity, and storage and their various combinations are given in the present work as yield stress and yield elongation. The yield stress corresponds to the yield strength, which is also referred to as the yield point. It refers to the mechanical stress at which a material can still be stretched reversibly. During flow, the plastic material is permanently deformed (Wiley Information Services). The yield elongation is the elongation in percent at the yield point (Grellmann and Seidler, 2011).

13.10.1 Artificial and natural weathering

The 18-month outdoor weathering according to DIN EN ISO 877: 2010 shows that the yield stress of the PLA-APP compounds increases. This is because the material continues to be postcrystallized by the sun's heat, thereby initially improving its mechanical properties. The yield elongation is reduced by 12%, since the material is more rigid.

After 855 h of artificial aging by global radiation, (ie, a simulation of approximately 10 m of outdoor storage), the PLA-APP compound without flame retardants exhibited a 78% lower yield elongation. With petroleum-based PC, which has both flame and UV protection, the yield elongation is reduced by 70%. The addition of UV stabilizers to the PLA-APP compound would probably, under the same aging conditions, lead to a similar yield strain as in PC, which is also suitable for outdoor applications.

In composite consisting of cellulose and TPP, the mechanical properties show insignificant change under both artificial and natural aging. Only the yield elongation decreases during the half year of natural weathering by 33%.

The test specimens composed of PLA and RP remain unchanged in terms of the yield elongation, whereas the yield stress decreases by 11%. This composite exhibits the greatest optical change (see Table 13.4; also Fig. 13.4, specimens 7 and 8). PLA,

Table 13.4 Specimen before and after 18 months of natural weathering

Composition	PLA	PLA + 10 wt% APP	CA + 10 wt% TPP	PLA + 25 wt% RP
Before natural weathering DIN EN ISO 877	Specimen 1	Specimen 3	Specimen 5	Specimen 7
After 16 month natural weathering DIN EN ISO 877	Specimen 2	Specimen 4	Specimen 6	Specimen 8

PLA, polylactide; *APP*, ammonium-polyphosphate; *TPP*, triphenylphosphate; *RP*, red phosphorus.

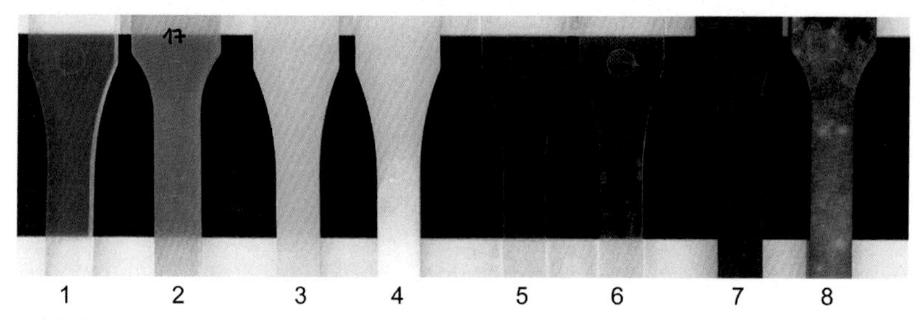

Figure 13.4 Specimen before and after 18 months of natural weathering.

which was compounded with the flame-retardant TPP or APP, shows no yellowing after natural weathering.

Natural weathering is preferable to the artificial aging, since the radiation for accelerated aging is often too energetic.

13.10.2 Absorption of water

For facade applications, the amount of moisture that a material absorbs should be determined, as it might be exposed to high humidity or rainwater.

The water absorption (DIN EN ISO 62) of PLA after 10 days of water storage at 21–23°C is 0%. Both the stretching elongation and the yield stress increase.

13.11 Further opportunities to improve properties

The heat resistance and the mechanical properties of PLA can be further improved by the addition of natural fibers or other fillers such as talc. These act as nucleating agents, whereby the crystallization of PLA is accelerated. The more crystallization of the polymer, the better the heat resistance and the stiffness of the material.

Durability can be improved by UV stabilizers. They convert, for example, UV radiation into heat. Therefore, oxalanilides (Affolter, p. 9) and hydroxyphenyltriazines, which have a particularly long functional life, or cinnamon acid esters/cinnamates can be used.

The properties of conventional petroleum-based plastics can be improved with the help of additives. Depending on the additive, the dosage is 0.1 to a maximum of 2% by weight and minimizes the proportion of renewable raw materials in most cases.

13.12 Biopolymers: scope for design

Bio-polymers combine the advantages of plastics, (ie, the formability), with those of naturally occurring materials that are particularly resource-efficient and sustainable. Bio-based polymers can be formed by processes including injection molding, thermoforming, and melt spinning into various forms and with diverse appearances.

Compared with wood, light-transmitting components can be produced from bio-based polymers. Therefore, in component design, focusing on these advantages appears useful for generating unique features.

13.12.1 Transparent and translucent components

PLA, regenerated cellulose, bio-based PET (PEF), or other bio-based plastics can be processed into solid boards, moldings, or hollow chamber or sandwich panels (Fig. 13.5). Dried natural fibers or other plant components can be integrated for design variety (Fig. 13.6).

13.12.2 Opaque components

The design possibilities for opaque components are virtually unlimited. Papers, dried leaves, and natural fibers, among others, can be integrated. Color master batches for bioplastics allow coloring in any hue. Furthermore, printed decorative foils can be laminated onto the surfaces of plastics (Figs. 13.7 and 13.8).

13.12.3 Molded components

In addition, the deep drawing of transparent or opaque semifinished products (sheets, plates) can produce diverse surface structures or moldings. The processing by CNC milling, drilling, and lasers is also possible. The design options for bioplastics could be well demonstrated on the facade of the mock-up.

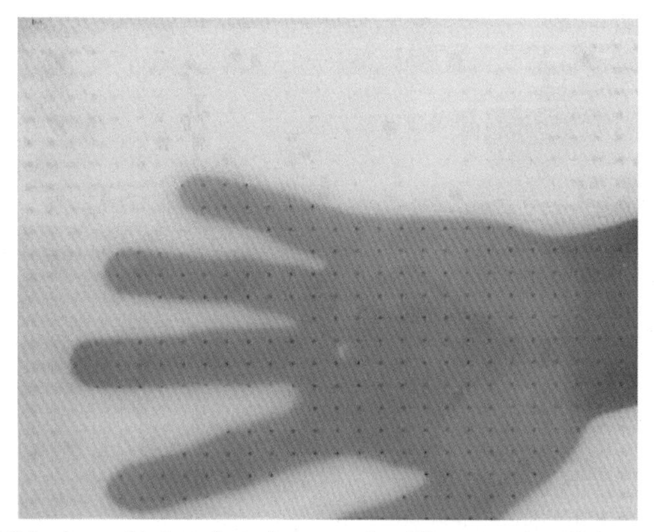

Figure 13.5 Injection-molded, perforated plate made of flame-retardant PLA. © M.R. Hammer/ITKE.

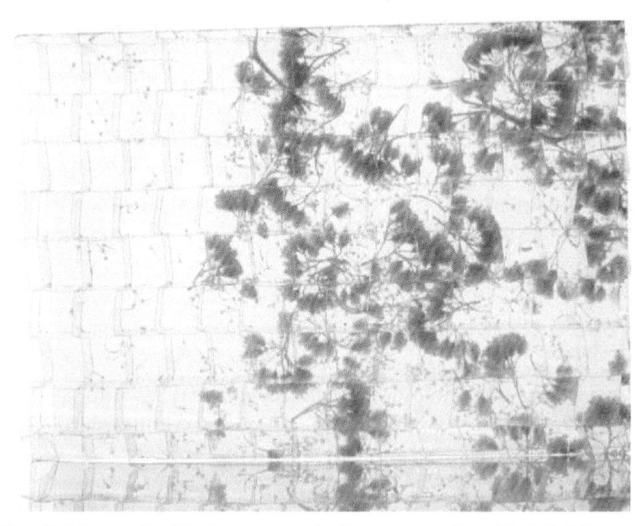

Figure 13.6 Sandwich panel with thermo-pressed cover panels with dried flower inlays.
© Damböck/Pixelcatcher.

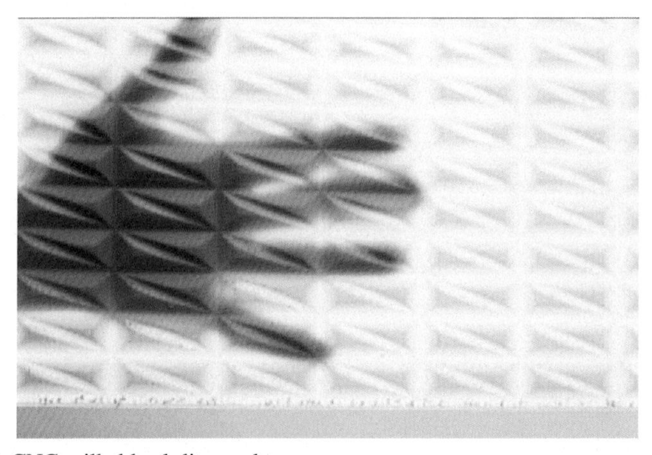

Figure 13.7 CNC-milled back-lit panel.
© Damböck/Pixelcatcher.

Figure 13.8 thermoformed PLA panel with surface structure.
© P. Siedler/ITKE.

13.12.4 Function integration

Facade moldings of resource-efficient biopolymers in combination with integrated moss demonstrate a possible application of thermoformable semifinished products based on renewable resources (Fig. 13.9). The three-dimensional element provides recesses in front for the moss and is formed by the vacuum forming of a plate. During the forming process, a surface structure can be obtained at defined points, which forms an adhesion base for moss mats. The moss takes fine dust from car exhausts and uses it as a fertilizer.

13.13 Facade mock-up

Based on the mock-up ArboSkin produced in October 2013, within the EFRE-research project bio-plastic façade, we could show that one or two differently sized pyramidal tools are sufficient to equip a free-form surface with over 380 different facade moldings (Fig. 13.10) (Knippers et al., 2013). Usually, in the production of free-form surfaces, many different tool shapes are relevant, such as uneven formed parts.

Figure 13.9 Bio-plastic moss facade, Hannover Exhibition 2013.
© M.R. Hammer/ITKE.

Figure 13.10 Bioplastic facade: facade mock-up.
© M.R. Hammer/ITKE.

13.13.1 Material, shaping, further processing, and recycling

Arboblend® modified with flame retardant by project partner Tecnaro was extruded into sheets and then thermoformed into pyramids (Fig. 13.11). The necessary post-CNC milling to remove the thermoforming edge allowed countless form variations in which the same thermoformed parts were trimmed unequally by different milling paths. A large amount of residual material was left over but this was granulated once again and was added as a regrind to the sheet extrusion.

13.13.2 Construction

The free-form bio-based plastic facade functions as a sheet-material-based shell structure with additional load-bearing and bracing-ring carriers and joists (Knippers et al., 2013). Contrary to common non-load-bearing facade constructions, this construction involves the load-bearing properties of the double-curved skin (made of 3.5-mm-thick biopolymer pyramids that are mechanically coupled) in the load-bearing and bracing processes of the entire system. On the one hand, this measure shows the potential of modified biopolymers. They are made primarily (>90%) from renewable resources, such as the bracing material (up to $E \approx 4000\ \mathrm{N/mm^2}$). This is suitable for exterior applications as it adds only a minor load attributable to its own weight ($13\ \mathrm{kN/m^3}$); on the other hand, it allows for the construction of a facade that utilizes a minimized number of points of support and/or mounting brackets on the structural work behind it.

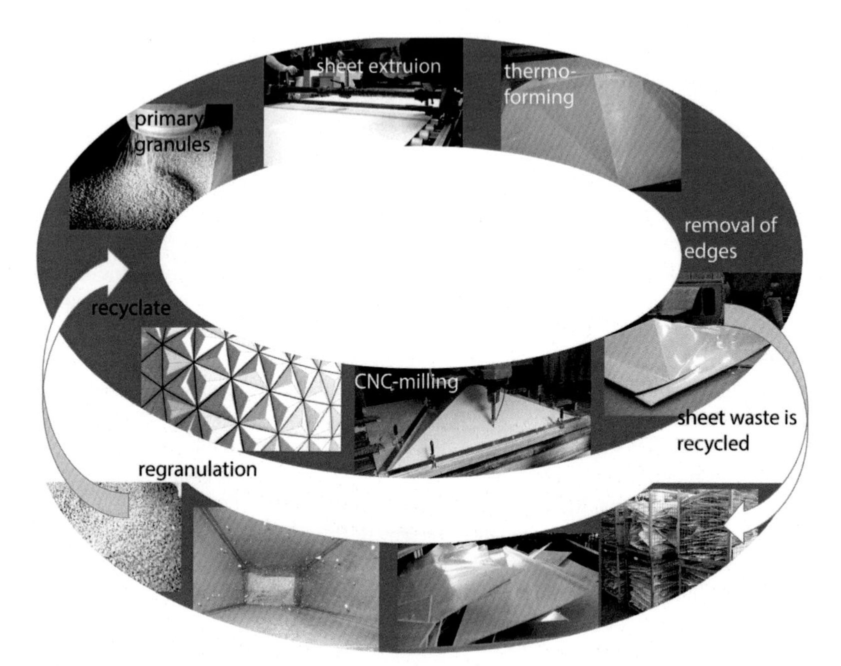

Figure 13.11 Scheme of production and recycling (Knippers et al., 2013).

13.14 Conclusion

A compound of PLA, APP and nucleating agents have shown to be suitable for outdoor applications on buildings. This also applies to PLA and TPP, for PLA and RP and for cellulose acetate and TPP.

With the flame-retardant APP, compounded PLA is based on about 88 wt% of renewable raw materials.

Currently, PLA appears the most interesting bioplastic because of its availability, good mechanical properties, UV resistance, and low price. A disadvantage is its low resistance to heat and hydrolysis, both of which, however, can be improved by the addition of additives and fillers (eg, natural fibers, talc, acid blockers).

Finally, we should mention once again that the performance of bio- and petroleum-based plastics does not depend on their origin but rather on their chemical structure.

The examples of modifications discussed in this chapter have shown that it is possible to improve the performance of bioplastics and to adopt them for architectural applications.

The use of bio-based plastics as a new building material for start-up creation will be a good way of saving resources. The plasticity of this material opens up various avenues for design, especially as bioplastics can be tailored to the requirements for their application in buildings. The mentioned aspects are not possible if wood is used, because wood is a building material that grows with specific properties and has an appearance that is difficult to reshape.

References

Arkema, 2014. Biobased Polyamide Block Based Elastomer Pebax RNew 80R53. Press picture information from 22 April 2014. http://en.european-bioplastics.org/press/press-pictures/other-applications/.

Brockmann, Mirjam, 2000. Plastics to Know: Cellulose Acetate. Principles of Chemistry, FU Berlin. http://www.chemie.fu-berlin.de/chemistry/kunststoffe/acetat.htm. Only available in German.

Bund Technischer Experten e.V. BTE (publisher), 2008. Worksheet of the BTE – Task Group: Durability of Components; Time Standards. BTE, Essen, p. 7. Only available in German.

Chemgapedia, Wiley Information Services GmbH: Streckgrenze. http://www.chemgapedia.de/vsengine/glossary/de/streckgrenze.glos.html.

DSM, 2014. Mercedes-Benz A-Class Engine Cover in EcopaXX, 70% Bio-Based PA 410. Press picture information from 23 January 2014. http://en.european-bioplastics.org/press/press-pictures/other-applications/.

DuPont, 2011. Hytrel® RS for Airbag Systems. Press picture information from 19 September 2011. http://en.european-bioplastics.org/press/press-pictures/other-applications/.

Eerhart, A., Faaija, A., Patela, M.K., 2012. Replacing fossil based PET with biobased PEF; pro cess analysis, energy and GHG balance. Energy & Environmental Science 5, 6407–6422. http://pubs.rsc.org/en/Content/ArticleLanding/2012/EE/c2ee02480b#!divAbstract.

European Flame Retardants Association (EFRA), 2004. Flame Retardants – Frequently Asked Questions. EFRA, p. 25. http://www.allinova.de/products/flame/explanation/efra06du.pdf. Only available in German.

Endres, Hans-Josef, Siebert-Raths, Andrea, 2011. Engineering Biopolymers – Markets, Manufacturing, Properties and Applications. Hanser, München. Page 7, 71, 182, 194.

European Bioplastics, Institute for Bioplastics and Biocomposites, December 2013a. Land Use for Bioplastics 2013 and 2018. Picture 4. http://en.european-bioplastics.org/press/press-pictures/labelling-logos-charts/.

European Bioplastics, Institute for Bioplastics and Biocomposites, December 2013b. Global Production Capacities of Bioplastics. http://en.european-bioplastics.org/press/press-pictures/labelling-logos-charts/.

European Bioplastics, 2015. Fact Sheet - European Bioplastics. What Are Bioplastics? Berlin, pp. 1–2. http://en.european-bioplastics.org/wp-content/uploads/2011/04/fs/Bioplastics_eng.pdf.

Eyerer, P., Elsner, P., Hirth, T., 2005. The Plastics and Their Properties, sixth ed. Springer, Berlin, p. 122. Only available in German.

European Union, 2011. Regulation (EU) No 305/2011 of the European Parliament and of the Council of 9 March 2011 Laying Down Harmonised Conditions for the Marketing of Construction Products and Repealing. Council Directive 89/106/EEC. Annex I - basic requirements for construction works. http://eur-lex.europa.eu/LexUriServ/LexUriServ.do?uri=OJ: L:2011:088:0005:0043:EN: PDF.

Farrington, D.W., Lunt, J., Davies, S., Blackburn, R.S., 2005. Poly(lactic Acid) Fibers. Table 6.1. http://www.jimluntllc.com/pdfs/polylactic_fibers.pdf.

German Standard, DIN 18516-1:2010-06, June 2010. Cladding for External Walls, Ventilated at Rear - Part 1: Requirements, Principles of Testing – Part 1. 5.2.2. Temperature Effects, Swelling and Shrinkage. Only available in German.

German Standard, DIN 50035:2012–09, September 2012. Terms and Definitions Used on Ageing of Materials – Polymeric Materials, p. 1. Only available in German.

Grellmann, Wolfgang, Seidler, Sabine, 2011. Plastics Testing, second ed. Carl Hanser Verlag, Munich. 11.1.1. Principles and Definitions. Page 126, 591, 593–595.

Kern GmbH technical plastic parts, 2014. Acrylnitril/Butadien/Styrol (ABS). www.kern-gmbh.de/cgi-bin/riweta.cgi?nr=2110&;lng=1. Only available in German.

Köhler-Hammer, C., 2015. Potential Applications for Bio-based Plastics in Buildings: Exemplary Development. Institute of Building Structures and Structural Design. University of Stuttgart, p. 88. Research report 38.

Köhler, C., Fischer, K., Kappler, E., Schweizer, M., Braun, F., 2013. Free-form Semi-finished Products Primarily Made of Renewable Raw Materials, for Use in Indoor and Outdoor Spaces – Exemplary Development. Final report on the funding program “EFRE environmental technology” of the Ministry of the Environment, Climate Protection and the Energy Sector Baden Württemberg. Only available in German.

K-Zeitung Online, 2013. Bio-Polyamide for Wiring, 18.06.2013. Only available in German. www.k-zeitung.de/bio-polyamide-fuer-leitungen/150/1195/67201/.

Mazzuccelli 1849, 2010. Plastiloid 22 MGP. Data sheet. Mazzuccelli 1849.

NatureWorks LLC, Ingeo™ Biopolymer 3251D Technical Data Sheet. Injection Molding Process Guide. http://www.natureworksllc.com/~/media/Technical_Resources/Technical_Data_Sheets/TechnicalDa taSheet_3251D_injection-molding_pdf.pdf, p. 1.

NatureWorks LLC, 2014. Ingeo resin product guide. Naturally Advanced Materials 2. www.natureworksllc.com/~/media/Technical_Resources/one-pagers/ingeo-resin-grades-brochure_pdf.pdf.

Natureworks LLC, 2015. From Plants to Plastics. www.natureworksllc.com/The-Ingeo-Journey/Eco-Profile-and-LCA/How-Ingeo-is-Made.

Tozzi-Spadoni, F., Seghizzi, A., Patritti, G., January 09, 2009. Process for the Production of Flame-Retardant Cellulose Acatate Fibers. Notification Number: 08100240.4. Publication Number: EP1944394 B1, (0062).

Underwriters Laboratories, August 23, 2013. UL 94 – Tests for Flammability of Plastic Materials for Parts in Devices and Appliances.

Vink, E., NatureWorks BV, Davies, S., NaturWorks LLC, Kolstad, J., August 2010. The eco-profile for current Ingeo® polylactide production. Biocatalytic vs. chemical production of alkanolamide biosurfactants. Industrial Biotechnology – The Journal of Biobased Industries 6 (4), 13. http://www.natureworksllc.com/~/media/The_Ingeo_Journey/EcoProfile_LCA/EcoProfile/NTR_Eco_Profile_Industrial_Biotechnology_082010_pdf.pdf.

Windsperger, A., Thurner, M., Brandt, B., Pilz, H., Fehringer, R., 2010. New Energies 2020. Climate Protection Potential by Forced Use of Biogenic and Conventional Plastics, p. 73. http://www.indoek.at/downloads/arbeitsbereich_1_klimaschutzpotenziale_biogene_kunststoffe.pdf. Only available in German.

Part Three

Algorithms, big data and IoT for eco-efficient and smart buildings

Development of algorithms for building retrofit

14

F. Cappelletti[1], *P. Penna*[2], *A. Prada*[2], *A. Gasparella*[2]
[1]University Iuav of Venice, Venice, Italy; [2]Free University of Bozen-Bolzano, Bolzano, Italy

14.1 Introduction

This introduction provides an overview of the building energy retrofit problem and of the main issues to be solved.

The European market has seen the increasing presence of energy service companies (ESCOs), which offer the provision of energy services to achieve energy and environmental goals. In particular these new companies are often energy consultants and managers for public administration supplying and installing energy-efficient equipment and scheduling building refurbishment actions. The peculiarity of ESCOs is the fact that they can also finance or arrange financing for the operation and their remuneration is strictly dependent on the energy saving achieved. Energy analysis and audits, monitoring, and evaluation of savings until project design and implementation are some of the activities included in the energy services. For these companies it is then especially important to evaluate a wide set of energy-efficiency measures (EEMs) to identify the most effective solutions, in terms of combination of measurements, in fulfilling different requirements and targets. As highlighted by Kolokotsa et al. (2009), the main criteria and aims for energy-efficiency projects are energy, costs, global environment, and indoor thermal comfort. In particular, retrofit interventions should aim at enhancing the overall building performance, not just the single energy aspect. In this framework the problem of choosing among different EEMs so as to achieve the best balance between each criteria can be tackled mainly through two different approaches (Diakaki et al., 2008).

The first approach is multicriteria analysis. This method requires a list of predefined and preevaluated alternative options and a given scale of priorities for the evaluation criterion defined by the stakeholders. The second approach, which is gaining growing attention in several fields, is optimization. This method, based on different optimization techniques often combined with simulation software, allows the evaluation of all the possibilities and the combination among them, setting one or more objective functions to be optimized. The application and limitations of the two approaches have been discussed in the following sections. It is worth saying that the implementation of these techniques is carried out today in research fields and not in the professional sectors, due to the lack of a specific, easy-to-use tool able to manage a

Start-Up Creation. http://dx.doi.org/10.1016/B978-0-08-100546-0.00014-5

multicriteria or multiobjective analysis associated with the building energy simulation. For larger applications, on the other hand, practitioners and ESCOs can find useful support in new entrepreneurship initiatives and start-ups focused on providing calculation and optimization services, with an in-depth knowledge of building physics, economics and finance, and fiscal aspects. High initial investments and entering barriers, both cultural and technical, are overcome thanks to an externalization strategy that can justify the settlement of new companies.

14.2 Methods for the choice of energy-efficiency measures for building retrofit

This section will explain the two main approaches discussed in literature. The aim is to give an overview of the potentials and the limits of the two methods in building retrofitting.

14.2.1 Multicriteria analysis

Multicriteria techniques, as well as optimization approaches, can efficiently cope with problems characterized by conflicting and multiple objectives. Using a multi-criteria approach, the possible solutions are evaluated and ranked according to criteria (quantitative or qualitative), fixed by professionals and users, involved in the decision-making process, by means of different weights. According to Wang et al. (2009) and to Asadi et al. (2013), this approach requires four main stages:

- Alternatives formulation and criteria selection
- Criteria weighting on the base of stakeholders' needs and preferences
- Evaluation of each alternative performance
- Final aggregation and sorting of the different solutions

The first step involves the definition of the possible alternative solutions and the criteria that have to be satisfied. In the case of building retrofits, this means defining the possible EEMs and the requisites to fulfill with the refurbishment project, such as energy savings, indoor thermal comfort, investment costs, CO_2 emissions, and so on. The right selection of the evaluation criteria, which should be consistent with the stakeholders' objectives and needs, is crucial to effectively find the best solutions.

The second stage requires the setting of weights for the selected criteria. Given that the objectives are often conflicting, the best solution highly depends on the stakeholders' preferences. The weights establish the scale of importance of the criteria and can be assessed using subjective weighting methods, objective weighting methods, and a combination of weighting methods. Subjective weighting methods rely only on the importance scale defined by stakeholders. On the other hand, the objective methods establish weights based on the analysis of the initial data. The third method combines subjective and objective weighting.

In the third step, the performance of each possible solution is evaluated, according to the allocated weights. In case of energy refurbishment design, Building Energy Simulation (BES) tools are often used to analyze the energy savings' potential.

Finally, the fourth stage of the multi-criteria approach involves the aggregation of the results and the ranking of the solutions. There are different approaches adopted in this last phase. According to Pohekar and Ramachandran (2004), the most used are weighted sum, weighted product, analytical hierarchy process (AHP), Preference ranking organization method for enrichment evaluation (PROMETHEE), and Elimination et choice translating reality (ELECTRE) method. The weighed sum and weighed product methods develop an overall synthesis value, based on the fulfillment of each criteria and their weights. AHP is a descriptive decision analysis methodology and involves the decomposition of the problem into goals, criteria, and subcriteria organized by a ratio-scale of importance. The hierarchy is defined through a pair-wise comparison of evaluation criteria and alternatives. PROMETHEE and ELECTRE are two outranking methods, based on the principle to rank the alternatives according to the decision-makers' preferences.

As stated by different authors (Wang et al., 2009; Pohekar and Ramachandran, 2004; Loken, 2007), the multi-criteria approaches are widely used, especially in the field of sustainable energy planning, where different aspects should be taken into account. Vucicevic et al. (2014) developed indicators of sustainable development, based on economic, social, and environmental aspects, to evaluate the residential building stock of Belgrade. The results highlighted how prioritizing different aspects influences the assessment of building stock sustainability. Georgopoulou et al. (1997) used the ELECTRE approach to evaluate energy policy for the exploitation of renewable energy to enhance the electricity production in the Greek island of Crete. Eight different strategies, from conservative options to innovative technologies based on renewable energy, have been evaluated considering economic, technical, political, and environmental criteria. The priority ranking of the different aspects have been established by stakeholders involved in the decision-making process.

In the field of building design, Hopfe et al. (2013) applied the AHP approach to optimize building design, taking into account several aspects such as energy efficiency, thermal comfort, and architectural layout, and considering uncertainty involved in design phase. The methodology has been applied to evaluate two different design options for a case study, applying to the considered performance aspects the weights defined by the project design team. Yongjun et al. (2015) proposed using a multi-criteria system design optimization to design nearly zero-energy buildings, optimizing the system's initial cost, indoor comfort, and grid stress caused by power mismatch. The proposed method proved to be effective in helping the professional in the design phase, identifying optimal building system sizes based on the weight defined for each performance criterion by users.

Concerning the effectiveness of multicriteria analysis to assist the choice among EEM in building retrofitting, this approach has proven to be effective. In particular, some authors (Flourentzou and Roulet, 2002; Caccavelli and Gugerli, 2002; Jaggs and Palmar, 2000) developed multicriteria tools to assist the decision-maker during the retrofit choice. Using this approach it is possible to identify the most effective retrofit strategies to satisfy the energy requirements and to have a general overview of the retrofit project. Also Rey (2004) proved the benefits of using a structured multicriteria assessment methodology to identify the best strategies in the refurbishment of an office building. The multivariate design method proposed by Kaklauskas et al. (2005) adopted a set of indexes based on significance, priorities, and utility degree of building retrofit alternatives; those indexes allow speeding up the designing process. Alanne (2004) develops a multicriteria knapsack model to help designers select the most feasible renovation actions in the conceptual phase of a renovation project, optimizing functionality, environmental value, and energy savings.

The main limits of these approaches are related to the necessity to know a priori the significance and priorities of the attainments. Often the criteria adopted by the decision-makers can be arbitrary and sometimes it could be difficult to define a scale of importance for the different objectives. Moreover, the judgment of the decision-makers typically depends on their knowledge or information about the problem, which could lead to some weighting errors or bias.

14.2.2 Optimization

The main drawback of the multicriteria approach, represented by the necessity to know in advance the priorities of the objectives, has been overcome using the optimization approach. By this method all the solutions are on the table and are evaluated using the same priorities, through the assessment of some objective functions quantifying the different performance aspects. The potential of this method is related to its ability to cope with complex problems, where one objective or multiple competing objectives need to be fulfilled.

This method requires the execution of the following four steps:

- Definition of retrofit options (variables) and constraints
- Definition of main objectives and conversion of them into objective functions
- Evaluation of all the possible options in the solution space
- Definition of the solutions set able to optimize the problem

In the first step a set of variables, their variation range, and the boundary conditions (constraints) have to be defined. In the case of building refurbishment these variables could be the different EEM and the constraints could be, for example, the weather data, the geometry of the building, and the air change rate.

The second stage is the definition of the objective function, also called cost or fitness function, which is normally calculated by means of simulation tools. Obviously the objective functions have to depend on the chosen variables, otherwise the problem would not be consistent. In the optimization the cost function can be more than one, if the aim is the research of a trade-off between competing objectives. In this case the optimization is called multiobjective optimization.

The third step of the optimization process is the calculation of the objective functions (also known as dependent variable) for all the possible combination of variables.

Finally the fourth stage is the selection of the optimal solution or the set of solutions in the case of multiobjective problems.

One of the main issues in this approach is related to the size of the decision problem: when there are several options to evaluate, the optimization process could become time-consuming and extremely difficult to handle. For this reason, computationally efficient solution techniques are more and more used. In literature, optimization approaches received increasing attention and, always more frequently, authors use these methods to deal with complex problems. In the renewable and sustainable energy field several works use optimization techniques, as reported by Iqbal et al. (2014) and Baños et al. (2011). The same occurs for building design, where Pacheco et al. (2012) provides a complete overview of the optimal design options for residential buildings, to reduce energy consumption for heating and cooling. Regarding the building retrofit design, De Boeck et al. (2015) have recently highlighted the use of an optimization approach in retrofit projects. Especially after the publication of the EU Directive 31/2010 (European Commission, 2010), the so-called EPBD-recast, which introduces the cost-optimal approach to define new energy requirements, the necessity of optimizing at the same time different aspects such as the energy and economic ones, has become even more important. Asadi et al. (2012a) developed a method to assist stakeholders in defining EEMs able to minimize energy consumption and the investment costs. Kurnitski et al. (2011) optimized the building energy performance and the net present value (NPV) for a typical Estonian detached house, to determine the cost-optimal approach and nearly zero-energy levels. Hamdy et al. (2013) and Ferrara et al. (2014) applied the cost-optimal approach proposed by EPBD-recast, to define cost-optimal retrofit solutions and nZEB energy performance levels for a standard single-family house located in Finland and in France, respectively. Several retrofit options to enhance the building envelope performances and the efficiency of the HVAC system have been evaluated to optimize energy savings and global costs during the lifespan of the building.

One aspect that only a few works (Asadi et al., 2012b, 2014; Chantrelle et al., 2011; Penna et al., 2015) considered in defining optimal retrofit solutions is thermal comfort. Asadi et al. (2012a,b, 2014) identified the retrofit options that minimize the energy consumption, investment costs, and percentage of discomfort hours for a residential building (Asadi et al. 2012a,b) and of a school building (Asadi et al., 2014) in Portugal. Chantrelle et al. (2011) presented a tool, MultiOpt, for the optimization of renovation operations. The tool was applied to a case study, a school building, to minimize energy consumption, investment costs, environmental impact, and occupants' discomfort. Penna et al. (2015) analyzed different EEMs, both on the building envelope and on the HVAC system, for a set of residential building modules in order to generalize the result's representativeness to the existing building stock. The solutions found are optimal in terms of maximum economic performance (NPV) and minimum energy consumption, for the lowest achievable thermal discomfort.

In the following section, the most common optimization techniques have been described in detail.

14.3 Algorithms for the implementation of multiobjective optimization

Optimization is the process of adjusting the input to find the minimum or the maximum of one or more objective functions (Haupt and Haupt, 2004). The input, in a retrofit project, are represented by the possible EEMs and the objective functions describe the aims to achieve, such as energy savings, costs, indoor environmental quality, and so on.

Generally, an optimization problem can be represented in mathematical form as (Machairas et al., 2014):

$$\min[f_1(x_1, x_2, \ldots, x_n)] \tag{14.1}$$

where f_1 represents the objective function to minimize (eg, life-cycle costs or NPV), $x_1, x_2, \ldots, x_n$ are the values of the evaluated retrofit measures (eg, insulation thickness, glazing type, boiler type, etc.), each of which can vary in a range and according to a step of variation typical for the selected retrofit (different insulation thickness, different glazing system U-values and solar heat gain coefficients, different boiler efficiency, etc.).

In multiobjective problems there are normally more than one objective function to minimize:

$$\begin{cases} \min[f_1(x_1, x_2, \ldots, x_n)] \\ \min[f_\mathrm{m}(x_1, x_2, \ldots, x_n)] \end{cases} \tag{14.2}$$

where f_1 and f_m are the two objectives, such as, for example, energy consumption and investment costs or energy consumption and discomfort time.

When more objectives occur, there are two common ways to approach the problem: the weighted-sum function method or the Pareto optimality.

Using a weighted-sum function, the objectives are normalized and summed up as:

$$\min[f_1(w_1 \cdot f_1 + w_\mathrm{m} \cdot f_\mathrm{m})] \tag{14.3}$$

where w_1 and w_m are the weight factors for each objective function. They are defined according to the importance and priority of the objectives, and their sum (from w_1 to w_m) is generally equal to 1.

The second approach, which is named after Pareto because of his works on economic efficiency (Pareto, 1896), does not identify just a single solution as the result of the optimization process, but a set of optimal solutions, which represent the nondominated ones. Those solutions are the best trade-offs between objectives. The benefits and the drawbacks of those approaches are described in the next paragraph.

14.3.1 Weighted sum function versus multiobjective approach through Pareto optimality

The weighted-sum function approach is a method used to simplify a multiobjective problem, lumping the objectives into one function by using weighted sum factors, as shown in Eq. [14.3]. The combined function f is used to evaluate and define the optimal solution. The weighted sum method is particularly efficient and easier to implement compared to the Pareto optimality. However, it requires prior knowledge on the compromise between the objectives and it does not give any information on the effect of the variable to be optimized on the different design targets. Some information can be derived testing different weighting factors, at the expense of increasing the number of optimization runs, lengthening the computational time.

The other popular approach to solve multiobjective optimization problems is Pareto optimality. This approach assigns the same weight to each objective and optimizes them simultaneously. According to Pareto, a solution X is said to dominate the other solution Y if both the following conditions are true:

- The solution X is no worse than Y in all objectives.
- The solution X is strictly better than Y in at least one objective.

Thus passing from Y to X produces an improvement for all the objectives, or an improvement for some, without the other ones being harmed. The Pareto optimum or optima represent the best trade-off among objectives. These solutions, which represent the Pareto Frontier, are also defined as nondominated, because no alternatives exist that increase the fulfillment of an objective without hampering the attainment of another. Fig. 14.1 reports an example of a Pareto frontier, which is a curve for an optimization problem with two objectives. The gray dots represent all the possible solutions of the problem that are dominated because solutions with better performance exist. The blue dots are the Pareto optimal solutions, the nondominated

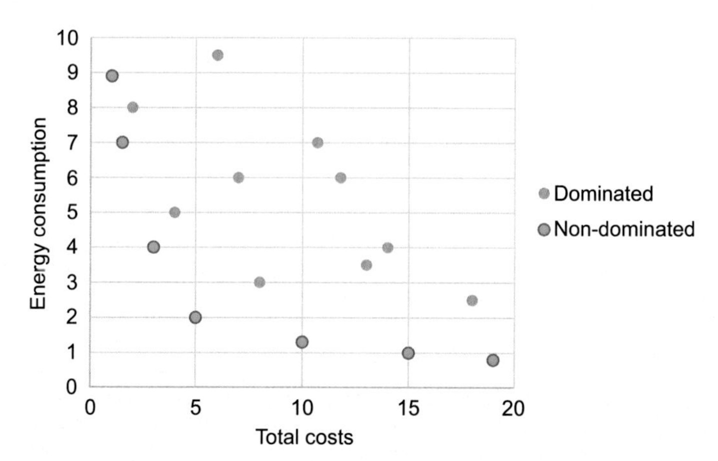

Figure 14.1 Example of optimization problem solved with Pareto optimality.

ones, which outline the Pareto frontier. If the objectives are three, the result is a Pareto surface.

One of the perceived disadvantages of this method is that it does not provide a single optimum solution, but a set of equally efficient configurations. Although it may seem that the final decision still must be made, identifying the Pareto optimal solution makes it possible to analyze the relationships among the objectives of the optimization problem and work on them excluding the largest part of inefficient configurations. Moreover, the Pareto optimality provides a better understanding of how each in a set of variables affects the attainment of the different objectives.

In literature some authors (Palonen et al., 2009; Lu et al., 2015) highlighted the benefits of using a multiobjective approach to find the nondominated solutions.

Palonen et al. (2009) compared the results of single and multiobjective optimizations in defining optimal design solutions for building envelopes and HVAC systems. The advantage of a multiobjective approach on generating a diverse set of Pareto front values and capturing extreme solutions to all objective functions has been proved. Lu et al. (2015) presented a study on the optimal design of renewable energy systems in buildings aiming at reducing the total cost, CO_2 emissions, and the stress on the grid. From the comparison of the results obtained with the two methods, the Pareto optimality was able to provide valuable information on the interaction among variables and objectives, allowing the decision-makers to find one or more appropriate solutions from the sets of Pareto solutions obtained.

The actual limitation of the approach is probably related to the computational time to explore the solution space and identify the frontier, which in most of the cases is much larger than what is needed by the weighted sum function approach.

14.3.2 Optimization solution techniques

If the long computational time required to get the nondominated solutions in the multiobjective approach of the Pareto optimization is the weakness of the method, when simulation tools are needed to assess the objectives of building retrofits, the optimization process could take an unsustainable long time, from several hours to some days, depending on the dimension of the problem (number of variables and their ranges and steps of variability) and on the complexity of the building.

Some optimization solution techniques exist that help reduce the time needed and allow a good approximation of the Pareto frontier. The choice of the most appropriate among the possible ones can significantly decrease the optimization time while preserving the possibility to approach the true optimal solutions. The performance of the algorithm, in terms of effectiveness (the capability to provide a good estimation of the frontier) and efficiency (the capability of approaching the frontier with a small number of configurations evaluated), depends on the type of problem and its mathematical description. This is the reason why so many methods are still in use. Brief details on common optimization techniques used for sustainable building optimization are given in the following.

14.3.2.1 Brute-force approach

This is not actually a solution technique since all the configurations attainable from any combinations of retrofit options are considered and evaluated. When the decision space is not wide, this method gives reliable results and may be as efficient as or even more efficient than other methods, when not properly implemented. Increasing the number of options makes the number of assessed configurations to increase geometrically, requiring an extremely long computational time to find the optimum. Especially in building retrofitting this method is often not feasible because the solution space is generally too large.

14.3.2.2 Direct search methods

Direct search methods are based on the explicit mathematical expression of objectives. As an example of direct search method, Fig. 14.2 illustrates the principle behind the sequential search optimization approach. In this work (Ihm and Krarti, 2012), the authors' aim was to minimize the life-cycle costs and the energy consumption for an existing building, finding a path that reaches the optimal package of EEMs.

The most cost-effective EEM options are chosen based on the steepest slope consisting of the life-cycle costs to energy savings ratio. In general, the direct search methods are effective in defining optimal solutions, because of the low computational time. The main drawback of these methods is that the possibility of convergence depends on the regularity of the objective functions. Indeed, if the domain of the optimization process is not continuous, the algorithm can fail or get trapped in local minima. Unfortunately, this issue is quite likely to occur in the refurbishment design, which is characterized by a discontinuous domain space, since most of the retrofit options assume discrete values.

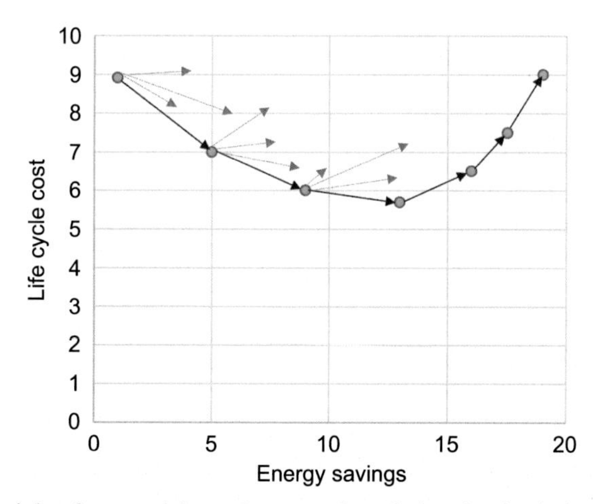

Figure 14.2 Principle of sequential search approach to find optimal solutions.
Adapted from Ihm, P., Krarti, M., 2012. Design optimization of energy efficient residential buildings in Tunisia. Building and Environment 58, 81–90.

14.3.2.3 Random methods

Using random methods, the problem of the regularity of objective function is overcome. By this approach the solutions to be evaluated are randomly defined and the nondominated solutions are identified applying stochastic operators. These methods reproduce natural phenomena and seems to be particularly successful for optimization problems. The first generation of solutions is randomly defined, and then new points in the search space are defined by applying operators to the current points and statistically moving toward more optimal places. The main drawbacks of these methods are related to the computational time, which is higher compared to the direct search method. In those approaches considering cost function derivatives is not required, which allows dealing with discrete variables and noncontinuous cost functions (Haupt and Haupt, 2004). In literature several random methods have been developed based on different natural phenomena. Some of the most important ones are reported in the next section.

14.3.3 Natural optimization algorithms

Natural optimization algorithms are able to solve and optimize complex problems by means of stochastic operators that mimic natural phenomena. The most adopted one is the evolutionary algorithm, based on the survival Darwinian principle. However, particle swarm optimization (PSO), ant colony optimization (ACO), harmony search (HS), and simulated annealing (SA) are also increasing their relevance.

14.3.3.1 Evolutionary algorithms

Evolutionary algorithms are based on the current scientific understanding of the natural selection process. According to Darwin's theory, only the organisms best adapted to their environment tend to survive and transmit their genetic pool in increasing number to succeeding generations. According to Haupt and Haupt (2004), this means that the organisms of today's world can be seen as the result of many iterations in a grand optimization algorithm, where the cost function is the survivability of the species.

Using the same principle, it is possible to address the solution of a multiobjective problem, such as a building refurbishment. The individuals of the population represent here the possible retrofit configurations, whose measures are the different genes, which assumes different values according to the specific intervention considered. The cost functions can be set as energy consumption, investment, operational or overall costs, environmental impact, and/or indoor thermal discomfort, and, by means of Darwinian principle of survival of the fittest, by maintaining the best individuals, the poorest solutions are eliminated at each generation. Genetic operators, such crossover (switching elements from different solutions) and mutation (introducing random changes), are applied to generate new solutions and to progressively define the best ones. According to Evins (2013) the most commonly used evolutionary algorithms are:

- Genetic algorithms (GA), the most commonly used in multiobjective optimization problems, especially the nondominated sorting genetic algorithm (NSGA-II) developed by Deb et al. (2002).

- Genetic programming, where the variables are various programming constructs and the output is a measure of how well the program achieves its objectives. It is in that case a computer program that writes other computer programs (Sette et al., 2001).
- Differential evolution, in which variable values are iteratively varied, introducing variable values of other best solutions (Storn and Price, 1997).

14.3.3.2 Genetic algorithm

The most popular evolutionary algorithm is the GA. It was first developed by John Holland (1975) and then popularized by David Goldberg (1989), one of his students, who applied it to the control of gas-pipeline transmission.

As previously explained, the GA uses the principle of natural selection to evolve a set of solutions toward an optimum solution. According to this method, each individual of a population can be seen as a possible solution of the optimization problem. The gene sequence, written into a single chromosome, defines the characteristics of the individual.

The GA starts with an initial population, where a group of chromosomes, randomly defined, represents the individuals of the population. Then, each individual is evaluated and ranked according to the objective function, and on the need of maximizing or minimizing it. The selection rate defines the fraction of individuals that survive and are kept as parents to create the next generations. Only the best solutions are selected to continue, while the rest are deleted. Defining the selection rate, besides the size of the original population, is somewhat arbitrary: keeping few chromosomes could limit the available genes in the offspring, but keeping too many chromosomes does not allow changes in the definition of the next generations and tends to increase the computational effort. Generally, the selection rate is fixed to 0.5 (Haupt and Haupt, 2004). Afterward, the selected individuals are used to define the new individuals of the next generation. The mating process involves two parents that produce two offspring by means of two operators: crossover and mutation (Fig. 14.3). Crossover swaps part of the genes of the parents' chromosomes, starting from a crossover point. In this way the offspring contain portions of the genetic code of both parents. Mutation is a random alteration of a gene contained in the chromosome.

The percentage of mutation should be carefully fixed; in fact, increasing the number of mutations increases the algorithm's freedom to search outside the current region of variable space. However, it also tends to distract the algorithm from converging to a popular solution. At the end of each generation, the process is repeated, the performance of the solutions are evaluated by means of the objective function, and the best ones are selected and used for the next generation.

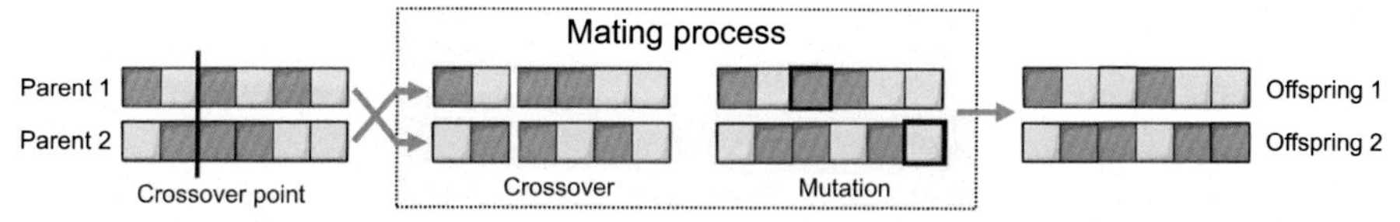

Figure 14.3 Mating process performed by genetic algorithms by means of the two operators, crossover and mutation.

This iterative process is ended when the convergence is reached or a limited number of iterations is exceeded. The nondominated solutions of the last generation are assumed to be the optimal ones.

A widely used variant of the GA is the NSGA-II, developed by Deb et al. (2002). Handling real, discrete, or even discontinuous variables, NSGA-II is particularly useful when dealing with multiobjective problems related to the field of building design optimization. NSGA-II has proved to be one of the most effective algorithms, showing good performance in exploring the search space, without being trapped in local minima and quickly reaching the Pareto frontier.

14.3.3.3 Particle swarm optimization

The PSO algorithm (Kennedy and Eberhart, 1995) is based on the social behavior of some animals, such as birds or fish. For example, when a fish flocking is searching for food around the surrounding space, none of the fish know where the food is, however at each iteration they know how far they are from the food. The most effective strategy is then to try to follow the fish closer to the food. The algorithm works according to the same principle; each fish, called a particle, moves around the space with a certain velocity. At each iteration the particles update their velocity and position on the base of the global optima and local optima, where the global optima is the best solution among the solutions of all the iterations and the local optima is the best solution according to the current iteration. Besides GA, PSO is the most common algorithm and it has been implemented in GenOpt (Wetter, 2001), which is a generic optimization program with ready-made routines offering coupling with building energy-simulation software.

14.3.3.4 Ant colony optimization

The ACO algorithm is inspired by the ants' behavior, in particular by their ability to find the shortest path to reach the food. In fact, ants leave on their path pheromones to encourage the others to follow them (Dorigo et al., 1996). Ants that are able to find the shortest path to reach the food leave a stronger trail of pheromone. Therefore, the others ants are always more attracted by the stronger trail and, at each iteration, a greater number of ants will chose the shortest path, until all the ants follow the shortest path. Among the presented algorithms, in literature there are still no works that are using it to optimize building design problems.

14.3.3.5 Harmony search

The HS optimization algorithm (Geem et al., 2001) is based on the musical process of searching a perfect state of harmony, the same way the optimization process tries to define the optimality. During an improvisation, each musician of an orchestra plays different notes with a different musical instrument; after a certain amount of time the musicians are able to find the best combination of frequency for the best tune. With a similar approach, the HS algorithm recombines the variable values, perturbing the neighbors' values, to find the best combinations. The process is repeated until the solutions stored in the harmony memories approach the optimum solutions.

The algorithm runs until the maximum iterations are overcome and the objective function is optimized (Patil Sachin and Patel, 2013).

14.3.3.6 Simulated annealing

The SA optimization algorithm has been developed by Kirkpatrick et al. (1983). The algorithm is inspired by the thermodynamic process of metal annealing, where the metal is melted and then gradually cooled. The product of this process is a crystalline lattice, composed by a regular atom structure, which minimizes the energy probability distribution. The optimization algorithm works in the same way: the minimization of internal energy represents the objective, and the heating process is represented by the random alteration of the variable values. At each iteration the algorithm tries to find the minimal energy states that represent the cost function. The higher the heat input, the more amplified are the random fluctuations.

14.4 Application example: optimization of energy retrofit measures applied on residential buildings

Here we present an application of the GA for the optimization of retrofit measures considering two (energy, overall cost) or three objectives (energy, overall cost, thermal comfort), in order to give an example of how to implement the algorithms and how to couple them with dynamic simulation, in order to show which can be the results of this approach.

In retrofit design, optimization techniques can be used successfully to drive the decision process toward solutions that are optimal not only in terms of energy savings but also of other aspects, first of all the economic one. As BES tools are needed to properly assess either energy or nonenergy aspects, which comes at a high computational cost, optimization algorithms can be used to speed up the otherwise extensive evaluation process of the possible alternatives.

In this section the potentialities of coupling the elitist NSGA-II developed by Deb et al. (2002), with the dynamic tool, TRNSYS (Solar Energy Laboratory, 2012), have been investigated in a case study in order to show the main differences between the brute-force approach versus the GA. Concerning the objectives to pursue, in this example a comparison between a two-objective optimization and three-objective optimization is presented in order to show the different results that can be obtained.

14.4.1 The case study

The analysis focuses on a residential building module, which is intended to be representative of a single unit inside a multifamily building, built prior to the first Italian energy legislation, in 1976 (Italian Parliament, 1976). The reference building is intentionally simplified as a shoebox-like module with realistic geometrical dimensions, window area, and an envelope's thermal resistance. The climatic context considered

Table 14.1 Linear thermal transmittance of the thermal bridges

Type	Ψ, Linear thermal transmittance (W/m/K)
Corners	0.098
Intermediate floor and walls	0.182
Window perimeters	0.060

in the example is that of Milan, which may be considered as a good population-based representative of Northern Italy, with heating degree days of 2404K d (4327°F d) calculated on a base temperature of 20°C, which corresponds to a climatic zone Cfa, according to the Köppen's classification (Köppen, 1884).

Concerning the geometry, the module is a single-story module with a square floor of 100 m^2 and an internal height of 3 m. The floor area is close to the weighted average surface for European residential buildings as from the data provided by the UNECE technical report (UNECE, 2006). The vertical walls are oriented toward the main cardinal points and the window-to-floor ratio is equal to 14.4%. For sake of simplicity the entire window surface is assumed to be south-oriented. The linear thermal transmittances of thermal bridges (Table 14.1) in the reference cases have been calculated according to the EN ISO 10211 (CEN, 2007a) by means of the finite element tool Therm (LBNL, 2013).

The exposed surface is that of the windowed façade and the two side walls. Floor, ceiling, and back wall are in contact with indoor spaces at the same temperature. This leads to a compactness ratio (ie, the ratio between externally exposed surface and volume $S/V = 0.3\ \mathrm{m}^{-1}$) typical of an intermediate flat in a block (Fig. 14.4).

Considering the common characteristics and/or the requisites in force in the considered construction period, the opaque envelope's resistance for the starting reference

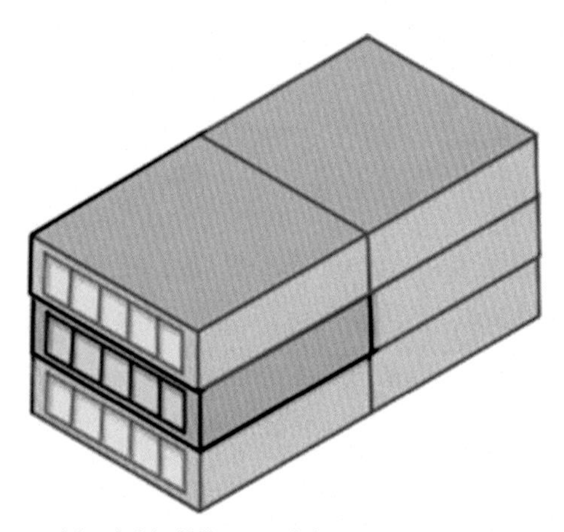

Figure 14.4 Reference residential building module.

condition is 0.97 $m^2/K/W$. A single pane glass of transmittance $U_{gl} = 5.7$ $W/m^2/K$ with a timber frame of transmittance $U_{fr} = 3.2$ $W/m^2/K$ are assumed. The heating system includes a standard gas boiler, coupled with radiators and an on-off control system, with a nominal global efficiency of 89%.

14.4.2 Energy-efficiency measures

The kind of retrofit options considered for refurbishing the building are conventional EEMs, applied either to the envelope components or to the heating and ventilation system:

1. External insulation of the opaque envelope with an expanded polystyrene layer, (conductivity $\lambda = 0.04$ W/m/K, specific heat $c = 1470$ J/kg/K, density $\rho = 40$ kg/m^3) in a range of thicknesses between 0 and 20 cm in steps of 1 cm.
2. Replacement of existing window systems with higher thermal performance glazing systems (Table 14.2), such as double or triple plane with either high or low solar heat gain coefficients, and frame replacement with an improved solution in aluminum with thermal break ($U_{fr} = 1.2$ $W/m^2/K$).
3. Substitution of heating generator with either modulating (nominal efficiency $\eta = 96\%$) or condensing boiler (nominal efficiency $\eta = 104\%$, calculated on the fuel lower heating value) both with a climatic control system.
4. Installation of a mechanical ventilation system with heat recovery to control the air exchange (ventilation rate $q_v = 150$ m^3/h, electric power $P = 59.7$ W).

The listed EEMs bring some extra energy performance improvements without extra costs:

- The linear thermal transmittance of thermal bridges is reduced depending on the insulation thickness and window type.
- The air tightness of the building is assumed to improve in case of substitution of the windows. The value of the infiltration rates is considered half the original values, reported in Table 14.2.
- The substitution of the radiators as emission system is not planned, and their nominal capacity does not change. However, when the heating peak load is reduced because of some intervention, the design capacity can be reduced accordingly, reducing the supply temperatures. This adds to the use of the climatic control, which is assumed when the boiler is replaced.

Table 14.2 Characteristics of the analyzed glazing systems

	Glazing system	U_{gl} ($W/m^2/K$)	SHGC (−)
DH	Double, high SHGC (4/9/4, krypton, low-e)	1.140	0.608
DL	Double, low SHGC (6/16/6, krypton, low-e)	1.099	0.352
TH	Triple, high SHGC (6/12/6/12/6 krypton, low-e)	0.613	0.575
TL	Triple, low SHGC (6/14/4/14/6 argon, low-e)	0.602	0.343

SHGC, solar heat gain coefficients.

14.4.3 Coupling NSGA-II with dynamic simulation

Before implementing the optimization, the EEMs useful to refurbish the building have to be codified consistently. According to this methodology, a possible combination of retrofit alternatives is individual, identified by a chromosome. The chromosome is composed of a genetic sequence and each gene corresponds to a single retrofit option, with the corresponding value. In the example shown in Fig. 14.5, the four genes contain information on the glazing typology, the insulation thickness, the presence of a mechanical ventilation with heat recovery, and the heat generation system, respectively.

As represented in Fig. 14.6, the first step in the GA procedure is the selection of the initial population, which in this example is set to 128 individuals/chromosomes. Usually, for each chromosome, the values of the genes are randomly defined. However, to improve the performance of the GA, the initial population and the genes values are defined through the Sobol's sequence sampling. This pseudo-random number generator avoids the oversampling of same regions that can occur with random sampling (Saltelli et al., 2004), giving a good individuals collection as the initial population.

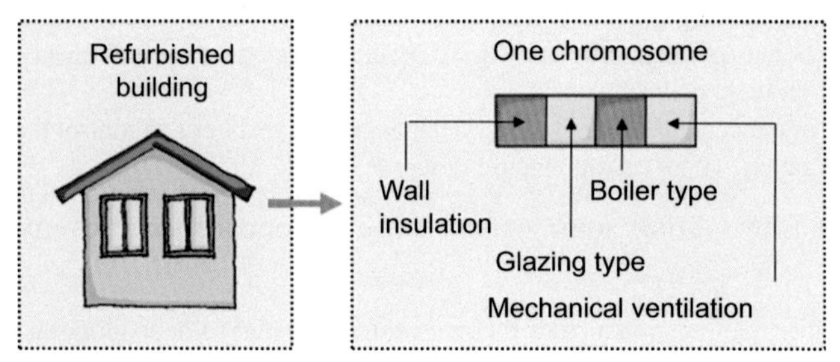

Figure 14.5 A possible solution of the optimization problem, represented by a chromosome.

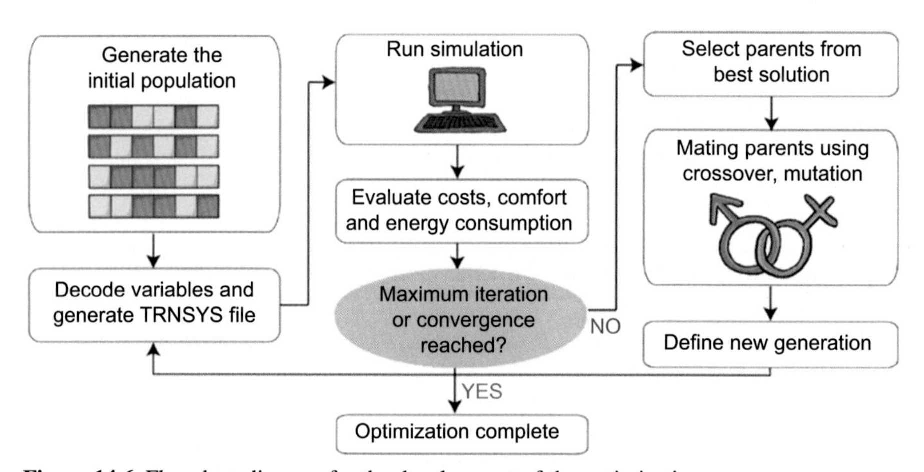

Figure 14.6 Flowchart diagram for the development of the optimization process.
Penna, P., Prada, A., Cappelletti, F., Gasparella, A., 2015. Multi-objectives optimization of energy efficiency measures in existing buildings. Energy and Buildings 95, 57–69.

Afterward, the genetic sequence of each chromosome is decoded and translated in a TRNSYS building description file by means of Matlab environment (Matlab, 2008). The annual simulation is then run and the objectives for the chosen retrofit combinations (ie, the individuals) are evaluated. The performance is then expressed by the value of the considered fitness functions. The nondominated solutions are selected as parents for the next generation.

The code combines the genetic characteristics of parents, giving rise to a number of newly generated individuals to complement the parents' number up to the original number of 128 individuals. The implemented recombination procedure is based on the operators of crossover and mutation, with an arithmetic crossover fraction of 0.8. The mutation is done by means of a Mersenne-Twister pseudo-random generator (Matsumoto et al., 1998), then a randomly selected gene is replaced by a uniformly distributed random value that meets the gene range.

The new population undergoes the selection process again, and the iterative process is repeated until the maximum number of iteration or the convergence level is reached. The convergence criteria are met when diversity of the optimal individuals is ensured, which is calculated in terms of Euclidean distance between the nondominated solutions (crowding distance), and when the number of nondominated solutions is a given percentage (75% in the example) of the entire population. The nondominated solutions in the final population are the GA Pareto frontier.

14.4.4 Evaluation objectives

In building retrofit the main goal to be pursued is usually the best trade-off between the investment costs and the energy savings during the building's operational life. This is what is intended by the cost-optimal analysis described by the EU regulation 244/2012 (European Commission, 2012). The cost-optimal analysis compares alternatives in terms of primary energy demand and global cost (life cycle) of the building, considering the investment for refurbishment and the operational costs that may occur during a 30-year life in case of residential buildings (20 years in other cases), expressed by its NPV. This analysis method, proposed as a procedure to establish new cost-optimal energy requirements for new and existing buildings, is useful to find not only cost-optimal levels and the combination of EEMs that guarantee its achievement, but also a set of nondominated solutions that are characterized by the maximum efficiency from the energy and economic point of view.

Next to this double-objective approach, when the building retrofit has to be pursued, the importance of implementing those interventions improving, or at least not deteriorating the indoor thermal comfort conditions, is clearly stated in the regulation and assessed by some studies (Penna et al., 2015). In this case the thermal comfort condition has been quantified in terms of an objective function, the weighted discomfort time (WDT), calculated according to the Standard EN 15251 (CEN, 2007c).

The multiobjective optimization approach can then be repeated, to also include the indoor thermal discomfort minimization.

For the optimization the objective functions defined are the energy performance for heating (EP_H), the total cost evaluated as the NPV and the WDT.

Therefore, the fitness functions to minimize are defined, according to Eqs. [14.4] and [14.5], as:

$$\begin{cases} \mathrm{EP_H} = f_1(x_1, x_2, x_3, x_4); \\ \mathrm{NPV} = f_2(x_1, x_2, x_3, x_4); \end{cases} \qquad [14.4]$$

in the two-objective optimization, or

$$\begin{cases} \mathrm{EP_H} = f_1(x_1, x_2, x_3, x_4); \\ \mathrm{NPV} = f_2(x_1, x_2, x_3, x_4); \\ \mathrm{WDT} = f_3(x_1, x_2, x_3, x_4); \end{cases} \qquad [14.5]$$

in the three-objective optimization, where x_1 represents the level of insulation of external walls; x_2, the adopted glazing system; x_3, the boiler type; and x_4, the presence of the mechanical ventilation system.

According to the EN ISO 15217:2007b (CEN, 2007b), the $\mathrm{EP_H}$ is the primary energy annually used by the building for heating divided by heated floor area. It is calculated by means of the simulation tool TRNSYS, considering the energy consumption for heating and for the electric devices, such as the circulation pump and the mechanical ventilation system, when present.

The economic evaluation of the different EEMs has been conducted according to the comparative framework methodology of cost-optimal level, proposed by the cited Regulation 244 (European Commission, 2012), by means of the calculation of the total cost evaluated as the NPV.

The NPV is a utility function that accounts not only for investment cost related to the retrofit, but also for the operational costs along the lifespan of the renovated building.

This function allows evaluation of the economic benefits related to the possible retrofit solutions, analyzing different time series of cash flows related to each intervention.

The NPV takes into account:

- The initial investment costs for the retrofits. The costs in the example are defined from the comparison of different regional databases (Regional Price List of Lombardia, Lazio, and Sicilia).
- The annual running costs, composed of the annual energy cost and the maintenance cost for preserving and restoring the building and its elements. The energy cost is calculated considering the fuel and electricity price rising.
- The replacement cost, for the periodic substitution of building/system elements.
- The residual value for the pieces of equipment with longer lifespan according to EN 15459 (CEN, 2009).

The evaluation of the long-term comfort performance is conducted by means of the calculation of the WDT index, as proposed by annex F of the Standard EN 15251

(CEN, 2007c) through the Degree Hours Criteria. The WDT accounts for how much and how long the operative temperature lies outside the comfort range, so it gives a good estimation of long-term comfort performance. With this approach the occupied hours, during which the actual operative temperature lies outside the specified comfort range, are weighted by a weighting factor that depends on the entity of the deviation from the range (Eqs. [14.6] and [14.7]).

$$\text{WDT} = \sum wf \cdot \text{time} \quad [14.6]$$

$$wf = \Theta_o - \Theta_{o,\text{limit}}$$

$$\text{when } \Theta_o < \Theta_{o,\text{limit,lower}} \text{ or } \Theta_o > \Theta_{o,\text{limit,upper}} \quad [14.7]$$

The evaluation of the WDT has been tightly integrated into the simulation model in TRNSYS.

14.4.5 Results

To assess the performance and characteristics of the GA, the optimal solutions for the refurbishment of the described building module have been evaluated using the brute-force approach, calculating all the possible combinations, and by means of NSGA-II for either the bioptimization (Fig. 14.7) or the trioptimization (Fig. 14.8).

The brute-force approach requires the calculation of all the possible configurations (in gray in the figures), among which the nondominated solutions are the real Pareto front (blue empty circles). The GA is able to identify most of the real front solutions (the light blue dots) even if in some cases some optimal are missing (in the dotted line circle).

More in detail, it is possible to observe that the optimal solutions found by the GA are slightly less than the ones in the real front (23 vs 24). This is an intrinsic limitation of the GA, which is only able to approximate the Pareto frontier, depending on the

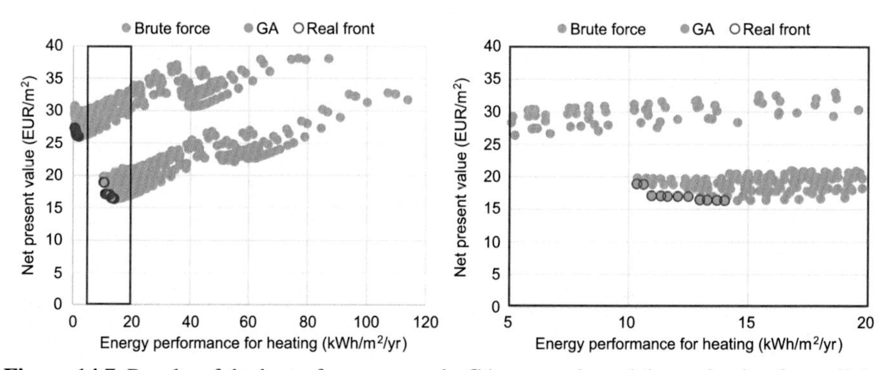

Figure 14.7 Results of the brute-force approach, GA approach, and the evaluation from all the evaluated combinations of the real Pareto front.

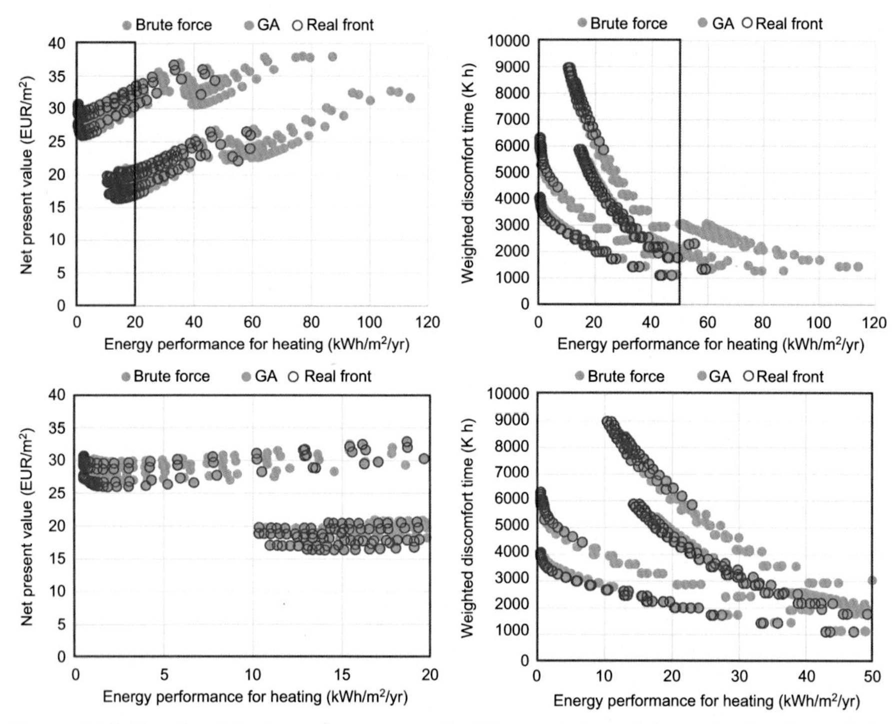

Figure 14.8 Results of the brute-force approach, GA approach, and the evaluation from all the evaluated combinations of the real Pareto front.

selected size of population and on the number of iterations or the convergence criterion. In this case, the convergence check has stopped the process before having found all the optimal solutions. However, the cost-optimal solution (ie, the one with the minimum NPV) found by calculating the real front and by the GA is the same, which means that the GA is actually able to define the best solutions in term of energy and costs.

Fig. 14.8 reports the results in which the comfort was also included into the optimization process. The relationship between energy and global costs has been reported on the left side, while on the right side, the results are reported in function of energy and comfort. First, it is possible to observe that increasing the number of objectives increases also the number of optimal solutions. In fact, the solutions composing the Pareto front is much higher than the ones of the previous optimization. However, the solutions selected by the GA are less than the ones of the real Pareto front (87 vs 235). This is due to the number of individual populations that is fixed for each generation; in this case it is smaller than the size of the actual optimal solutions set. Moreover, a second issue appears because of that, which is the inclusion within the 87 of 3 suboptimal solutions. In fact, the larger the difference between the number of the real Pareto front solutions and the number of the GA generation, the higher the possibility of selecting suboptimal solutions in the final step. On the other hand, it is also true that the optima selected not only have a very small differences in terms of energy and comfort performances, but they are very similar in terms of total cost, thus confirming that the GA algorithm maintains a good performance for the best EEMs selection.

14.5 Conclusion

In this chapter an overview of different methodologies to assist and drive the decision process in building energy retrofit has been presented. Designing the retrofit can become a complex situation because decision-makers are in front of a large number of innovative technologies and EEMs, each of them with a different energy and cost impact. Moreover, decision-makers, would they be the policy maker, a private owner, or an ESCO, probably can give different objectives and priorities to their investment. Therefore one of the main issues is to identify those measures able to fulfill energy, financial, legal regulation, environmental requirements, and occupant satisfaction, reaching the best compromise among these objectives. In literature, several approaches to cope with multiobjective problems exist, even though one of the most promising is the multiobjective optimization through Pareto optimality. By identifying the so-called Pareto optimal solutions, it is possible to find solutions that represent the best trade-off among the objectives of the refurbishment project. Pareto optimality also allows for investigation of the effect of different retrofit strategies on the attainment of different objectives, because not just one best solution is found, but it can be a set of nondominated ones. Otherwise, the efficacy of optimization results is strictly connected to the solution space investigated. The wider the space, the more reliable the results. Optimization solution algorithms based on the mimicking of natural phenomena are particularly effective to deal with the field of building optimization because of their ability in selecting real optima, not being trapped in local minima. The optimization approach most widely adopted in literature is the GA, inspired by evolutionary biology such as inheritance, mutation, selection, and crossover, and able to identify the Pareto optimal solutions.

In the last part of the chapter the GA has been applied in a case study in order to show how this algorithm can be coupled to building energy simulation and used in the retrofit design. The ability of the GA in finding the optimal solutions has been proved, analyzing the differences between the solutions laying on the real Pareto front of all the calculated solutions (brute-force approach) versus the solutions selected by the GA. The case study has highlighted the importance of properly selecting the number of the individual populations that should be large enough to identify the optimal combinations. As shown in the analysis, coupling optimization algorithms with simulation codes is a good instrument to automatically investigate the decision space and to quickly define optimal solutions. Concerning the objectives to be reached, the importance of considering people's thermal comfort in the optimization process has been highlighted. In fact, neglecting this aspect leads to select those EEMs that, despite a low total cost, would hamper the thermal indoor environment; to reject some EEMs presents a higher investment cost but does not compromise the indoor thermal comfort.

Start-up companies operating in the field of computational software should consider the lack of tools developing this sort of analysis as a great challenge and opportunity. They should try to propose some applications able to assist the stakeholders during the evaluation phases and able to guide the retrofit decision of the investors. On the other side, energy consultant agencies and ESCOs are the main subjects to be interested in using new tools in order to support policy makers, public administrators, or private owners in the energy retrofit of buildings.

References

Alanne, K., 2004. Selection of renovation actions using multi-criteria "knapsack" model. Automation in Construction 13, 377–391.

Asadi, E., Da Silva, M.G., Antunes, C.H., Dias, L., 2012a. Multi-objective optimization for building retrofit strategies: a model and an application. Energy and Buildings 44, 81–87.

Asadi, E., Da Silva, M.G., Antunes, C.H., Dias, L., 2012b. A multi-objective optimization model for building retrofit strategies using TRNSYS simulations, GenOpt and MATLAB. Buildings and Environment 56, 370–378.

Asadi, E., Da Silva, M.G., Antunes, C.H., Dias, L., 2013. State of the art on retrofit strategies selection using multi-objective optimization and genetic algorithms. In: Pacheco-Torgal, P., Mistretta, M., Kakliauskas, A., Granqvist, C.G., Cabeza, L.F. (Eds.), Nearly Zero Energy Building Refurbishment, pp. 279–297.

Asadi, E., Da Silva, M.G., Antunes, C.H., Dias, L., Glicksman, L., 2014. Multi-objective optimization for building retrofit: a model using genetic algorithm and artificial neural network and an application. Energy and Buildings 56, 370–378.

Baños, R., Manzano-Agugliaro, F., Montoya, F.G., Gil, C., Alcayde, A., Gómez, J., 2011. Optimization methods applied to renewable and sustainable energy: a review. Renewable and Sustainable Energy Reviews 15, 1753–1766.

Caccavelli, D., Gugerli, H., 2002. TOBUS – a European diagnosis and decision-making tool for office building upgrading. Energy and Buildings 34, 113–119.

CEN European Committee of Standardization, 2007a. EN ISO 10211, Thermal Bridges in Building Construction – Heat Flows and Surface Temperatures – Detailed Calculations.

Chantrelle, F.P., Lahmidi, H., Keilholz, W., El Mankibi, M., Michel, P., 2011. Development of a multicriteria tool for optimizing the renovation of buildings. Applied Energy 88, 1386–1394.

De Boeck, L., Audenaert, A., De Mesmaeker, L., 2015. Improving the energy performance of residential buildings: a literature review. Renewable and Sustainable Energy Reviews 52, 960–975.

Deb, K., Pratap, A., Agarwal, S., Meyarivan, T., 2002. A fast and elitist multi-objective genetic algorithm: NSGA-II. IEEE Transactions on Evolutionary Computation 6 (2), 182–197.

Diakaki, C., Grigoroudis, E., Kolokotsa, D., 2008. Towards a multi-objective optimization approach for improving energy efficiency in buildings. Energy and Buildings 40, 1747–1754.

Dorigo, M., Maniezzo, V., Colorni, A., 1996. The ant system: optimization by a colony of cooperating agents. IEEE Transactions on Systems, Man, and Cybernetics—Part B 26 (1), 29–41.

European Commission, 2010. Directive 2010/31/EU of the European parliament and of the council. Official Journal of European Union. L-153/13, 18/06/2010.

European Commission, 2012. Commission Delegated Regulation (EU) No 244/2012 of 16 January 2012 supplementing Directive 2010/31/EU.

European Committee for Standardization (CEN), 2007b. EN 15217:2007. Energy Performance of Buildings. - Methods for ex- pressing energy performance and for heating certification of buildings. CEN, Bruxelles, Belgium.

European Committee for Standardization (CEN), 2009. EN 15459:2009. Energy performance of buildings - Economic evaluation procedure for energy systems in buildings. Bruxelles, Belgium, CEN.

Evins, R., 2013. A review of computational optimization methods applied to sustainable building design. Renewable and Sustainable Energy Reviews 22, 230–245.

Ferrara, M., Fabrizio, E., Virgone, J., Filippi, M., 2014. A simulation-based optimization method for cost-optimal analysis of nearly Zero Energy Buildings. Energy and Buildings 84, 442–457.

Flourentzou, F., Roulet, C.-A., 2002. Elaboration of retrofit scenarios. Energy and Buildings 34, 185–192.

Geem, Z., Kim, J., Loganathan, G., 2001. A new heuristic optimization algorithm: harmony search. Simulation 76 (2), 60–68.

Georgopoulou, E., Lalas, D., Papagiannakis, L., 1997. A multi criteria decision aid approach for energy plannings problems: the case of renewable energy option. European Journal of Operational Research 103, 38–54.

Goldberg, D.E., 1989. Genetic Algorithms in Search, Optimization, and Machine Learning. Addison-Wesley, Reading, MA.

Hamdy, M., Hasan, A., Siren, K., 2013. A multi-stage optimization method for cost-optimal and nearly-zero-energy building solutions in line with the EPBD-recast 2010. Energy and Buildings 56, 189–203.

Haupt, R.L., Haipt, S.E., 2004. Algorithms Practical Genetic Algorithms, second ed. John Wiley & Sons, Hoboken, New Jersey.

Holland, J.H., 1975. Adaptation in Natural and Artificial Systems. University of Michigan Press, Ann Arbor.

Hopfe, C.J., Augenbroe, G.L.M., Hensen, J.L.M., 2013. Multi-criteria decision making under uncertainty in building performance assessment. Building and Environment 69, 81–91.

Ihm, P., Krarti, M., 2012. Design optimization of energy efficient residential buildings in Tunisia. Building and Environment 58, 81–90.

Iqbal, M., Azam, M., Naeem, M., Khwaja, A.S., Anpalagan, A., 2014. Optimization classification, algorithms and tools for renewable energy: a review. Renewable and Sustainable Energy Reviews 39, 640–654.

Italian Parliament, 1976. Law 373 Norme per il contenimento del consume energatico per usi termici negli edifici.

Jaggs, M., Palmar, J., 2000. Energy performance indoor environmental quality retrofit—a European diagnosis and decision making method for building refurbishment. Energy and Buildings 31, 97–101.

Kaklauskas, A., Zavadskas, E., Raslanas, S., 2005. Multivariant design and multiple criteria analysis of building refurbishments. Energy and Buildings 37, 361–372.

Kennedy, J., Eberhart, R., 1995. Particle swarm optimization. Proceedings of the IEEE International Conference on Neural Networks 4, 1942–1948.

Kirkpatrick, S., Cd, G., Mp, V., 1983. Optimization by simulated annealing. Science 220 (4538), 671–680.

Köppen, W., 1884. Die wärmezonen der erde, nach der dauer der heissen, gemässigten und kalten zeit und nach der wirkung der wärme auf die organische welt betrachtet.

Kolokotsa, D., Diakaki, C., Grigoroudis, E., Stavrakakis, G., Kalaitzakis, K., 2009. Decision support methodologies on the energy efficiency and energy management in buildings. Advances in Building Energy Research 3, 121–146.

Kurnitski, J., Saari, A., Kalamees, T., Vuolle, M., Niemelä, J., Tark, T., 2011. Cost optimal and nearly zero (nZEB) energy performance calculations for residential buildings with REHVA definition for nZEB national implementation. Energy and Buildings 43, 3279–3288.

LBNL, 2013. Lawrence Berkeley National Laboratory, Therm 6.3/Window 6.3 NFRC. Simulation Manual, July 2013. http://windows.lbl.gov/software/window/6/.

Loken, E., 2007. Use of multicriteria decision analysis methods for energy planning problems. Renewable and Sustainable Energy Reviews 11, 1584–1595.

Lu, Y., Wang, S., Zhao, Y., Yan, C., 2015. Renewable energy system optimization of low/zero energy buildings using single-objective and multi-objective optimization methods. Energy and Buildings 89, 61–75.

Machairas, V., Tsangrassoulis, A., Axarli, K., 2014. Algorithms for optimization of building design: a review. Renewable and Sustainable Energy Reviews 31, 101–112.

MATLAB 7.7.0 R, September 2008.

Matsumoto, M., Nishimura, T., 1998. Mersenne Twister: a 623-dimensionally equidistributed uniform pseudo-random number generator. ACM Transactions on Modeling and Computer Simulation 8, 3–30.

Pacheco, R., Ordóñez, J., Martínez, G., 2012. Energy efficient design of buildings: a review. Renewable and Sustainable Energy Reviews 16, 3559–3573.

Palonen, M., Hasan, A., Siren, K., 2009. A genetic algorithm for optimization of building envelope and HVAC system parameters. In: IBPSA Conference on Eighth International Building Performance Simulation Association, pp. 159–166.

Pareto, V., 1896. In: Rouge, F. (Ed.), Cours D' Economie Politique.

Patil Sachin, A., Patel, D.A., 2013. An overview: improved harmony search algorithm and its applications in mechanical engineering. International Journal of Engineering Science and Innovative Technology (IJESIT) 2 (1), 433–444.

Penna, P., Prada, A., Cappelletti, F., Gasparella, A., 2015. Multi-objectives optimization of energy efficiency measures in existing buildings. Energy and Buildings 95, 57–69.

Pohekar, S.D., Ramachandran, M., 2004. Application of multi-criteria decision making to sustainable energy planning – a review. Renewable and Sustainable Energy Reviews 8, 365–381.

Rey, E., 2004. Office building retrofitting strategies: multicriteria approach of an architectural and technical issue. Energy and Buildings 36, 367–372.

Saltelli, A., Tarantola, S., Campolongo, F., Ratto, M., 2004. Sample Generation, in Sensitivity Analysis in Practice. A Guide to Assessing Scientific Models. John Wiley & Sons, Chichester, UK, pp. 193–204.

Sette, S., Boullart, L., 2001. Genetic programming: principles and applications. Engineering Applications of Artificial Intelligence 14 (6), 727–736.

Solar Energy Laboratory, TRNSYS 17, 2012. A Transient System Simulation Program. http://sel.me.wisc.edu/trnsys.

Storn, R., Price, K., 1997. Differential evolution a simple and efficient heuristic for global optimization over continuous spaces. Journal of Global Optimization 11 (4), 341–359.

UNECE, 2006. Bulletin of Housing Statistics for Europe and North America, Technical Report. United Nations Economic Commission for Europe.

Vucicevic, B., Jovanovic, M., Afgan, N., Turanjanin, V., 2014. Assessing the sustainability of the energy use of residential buildings in Belgrade through multi-criteria analysis. Energy and Buildings 69, 51–61.

Wang, J.J., Jing, Y.Y., Zhang, C.F., Zhao, J.H., 2009. Review on multi-criteria decision analysis aid in sustainable energy decision-making. Renewable and Sustainable Energy Reviews 13, 2263–2278.

Wetter, M., 2001. GenOpt – a generic optimization program. In: Seventh International IBPSA Conference Rio de Janeiro, Brazil, August 13–15, 2001.

Yongjun, S., Pei, H., Gongsheng, H., 2015. A multi-criteria system design optimization for net zero energy buildings under uncertainties. Energy and Buildings 97, 196–204.

The use of algorithms for light control

C. Cristalli[1], L. Standardi[1], D. Kolokotsa[2], S. Papantoniou[2]
[1]Loccioni Group, Angeli di Rosora, (An), Italy; [2]Technical University of Crete, Chania, Greece

15.1 Introduction

Increase of CO_2 level in the atmosphere, elimination of fossil fuels, and economic and political reasons are among the causes of reduction in energy consumption and growth of the energy-efficiency product market. In particular, part of such a market is meant to exponentially grow over the next few years, so it represents a fertile ground for innovative start-ups. Moreover, energy consumption in buildings measures up to 40% of the worldwide energy consumption according to Pérez-Lombard's work and this clearly causes interest in such a sector (Pérez-Lombard et al., 2008). Hereafter, the importance of the lighting control market is explained by introducing the compelling role of light for human beings, the light control systems, and the features of such a market.

15.1.1 Importance of light

The availability and intensity of light both indoors and outdoors has a direct impact on physical health and emotional well-being according to Hraska's research (Hraska, 2014). The ways in which light affects the human body are two-fold: through the visual system and by interacting with the skin.

The light that enters the human body through the retina of the eye affects our metabolism and our endocrine and hormone systems. When the light levels are low, melatonin secretion increases and the activity levels of the human body, regulated by this hormone, drop. Great availability of daylight suppresses the production of melatonin and fosters an alert state of mind by secreting serotonin. Long-term stay in environments with little or no daylight may cause deregulation of the internal clock of humans.

An emotional disorder is SAD, or Seasonal Affective Disorder. It is common in the populations of northern latitudes, where daylight levels are low, and is linked to the melatonin and serotonin levels in the body. People suffering from SAD experience drastic mood swings, lowered energy, and depression. Artificial or natural light therapy has been found to be an effective antidepressant, but only when the light is bright enough as reported by Boubekri (2008).

Light affects the body interacting with our skin by way of photosynthesis and production of vitamin D. The ultraviolet light emitted by the sun is divided into three

Start-Up Creation. http://dx.doi.org/10.1016/B978-0-08-100546-0.00015-7

wavelength spectra: UV-A, UV-B, and UV-C. UV-C (<280 nm), which can burn the skin even at small doses, is completely absorbed by the ozone layer. UV-A (320–400 nm) is responsible for skin darkening and pigmentation. UV-B (290–315 nm) is responsible for photosynthesis, for skin burning, and aging. Overexposure to UV radiation over long periods of time might contribute to the risk of developing skin cancer (melanoma) as well as squamous cell carcinoma and basal cell carcinoma.

However, UV-B radiation also stimulates our skin to produce vitamin D. Vitamin D helps maintain serum calcium and phosphorus concentrations within the normal range, thereby enhancing the capability of the small intestine to absorb these minerals from the diet. Lack of vitamin D is linked to bone loss and fracture and to multiple harmful effects on the cardiovascular system. There is also evidence that there is correlation between multiple sclerosis and UV-B exposure.

Visible and infrared radiations from artificial lights are unlikely to have any effect on health, unless they are extremely intense and used at close range. However, people with conditions affected by light should be cautious even when being inside. Some people with skin conditions may be sensitive even to artificial lighting and should avoid light sources with UV emissions. For instance, if they use compact fluorescent lamps (CFLs) it would be better if they used those with a double envelope. An even better option for some people might be light-emitting diodes (LEDs), because they do not emit UV. Patients with eye conditions that are intolerant to high levels of light should wear glasses with protective filters. Finally, there is no scientific evidence to evaluate whether contemporary artificial light sources, like CFLs and LEDs, have any effect on conditions such as Irlen-Meares syndrome, myalgic encephalomyelitis, fibromyalgia, dyspraxia, autism, and HIV, as they are essentially flicker-free according to the European Commission (2012).

To sum up, light directly influences human health in many ways. However, artificial lights can be controlled in the view of fulfilling the requirements for lighting levels, so, ensuring visual comfort in buildings.

15.1.2 Light control algorithms

Among the most common internal gains available in the majority of buildings are those of artificial lights. Artificial lights in office buildings consume a significant amount of energy all around the world compared to the whole building's consumption as presented by Santamouris et al. (1994) and Lam et al. (2003), and this affects the cooling loads of buildings as reported by Franzetti et al. (2004). Moreover, according to Santamouris's research based on measurements in buildings of Greece, artificial lights consume 10% of the total energy consumption.

Many researchers have raised the issue of energy savings from artificial lights maximizing the benefits from natural daylight. The first approach is done by Knight (1998), who considers controllers that switch artificial lights on/off based on the indoor illuminance level. As Knight mentioned in his research, this model could only adjust the illuminance set point at 500 lux, forbidding the user to set a different set point based on the usage of each workstation. Furthermore, an example of a wireless communication between sensors and actuators involving on/off controllers is designed and applied by Nippun Kumaar et al. (2010). In this installation, wireless sensors are located in various areas in a room, then the wireless actuators, located next to the light fixtures, receive the control signals and switch the lights on/off. This causes savings up to 14.4 kWh per month, which is the 20% of the energy consumption of the artificial lights in the selected case study.

Furthermore, light control in buildings may also involve other comfort parameters, such as indoor air quality and thermal comfort in the view of achieving an overall optimization of the building's energy consumption while guarantying indoor comfort. This strategy is described by Dounis et al. (2011) and Kolokotsa et al. (2009): the former research considers simulation results while the latter is based on real measurements.

Similarly, more sophisticated controllers can be used especially for lighting systems that integrate dimmers. A wireless control system is designed and tested by Wen and Agogino (2011). According to their research, if a photo sensor and a controller are located above each workstation, energy savings can be up to 60.8% considering a specific occupancy profile for an office and comparing it to the initial state where all the lights were switched on/off simultaneously. The advantage of the latter control system compared to the former one is the capability to dim artificial lights separately. Another comparison between automated on/off systems and fully dimmable systems is presented by Frattari et al. (2009) where it can be noticed that a fully dimmable system can save up to 68% during autumn and up to 43% during winter, while an automated on/off system saves 56% and 20%, respectively. Furthermore, Knight's work shows that dimmable systems enable higher energy savings during the night since they adjust the provided illuminance to the required set point (Knight, 1998).

The development of the fuzzy control for dimming the artificial lights is presented by Kurian et al. (2005), and its application including the achieved results is published by Colaco et al. (2012). Although fuzzy technology has been developed since 1965, its application is continuously increasing. Their main advantage is the users' knowledge inserted in the controller in the form of rules. Another advantage of the fuzzy technology is its adaptability for actual measurements using ANFIS (Adaptive Neuro Fuzzy Inference System) architecture introduced by Jang's work, in which the fuzzification and defuzzification parameters are updated based on measurements collected on-site (Jang, 1993). Additional work is presented by Papantoniou et al. (2014) in which a fuzzy-based controller is applied to artificial lights in cooperation with dedicated smart sensors. Similarly, an automated controller can also be combined with a fault detection system in order to inform the energy managers that a sensor might have sent some fault measurements. Such a fault detection system is developed by Kolokotsa et al. (2005), showing remarkable results despite its simplicity.

15.1.3 Lighting control market

In the massive market of energy-efficiency products, the lighting control system sector is attracting much attention worldwide. While buildings' owners are willing to maximize their energy savings while minimizing costs, Curwin's work reports that only 7% of buildings in the United States have installed lighting control systems of any kind, including retrofitted lighting fixtures, dimmable lighting systems, or building automation networks (Curwin, 2010). However, these actions are necessary in view of being in line with the consumption reduction targets: this yields to billions of dollars per year that must be invested in energy efficiency in the building industry by 2030 according to Kachan's research (Kachan & Co, 2013; Green Growth Alliance, 2013). Because of this, the market of smart lighting control systems in the building industry is one of the most attractive as it is offering the greatest cost-saving potential and reduced energy use, while showcasing demonstrable return on investment. In particular, Navigant Research reports that, in the United States, global networked lighting control revenue is expected to grow from $2.2 billion in 2015 to $4.8 billion in 2024 (Navigant Research Report, 2015).

In this attractive market, information and communication technology (ICT) is playing a key role as the innovative lighting systems are networked, smart, and digital. As a consequence, differentiation, which is a key requirement for a successful start-up rollout, can be achieved through the development of customized light control algorithms to be integrated in the lighting project.

Therefore, the light control market has significant potential to grow exponentially and become a key driver to the building automation market segment. However, low costs, ease of installation, and scalability of the proposed solutions are crucial. Nowadays, the lighting control market consists mainly of firms that are less than 3 years old: it is a new market but its size is substantial and the best of these start-ups have a good chance of getting acquired by larger players. The cheaper equipment, installations, and labor and lost revenue due to disruption are the key of success for a firm. Moreover, smaller vendors are able to provide customized light control systems and this is an advantage over larger firms that include light controllers embodied in their larger offering.

The innovative lighting control systems mostly include LEDs despite the higher costs: this is due not only to their high energy efficiency but also to their high life expectancy. LEDs combined to control algorithms involve sensors too and, eventually, dimmable systems. Furthermore, both wireless and cabled communications enable data from the sensors to be collected for analysis.

In the view of showing the potential of start-ups in the lighting control market, few examples are reported. A silicon valley start-up called Enlighted has developed and installed lighting control systems for the biggest hi-tech companies, such as HP, Google, and LinkedIn. For example, in HP the installed control system tracks the number of people in a given space in order to avoid an employee waving his arms or walking around to get the lights turned back on again if he sits far from the motion detection sensor that controls the lights. In addition, Ketra, a start-up based in Austin, has developed digital bulbs that enable LEDs not only to emit lights but also to receive it in order to deliver quality control on the color. Comfy is a newborn Swiss start-up that provides a lighting control system that also includes flashing lights in case of intruders, notifications on mobile phone when the sensors detect motion, and simulates presence by switching on and off on a natural rhythm. Moreover, Brightup is a German start-up of a light control system that enables the user to set his or her preferences in terms of lights and then the controller regulates all the lamps in the house.

The purpose of this chapter is to introduce lighting control systems that implement control algorithms tailored to specific types of buildings, like hospitals and offices: these solutions have low up-front costs and they ensure substantial energy savings. Moreover, the proposed lighting control systems can be implemented in other types of buildings, such as the residential ones. Finally, three real-case implementations show the efficacy of the proposed strategy. A brief description of the buildings used as case studies is hereafter reported. The work presented in this chapter is partly part of the research project GREEN@Hospital (2015), a 3-year European Project cofunded by the ICT Policy Support Program as part of the Competitiveness and Innovation framework Program (CIP). Loccioni Group, both hospitals and the Technical University of Crete, were among the partners of the project concluded in February 2015.

15.1.3.1 Azienda Ospedaliera Universitaria Ospedali Riuniti in Ancona

The Azienda Ospedaliera Universitaria Ospedali Riuniti Umberto I—G.M. Lancisi—G. Salesi (AOR) of Ancona, shown in Fig. 15.1, is a university hospital and it is the biggest hospital of Italy's Marche region.

In this hospital, the consumption given by artificial lights is about 14%, compared to the total energy consumption of the hospital as presented by Kolokotsa et al. (2012). Three areas are selected to improve energy performances by implementing control and optimization algorithms: the oncology and hematology departments and the analysis lab. In these three areas of the hospital, the lighting system consisted of fluorescent artificial lights that were manually controlled. Accordingly, the lights were often left on during the absence of persons in the rooms or when daylight was sufficient and these caused energy losses. In these three areas of the hospital, the existing lighting structure was replaced by new dimmable LED lights that implement DALI protocol and advanced control algorithms, which include an indoor illuminance level and

Figure 15.1 Azienda Ospedaliera Universitaria Ospedali Riuniti in Ancona, Italy.

presence. The upgrade of the artificial lights combined with the development of the control algorithms significantly reduces energy consumption.

15.1.3.2 General hospital Chania Saint George

The Saint George Hospital, shown in Fig. 15.2, is in the city of Chania in Crete; it was built in 1997–2000 and first operated in 2000.

In the pediatric department of the Saint George Hospital, some rooms are selected to improve the performance of the artificial lights that were manually controlled by the users. In all the selected rooms, two light fixtures operate based on different circuits as they are connected to the central power line and the generator line, respectively. Also in the Saint George Hospital, advanced control algorithms that include indoor illuminance level and presence are implemented, and this yields to significant reduction in the energy consumption. However, the lighting system in this hospital differs from the one installed in Ancona: in the Saint George hospital, the artificial lights are not dimmable.

Figure 15.2 General hospital Chania Saint George, in Greece.

15.1.3.3 Loccioni Group building

Loccioni Group headquarters is located in Angeli di Rosora, in Italy's Marche region, and the related industrial microgrid has extended through two new buildings called Leaf Lab, shown in Fig. 15.3 and Kite Lab. The Leaf Lab was built in 2013 and it was fully operative in 2014; it has been built in view of maximizing its energy efficiency and it is indeed a class A+ building. Moreover, internal and external lights are LEDs and they are compliant with the DALI protocol, while dedicated control algorithms are applied to internal and external lighting systems together with exterior shades. These controllers are tailored to the users' comfort and they utilize a huge variety of sensors in the view of minimizing energy consumption. It is worth mentioning that the building automation technology applies to the overall lighting system and all exterior shades in both buildings; for example, Leaf Lab and Kite Lab: every room, external lights, and shades are controlled in a fully or semiautomatic way. Users can easily set the desired set-points and switch from one operation mode to another.

Figure 15.3 Leaf Lab of the Loccioni Group in Angeli di Rosora, Italy.

15.2 Lighting requirements in buildings

Most of the lighting systems are based on fluorescent lamps and only the more recent ones allow dimming by using electronic ballasts that replace the older magnetic ones. Before the mandatory installation of fluorescent light bulbs in Europe (European Commission, 2009), most buildings had incandescent light bulbs which were very inefficient. Lately, electronic ballasts combined with LED lights, which can dim up and down the artificial lights linearly, adjust the indoor light level to comfort levels, which causes an increase in energy savings. The luminous efficiency of each aforementioned system can be seen in Table 15.1. Accordingly, the design and the power of luminaries are based on the standard EN12461.1 2002: European Standard for Interior Lighting.

Moreover the illuminance and its distribution on the task area have a great impact on how quickly, safely, and comfortably a person perceives and carries out a visual task. The illuminance category and lux levels are tabulated in Table 15.2 (Alrubaih et al., 2013).

Table 15.1 Luminous efficiency of different types of light

Type of light	Luminous efficiency (lm/W)
Incandescent	10–18
Halogen	15–25
Compact fluorescent	50–60
Linear fluorescent	50–60
LED	30–60

Based on EN 12464.1 2002: European standard for interior lighting.

Table 15.2 Illuminance categories and illuminance values for different types of activities

Activity	Illuminance category	Illuminance (lux)
Visual tasks of high contrast or large size	D	200–300–500
Visual tasks of medium contrast or small size	E	500–750–1000
Visual tasks of low contrast or large size	F	1000–1500–2000

Adapted from Alrubaih, M.S., Zain, M.F.M., Alghoul, M.A., Ibrahim, N.L.N., Shameri, M.A., Elayeb, O., 2013. Research and development on aspects of daylighting fundamentals. Renewable and Sustainable Energy Reviews 21, 494–505.

However, the efficiency of such energy saving techniques is also affected by the behavior of occupants inside the buildings. Environmental culture and awareness have been improved only in recent years, thus smart systems aimed to maximize energy savings from the usage of artificial lights are now largely applied.

15.3 Development of light-control algorithms

The operations of the luminaries can be classified as:

- Manual: Occupants in the room switch on/off the lights
- Semiautomatic: Users activate the artificial light that shuts automatically based on a time delay or a presence detector
- Fully automatic: The turning on and off operations are fully based on presence detection and indoor illuminance sensors

Control algorithms apply to artificial lights in order to automatically maintain the illuminance level that fulfills the lighting requirements by including a presence detector and/or an indoor illuminance sensor. A possible structure of such a control algorithm is depicted in Fig. 15.4: the control architecture is closed-loop, thus the system continuously monitors indoor conditions that are fed back to the controller.

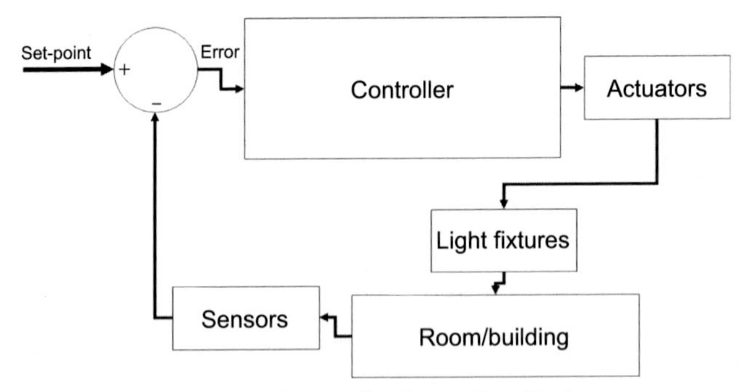

Figure 15.4 Structure of the proposed controller for artificial lights.

The input of the controller is the error that is defined as the difference between the current illuminance level, provided by the related sensor, and the required set point depending on the usage of each room.

$$\text{error}_{\text{illuminance}} = \text{Set point illuminance} - \text{measured illuminance} \quad [15.1]$$

The output of the controller is linked to the actuators of the light fixtures. The controller chosen for the proposed applications is a fuzzy controller that estimates the required change in the dimming level of the artificial lights from the error given as an input. Hereafter, the proposed fuzzy-based control is described in the view of showing its ease of replicability in other applications: naturally, this represents a strength in case of a start-up because it enables lighting control projects to be easily reproduced and implemented in other applications while maintaining low costs.

Table 15.3 reports its architecture, the fuzzification and defuzzification parameters. Moreover, the Sugeno architecture is selected because the control algorithm has only one output.

Furthermore, the input of the controller ($\text{error}_{\text{illuminance}}$) is fuzzified by using trapezoidal membership functions as reported in Fig. 15.5.

The membership functions are designed in order to maintain the desired value of illuminance and to simultaneously prevent recurring changes of the artificial lights that will reduce their life expectancy. Finally, Table 15.4 lists the rules used in the fuzzy controller.

Fig. 15.6 illustrates the architecture of the proposed control algorithm: fuzzy controller switches on the light in the controlled room if a presence is detected.

Table 15.3 Properties of fuzzy control algorithm for artificial lights in the two hospitals used as case studies

Type of fuzzy control algorithm	'Sugeno'	
Number of inputs	1: Error between current and desired light level	
Number of outputs	1: Change in the artificial lights state	
Fuzzification membership functions	5	
	'NE'	−0.7
	'SNE'	−0.3
	'ZERO'	0
	'SPO'	0.3
	'PO'	0.7
Further fuzzy parameters	AndMethod: 'prod'	OrMethod: 'probor'
	ImpMethod: 'prod'	AggMethod: 'sum'
	DefuzzMethod: 'wtaver'	

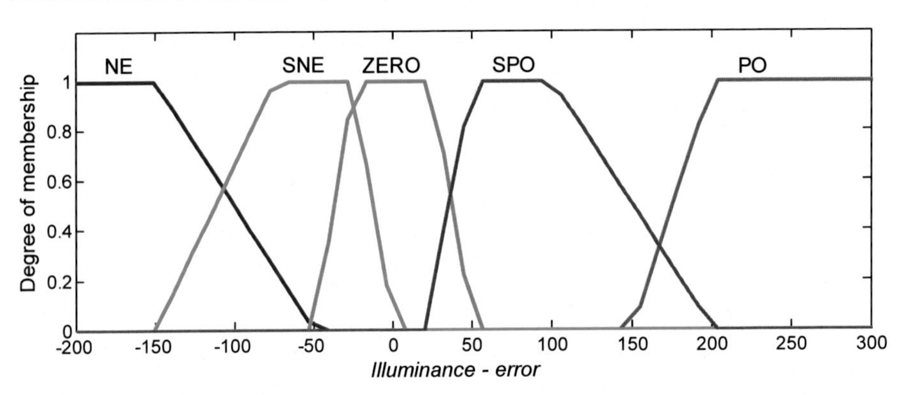

Figure 15.5 Membership functions for the fuzzification of illuminance error in the developed controller.

Table 15.4 Embedded rules of the proposed fuzzy controller

Input	Output
Illuminance difference from set point (error)	Artificial lights change
NE	PO
SNE	SPO
ZERO	ZERO
SPO	SNE
PO	NE

The lighting system in the hospital in Chania is not dimmable, so the fuzzy control output must be converted for such light fixtures.

15.4 Integration and implementation of the control algorithms in pilot buildings

In the following paragraph, the results obtained from the integration of the described control algorithm in the different real case studies are presented. For each pilot there is also a description of the hardware installed in order to maximize the energy savings. The control algorithms run as part of an overall system that also includes communication with a central computer, user interface, and required drivers for communicating with different systems. Although the control system has the capacity to meet the

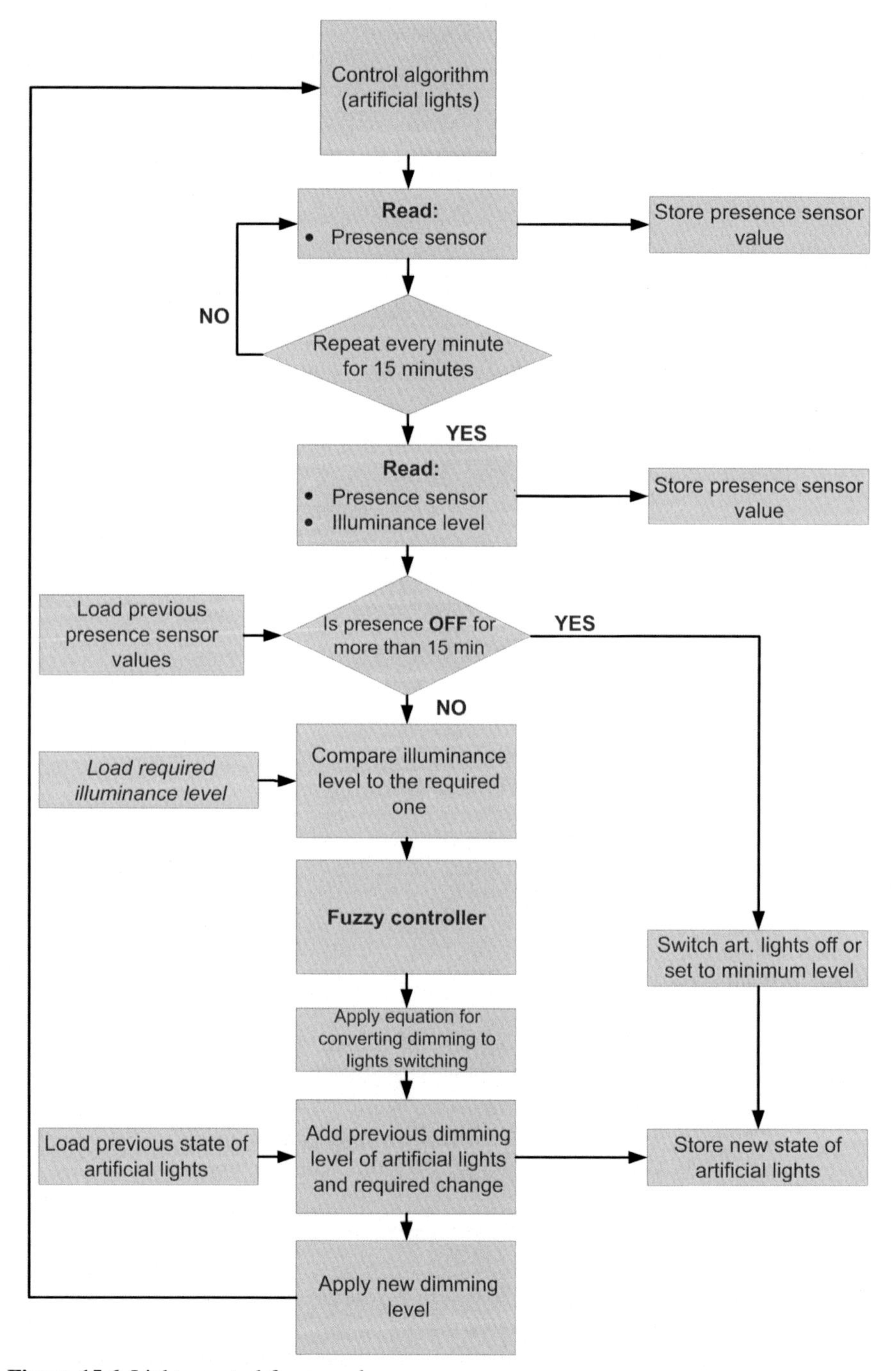

Figure 15.6 Lights control framework.

restrictions and preferences imposed by the user, additional safety features are incorporated to avoid any possibility of circumvent conditions. The special requirements given by hospitals, combined with the need of constantly ensuring comfort conditions, has led to the development of a software-based override system for all the aforementioned systems. The override system enables authorized personnel through a user-friendly interface to switch the system back to manual operation in the unlikely event of a failed sensor or a communication problem between the control algorithm.

Clearly, all these features for lighting control systems require flexibility in the solutions and this is achieved through dedicated control algorithms. Moreover, there is a huge variety of features to exploit linked to the applications of the lighting control systems: because of this, vendors should either be flexible, by developing customized light control algorithms, or address a specific application and master the connecting area of the market.

With reference to the control algorithm, input and output variables are connected with the measurements from the sensors and the actuators of the lighting systems in the selected rooms. The input variables sent to the artificial lights control system are:

- Override indication
- Presence detection
- Illuminance level
- Input definition

Hereafter, the proposed lighting control algorithm applied to the hospital in Ancona (AOR) and the one in Chania (SGH) is described as well as the obtained energy saving results. Furthermore, a lighting control system applied to office buildings is reported.

15.4.1 Azienda Ospedaliera Universitaria Ospedali Riuniti

Hospital wards and corridors were equipped with traditional T8 fluorescent lights. Main corridors are usually lit 24 h per day, ward corridors are manually controlled from the ward switchboard, while room lighting is controlled from dedicated switches installed in each room. In this pilot, LED DALI dimmable luminaries, presence sensors, and controllers are installed in order to implement control strategies based on presence detection, luminance level optimization, natural/artificial light mix, time schedule, and variable luminance set point. The control algorithms implemented in the field controllers continuously monitor the presence and the indoor illuminance and calculate the new dimming level that maintains indoor illuminance at the required set point. The performance of the control algorithm during 2 days is illustrated in Fig. 15.7.

When the artificial lights dim below the maximum capacity, savings are achieved compared to the initial case when lights were working at their maximum power. Furthermore, the presence detector switches off the artificial lights if the room is unoccupied. Fig. 15.7 also illustrates that energy is saved during the biggest part of the day when outdoor daylight contributes to indoor illuminance level.

Estimation of the annual savings is performed by comparing the baseline period, which is before and after the integration of the proposed control algorithm with the

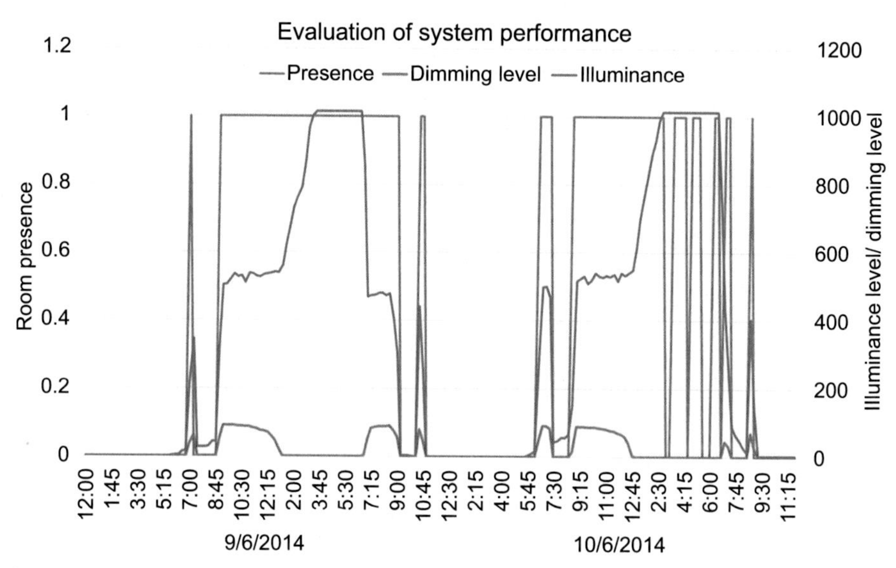

Figure 15.7 Evaluation of the control algorithms for dimmable artificial lights in AOR.

light fixture updates. The baseline period is 1 week per room. Moreover, the weekly energy consumption is calculated by multiplying the power consumption of the artificial lights by the time of operation. It should be noticed that all final energy values refer to electricity; energy efficiency (kWh) is finally converted into € using as conversion factor 0.15 €/kWh.

Yearly savings in terms of percentage, electricity, and € are presented in Table 15.5.

Accordingly, Table 15.5 illustrates that in all the retrofitted areas (except Archives PC) at least 50% energy efficiency is reached. Three areas are underperforming compared to the others: oncology nurse office, doctor office, and archive PC area. However, a higher energy efficiency in terms of percentage does not imply a lower payback time (PBT) of the solution in the selected area. For example, comparing the oncology nurse office and the oncology patient's waiting room (where the same peak power for lighting is installed) even if the former room presents higher savings in terms of percentage, the latter has higher savings in terms of energy and money. This is clearly due to the different light switch patterns of the two rooms.

Furthermore, Fig. 15.8 shows the evolution of the energy efficiency obtained in the hematology department during the test period. Generally speaking, different tendencies can be highlighted:

- Saving increase in the first month of the monitoring campaign due to algorithm optimization
- A seasonal tendency can be highlighted: the higher availability of daylight during spring and summer increases the savings achieved thanks to dimming in rooms equipped with windows

With reference to the PBT, Table 15.6 lists, for each room typology and for the overall facility, the solution PBT.

Table 15.5 Smart lighting system yearly results in AOR

Dept.	Room	Saving (%)	Saving (kWh)	Saving (€)
Oncology	Visitors waiting room corridor	80	379	57
	Visitors waiting room middle	73	688	103
	Visitors waiting room window	82	389	58
	Patients waiting room	76	378	57
	Nurse office	56	572	86
	Doctor office	52	59	9
	Archives shelves	94	592	89
	Archives PC	47	56	8
Hematology	Warehouse door	63	17	3
	Warehouse window	99	622	93
	Nurse office	82	746	112
	Doctor office	74	3.3	0
Analysis lab	Waiting room queue	76	1910	286
	Waiting room chairs	79	2972	446
	Oncology lab window	75	1903	286
	Oncology lab door	64	1626	244

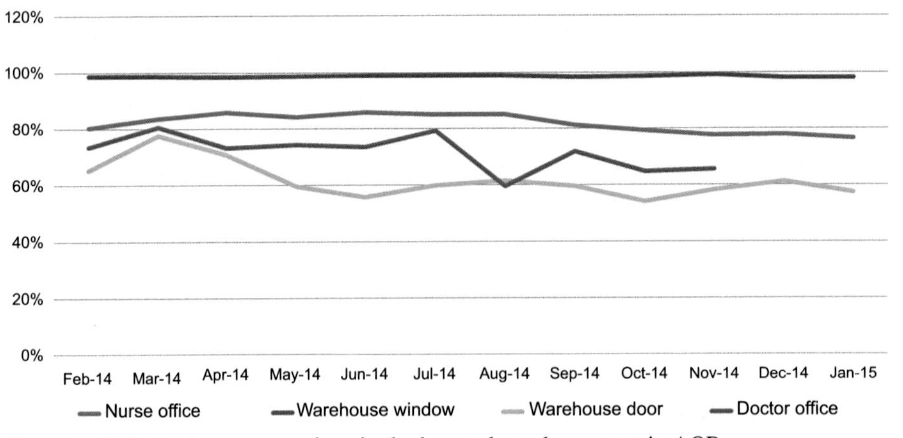

Figure 15.8 Monthly energy savings in the hematology department in AOR.

Table 15.6 Payback time in each room typology and in the overall facility

Room typology	Saving (%)	Saving (€)	Implementation cost (€)	PBT
Archives and rest rooms	87	23,136	274,362	11.9
Warehouse	97	157,415	650,676	4.1
Corridor	77	136,235	518,054	3.8
Ambulatory	52	4514	240,377	53.3
Nurse office and other offices	56	17,032	186,500	10.9
Waiting room	76	4186	103,611	24.8
Laboratory and technical areas	70	487,587	2,080,506	4.3
TOTAL	**75**	**830,104**	**4,054,085**	**4.9**

The result achievable of 75% of energy saved is due to two different factors:

- Retrofitted light features (LED lights are more efficient than fluorescent lights)
- Lighting control systems (fuzzy algorithms)

Comparing old and new lights datasheets it can be noted that new lights can save 52% of energy due to higher efficiency. As a consequence, a contribution of 48% can be assigned to improved control algorithms. It can be noticed that the PBT in case of implementation of the solution in the overall facility is very interesting. However retrofitting in some areas is particularly convenient: areas where lights remain switched on for many hours during the day (warehouses, corridors, laboratories, and technical areas) are more interesting. Furthermore, it should be noticed that PBT can be reduced in those areas where a retrofitted activity is already planned.

15.4.2 Saint George hospital

In the selected rooms of the pediatric department in the Saint George hospital, the artificial lights are not dimmable. However, the control algorithms implemented in the field controllers continuously monitor presence and indoor illuminance and calculate the new operation level of the artificial lights that maintains indoor illuminance at the required set point. The performance of the control algorithm during one day is illustrated in Fig. 15.9. Clearly, when the artificial lights switch off due to the lack of occupancy, savings are achieved.

The annual saving results are presented, per type of room, in Table 15.7 in order to point out the significant role of the room's type. The energy efficiency (kWh) is converted to € using as conversion factor 0.07€/kWh.

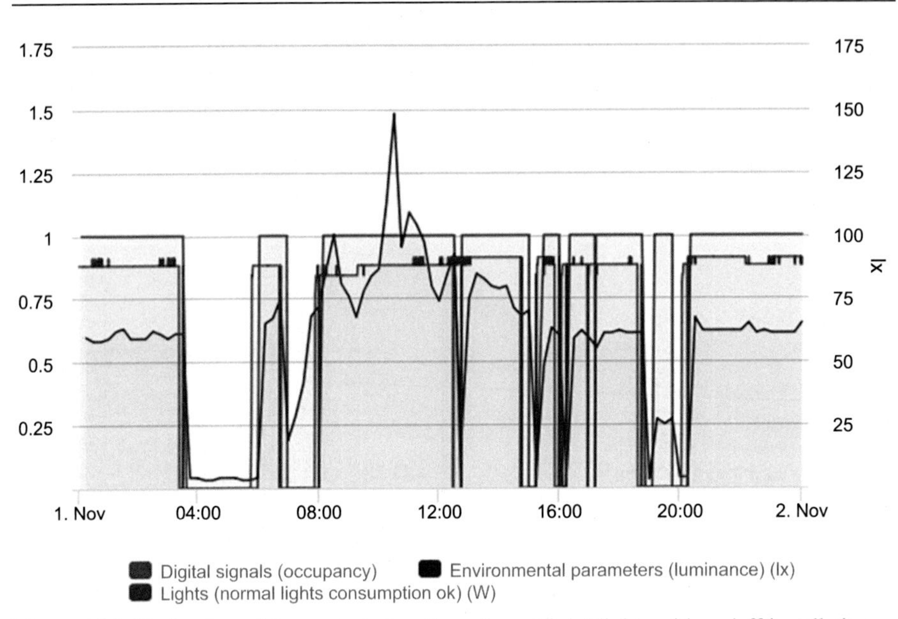

Figure 15.9 Evaluation of the control algorithms for artificial lights with on/off installation in SGH.

Table 15.7 Smart lighting system yearly results in Saint George hospital

Dept.	Room	Saving (%)	Saving (kWh)	Saving (€)
Pediatric	Doctors rest room	26	27.73	1.94
	Patients room	60	123.55	8.65
	Doctors office	51	803.81	57.85

It can be seen that savings in the doctors room are much higher compared to the other rooms.

The solution sets required the installation of new hardware, such as lights and sensors, and an update of the Building Management System through the implementation of lighting control algorithms. The following investment costs for the three rooms are analyzed:

- 600€ implementation of control algorithm and hardware installation
- 892.56€ equipment cost

Thus, a PBT of 21.81 years is estimated.

In the case of replication of the solution sets in the hospital wards, a controller can be applied to the artificial lights of three rooms: accordingly, the cost of such a controller is divided among three rooms. Furthermore for each room the sensor required is just a presence detector. Consequently, in case of replication in the overall building, 180 rooms will be retrofitted with a total capital cost estimated at 50.400€.

15.4.3 LOC and Its Energy Department: A Well Established Start-up

In the Loccioni Group, the energy business unit (BU) can be seen as a start-up operating in the lighting control market too. In fact, this department of the group addresses the broader market of energy efficiency in buildings, also including lighting control systems. The energy BU has developed and implemented customized lighting systems to the two aforementioned buildings in the Loccioni Group headquarters, Leaf Lab and Kite Lab. It is worth noticing that data related to energy savings are available only for the Leaf Lab because the Kite Lab has started to be fully operational in July 2015. However, the overall control system is developed and implemented in a proprietary software platform by Schneider.

Ninety percent of the indoor lights are LEDs and they implement DALI technology; this enables the control of both one single lamp and a group of lamps (eg, in a room). Moreover, the DALI protocol adds flexibility to the lighting system because it allows a direct communication between each lamp and the user through feedback: as a result, the user knows precisely which lamp is not working properly.

The lighting system integrates presence detectors and illuminance sensors located in each room. Furthermore, sensors are calibrated only when the room is fully furnished in order to include lights reflections.

With reference to the presence detector, it switches on the lights in the room after the detection and it switches off 30 min after no movements are detected.

Similarly, due to regulations, the illuminance level in a room must be at least 500 lux and the control system dimmer lights in the view of achieving such a value.

The Leaf Lab building embodies offices, laboratory, and warehouse and they all have different needs with reference to light: consequently, specific control strategies are deployed for each area. For example, in the laboratory and in the warehouse, timers switch the lights on at 8:00 am and off at 7:30 pm. Furthermore, the control strategy dims the lights in order to achieve 500 lux of illuminance. Finally, through a touch screen, the user can manually switch on and off the lights and dim them both in the warehouse and laboratory.

These light-control algorithms are implemented on the Schneider software platform and Fig. 15.10 shows their performance from October 2014 to September 2015. The dimming action on the LED lights causes a decrease in the related energy consumption that is measured and reported every 15 min. In addition, in spring and in summer, as expected, the consumption is even less due to the sunlight. However, the blank white spaces in the plot indicate errors or faults in the monitoring systems due to external reasons.

Finally, Fig. 15.11 illustrates the monthly energy consumption and the related savings given by a comparison with a reference baseline: this baseline is given by the consumption that would have been given by the traditional lighting systems that are not dimmable and do not include data from dedicated sensors such as presence detectors and illuminance sensors.

The proposed lighting system causes a decrease in energy consumption up to 72%, which corresponds to almost 13.000€ in 11 months.

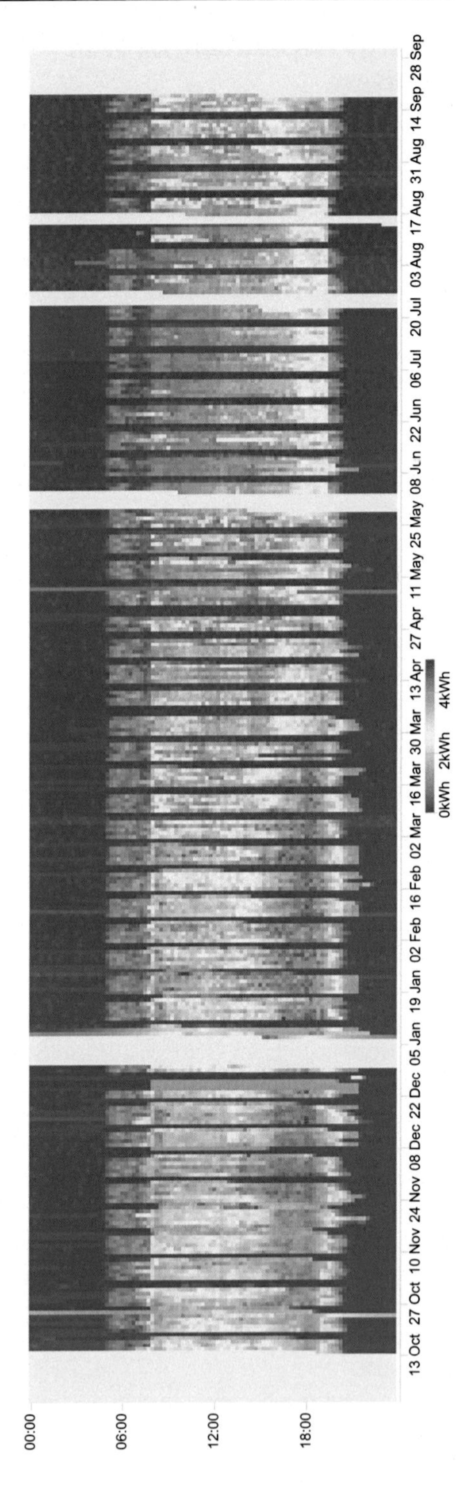

Figure 15.10 Energy consumption (data collected every 15 min) of the internal lighting system due to proposed control strategy in LOC.

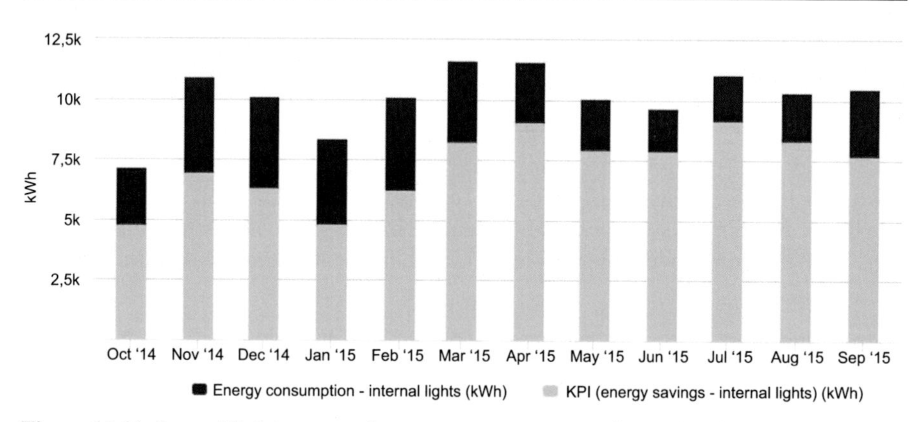

Figure 15.11 Internal lighting control system: energy consumption and savings in the Leaf Lab building (LOC).

Additionally, such lighting solutions are designed by the energy BU and tailored to other customers based on their needs. Hence, the development and the implementation of light control algorithms are products that the energy BU sells to customers.

15.5 Conclusion

This chapter has presented the lighting control market and has explained how innovative start-ups should exploit such a scenario. First, the light influence on human health has been introduced. Then, the key factors of success for new and small vendors (eg, start-ups) have been denoted as low costs, long life expectancy of solutions, and flexibility. Furthermore, the importance of light control algorithms has been introduced as they enable start-ups to develop solutions tailored to the customer needs while satisfying all the requirements for success. In addition, this chapter has introduced the connection between lighting requirements and buildings.

Finally, real case operations have been reported in three phases. First, a fuzzy-based controller has been developed and integrated in an artificial light-control framework that has included also illuminance sensors and presence detectors. In the view of implementing the proposed control system in the two pilot hospitals of Ancona and Chania, a refurbishment of the existing lighting systems was necessary by utilizing only LED lights, possibly dimmable, and by installing dedicated indoor sensors. Second, the results given by the combination of the proposed control systems and the refurbishment of the existing lighting systems have shown significant energy savings for both hospitals. It should be noticed that the results are related to the period of time of 1 year: this demonstrates that the proposed control system ensures indoor comfort. Third, this chapter has also emphasized that energy savings are achieved in an office building: the Leaf Lab, located in Italy, has implemented smart light control algorithms, developed by the Loccioni energy BU, to dimmable LED lights and integrated illuminance sensors and presence detectors too. The control strategy implemented in this building has caused significant energy savings and it has not affected the indoor comfort: such a control system has been continuously operating in the Leaf Lab. Moreover, the developed control strategy has considered the room's purpose; thus, it has implemented dedicated control systems according to these needs.

Consequently, the lighting control market has been demonstrated to be a fertile ground for innovative start-ups; in such a scenario, through customized light-control algorithms the start-up solutions enable increase in energy savings while minimizing costs and ensuring both internal comfort and the fulfillment of the lighting requirements.

Acknowledgments

This work is partly funded by the EU Commission, within the research contract GREEN@Hospital a three years European Project co-funded by the ICT Policy Support Program as part of the Competitiveness and Innovation framework Program (CIP).

References

Alrubaih, M.S., Zain, M.F.M., Alghoul, M.A., Ibrahim, N.L.N., Shameri, M.A., Elayeb, O., 2013. Research and development on aspects of daylighting fundamentals. Renewable and Sustainable Energy Reviews 21, 494–505.

Boubekri, M., 2008. Daylighting, Architecture and Health. Building Design Strategies. Architectural Press.

Colaco, S., Kurian, C., George, V., Colaco, A., 2012. Integrated design and real-time implementation of an adaptive, predictive light controller. Lighting Research and Technology 44, 459–476.

Curwin, T., 2010. Report: An Assessment of the Lighting Control Market Segment.

Dounis, A.I., Tiropanis, P., Argiriou, A., Diamantis, A., 2011. Intelligent control system for reconciliation of the energy savings with comfort in buildings using soft computing techniques. Energy and Buildings 43 (1), 66–74.

European Commission, 2009. Implementing Directive 205/32/EC of the European Parliament and of the Council with regard to ecodesign requirements for fluorescent lamps without integrated ballast, for high intensity discharge lamps, and for ballasts and luminaires able to operate such. Official Journal of the European Union 17–44.

European Commission, 2012. Health Effects of Artificial Light. http://ec.europa.eu/health/scientific_committees/opinions_layman/artificial-light/en/index.htm#3.

Franzetti, C., Fraisse, G., Achard, G., 2004. Influence of the coupling between daylight and artificial lighting on thermal loads in office buildings. Energy and Buildings 36, 117–126.

Frattari, A., Chiognm, M., Boer, J., 2009. Automation system for lighting control: a comparison between data recorded and simulation model. International Journal for Housing Science 33 (1), 45–56.

GREEN@Hospital, 2015. Final Report. http://www.greenhospital-project.eu/wp-content/uploads/2015/08/Green@Hospital-Final-report.pdf.

Green Growth Alliance, 2013. The Green Investment Report: The Ways and Means to Unlock Private Finance for Green Growth. World Economic Forum.

Hraska, J., 2014. Chronobiological aspects of green buildings daylighting. Renewable Energy 73, 109–114.
Jang, J.-S., 1993. ANFIS: adaptive-network-based fuzzy inference system. IEEE Transactions on Systems, Man, and Cybernetics 23, 665–685.
Kachan & Co, 2013. Cleantech Redefined.
Knight, I.P., 1998. Measured energy savings due to photocell control of individual luminaires. Renewable Energy 15, 441–444.
Kolokotsa, D., Pouliezos, A., Stavrakakis, G., 2005. Sensor fault detection in building energy management systems. In: 5th International Conference on Technology and Automation, pp. 1–5.
Kolokotsa, D., Pouliezos, A., Stavrakakis, G., Lazos, C., 2009. Predictive control techniques for energy and indoor environmental quality management in buildings. Building and Environment 44, 1850–1863.
Kolokotsa, D., Tsoutsos, T., Papantoniou, S., 2012. Energy conservation techniques for hospital buildings. Advances in Building Energy Research 6, 159–172.
Kurian, C., Kuriachan, S., Bhat, J., Aithal, R., 2005. An adaptive neuro-fuzzy model for the prediction and control of light in integrated lighting schemes. Lighting Research and Technology 37, 343–352.
Lam, J.C., Li, D.H.W., Cheung, S.O., 2003. An analysis of electricity end-use in air-conditioned office buildings in Hong Kong. Building and Environment 38, 493–498.
Navigant Research Report, 2015. Intelligent Lighting Controls for Commercial Buildings.
Nippun Kumaar, A.A., Kiran, G., Sudarshan, T.S.B., 2010. Intelligent lighting system using wireless sensor networks. International Journal of Ad Hoc, Sensor & Ubiquitous Computing (IJASUC) 1 (4), 17–27.
Papantoniou, S., Kolokotsa, D., Kalaitzakis, K., Cesarini, D.N., Cubi, E., Cristalli, C., 2014. Adaptive lighting controllers using smart sensors. International Journal of Sustainable Energy.
Pérez-Lombard, L., Ortiz, J., Pout, C., 2008. A review on buildings energy consumption information. Energy and Buildings 40, 394–398.
Santamouris, M., Argiriou, A., Dascalaki, E., Balaras, C., Gaglia, A., 1994. Energy characteristics and savings potential in office buildings. Solar Energy 52, 59–66.
Wen, Y.J., Agogino, A.M., 2011. Control of wireless-networked lighting in open-plan offices. Lighting Research and Technology 43 (2), 235–248.

16 Big data analytics and cloud computing for sustainable building energy efficiency

J.-S. Chou[1,2], N.-T. Ngo[1], W.K. Chong[2], G.E. Gibson Jr.[2]
[1]National Taiwan University of Science and Technology, Taipei, Taiwan; [2]Arizona State University, Tempe, AZ, United States

16.1 Introduction

According to the International Energy Agency, global energy consumption is expected to increase by 53% from 505 quadrillion Btu in 2008 to 770 quadrillion Btu in 2035 (IEA, 2011). The building sector is one of the largest energy consumers, accounting for approximately 40% of the global energy usage while generating 30% of all CO_2 emissions (Costa et al., 2012); energy consumption by this sector is continuously increasing because of urbanization.

In Europe, buildings account for 40% of energy consumption and 36% of CO_2 emissions. The efficient usage of building energy plays a vital role in controlling energy costs, reducing environment impacts, and increasing the value and competitiveness of buildings. Energy efficiency is a cost-effective method for curbing energy use and carbon emissions from buildings (Mathew et al., 2015). Managing and utilizing building energy usage data are critical for the successful deployment of energy efficiency.

Smart grids are a promising solution to the drastic increase in power demand (Yuan and Hu, 2011). They can potentially enhance reliability, power quality, and energy efficiency; reduce peak demand and transmission congestion costs; realize environmental benefits through increased asset utilization and improved capability for accommodating renewable energy; and increase security, durability, and ease of repair in response to malicious attacks and adverse natural events (El-hawary, 2014). The development of smart grids is one of the driving forces for the adoption of big data management technology (Schuelke-Leech et al., 2015). However, processing the large volume of data generated using smart grids is difficult.

This challenge can be overcome by using big data analytics and cloud computing technology, which are powerful tools for analyzing such energy data. Advances and innovations are crucial for a sustainable energy system with big data techniques and cloud computing. Big data analytics and cloud computing are two initiatives of information technology (IT), which are highly potential in building energy management

Start-Up Creation. http://dx.doi.org/10.1016/B978-0-08-100546-0.00016-9

analysis. A start-up IT application is needed to explore a huge amount of building energy data. Big data analytics provides valuable insights that can facilitate efficient energy usage by users. For instance, data mining (DM) and optimization algorithms can be used for building energy consumption analysis to generate energy-saving solutions for users.

Cloud computing is one of the most substantial innovations in modern information and communication technology. Cloud computing is advantageous because of its virtualized resources, parallel processing, security, and data service integration with scalable data storage. Cloud computing can enhance the agility and productivity of big data analytics, increase efficiencies, and reduce costs. The computing infrastructures in conjunction with fast communication networks, data-intensive programming paradigms (eg, MapReduce), distributed storage system, Web service, machine learning algorithms, and metaheuristic optimization algorithms can form the basis for designing and developing big data analysis-based innovation framework in the building energy domain.

This chapter presents the conceptual framework of a smart decision support system (SDSS) that integrates smart grid big data analytics and cloud computing for building energy efficiency. The framework serves as a start-up creation in an energy management application of smart eco-efficient built environment. The framework is based on a layered architecture that includes smart grid and data collection, an analytics bench, and a web-based portal. Notably, advanced artificial intelligence techniques and a dynamic optimization algorithm are designed along with the analytics bench to enable accurate prediction and optimization of energy consumption.

A real-world smart metering infrastructure was installed in a residential building for the prototype experiments. The SDSS integrates data analytics and dynamic multiobjective optimization models to generate energy consumption patterns and alternative energy-saving solutions at the appliance level. The SDSS is expected to identify consumer usage patterns, facilitate energy usage efficiency, and accurately estimate future energy demands. End users can reduce electricity costs through the automatic and optimal operation of appliances by using the SDSS.

16.2 Literature review of building energy management systems

Recently developed energy consumption management systems facilitate effective usage of electricity by end users (Hao et al., 2012; Fróes Lima and Portillo Navas, 2012; Zhou et al., 2014). For instance, Zhou et al. (2014) investigated a real-time energy control approach for a home energy management system in the United Kingdom. Combining half-hour-ahead rolling optimization and a real-time control strategy realized household economic benefits in complex operating environments. Simulation test results indicated that the proposed control approach can optimize the schedule for home appliances and battery charging/discharging behavior, even if the forecasting data are inaccurate.

Aghemo et al. (2014) proposed a building automation and control system that can control the functioning of plants in buildings to increase user comfort and reduce operation and maintenance costs (Aghemo et al., 2014). Arghira et al. (2012) proposed a stochastic method for predicting energy consumption in the next 24 h. Basic predictors that were presented and tested in the available historical data included "will always consume," "will never consume," and auto regressive moving average parameters.

Chen et al. (2012) presented a smart appliance management system for recognizing electric appliances in a home network by measuring the energy consumption of appliances using a current sensing device (Chen et al., 2012). This system can search the corresponding cluster data and eliminate noise for recognition functionality and error detection mechanism of electric appliances by applying the current clustering algorithm. Radulovic et al. (2015) presented guidelines for linked data generation and publication, together with one complete example in the domain of energy consumption in buildings. This helped researchers and practitioners interested in energy consumption in buildings to exploit linked data technologies.

Bapat et al. (2011) developed a Yupik system that enables users to respond to real-time changes in electricity costs. Yupik combined sensing, analytics, and an integer linear program for generating appliance usage schedules, which can be used by households to minimize their electricity costs and potential lifestyle disruptions. Similarly, Fróes Lima and Portillo Navas (2012) integrated automated remote metering and submetering of electricity into a structured knowledge tool. This integration environment received electricity meter measurements.

Lach and Punchihewa (2007) proposed an automatic monitoring system for reducing the energy usage of a typical home by using WiFi-enabled smart switches. Multiple sensors were used to enable automatic monitoring and control of the environment according to user preference profiling. Reinisch et al. (2011) proposed ThinkHome, a home system concept that uses multiagent techniques to reduce energy consumption. This system includes a wide knowledge base of information used to meet the goals of energy efficiency and user comfort.

Lee et al. (2014) proposed a green construction hoist that uses an energy regeneration system (ERS) to reduce the operating energy requirements of hoists in construction sites. The ERS was customized to improve energy-saving efficiency. Zeng et al. (2014) investigated applications of several energy management strategies in hybrid electric wheel loaders, including engine optimal control, minimum motor power control, motor optimal control, and instantaneous optimal control strategies.

For time-series energy data prediction, Chou and Telaga (2014) proposed a novel approach for using large data sets to identify anomalous power consumption in building office spaces. This approach comprised two stages: consumption prediction and anomaly detection. Daily real-time consumption was predicted using a hybrid neural network (NN) autoregressive integrated moving average (ARIMA) model. Anomalies were identified by applying the two-sigma rule to compare actual and predicted consumption. Their research contributed to the development of a methodology for real-time detection of anomalous patterns in large data sets. The prediction module helps building managers to plan energy consumption, whereas the anomaly detection module helps building managers to identify unusual consumption of electricity by tenants.

Until now, most studies have focused on establishing a system for predicting the baseline of future electricity usage through stochastic methods and regression analysis. No study has proposed dynamic operating strategies for home appliances in which the energy costs are minimized and energy is saved effectively with consideration of user comfort. This study therefore describes a procedure and the associated components for an energy-saving decision support system (DSS).

The proposed smart-metering infrastructure collects real-time data for building energy use. Bluetooth and WiFi are used in home area networks for data communication. Then real-time data are dynamically analyzed using advanced artificial intelligence and a multiobjective optimization algorithm to provide a list of energy-saving alternatives. Finally, a web-based technology is used to visualize the optimal operating schedules for home appliances.

16.3 Overview of big data and cloud computing technologies

Big data analytics and cloud computing are currently being developed and employed for managing and mining large volumes of data. A brief introduction is given in the following subsections.

16.3.1 Big data analytics

Big data refers to very large datasets with an increasing volume of data; conventional database technologies cannot capture, store, manage, and analyze such large volumes of data (Hashem et al., 2015). Several researchers have defined big data. Cox and Ellsworth (1997) defined big data as a large volume of scientific data for visualization. Manyika et al. (2011) defined big data as the amount of data that is beyond technology's capability to store, manage, and process efficiently. Gartner (2015) argued that big data, in general, is defined as high-variety information assets that demand cost-effective innovative forms of information processing for enhanced insight, decision-making, and process automation.

Similarly, Zikopoulos et al. (2012) defined big data in the form of three Vs: volume, variety, and velocity. Gantz and Reinsel (2011) extended the definition of big data to four Vs: volume, variety, velocity, and value. This definition is commonly used because it emphasizes the meaning and necessity of big data.

- **Volume** refers to the amount of data generated using various sources such as smart meters and sensors.
- **Variety** represents the data types that are collected through sensors, smart meters, smartphones, or social networks. Such data can be structured, unstructured, or semistructured.

- **Velocity** refers to the rate at which the data are produced and processed. For example, the data of energy consumption and carbon dioxide emissions from the building are retrieved at 1-min intervals.
- **Value** refers to the process of discovering hidden values from large datasets.

The purpose of big data analytics is to obtain useful information. The real value of big data is obtained through big data analytics: a set of advanced technologies assigned to process large volumes of data. Sophisticated quantitative methods, such as machine learning, artificial intelligence, and metaheuristic optimization algorithms, can be utilized to explore the energy consumption pattern, predict future energy consumption, and minimize energy costs for users. Predictive analytics, a part of big data analytics, enables users to visualize trends and offers solutions for deriving value from big data.

Advanced computing technologies can improve the reliability of electrical systems and its energy efficiency in addition to reducing costs to the consumer. Big data analytics allows large volumes of data generated using electronic sensors, smart grid technologies, electricity supply, grid operations, and customer demands to be coordinated, analyzed, and effectively utilized (Schuelke-Leech et al., 2015). Thus, big data analytics is a highly potential technique and start-up creation for analyzing usage patterns of building energy consumption.

16.3.2 Cloud computing

Cloud computing is one of the most substantial innovations in modern information and communication technology because of its virtualized resources, parallel processing, security, and data service integration with scalable data storage. Cloud computing is a rapidly growing technology, whose architecture enables it to perform large-scale and complex computing (Hashem et al., 2015). This technology promises reliable software, hardware, and infrastructure as a service (IaaS) delivered over the Internet and remote data centers (Armbrust et al., 2010). Cloud technology addresses barriers to adoption with enhanced security and data integration. Data are being increasingly stored in cloud environments, which are immense and valuable sources of information.

The demand for storing, processing, and analyzing large amounts of datasets has driven organizations to adopt cloud computing (Huan, 2013). It allows organizations to concentrate on the core business without concerns of infrastructure, flexibility, and availability of resources (Aceto et al., 2013). Cloud computing is an extremely successful paradigm of service-oriented computing that has revolutionized computing infrastructure since it eliminates the need for maintaining expensive computing hardware, dedicated space, and software (Chandrashekar et al., 2015). The basic cloud service types for analytics as a service consist of IaaS, platform as a service (PaaS), and software as a servicc (SaaS).

- IaaS refers to hardware equipment operating on a cloud and is used by end users on demand. IaaS enables users to allocate time for shared server resources to meet computing and storage demands for big data analytics. IaaS is the foundation for cloud services. It requires a high investment of IT resources, such as a Hadoop framework and NoSQL database, which implement big data analytics. Amazon Web Services, Citrix CloudPlatform, Microsoft Azure, and Microsoft System Center are examples of IaaS solutions in the cloud technology ecosystem.

- PaaS refers to various resources operating on a cloud to provide platform computing for end users. PaaS assists developers with tools and libraries to develop and run applications on a cloud infrastructure. PaaS serves as a development platform for advanced analytics applications. Force.com, Google App Engine, and Microsoft Azure are examples of PaaS solutions in the cloud technology ecosystem.
- SaaS refers to applications operating on a remote cloud infrastructure accessible through the Internet. Specific applications for big data analytics in cloud can be provided by SaaS. Amazon Elastic MapReduce and Google BigQuery are examples of SaaS solutions in the cloud technology ecosystem.

Cloud computing uses networks of a large group of servers with specialized connections to distribute data processing among servers. Instead of installing a software suite for each computer, this technology requires installing a single piece of software, which allows users to log into a web-based service, in each computer; it also hosts all the programs required by users (Chandrashekar et al., 2015). Users can access applications in the cloud network from any location by connecting to the cloud through the Internet. Some real-time applications that use cloud computing are Gmail, Google Calendar, Google Docs, and Dropbox.

16.3.3 Benefits of big data analytics with cloud computing

The benefits of online analytical processing with cloud computing include the following:

- Cost reduction: Cloud computing offers a cost-effective approach to support big data technologies and advanced analytics applications that can drive business value. Enterprises want to unlock the hidden potential of data, which can yield a competitive advantage. Big data environments require clusters of servers for supporting the tools that process large volumes, high velocity, and varied formats of big data. IT organizations must use cloud computing as the structure to save costs through the cloud's pay-per-use model.
- Reduced overhead: Various components and integration are required for the implementation of any big data solution. With cloud computing, these components can be automated, thus reducing complexity and improving the productivity of the IT team.
- Rapid provisioning/time to market: Provisioning servers in the cloud is as easy as shopping over the Internet. Big data environments can be scaled up or down easily according to the processing requirements. Rapid provisioning is crucial for big data applications because the value of data reduces quickly with time.
- Flexibility/scalability: Big data analysis, particularly to meet the power demand, requires considerable computer power for a brief amount of time. For such analysis, servers must be provisioned in few minutes or in desired intervals. Such scalability and flexibility can be achieved in the cloud, reducing huge investments on supercomputers by only paying for computing on an hourly basis.

16.4 Framework design of the proposed smart decision support system for building energy efficiency

The framework of the proposed SDSS is based on layered architecture, which contains a data access layer, an integrated analytics bench, and a web-based portal.

16.4.1 Real-time data access layer

The data layer is a database management system containing real-time electricity data, appliance information, electricity costs, data from temperature and humidity sensors, and analytical results data. Data on the amount of electricity consumed is retrieved from smart meters and transferred to dedicated servers through the communication network. In addition, electrical parameters, such as voltage, current, power, frequency, and power factors, are saved in the database management system. The data streams from the smart meters arrive at 1-min intervals, which results in 1440 data points every 24 h for each smart meter.

Schematically, the IT infrastructure can be physically and functionally divided into (1) a smart grid, (2) a communication network, and (3) a data management infrastructure, as illustrated in Fig. 16.1.

1. Smart grid
   The residential smart grid includes meters, submeters, and network devices necessary to organize data collection and transfer data through the communication network to the database management system. Bluetooth is one of the most common wireless communications systems used for exchanging data over short distances (Davies, 2002; Bisdikian, 2001), and the IEEE standard for Bluetooth is IEEE 802.15.1 (Mahmood et al., 2015). The wireless Bluetooth network is connected to a smart IP power controller to transmit the data and provide Internet access.

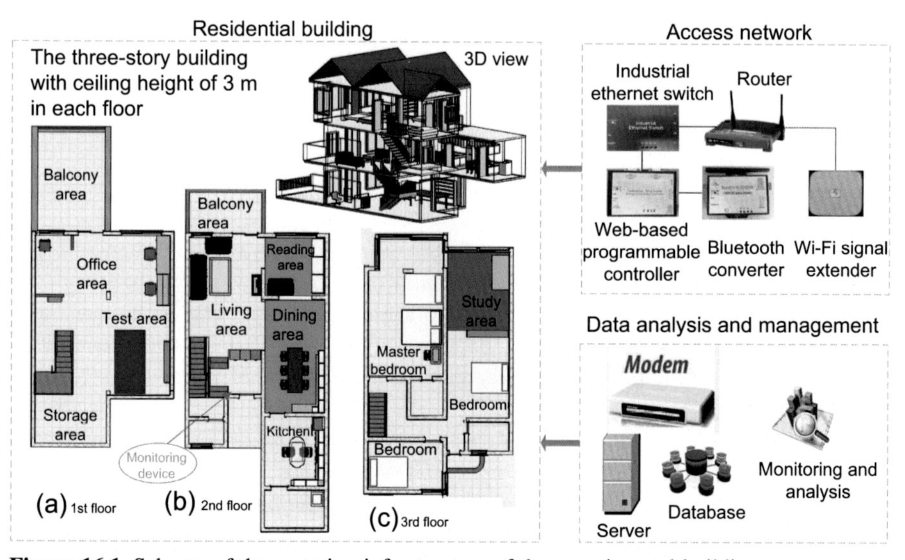

Figure 16.1 Scheme of the metering infrastructure of the experimental building.

2. Communication network
   The communication network transfers the information collected by each of the metering devices to the database management system. The elements of the communication network comprise a router, an industrial Ethernet switch, a web-based programmable controller, a Bluetooth converter, and a WiFi signal extender.
3. Infrastructure for data management
   New methods of managing data, such as Hadoop, offer alternatives to conventional data warehousing. Apache™ Hadoop® is an open-source software that enables the distributed processing of large data sets across clusters of commodity servers. Fig. 16.2 illustrates a Hadoop system integrated with an advanced analytics module. This environment system is highly beneficial and effective for solving the metadata problem. Apache Hadoop comprises two main components: MapReduce, which provides fault tolerance for distributed processing, and Hadoop distributed file system (HDFS), which is a file system spanning all nodes in a Hadoop cluster for data storage.
   MapReduce is a programming model and an associated implementation for processing and generating large data sets. Users specify a map function that processes a key/value pair to generate a set of intermediate key/value pairs and a reduce function that merges all intermediate values associated with the same intermediate key (Dean and Ghemawat, 2008). The HDFS stores large data sets reliably and streams them at a high bandwidth for user applications. Big data environments require clusters of servers to support the tools that process large volumes, high velocity, and varied formats of big data. Cloud computing supports big data analytics for building energy consumption.

16.4.2 Integrated analytics bench

The analytics layer creates computational models by integrating big data analytics and dynamic multiobjective optimization modules to continuously analyze the real-time data received through smart meters and generate energy consumption patterns, alternative optimal schedules of home appliances, and the electricity cost-saving percentage.

16.4.2.1 Hybrids of machine learners and time-series data mining techniques

The aim of analytics layer is to identify energy usage patterns at the appliance level. The scheme can help users understand and monitor their appliances so that they can take further action to reduce energy costs. To achieve this goal, DM techniques and time-series analysis are adopted for data analytics. The various time-dependent methods of model fitting include ARIMA and nonlinear Neural Network Auto Regressive (NNAR) model.

DM techniques have attracted considerable scientific attention and have become a crucial research area. Several major types of DM methods, such as generalization, characterization, classification, clustering, association, evolution, pattern matching, data visualization, and metarule-guided mining, exist (Liao et al., 2012). The DM techniques typically used to solve real-world problems include artificial NNs (ANNs),

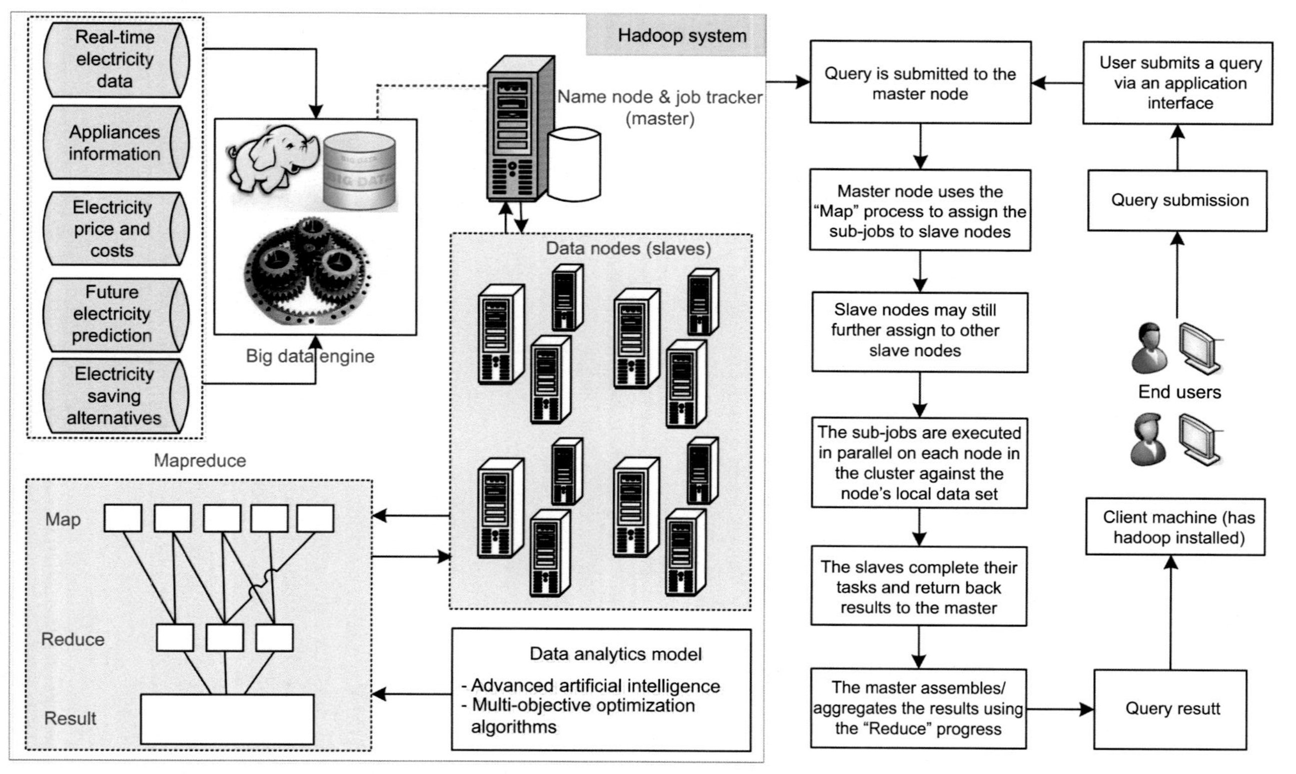

Figure 16.2 Hadoop system with integrated advanced data analytics model.

support vector regression (SVR), classification and regression trees, chi-squared automatic interaction detector, and generalized linear model.

Energy consumption data are time-series data for both linear and nonlinear components. Therefore, hybrid linear and nonlinear models effectively capture energy consumption patterns. ANNs and SVR, for instance, are two popular nonlinear models that can address nonlinear components, whereas ARIMA, seasonal ARIMA, Auto Regressive (AR), exponential smoothing, and moving average are used to fit linear components in linear and nonlinear models.

Models integrating ARIMA and ANN have been extensively used (Sánchez Lasheras et al., 2015; Khashei and Bijari, 2011; Khandelwal et al., 2015; Zhang, 2003; Hansen and Nelson, 2003; Sallehuddin and Shamsuddin, 2009). Khashei and Bijair (2011) proposed a novel hybridization of ANN and ARIMA models for time-series forecasting. The unique advantage of the ARIMA model in linear modeling is its effectiveness in identifying and magnifying the existing linear structure in data, following which the ANN model is used to capture the nonlinear part, which is the residual from the first stage of ARIMA modeling (Khashei and Bijari, 2011). However, sufficient data are required for obtaining an efficient hybrid model. Other problems with ANN models are their numerous parameters, uncertain solutions, and potential for overfitting.

SVR was proposed by Vapnik (1995) to overcome the drawbacks of ANN. SVR is a nonlinear alternative to ANN. Researchers have successfully used SVR to solve many regression problems (Pai and Lin, 2005; Alwee et al., 2013; Che and Wang, 2010). A hybridization of ARIMA and SVR has been successfully applied in time-series forecasting for stock market prediction (Pai and Lin, 2005; Da-yong et al., 2008) and reducing electricity costs (Chen and Lin, 2010).

The main problem with SVR is setting the appropriate hyperparameters, which requires practitioner experience. Unfavorable kernel functions or hyperparameter settings may lead to considerably inferior performance. Therefore, a novel hybrid model combining ARIMA and nature-inspired metaheuristic optimization in SVR is a potential solution for forecasting energy consumption of residential buildings. Hybrid models, such as the fine-tuned SVR model, are used to address nonlinearity, and the ARIMA model is used to address the nonstationary linear component.

Fig. 16.3 illustrates the generic process for building and evaluating the proposed hybrid time-series forecasting system. The first stage of the linear time-series prediction model captures the linear component. In the second stage, the residual from the first stage along with meteorological conditions, such as outdoor temperature, is used as input for the nature-metaheuristic optimized artificial intelligence model. By separating the linear and nonlinear components, the forecasting accuracy of the system can be improved.

16.4.2.2 Dynamic optimization algorithm for energy-saving alternatives

A dynamic multiobjective optimization algorithm is used to optimize appliance operating schedules. Users can compare alternatives to determine when appliances should

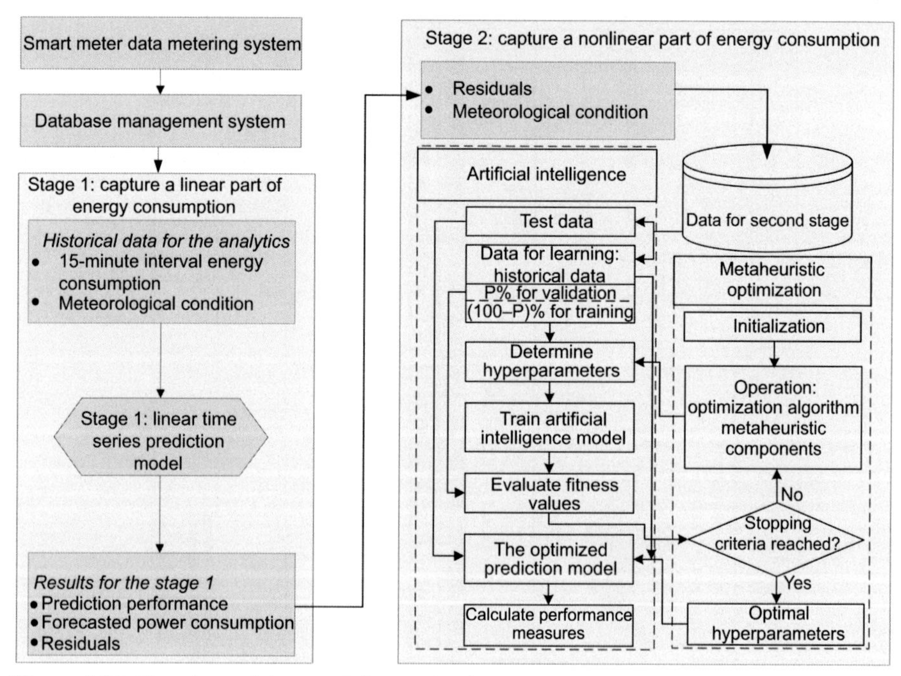

Figure 16.3 Flowchart of the novel time-series forecasting system.

be switched on or off. Each solution is an alternative (nondominant) energy-saving strategy.

The aim is to optimize the objective functions; that is, appliance electricity consumption and electricity cost. The first constraint is the electricity pricing policy, which is categorized by season, day (weekday, weekend, or holiday), and time of day (peak and off-peak time). The second constraint is the available operating time for the appliances and the status of occupants of the residential building. Fig. 16.4 illustrates the dynamic multiobjective optimization module used to optimize energy usage.

Swarm intelligence (SI) and bioinspired computation have attracted considerable attention. In optimization, computational intelligence, and computer science, bioinspired algorithms, particularly the SI-based algorithms, are commonly used (Yang, 2014). Examples of SI-based algorithms are bee algorithm (BA), particle swarm optimization (PSO), cuckoo search, and firefly algorithm (FA). Optimization algorithms used for allocating energy resources include genetic algorithm, PSO, FA, and BA.

16.4.3 Web-based smart decision support system

The web-based portal is the presentation layer that enables the user to interact with the energy-saving DSS. Fig. 16.5 illustrates a sample SDSS interface. The functions are defined to support consumers' requirements, including (1) real-time electricity consumption; (2) monthly consumption records; (3) monthly comparisons; (4) maximum, average, and minimum consumption; (5) consumption forecasts for the current month

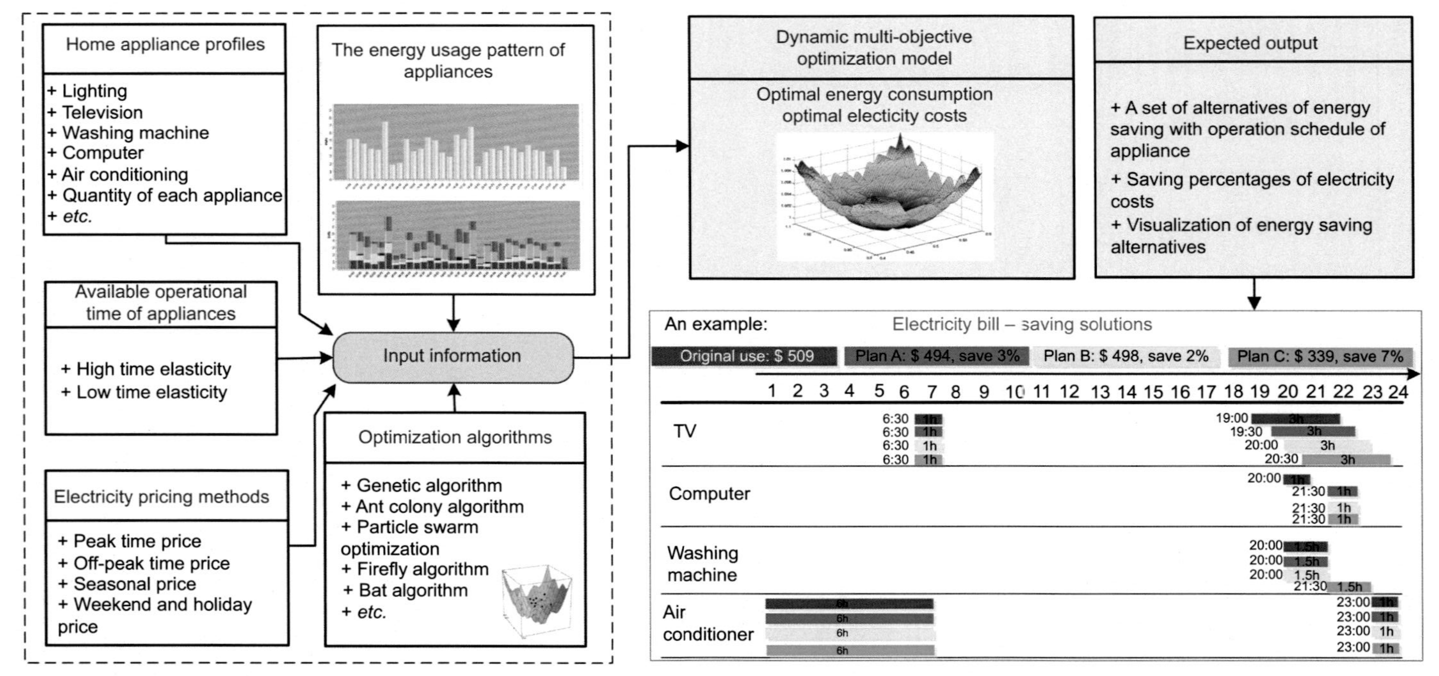

Figure 16.4 Dynamic multiobjective optimization model.

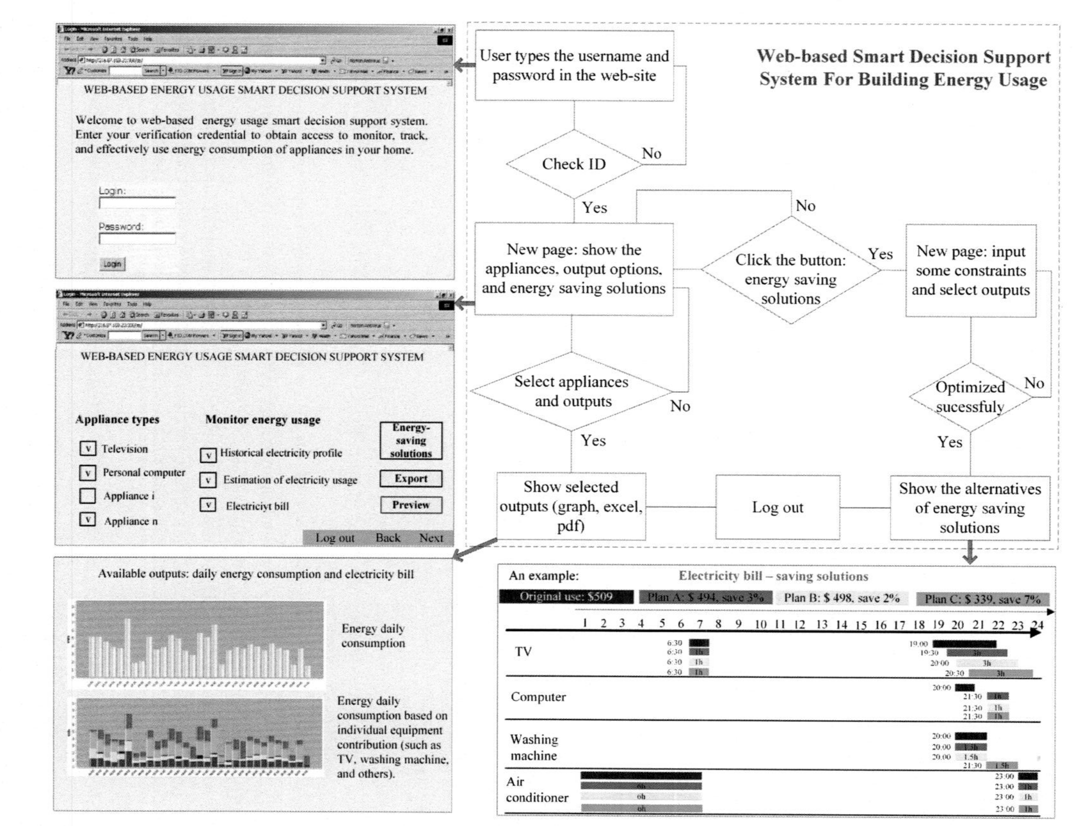

Figure 16.5 Sample interface of the web-based SDSS for building energy-usage tracking.

and the resulting expenditure; (6) alternative operation schedules of home appliances with optimal electricity costs; and (7) saving percentage of electricity costs by using alternative operation schedules.

16.5 Conclusions

Big data and cloud computing are emerging practices for building energy efficiency. This chapter presented the framework for an energy-saving DSS that integrates smart grid big data analytics and cloud computing for building energy efficiency. The layered architecture of the proposed framework includes a real-time data access layer, an integrated analytics bench, and a web-based portal. In particular, a novel hybrid nature-inspired metaheuristic time-series system and dynamic optimization algorithm are potentially integrated behind the analytics bench to enable accurate prediction and optimization of energy consumption. The framework serves as a start-up creation in an application of big data analytics and cloud computing technology for sustainable building energy efficiency.

A real-world smart metering infrastructure was installed in a residential building for the prototype experiments. By identifying energy consumption patterns, the SDSS can help improve energy usage efficiency and accuracy in estimating future energy demand. In addition, end users can reduce electricity costs by using the system to optimize operation schedules of appliances, lighting systems, and heating, ventilation, and air conditioning. The proposed framework is expected to attract an investor company to start-up and develop a real-time monitoring system for building energy efficiency.

References

Aceto, G., Botta, A., de Donato, W., et al., 2013. Cloud monitoring: a survey. Computer Networks 57 (9), 2093–2115.

Aghemo, C., Blaso, L., Pellegrino, A., 2014. Building automation and control systems: a case study to evaluate the energy and environmental performances of a lighting control system in offices. Automation in Construction 43, 10–22.

Alwee, R., Shamsuddin, S.M. Hj., Sallehuddin, R., 2013. Hybrid support vector regression and autoregressive integrated moving average models improved by particle swarm optimization for property crime rates forecasting with economic indicators. The Scientific World Journal 2013, 11.

Arghira, N., Hawarah, L., Ploix, S., et al., 2012. Prediction of appliances energy use in smart homes. Energy 48 (1), 128–134.

Armbrust, M., Fox, A., Griffith, R., et al., 2010. A view of cloud computing. Communications of the ACM 53 (4), 50–58.

Bapat, T., Sengupta, N., Ghai, S.K., et al., 2011. User-sensitive scheduling of home appliances. In: Proceedings of the 2nd ACM SIGCOMM Workshop on Green Networking. ACM, Toronto, Ontario, Canada, pp. 43–48.

Bisdikian, C., 2001. An overview of the bluetooth wireless technology. Communications Magazine, IEEE 39 (12), 86–94.

Chandrashekar, R., Kala, M., Mane, D., 2015. Integration of big data in cloud computing environments for enhanced data processing capabilities. International Journal of Engineering Research and General Science 3 (3), 240–245.

Che, J., Wang, J., 2010. Short-term electricity prices forecasting based on support vector regression and auto-regressive integrated moving average modeling. Energy Conversion and Management 51 (10), 1911–1917.

Chen, J.-H., Lin, J.-Z., 2010. Developing an SVM based risk hedging prediction model for construction material suppliers. Automation in Construction 19 (6), 702–708.

Chen, S.-Y., Lu, Y.-S., Lai, C.-F., 2012. A smart appliance management system with current clustering algorithm in home network. In: Rodrigues, J.P.C., Zhou, L., Chen, M., et al. (Eds.), Green Communications and Networking. Springer Berlin Heidelberg, pp. 13–24.

Chou, J.-S., Telaga, A.S., 2014. Real-time detection of anomalous power consumption. Renewable and Sustainable Energy Reviews 33 (0), 400–411.

Costa, A., Keane, M.M., Raftery, P., et al., 2012. Key factors methodology—A novel support to the decision making process of the building energy manager in defining optimal operation strategies. Energy and Buildings 49 (0), 158–163.

Cox, M., Ellsworth, D., 1997. Managing Big Data for Scientific Visualization.

Da-yong, Z., Hong-wei, S., Pu, C., 2008. Stock market forecasting model based on a hybrid ARMA and support vector machines. In: Management Science and Engineering, 2008. ICMSE 2008. 15th Annual Conference Proceedings., International Conference on, pp. 1312–1317.

Davies, A.C., 2002. An overview of bluetooth wireless technology and some competing LAN standards. In: ICCSC '02. 1st IEEE International Conference on Circuits and Systems for Communications, pp. 206–211.

Dean, J., Ghemawat, S., 2008. MapReduce: simplified data processing on large clusters. Communications of the ACM 51 (1), 107–113.

El-hawary, M.E., 2014. The smart grid—state-of-the-art and future trends. Electric Power Components and Systems 42 (3–4), 239–250.

Fróes Lima, C.A., Portillo Navas, J.R., 2012. Smart metering and systems to support a conscious use of water and electricity. Energy 45 (1), 528–540.

Gantz, J., Reinsel, D., 2011. Extracting Value from Chaos. IDC Review, Framingham, MA.

Gartner, 2015. http://www.gartner.com/it-glossary/big-data.

Hansen, J.V., Nelson, R.D., 2003. Time-series analysis with neural networks and ARIMA-neural network hybrids. Journal of Experimental & Theoretical Artificial Intelligence 15 (3), 315–330.

Hao, X.H., Wang, Y.C., Wu, C.Y., et al., 2012. Smart meter deployment optimization for efficient electrical appliance state monitoring. In: 2012 IEEE Third International Conference on Smart Grid Communications (SmartGridComm), pp. 25–30.

Hashem, I.A.T., Yaqoob, I., Anuar, N.B., et al., 2015. The rise of "big data" on cloud computing: review and open research issues. Information Systems 47 (0), 98–115.

Huan, L., 2013. Big data drives cloud adoption in enterprise. Internet Computing, IEEE 17 (4), 68–71.

IEA, 2011. World Energy Outlook 2011 Executive Summary. International Energy Agency.

Khandelwal, I., Adhikari, R., Verma, G., 2015. Time series forecasting using hybrid ARIMA and ANN models based on DWT decomposition. Procedia Computer Science 48, 173–179.

Khashei, M., Bijari, M., 2011. A novel hybridization of artificial neural networks and ARIMA models for time series forecasting. Applied Soft Computing 11 (2), 2664–2675.

Lach, C., Punchihewa, A., 2007. Smart home system operating remotely via 802.11b/g wireless technology. In: Proceedings of the Fourth International Conference Computational Intelligence and Robotics and Autonomous Systems.
Lee, M., Kim, T., Jung, H.-K., et al., 2014. Green construction hoist with customized energy regeneration system. Automation in Construction 45, 66–71.
Liao, S.-H., Chu, P.-H., Hsiao, P.-Y., 2012. Data mining techniques and applications – a decade review from 2000 to 2011. Expert Systems with Applications 39 (12), 11303–11311.
Mahmood, A., Javaid, N., Razzaq, S., 2015. A review of wireless communications for smart grid. Renewable and Sustainable Energy Reviews 41 (0), 248–260.
Manyika, J., Chui, M., Brown, B., et al., 2011. Big Data: The Next Frontier for Innovation, Competition, and Productivity. McKinsey Global Institute.
Mathew, P.A., Dunn, L.N., Sohn, M.D., et al., 2015. Big-data for building energy performance: lessons from assembling a very large national database of building energy use. Applied Energy 140, 85–93.
Pai, P.-F., Lin, C.-S., 2005. A hybrid ARIMA and support vector machines model in stock price forecasting. Omega 33 (6), 497–505.
Radulovic, F., Poveda-Villalón, M., Vila-Suero, D., et al., 2015. Guidelines for linked data generation and publication: an example in building energy consumption. Automation in Construction 57, 178–187.
Reinisch, C., Kofler, M., Iglesias, F., et al., 2011. Think home energy efficiency in future smart homes. EURASIP Journal on Embedded Systems 2011 (1), 104617.
Sallehuddin, R., Shamsuddin, S.M. Hj., 2009. Hybrid grey relational artificial neural network and auto regressive integrated moving average model for forecasting time-series data. Applied Artificial Intelligence 23 (5), 443–486.
Sánchez Lasheras, F., de Cos Juez, F.J., Suárez Sánchez, A., et al., 2015. Forecasting the COMEX copper spot price by means of neural networks and ARIMA models. Resources Policy 45, 37–43.
Schuelke-Leech, B.-A., Barry, B., Muratori, M., et al., 2015. Big data issues and opportunities for electric utilities. Renewable and Sustainable Energy Reviews 52, 937–947.
Vapnik, V.N., 1995. The Nature of Statistical Learning Theory. Springer-Verlag, New York.
Yang, X.-S., 2014. Chapter 2-Analysis of algorithms. In: Yang, X.-S. (Ed.), Nature-Inspired Optimization Algorithms. Elsevier, Oxford, pp. 23–44.
Yuan, J., Hu, Z., 2011. Low carbon electricity development in China—An IRSP perspective based on super smart grid. Renewable and Sustainable Energy Reviews 15 (6), 2707–2713.
Zeng, X., Yang, N., Peng, Y., et al., 2014. Research on energy saving control strategy of parallel hybrid loader. Automation in Construction 38, 100–108.
Zhang, G.P., 2003. Time series forecasting using a hybrid ARIMA and neural network model. Neurocomputing 50, 159–175.
Zhou, S., Wu, Z., Li, J., et al., 2014. Real-time energy control approach for smart home energy management system. Electric Power Components and Systems 42 (3–4), 315–326.
Zikopoulos, P., deRoos, D., Corrigan, D., et al., 2012. Harness the Power of Big Data – The IBM Big Data Platform. McGraw Hill Professional.

Intelligent decision-support systems and the Internet of Things for the smart built environment

17

A. Kaklauskas, R. Gudauskas
Vilnius Gediminas Technical University, Vilnius, Lithuania

17.1 Introduction

17.1.1 Definition, characteristics, and components

Today computers—and, therefore, the Internet—are almost wholly dependent on human beings for information. Nearly all of the roughly 50 petabytes (a petabyte is 1024 terabytes) of data available on the Internet were first captured and created by human beings—by typing, pressing a record button, taking a digital picture, or scanning a bar code. The problem is that people have limited time, attention, and accuracy—all of which means they are not very good at capturing data about things in the real world (Ashton, 2009).

The idea of an Internet of Things (IoT) started many years ago. Nikola Tesla in an interview with Colliers magazine in 1926 stated, "When wireless is perfectly applied the whole earth will be converted into a huge brain, which in fact it is, all things being particles of a real and rhythmic whole … and the instruments through which we shall be able to do this will be amazingly simple compared with our present telephone. A man will be able to carry one in his vest pocket" (Friess et al., 2012).

Before the Internet was developed in 1969, Alan Turing already proposed the question whether machines can think, in his 1950 article, Computing Machinery and Intelligence. He stated that "It can also be maintained that it is best to provide the machine with the best sense organs that money can buy, and then teach it to understand and speak English. This process could follow the normal teaching of a child." So, years before the first message was send across Internet, Alan Turing was already thinking about smart machines communicating with each other (Van Rijmenam, 2015).

Kevin Ashton is recognized for inventing the term Internet of Things. Ashton (2009) stated that "I could be wrong, but I'm fairly sure the phrase 'Internet of Things' started life as the title of a presentation I made at Procter & Gamble (P&G) in 1999."

There is no commonly recognized definition of IoT; it can be defined in several ways:

- The concept goal of the IoT is to enable things to be connected anytime, anyplace, with anything and anyone, ideally using any path/network and any service. IoT is a new revolution of

Start-Up Creation. http://dx.doi.org/10.1016/B978-0-08-100546-0.00017-0

the Internet. Objects make themselves recognizable and they obtain intelligence thanks to the fact that they can communicate information about themselves and they can access information that has been aggregated by other things. For example, alarm clocks will go off early if there's traffic; plants will communicate to the sprinkler system when it's time for them to be watered; running shoes communicate time, speed, and distance so that the wearer can compete in real time with people on the other side of the world; medicine containers tell your family members if you forget to take the medicine. All objects can play an active role thanks to their connection to the Internet (Friess et al., 2012).

- The IoT is an integrated part of the future Internet, including existing and evolving Internet and network developments, and could be conceptually defined as a dynamic global network infrastructure with self-configuring capabilities based on standard and interoperable communication protocols where physical and virtual things have identities, physical attributes, and virtual personalities; use intelligent interfaces; and are seamlessly integrated into the information network (Vermesan et al., 2011).
- International Telecommunication Union (ITU) defines the IoT as a "global infrastructure for the information society, enabling advanced services by interconnecting (physical and virtual) things, based on existing and evolving interoperable information and communication technologies." ITU's foundational definition, published on July 4, 2012, offers useful insight and a sound springboard for further analysis and research into the IoT. Importantly, ITU points out that the IoT is a "vision," not a single technology, and that it has "technological and societal implications" (Louchez, 2013).
- The IoT is the network of physical objects that contain embedded technology to communicate and sense or interact with their internal states or the external environment (Gartner).
- The Oxford Dictionary defines IoT as the interconnection via the Internet of computing devices embedded in everyday objects, enabling them to send and receive data.
- The IoT is a dynamic global network infrastructure with self-configuring capabilities based on standard and interoperable communication protocols where physical and virtual things have identities, physical attributes, and virtual personalities and use intelligent interfaces, and are seamlessly integrated into the information network (IERC definition, 2014).
- The IoT enables the objects in our environment to become active participants; that is, they share information with other members of the network or with any other stakeholder and they are capable of recognizing events and changes in their surroundings and of acting and reacting autonomously in an appropriate manner. In this context the research and development challenges to create a smart world are enormous. A world where the real, digital, and virtual are converging to create smart environments that make energy, cities, and many other areas more intelligent (Friess et al., 2012).

Comfort zone defines certain types of behaviors that do not cause a person anxiety or tension. A comfort zone involves certain attitudes of thought encouraging formation and use of psychological limitations inducing a sense of safety. A person who has created a comfort zone while creating IoT products tends to remain in this zone, never stepping beyond its bounds, from inertia. If there is a desire to step over the bounds of a comfort zone, a person must experiment with some new or differing IoT creation and, thereby, face new and differing reactions appearing in the IoT creation environment. The boundaries of a comfort zone determine the rigidity of inner thinking. Terms describing a comfort zone can be inflexibility, boundaries or walls, habit, and the like.

As one example, there is a list of information that a smart home investor needs to have explained about the comfort zone of an investment: geographical location, names of the streets, name of the region, the rules and regulations on zoning, smart home decrees applicable to the location, prices according to the region or to the streets, data on the market for leases, future road plans, future plans on utility services, school designations by region and ability to enter other schools, route schedules of buses and other local transportation means, and so forth. Investors will be better off remembering that most of these factors can change; therefore the existing situation needs to be constantly reviewed. It could become clear to an investor that an investment will be short-term, if it becomes necessary to make do with the existing accustomed smart home development and financing methods in forming an investment portfolio.

In Lenz (2014) opinion, creating IoT products can place businesses and their product teams in an uncomfortable place by pushing them outside of their comfort zone. Lenz (2014) explores six design characteristics to help guide IoT product teams as they set forth into uncharted territory:

- *Intelligence*. Together algorithms and computing (ie, software and hardware) provide the intelligent spark that makes a product experience smart. Consider Misfit Shine compared to Nest's intelligent thermostat. The Shine experience distributes compute tasks between a smartphone and the cloud. The Nest thermostat has more compute horsepower for the artificial intelligence that make them smart.
- *Connectivity*. Connectivity in the IoT is more than slapping on a WiFi module and calling it a day. Connectivity enables network accessibility and compatibility. Accessibility is getting on a network while compatibility provides the common ability to consume and produce data.
- *Autonomous behavior*. Virtual things will be able to act autonomously (following their own objectives or shared ones) depending on the context, circumstances, or environments.
- *Complexity*. The IoT will interconnect trillions of smart devices and be a huge, heterogeneous, and complex system.
- *Sensing*. We tend to take for granted our senses and ability to understand the physical world and people around us. Sensing technologies provide us with the means to create experiences that reflect a true awareness of the physical world and the people in it. This is simply the analog input from the physical world, but it can provide rich understanding of our complex world.
- *Expressing*. Expressing enables interactivity with people and the physical world. Whether it is a smart home, expressing provides us with a means to create products that interact intelligently with the real world. This means more than just rendering beautiful user interfaces to a screen. Expressing allows us to output into the real world and directly interact with people and the environment.
- *Energy*. Without energy we can't bring our creations to life. The problem is we can't create billions of things that all run on batteries. Energy harvesting, power efficiency, and charging infrastructure are necessary parts of a power-intelligent ecosystem that we must design. Today, it is woefully inadequate and lacks the focus of many product teams.
- *Safety*. As we gain efficiencies, novel experiences, and other benefits from the IoT, we must not forget about safety. As both the creators and recipients of the IoT, we must design for safety. This includes the safety of our personal data and the safety of our physical well-being. Securing the end points, the networks, and the data moving across all of it means creating a security paradigm that will scale.

In the IoT, smart things or objects are expected to become active participants in business, information, and social processes where they are enabled to interact and communicate among themselves and with the environment by exchanging data and information sensed about the environment, while reacting autonomously to the real/physical world events, and influencing it by running processes that trigger actions and create services with or without direct human intervention. By framing IoT design with these characteristics, multidiscipline teams can work across their domains to make tradeoffs in interaction design, software architectures, and business models (Vermesan et al., 2011).

The IoT concept refers to uniquely identifiable things with their virtual representations in an Internet-like structure and IoT solutions comprising a number of components such as (Friess et al., 2012):

- Module for interaction with local IoT devices (for example embedded in a mobile phone or located in the immediate vicinity of the user and thus contactable via a short-range wireless interface). This module is responsible for acquisition of observations and their forwarding to remote servers for analysis and permanent storage.
- Module for local analysis and processing of observations acquired by IoT devices.
- Module for interaction with remote IoT devices, directly over the Internet or more likely via a proxy. This module is responsible for acquisition of observations and their forwarding to remote servers for analysis and permanent storage.
- Module for application-specific data analysis and processing. This module is running on an application server serving all clients. It is taking requests from mobile and web clients and relevant IoT observations as input, executes appropriate data processing algorithms, and generates output in terms of knowledge that is later presented to users.
- Module for integration of IoT-generated information into the business processes of an enterprise. This module will be gaining importance with the increased use of IoT data by enterprises as one of the important factors in day-to-day business or business strategy definition.
- User interface (web or mobile): visual representation of measurements in a given context (for example on a map) and interaction with the user (ie, definition of user queries).

17.1.2 Design

Distributed hypermedia provides a uniform means of accessing services through the embedding of action controls within the presentation of information retrieved from remote sites. An architecture for the Web must therefore be designed with the context of communicating large-grain data objects across high-latency networks and multiple trust boundaries. The anarchic scalability requirement applies to all architectural elements. Clients cannot be expected to maintain knowledge of all servers. Servers cannot be expected to retain knowledge of state across requests. Hypermedia data elements cannot retain back-pointers, an identifier for each data element that references them, since the number of references to a resource is proportional to the number of people interested in that information. Security of the architectural elements, and the platforms on which they operate, also becomes a significant concern. Multiple organizational boundaries implies that multiple trust boundaries could be present in any communication. The participants in an application interaction should either assume that any information received is untrusted, or require some additional authentication before trust can be given (Fielding, 2000). According to Fielding (2000), above anarchic scalability can completely realize the potential of IoT solutions.

Well-designed devices must have accessible, intuitive interfaces that make it possible for users to communicate their intentions without having to learn any sophisticated programming languages. If users need to learn different interfaces for their vacuums, their locks, their sprinklers, their lights, and their coffeemakers, it's tough to say that their lives have been made any easier (Littman and Kortchmar, 2014). In Littman and Kortchmar (2014) opinion, effective implementation of the IoT requires consideration of the interface's usability (more user friendly and better integrated).

17.1.3 Technological trends

Advances in wireless networking technology and the greater standardization of communications protocols make it possible to collect data from sensors and wireless identifiable devices almost anywhere at any time. Massive increases in storage and computing power, some of it available via cloud computing, make number crunching possible at a very large scale, at a high volume, and at a low cost. It is possible to identify, for the years to come, a number of distinct macrotrends that will shape the future of IoT (Vermesan et al., 2011):

- The explosion in the volumes of data collected, exchanged, and stored by IoT interconnected objects will require novel methods and mechanisms to find, fetch, and transmit data.
- Research is looking for ultralow power autonomic devices and systems from the tiniest smart dust to the huge data centers that will self-harvest the energy they need.
- Miniaturization of devices is also taking place at a lightning speed, and the objective of a single-electron transistor, which seems to be (depending on new discoveries in physics) the ultimate limit, is getting closer.
- The trend is toward the autonomous and responsible behavior of resources. The ever-growing complexity of systems, possibly including mobile devices, will be unmanageable, and will hamper the creation of new services and applications, unless the systems will show "self" functionality, such as self-management, self-healing, and self-configuration.

IoT research and development is becoming more complex due to the already highly advanced level of technology; the global, intersectoral, and interdisciplinary collaboration needed; and the ever-increasing demands of society and the economic global marketplace. Development of certain enabling technologies such as nanoelectronics, communications, sensors, smartphones, embedded systems, cloud computing, and software technologies will be essential to support important future IoT product innovations affecting the different industrial sectors (Vermesan et al., 2011). Chamberlin (2014) takes a look at two important enabling technologies that are helping to make the future of IoT a reality today:

- Advances in connectivity and networks. To connect all the expected billions of devices to the Internet, more Internet addresses were needed than were available through the IPv4 protocol. Once connected, all these devices need networks to communicate with other devices and computer systems. There are many different types of networks that are available and each has different strengths for different applications. While all these different wireless

communications standards and technologies result in a fragmented approach, each of these networks has strengths for certain applications.
- Advances in sensor and microprocessor design. The following advancements are enabling new types of IoT systems and applications: smaller, more durable sensors, multiprocessor chips, increasing processor performance and efficiency, and lower costs.

17.2 Domain-specific examples and applications

Built environment can be defined in several ways:

- The term built environment refers to the human-made surroundings that provide the setting for human activity, ranging in scale from buildings and parks or green space to neighborhoods and cities that can often include their supporting infrastructure, such as water supply or energy networks. The built environment is a material, spatial, and cultural product of human labor that combines physical elements and energy in forms for living, working, and playing. It has been defined as "the human-made space in which people live, work, and recreate on a day-to-day basis" (Wikipedia, https://en.wikipedia.org/wiki/Built_environment).
- The "built environment encompasses places and spaces created or modified by people including buildings, parks, and transportation systems." In recent years, public health research has expanded the definition of built environment to include healthy food access, community gardens, walkability, and bikability (http://www.ieltsinternational.com/).

A built environment is developed in order to satisfy residents' requirements. Human needs can be physiological or social and are related to security, respect, and self-expression. People want their built environment to be aesthetically attractive and to be in an accessible place with a well-developed infrastructure, convenient communication access, and good roads, and the dwelling should also be comparatively cheap, comfortable, with low maintenance costs, and have sound and thermal insulation of walls. People are also interested in ecologically clean and almost noiseless environments, with sufficient options for relaxation, shopping, fast access to work or other destinations, and good relationships with neighbors.

It must be admitted that the most serious problems of built environments (eg, unemployment, vandalism, lack of education, robberies) are not always related to the direct physical structure of housing. Increasing investment into the development of social and recreational centers, such as athletic clubs, physical fitness centers, and family entertainment centers, the infrastructure, a good neighborhood and better education of young people, can solve such problems. Investment, purchase and sale of a property, and its registration have related legal issues. The legal system of a country aims to reflect its existing social, economic, political, and technical state and the requirements of the market economy. As illustrated, the life cycle of the built environment can be assessed taking into account many quantitative and qualitative criteria. Life cycle of the built environment quantitative and qualitative analyses aspects are presented in Fig. 17.1.

Each one of these Level 1 aspects subsystems (see Fig. 17.1) based on the principle of a tree diagram can be discussed in much greater detail. To illustrate, the life cycle of the built environment described according to an example of the ith level aspects subsystem could be energy. In view of the global practice, any analysis of various aspects

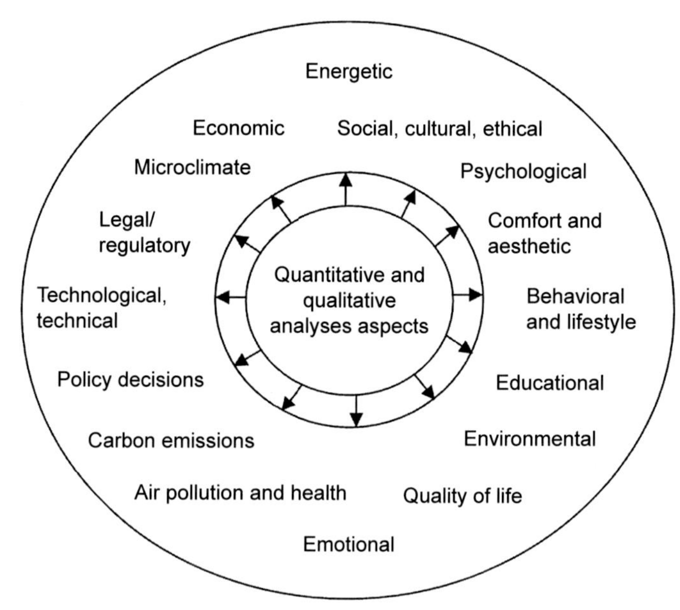

Figure 17.1 Life cycle of the built environment quantitative and qualitative analyses aspects.

characteristic to the life cycle of the built environment focuses on the analysis of energy. Life cycle of the energy-efficient built environment application areas are presented in Fig. 17.2.

Potential applications of the IoT for the built environment are many and various, fitting into almost all activities done by persons, organizations, and the community as a whole. Libelium (2014) has released the document "Top 50 Internet of Things Applications". Based on Libelium (2014), here is an overview of the applications used in the built environment:

- *Domotic and home automation:* Energy and water use (energy and water supply consumption monitoring to obtain advice on how to save cost and resources), remote control

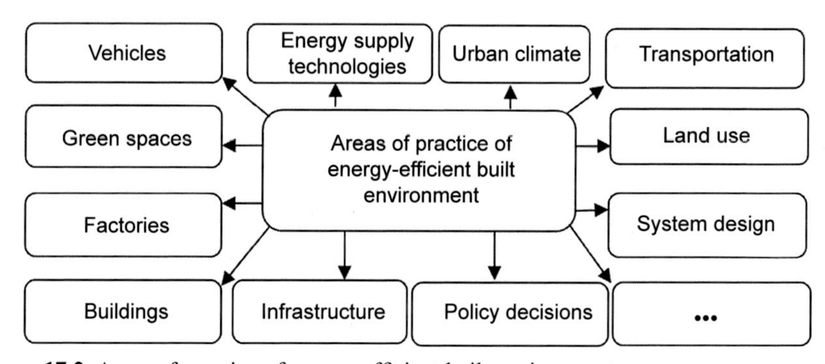

Figure 17.2 Areas of practice of energy-efficient built environment.

appliances (switching on and off appliances remotely to avoid accidents and save energy), intrusion detection systems (detection of window and door openings and violations to prevent intruders), art and goods preservation (monitoring of conditions inside museums and art warehouses).

- *Smart cities:* Smart parking (monitoring of parking spaces availability in the city), structural health (monitoring of vibrations and material conditions in buildings, bridges, and historical monuments), noise urban maps (sound monitoring in bar areas and centric zones in real-time), electromagnetic field levels (measurement of the energy radiated by cell stations and WiFi routers), traffic congestion (monitoring of vehicles and pedestrian levels to optimize driving and walking routes), smart lighting (intelligent and weather-adaptive lighting in street lights), waste management (detection of rubbish levels in containers to optimize the trash collection routes), smart roads (intelligent highways with warning messages and diversions according to climate conditions and unexpected events like accidents or traffic jams).
- *Smart environment:* Forest fire detection (monitoring of combustion gases and preemptive fire conditions to define alert zones), air pollution (control of CO_2 emissions of factories, pollution emitted by cars), snow level monitoring (snow level measurement to know in real time the quality of ski tracks and allow security corps avalanche prevention), landslide and avalanche prevention (monitoring of soil moisture, vibrations, and earth density to detect dangerous patterns in land conditions), earthquake early detection (distributed control in specific places of tremors).
- *Smart water:* Potable water monitoring (monitor the quality of tap water in cities), chemical leakage detection in rivers (detect leakages and wastes of factories in rivers), swimming pool remote measurement (control remotely the swimming pool conditions), pollution levels in the sea (control real-time leakages and wastes in the sea), water leakages (detection of liquid presence outside tanks and pressure variations along pipes), river floods (monitoring of water level variations in rivers, dams, and reservoirs).
- *Smart metering:* Smart grid (energy consumption monitoring and management), tank level (monitoring of water, oil, and gas levels in storage tanks and cisterns), photovoltaic installations (monitoring and optimization of performance in solar energy plants), water flow (measurement of water pressure in water transportation systems), silos stock calculation (measurement of emptiness level and weight of the goods).
- *Security and emergencies:* Perimeter access control (access control to restricted areas and detection of people in nonauthorized areas), liquid presence (liquid detection in data centers, warehouses, and sensitive building grounds to prevent breakdowns and corrosion), radiation levels (distributed measurement of radiation levels in nuclear power stations surroundings to generate leakage alerts), explosive and hazardous gases (detection of gas levels and leakages in industrial environments, surroundings of chemical factories, and inside mines).
- *Retail:* Supply-chain control (monitoring of storage conditions along the supply chain and product tracking for traceability purposes), NFC payment (payment processing based on location or activity duration for public transport, gyms, theme parks, etc.), intelligent shopping applications (getting advice in the point of sale according to customer habits, preferences, presence of allergic components for them, or expiring dates), smart product management (control of rotation of products in shelves and warehouses to automate restocking processes).

According to Friess (2012), the following list of aspects provides a wide but certainly not exhaustive compilation of the current IoT issues at stake:

- *Architecture:* Development and refinement of structural reference frameworks for the arrangement of physical and logical hardware and software components, including questions of object identification, virtualization, and decentralization; also ensuring interoperability across application sectors.

- *Security and trust issues:* Development of mechanisms and frameworks (by design) for ensuring that all users in business and private contexts trust the applications and maintain a certain power of control on their data across the full data and information life cycle.
- *Software and middleware platforms:* Support for analysis and processing of data flows from sensing devices and a high quantity of object instances, complemented with event filtering and management capabilities and including complexity management considerations.
- *Interfaces:* Integration of multimodal interface approaches for enriching all kinds of man–machine interaction for both changing the user experience and coping with the information density.
- *Smart sensors:* Integration of sensing and reasoning capabilities into networked and energy-harvesting devices.
- *Testing and standardization:* Current IoT dispositions are still ongoing and effects on mass deployments need to be much better understood. Testing and large-scale pilots are absolutely crucial and should also lead subsequently to standardization for ensuring interoperability and reducing complexity.
- *Business models:* A sound exploitation of the IoT business potential is still missing and new business models for the existing incumbents but also new and innovative players need to be developed.
- *Societal and ethical implications:* The IoT has already started to change our lives virtually but questions about the physical and logical usage coupled with considerations of needs for privacy, inclusiveness of the society, and evolution of social behavior remain very valid and only partly addressed.
- *IoT governance:* Often misunderstood, IoT governance is, in particular, about the governance of the IoT and their context of usage rather than Internet aspects. New models, mechanisms, and frameworks covering legal aspects too are necessary for guaranteeing proper trust, identity, and liability management.
- *International cooperation:* The IoT is a truly global subject that shows interesting application cases in different parts of the world. Moreover, as it will only work if a certain level of interoperability is maintained, a common understanding among the different nations involved is pivotal.
- *Integration of results from other disciplines:* Basic ICT (information and communications technology), robotics, nanotechnology, biomedicine, and cognitive sciences provide a rich source of inspiration and applications for developing the IoT further on.

Potential applications of the IoT for the built environment are many and various, fitting into almost all activities done by persons, organizations, and the community as a whole. These (smart home, real-time information about the city's environment, Oxford Flood Network, waste collection for smart cities, wireless monitoring systems in the field of civil engineering, urban intelligence platform, emotional gateway to Minneapolis, waste management, cyber security challenges in smart cities, smart environment monitoring system for pollution, health e-research system, negotiation in cyber-physical systems, real-time safety early warning system for cross-passage construction, RFID-plants in the smart city) are provided in brief next.

All of Samsung's products would be built on platforms that are open and compatible with other products and 90% of its products—which range from smartphones to refrigerators—would be able to connect to the Web by 2017. In 5 years, every product in the company's entire catalog is expected to be Internet-connected. In effect, Samsung is readying for the IoT, the term for the concept of using sensors and other technologies to hook just about anything you can think of into the Internet. Samsung introduced a new home-monitoring subscription service that will send immediate texts or calls to the smartphone of a user or designated contacts about problems or emergencies at their home, such as a flood, fire, plumbing leak, or a pet out in the yard when a storm is starting. The premium service also includes built-in DVR services for cameras (watch around your home for different issues), alert for different issues (for example, grandma did not get up this morning; my kid did not get home from school on time; my dog is out in the yard and there's a storm coming, etc.) (Tibken, 2015).

This summer, data scientists and architects in Chicago are working on a new form of civic infrastructure: highly visible, aesthetically pleasing, 1-foot-square boxes mounted on light poles that track environmental conditions around them. Those small boxes represent a big idea: Inside each one, about a dozen sensors measure heat, humidity, air quality, carbon monoxide, and carbon dioxide levels, and light and noise levels, and those data will be made publicly available so that they can be used by application developers and researchers as well as the city. About 50 will be installed this year in the Loop area of the city (Crawford, 2014).

Right now, cities collect information in the form of permit applications, inspection results, and other service-related inputs. Analysis of these data can help cities know how the city is doing and assist it in targeting its efforts. But information about the well-being of a city—the quality of lives lived on its streets—is harder to come by. The Array of Things, as Chicago's Urban Center for Computation and Data calls this project, will start providing real-time information about the city's environment. For example, sensors will be able to detect mobile devices that have Bluetooth turned on, so the city will have information about the level of pedestrian density in a particular area. The city, as well as any researcher, will know about fine-grained pollution levels in different neighborhoods for the first time. Now it's moving to understand its weather, pollution, and noise in a transparent, public-friendly way. This means that the city will be able to investigate reams of these data, combine it with other information, and make predictions about its future that inform how the city allocates its resources and changes its policies. It's crowded? Change the traffic light patterns. Pollution is a problem in particular neighborhoods? Find out why and fix it. Gathering these data will not solve all of Chicago's challenges, such as a shooting rate that remains among the nation's highest. But making a better city also means improving the quality of daily life at street level. Investing time and money in data makes sense, and it's changing how local government works. Chicago, the quintessential American city, is quickly becoming the nation's leading city for data analytics (Crawford, 2014).

Oxford Flood Network is installing sensors around Oxford. Network have several on the Thames and Castle Mill Stream area and some under floors to detect rising water when the time comes. The levels are very low at the moment, but we know how quickly that can change. Oxford Flood Network is collecting a list of people who are happy to host a sensor ($50 \times 50 \times 100$ mm) and/or gateway device ($90 \times 60 \times 26$ mm). There is no cost to the host for the device, but inhabitants will need to help keep it up and running by checking it periodically online and perhaps changing the battery once a year. Oxford Flood Network will use the sensors to create a detailed map of water levels around the city in higher detail than the Environment Agency's existing sensors. Oxford Flood Network involves communities and citizens, improving literacy in the IoT (Handsome, 2015).

Until now collecting waste has been done with static routes and schedules. Containers are collected every day or every week regardless of whether they are full or not. This causes unnecessary costs, poor equipment utilization, and the constant nuisance of container overfill. Enevo ONe uses smart wireless sensors to gather fill-level data from waste containers and sends it to a cloud-based analytics platform. The platform then generates accurate forecasts for ideal container pick-up schedules and routes that can be can be accessed directly by the driver through any cellular-enabled tablet or smartphone. The Enevo ONe service provides not only monitoring, scheduling, and optimized routes, but truly smart waste collection plans, which are the result of millions of complex calculations regarding fill-level trends and projections, scheduling constraints, and routing options. Collection based on Enevo's smart plans significantly reduces costs, emissions, road wear, vehicle wear, noise pollution, and work hours. Enevo ONe provides up to 50% in direct cost savings in waste logistics. And that's not all. Reducing the amount of overfull containers means less litter and happier customers (Enevo, 2015).

A long-term deployment has been set up to demonstrate the capabilities and the ease of use of wireless monitoring systems in the field of civil engineering. In this application tensile forces of cable stays of a cable-stayed bridge are monitored by tracking natural frequencies of cable vibrations. Wireless sensors (accelerometer, air temperature, air humidity), running on a single set of batteries were installed on six stays to measure cable acceleration. Since energy resources are limited and data communication is an energy-consuming task, the amount of transmitted data has to be kept small in order to extend system lifetime. In this case, the acceleration time series is processed on the node and reduced to one frequency value, which has to be transmitted over the air. The concept of data reduction by means of processing raw data on the sensor node level is demonstrated in the deployment at the Stork Bridge in Winterthur. The installation has been running since 2006 and is one of the first long-term wireless monitoring applications worldwide (Decentlab, 2015).

Founded in 2012 and based in New York City, Placemeter is an urban intelligence platform that quantifies the movement of modern cities, at scale. Placemeter ingests any kind of video to analyze pedestrian and vehicular movement, revealing hidden patterns and strategic opportunities. Placemeter (2015) platform leverages proprietary computer vision technology to gather data without identity detection from live streams and archival video. Placemeter is using feeds from hundreds of traffic video cameras to study 10 million pedestrian movements each day. It's using that data to help businesses learn how to market to pedestrian consumers. Placemeter also says it wants to use the data to help consumers with information such as when to visit your neighborhood coffee bar when the line is shorter. Placemeter says it does not store the video, nor does their analysis involve facial recognition (Patterson, 2014).

Placemeter is turning disused smartphones into big data. Measuring data about how the city moves in real time, being able to make predictions on that, is definitely a good way to help cities work better. That's the vision of Placemeter—to build a data platform where anyone at any time can know how busy the city is, and use that. City residents send Placemeter a little information about where they live and what they see from their window. In turn, Placemeter sends participants a kit to convert their unused smartphone into a street sensor, and agrees to pay cash as long as the device stays on and collects data. The more action outside—the more shops, pedestrians, traffic, and public space—the more the view is worth (Jaffe, 2014).

On the back end, Placemeter converts the smartphone images into statistical data using proprietary computer vision. The company first detects moving objects and classifies them either as people or as 11 types of vehicles or other common urban elements, such as food carts. A second layer of analysis connects this movement with behavioral patterns based on the location—how many cars are speeding down a street, for instance, or how many people are going into a store. Placemeter taking all measures to ensure anonymity. The smartphone sensors do not capture anything that goes on in a meter's home (such as conversations), and the street images themselves are analyzed by the computer, then deleted without being stored (Jaffe, 2014).

Efforts to quantify city life with big data are not new, but Placemeter's clear advance is its ability to count pedestrians. With its army of smartphone eyes, Placemeter promises a much wider net of real-time data dynamic enough to recognize not only that a person exists but also that person's behavior, from walking speed to retail interest to general interaction with streets or public spaces. The benefits could extend to both private and public entities alike. Investors might use Placemeter data to find the best location for a store, while retailers could learn things like their sidewalk-to-store conversion rate and how it compares to other stores on the block. Meanwhile, municipal agencies could detect the use of benches or near misses at intersections—and generally evaluate (and perhaps improve) public projects more quickly than they might otherwise. In the future people can use Placemeter data to know when a basketball court is free or when the grocery store will be least crowded. It's this grassroots approach to big data that could make Placemeter a powerful platform for government accountability (Jaffe, 2014).

It needs between 2500 and 2700 video feeds to properly cover the city. With high-resolution cameras Placemeter says it can detect the gender of pedestrians with between 75% and 80% accuracy. This opens the potential for advertisements to be targeted to a more appropriate audience. Better foot traffic data could let retailers know whether they're paying too much for a location. The location of a store can make a huge difference in its success. Placemeter wants to sell its foot traffic data to businesses to help them get a better opportunity. Placemeter is trying to find inefficiency in the market. Ultimately the price starts to increase to a level that accurately reflects how good or bad the area is. Placemeter is trying to effectively short circuit that (McFarland, 2014).

Minneapolis Interactive Macro Mood Installation (MIMMI) is an emotional gateway to Minneapolis, bringing residents and visitors together to experience and participate in the collective mood of the city. MIMMI is a large, air-pressurized sculpture suspended from a slender structure located at the Minneapolis Convention Center Plaza. Cloud-like in concept, the sculpture hovers 30 feet above the ground, gathering emotive information online from Minneapolis residents and visitors to the plaza. MIMMI analyzes this information in real time, creating abstracted light displays and triggering misting in response to this input, creating light shows at nighttime and cooling microclimates during the daytime. Whether the city is elated following a Minnesota Twins win or frustrated from the afternoon commute, MIMMI responds, changing behavior throughout the day and night. To understand the city's mood, MIMMI sources information from local Twitter feeds and uses textual analysis to detect the emotion of those tweets, a process developed by INVIVIA's technologists using open source technology. By aggregating the positivity and negativity of tweets in real time, MIMMI transmits the abstracted emotion of the city to a series of WiFi-enabled LED bulbs and an integrated water misting system. The low-energy lights, hung inside of the sculpture material and stretching throughout the entire shape, display the mood beginning at sunset. The color of the lights shifts from cool colors (negative) to warm and hot colors (positive) depending on the mood, with rate of the lights' change depending on the rate of tweets (Minneapolis, 2015).

If the city mood is particularly "sad" or emotional for any particular reason, visitors to the plaza can come together to lift MIMMI's (and the city's collective) spirits, as MIMMI can detect movement at the plaza and include this information in its analytics. The more people present and moving around under the cloud, the more active MIMMI will become, responding either with increased lighting or misting depending on the time of day. Dance, high activity, and movement will positively affect MIMMI's mood displays. The website, www.minneapolis.org/mimmi, will catalog the mood of the city generated by MIMMI over the summer and fall, allowing visitors to see daily and weekly trends in the city's emotions. When visitors using iOS (iPhones) arrive at the plaza, the app will transform into an augmented reality view of MIMMI, providing a wholly new way of looking at the installation with additional animations emphasizing the city's current mood (Minneapolis, 2015).

Homes, cars, public venues, and other social systems are now on their path to the full connectivity known as the IoT. Standards are evolving for all of these potentially connected systems. They will lead to unprecedented improvements in the quality of life. To benefit from them, city infrastructures and services are changing with new interconnected systems for monitoring, control, and automation. Intelligent transportation, public and private, will access a web of interconnected data from GPS locations to weather and traffic updates. Integrated systems will aid public safety, emergency responders, and in disaster recovery (Elmaghraby and Losavio, 2014). Elmaghraby and Losavio (2014) examine two important and entangled challenges: security and privacy. Security includes illegal access to information and attacks causing physical disruptions in service availability. As digital citizens are more and more instrumented with data available about their location and activities, privacy seems to disappear (Elmaghraby and Losavio, 2014).

Air pollution is a major environmental change that causes many hazardous effects on human beings, which needs to be controlled (Jamil et al., 2015). Jamil et al. (2015) deployed wireless sensor network (WSN) nodes for constant monitoring of the air pollution around the city and the moving public transport buses and cars. The data of the air pollution particles such as gases, smoke, and other pollutants is collected via sensors on the public transport buses and the data is being analyzed when the buses and cars reach back to the source destination after passing through the stationary nodes around the city (Jamil et al., 2015).

Clarke and Steele (2014) introduce a novel system to capture aggregate population health research data via utilizing smartphone capabilities while fully maintaining the anonymity and privacy of each individual contributing such data. A key and novel capability of this system is the support for customizable data collection, without the need to know specific details about an individual. The customized collection rules can be deployed on the local device based on detailed local data, and the resultant collection can be measured by the anonymous data collection network (Clarke and Steele, 2014).

In the near future, the IoT is expected to penetrate all aspects of the physical world, including homes and urban spaces. In order to handle the massive amount of data that becomes collectible and to offer services on top of this data, the most convincing solution is the federation of the IoT and cloud computing. Yet, the wide adoption of this promising vision, especially for application areas such as pervasive health care, assisted living, and smart cities, is hindered by severe privacy concerns of the individual users. Hence, user acceptance is a critical factor to turn this vision into reality (Henze et al., 2015).

With the rapid development of urbanization in China, the number and size of underground space development projects are increasing quickly. At the same time, more and more accidents are causing underground construction to increasingly become a focus of social attention. Therefore, this research presents a real-time safety early warning system to prevent accidents and improve safety management in underground construction, based on IoT technology. The proposed system seamlessly integrates a fiber Bragg grating sensor system and a radio frequency identification (RFID)-based labor tracking system. This system has been validated and verified through a real-world application at the cross passage construction site in the Yangtze Riverbed Metro Tunnel project in Wuhan, China (Ding et al., 2013).

A city may become smart and green through strategic deployment of information and communication technology infrastructure and services to achieve sustainability policy objectives in which trees have to be involved. Plants not only constitute green space useful to contrast urban pollution effects or provide ecosystemic benefits to residents but they can also be used as bioindicators and their involvement in communication networks can represent a significant contribution to build a smart, green city. RFID tags can be easily associated with plants, externally or internally. This latter approach is particularly indicated if the identification of trees needs to be secured since its production, eliminating the risk of tag losses or removal. Interesting applications may be derived by implementing RFID tags in biomonitoring systems in order to guarantee a real-time data communication in which tags may act as antennas for multifunctional green spaces (Luvisi and Lorenzini, 2014).

17.3 Machine-to-machine

Machine-to-machine (M2M) communication uses technologies to allow both wireless and wired systems to connect with devices of the same ability. It is also to enable applications that allow businesses to increase productivity and competitiveness through increased efficiencies, cost-savings, and improved levels of services. The machines use telemetry or telematics to connect to one another; this is accomplished over wireless networks. Wireless networks can transmit data to and from each machine. M2M technology is the means of communication of different types of mechanical devices for the exchange of data or information to one another. The M2M interface allows monitoring, control, and management of remote equipment or machines. Remote monitoring, control, and management of devices and machines allow businesses to address maintenance issues and restore functionality. The basic structure involves a central system that is able to connect with other systems at various locations via wireless networks. The central computer system can collect or send data or information to each remote machine or headquarters. Key applications are (M2M, 2015):

- Connecting machines to other machines: remote production environments, for example
- Connecting machines to services centers: cars notifying service centers of maintenance issues, for example
- Connecting service centers to machines: vending machines reporting stock status to a central inventory system, for example
- Connecting vehicles to machines: fleet management and location, for example

Not many M2M applications for the built environment are available in the world. These (M2M communications for smart cities, from M2M to the IoT in the built environment) are discussed briefly, next.

Vilajosana and Dohler (2015) have observed a steady increase in the deployment of M2M technologies in urban environments. It is the beginning of a paradigm referred to as the smart city, where smartness is added to city infrastructure through sensors, big data, and other capabilities. Vilajosana and Dohler (2015) review currently used smart city M2M technologies that are due to appear. It dwells in great detail on one of the most prominent deployment use cases, smart parking. The more holistic problem of handling data in advanced platforms, and how to offer it to the emerging ecosystem, is also discussed. Financial and governance issues are then outlined through 10 challenges that lie ahead for a successful explosion of the smart city market (Vilajosana and Dohler, 2015).

Digital technologies are often suggested as the panacea through the development of smart cities—cities that in some form integrate a digital infrastructure with the physical city in order to reduce environmental impact while improving quality of life and economic prospects. While these sorts of concepts have been around for several decades, the advent of smartphones and cheaper sensor technology means that digitally enabled, or smart, cities are fast becoming a real-world possibility (Höller et al., 2014).

Participatory sensing (PS), also known as urban, citizen, or people-centric sensing, is a form of citizen engagement for the purpose of capturing the surrounding environment in a city as a first step for contributing to the solution of specific issues such as public health and well-being. Either citizens on their own initiative, or citizens organized through a specific campaign initiated by city authorities, collect sounds, pictures, videos, and other sensor data using their mobile phones as the main tool to monitor the environment and transfer the collected data to a storage space. The collected data are analyzed by citizens or city authorities, conclusions and action plans are drawn, and actions are taken. Although this form of engagement was typical a few years ago, nowadays the PS concept has been enriched to include active citizen journalists or passive social media sensing in the sense that citizen engagement to social media such as Twitter can also be used as additional input to PS campaigns (Höller et al., 2014).

WSNs as a central part of cyber-physical systems are gaining commercial momentum in many areas, including building monitoring and intelligent home automation. Users wish to successively deploy hardware from different vendors. Interoperability is taken for granted by the customers who want to avoid the need for exhaustive configuration and set up. Therefore, the need for an interoperable and efficient application layer protocol for M2M communication in and across the boundaries of WSNs arises (Schmitt et al., 2014).

A building automation system (BAS) is a computerized, intelligent system that controls and measures lighting, climate, security, and other mechanical and electrical systems in a building. The purpose of a BAS is typically to reduce energy and maintenance costs, as well as to increase control, comfort, reliability, and ease of use for maintenance staff and tenants (Höller et al., 2014).

17.4 Intelligent decision-support systems for Internet of Things

Intelligent decision support systems (IDSSs) add artificial intelligence functions to traditional decision support systems (DSSs) with the aim of guiding users through some of the decision-making phases and tasks or supplying new capabilities. This notion has been applied in various ways (Phillips-Wren et al., 2009). The term IDSS describes DSSs that make extensive use of artificial intelligence (AI) techniques. Along with knowledge-based decision analysis models and methods, IDSSs incorporate databases, model bases, and intellectual resources of individuals or groups well to support effective decision making (Wan and Lei, 2009).

A DSS, as a kind of interactive computer-based information system, helps decision makers utilize data and models to solve mostly semistructured or unstructured decision problems in practice. IDSSs, along with knowledge-based decision analysis models and methods, incorporate databases, model bases, and intellectual resources of individuals or groups well to improve the quality of complex decisions. Multicriteria DSS, group DSS, and web-based customer recommender systems have had unimaginable developments and improvements in dealing with complex, uncertain, and unstructured decision problems under the support of computational intelligent technologies (Lu et al., 2010).

Some research in AI, focused on enabling systems to respond to novelty and uncertainty in more flexible ways has been successfully used in IDSS. For example, data mining in AI that searches for hidden patterns in a database has been used in a range of decision-support applications. The data mining process involves identifying an appropriate data set to mine or sift through to identify relations and rules for IDSS. Data mining tools include techniques like case-based reasoning, clustering analysis, classification, association rule mining, and data visualization. Data mining increases the intelligence of DSS and becomes an important component in designing IDSS (Yang et al., 2012).

Currently IDSS provides decision support via text analytics and mining-based DSSs; ambient intelligence and the IoT-based DSSs; biometrics-based DSSs; recommender, advisory, and expert systems; data mining, data analytics, neural networks, remote sensing and their integration with DSSs and other IDSSs. These other IDSSs include genetic algorithm (GA (genetic algorithm)-based) DSS, fuzzy sets DSS, rough sets-based DSS, intelligent agent-assisted DSS, process mining integration to decision support, adaptive DSS, computer vision-based DSS, sensory DSS, and robotic DSS.

The concepts of the IoT and their link to DSSs are briefly deliberated. In the opinion held by Höller et al. (2014), the IoT refers to the interconnection of uniquely identifiable embedded computing-like devices within the existing Internet infrastructure. Typically, IoT is expected to offer advanced connectivity of devices, systems, and services that goes beyond M2M communications and covers a variety of protocols, domains, and applications (Höller et al., 2014).

Very few IDSSs for IoT systems are available in the world. A few of them (integrated sensor and management system for urban waste water networks, real-time management system of urban water cycle, energy management based on IoT, cyber-based DSS for a low-carbon integrated waste management system in a smart city, individual customized services in symbiosis houses, environmental DSS (EDSS), DSSs for water utilities and consumers, DSSs based on water smart metering, cloud asset for urban flood control, DSSs for the management of sewer infrastructure, Internet-enabled prefabricated housing construction) are described here.

Sempere-Payá and Santonja-Climent (2012) describe the design and implementation of improvements to the monitoring system of an urban waste water network, resulting in more efficient management of the system. To achieve this objective, the latest communications technology has been incorporated into heterogeneous networks and sensor systems. This technology includes mobile systems, which take measurements and transmit images in real time; an intelligent platform for processing and management of variables; and the implementation of WSNs designed with specific protocols and tools that allow the rapid deployment of the network and allow measurements to be taken in emergency situations. The sensors in this type of installation are extremely important for the management of the system as they allow us to collect information and make decisions with sufficient time to deal effectively with critical situations such as flooding or overloading of the waste water system, or environmental problems such as dumping of possible pollutants, as well as to make the best use of the water cycle. The solution presented here automates large portions of the processes, minimizing the possibility of human error, and increasing the frequency and accuracy of the measurements taken, ensuring a robust communication system covering all the elements involved to provide ubiquity of information, and finally gives an application layer to manage the system and receive alerts (Sempere-Payá and Santonja-Climent, 2012).

A main feature of the upgrading of urban waste water networks is the application of DSSs to support in the fast reaction to a disaster state. The Neptune project (Morley et al., 2009) involves two cases of urban waste water (monitoring water quality, prevention, and early action in case of contamination) network integration with DSS. Pyayt et al. (2010) used a sensor network on dikes for the early prediction of structural breaks and floods, as well as management and decision-making assistance. COWAMA (Suńer et al., 2008) is a DSS for the management of bathing water. With the entry and flow of water into the sewers, direct discharges of sewage to the receiving environment, and sewage treatment and dispersal in the sea, this system can predict the magnitude and duration of pollution episodes (Suńer et al., 2008).

The growth in energy monitoring and management requires integration of energy consumption data in several tools to support energy-aware decision-making, such as energy efficiency KPIs (e-KPIs), visualization energy and visualization e-KPIs, energy simulation KPIs, energy-DSS (e-DSS) and optimization tools. The integration of energy data into production management decisions also requires an e-DSS to support energy-aware decision-making. Such systems provide several benefits for the factories. The first benefit is providing solutions and mechanisms to support production processes to be more energy and cost efficient. The second benefit is the rapid response to production process needs, such as faster response to changes in energy prices (ie, demand response) (Shrouf and Miragliotta, 2015).

In today's manufacturing scenario, rising energy prices, increasing ecological awareness, and changing consumer behaviors are driving decision-makers to prioritize green manufacturing. The IoT paradigm promises to increase the visibility and awareness of energy consumption, thanks to smart sensors and smart meters at the machine and production line level. Consequently, real-time energy consumption data from manufacturing processes can be easily collected, and then analyzed, to improve energy-aware decision-making (Shrouf and Miragliotta, 2015). Relying on a comprehensive literature review and on experts' insight, Shrouf and Miragliotta (2015) contribute to the understanding of energy-efficient production management practices that are enhanced and enabled by the IoT technology. In addition, it discusses the benefits that can be obtained thanks to adopting such management practices. Eventually, a framework is presented to support the integration of gathered energy data into a company's information technology tools and platforms. This is done with the ultimate goal of highlighting how operational and tactical decision-making processes could leverage on such data in order to improve energy efficiency, and therefore competitiveness, of manufacturing companies. With the outcomes of this research, energy managers can approach the IoT adoption in a benefit-driven manner, addressing those energy management practices that are more aligned with company maturity, measurable data, and available information systems and tools (Shrouf and Miragliotta, 2015).

Sustainable integrated waste management system (IWMS) is one of the big challenges of modern society. The topic is a priority since it contributes to the goals set by EU 2020 Strategy. New web-based technologies are available to monitor, manage, and elaborate information concerning spatially decentralized systems interacting with each other by cloud computing. A growing interest is paid to potential applications of these technologies to manufacturing systems and to municipal services ensuring both flexibility and efficiency. The authors propose a DSS for a Social–Cyber–Physical framework suitable for the strategic planning and operational phases, by supporting public administrators at a municipal level in design and plan an IWMS. A DSS allows public decision-makers and technical staff to assess technical options and local policies for an IWMS according to a what-if analysis. The main goal of the DSS is minimize the net carbon emissions of the IWMS. The model has been applied to a middle-sized city located in Southern Italy. A comparative analysis is carried out investigating the optimization of the standalone collecting phase and of the overall IWMS. Results in both cases suggest the adoption of multistream grouping systems for dry recycled fraction and of single stream grouping system for organic and glass fraction. At the same time a different mix of waste collection modalities is found in the two reference cases (Digiesi et al., 2015).

Smart residential spaces designed for residents departs from simply providing conventional standardized services, to go beyond and provide personalized experiences that take individual circumstances into account. Such services play a key role in enhancing the residents' quality of life (Kim et al., 2014). Kim et al. (2014) look into such personalized services that reflect different characteristics of a diverse range of residents who have different behavior patterns. Such services can increase the satisfaction of the residents by providing flexible services that take into account the lifestyle and circumstances of each resident. A problem with offering customized services, however, is that there is a dearth of data on individuals. Sufficient amounts of data must be collected in order to determine what a proper service for an individual is. The automatic ventilation system developed is a system that automatically opens and closes windows by sensing smoke, rain, hail, temperature, and wind using diverse sensors. For example, if there is fire or smoke outside the building, windows are automatically closed to prevent it from entering. If the indoor temperature is too high, windows are automatically opened for natural ventilation (Kim et al., 2014). As such, Kim et al. (2014) explain the life-log data and discuss its collection method. The life-log data collected serves as crucial grounds for decision-making. How decisions are made using the life-log data is an intriguing research topic. Kim et al. (2014) propose and discuss logics and processes of various decision-making methods that can be executed using the life-log data. In their homes, residents tend to regularly display certain behaviors in patterns, which allows for identifying the residents' behavior patterns, as well as predicting the residents' future behavior. In this aspect, the residents' location, needs, and current behavior must be recognized in order to provide personalized services. As such, Kim et al. (2014) propose a decision-making method by verifying in-house behavior of Korean elderly with companion dogs in symbiosis homes. Such dog's hair and foul smells cause indoor pollution that damage elderly heath. Kim et al. (2014) propose an automatic personalized window opening and closing service by using life-log data of the resident.

EDSSs have become an important research topic in the fields of water environment protection and emergency early warning. Most existing systems require plenty of measured data, which are typically unavailable in ungauged basins. In order to develop an early warning system that can work with few measured data, a mobile EDSS (MEWSUB) for early warning and emergency assessment in ungauged river basins was developed. In an emergency situation, the pollution sources/situation should be first selected by the users, then the system recommends the decision of pollutant discharge patterns for different pollution sources. A one-dimensional water quality model was built for the system and the terrain module and pollutant module were developed for terrain digitization and pollutant release process data generation. MEWSUB can quickly create water quality model input files, simulate the temporal release of pollutants, and display the results in a dynamically rendered map and trend line diagram and present recommendations via HTML5. The system has been used successfully to support early warning efforts during accidents on several ungauged rivers in China (Wang et al., 2015).

Water smart metering enables the measurement and reporting of water consumption at subdaily intervals. However, assuming that increased availability of consumption information will necessarily result in changed behavior is simplistic. The main scientific challenges for the iWIDGET project are the management and extraction of useful information from vast amounts of high-resolution consumption data, the development of customized information to influence awareness and support behavioral change, and the integration of iWIDGET concepts into a set of decision support tools for water utilities and consumers (Ribeiro et al., 2015).

The Water Framework Directive and the European Union carbon emission reduction target by 2020 set out the strategic context for the EU FP7 iWIDGET (improved water efficiency through ICT for integrated supply–demand-side management) project (http://www.i-widget.eu/) deployment. The goal of this project is to provide a web-based platform—the iWIDGET system—targeting both the household and water utilities end-users, capable of offering near real-time (at subdaily intervals) information about water consumption (and energy use, in specific conditions) and a set of decision support tools aimed to promote water and related energy-efficient use behaviors. The main scientific challenges of the iWIDGET system are the management and extraction of useful information from vast amounts of high-resolution consumption data, the development of customized interventions to influence behavioral change, and the integration of iWIDGET concepts into a set of decision-support tools for water utilities and consumers, applicable in differing local conditions (Ribeiro et al., 2015).

The performance of physical assets has become a major determinant success factor for urban flood control. However, managing these assets is always challenging as there are a huge number of diverse assets involved, which are distributed throughout the city, and owned by different agencies (Xu et al., 2015). Aiming at improving the management efficiency of these assets, and ensuring their performance, Xu et al. (2015) propose the concept of cloud asset based on cloud computing, mobile agent, and various smart devices. Through hardware integration and software encapsulation, cloud asset could sense its real-time status, adapt to varied working scenarios, be controlled remotely, and shared among agencies. It enables accurate real-time control of every asset, and thus improves the management efficiency and effectiveness. When flood occurs, decisions can be made immediately by cloud services according to the real-time situations, and demands could be distributed to specific assets directly in near real time through the cloud platform. Considering the specific emergency situation in urban flood control, corresponding real-time data analytics tools and decision-making models should be further explored to provide timely and accuracy decisions on both urban flood control and asset management processes (Xu et al., 2015).

Park and Kim (2013) developed a data warehouse-based DSS for the management of sewer infrastructure. To monitor the condition of water pipeline and networks, Lau and Dwight (2011) proposed a fuzzy-based decision support model. Various DSSs have also been built to support the different stages of urban flood control.

Prefabrication could provide various benefits for housing production, including lowering the cost, improving the management efficiency and quality, ensuring the safety, and realizing sustainability. However, to fully acquire these benefits in a real-world construction process, challenges still exist on real-time tracing and tracking of the prefabrication components, and getting real-time interactions among all the stockholders. Physical Internet (PI), as has been well adopted in logistics, could be an effective solution for real-time visibility. Based on the prefabrication practice in Hong Kong, this paper proposes the architecture of a PI-enabled prefabricated housing construction that upgrades and transforms the prefabrication construction so that the logistics echelons could be seamlessly integrated and synchronized, together with a PI-enabled DSS that uses IoT and cloud techniques for designing the architecture and a rich set of services and tools to assist different decision-makers (Zhong et al., 2015).

17.5 The trends and future of the Internet of Things

This section presents the main trends related to the IoT (Chamberlin, 2015; Schmitz, 2015; Harper, 2015) in the built environment. These trends can be further analyzed in more detail, which will be done using new strategic choices that companies face (Porter and Heppelmann, 2014) and four key ways the IoT will change the face of modern marketing (Rossi, 2015) as an example.

The IoT is in a stage of adoption very similar to what we saw in 2008–2010 around cloud computing and in 2011–2013 around big data analytics. Those two trends have been fully defined, and enterprises are actively deploying mission-critical enterprise applications utilizing those technologies. IBM understand that the IoT is an emerging and disruptive force that integrates devices, data, connections, processes, and people. As with any emerging technology area, there is a lot of activity going on in the IoT space (Chamberlin, 2015). Chamberlin (2015) describes 20 IoT trends to watch in 2015:

1. *Security and privacy:* Trust and authentication become critical across all elements of the IoT, including devices, the networks, the cloud, and software apps.
2. *Standards:* Competing proprietary and open-source standards efforts have continued to struggle to come together for the common good, but perhaps we will see breakthroughs in 2015.
3. *Hardware:* Advanced microcontrollers, systems on chips, and sensor technologies are enabling new types of IoT devices. These devices are also getting smaller, smarter, and cheaper.
4. *Software:* A huge rush of activity is expected as developers increase their focus on developing IoT platforms and solutions that upload data from sensors and perform the analytics necessary to deliver the insights required for business decisions.
5. *Edge analytics:* Increasingly, analytics capabilities will be pushed to the edge of networks. Advanced analytics and dashboards will be needed to provide insights from all the "things."

6. *M2M automation:* In the future, sensors, devices, and whole IoT systems will be talking to each other, providing insights and making decisions without human intervention.
7. *Platform-to-platform integration:* Expect movement from closed platforms toward open IoT platforms that support multiple applications, devices, and networks.
8. *Wearables:* This is a key subsegment within the overall IoT market. Watch developments in systems on chips, sensor hubs, sensor fusion, low-power wireless connectivity, battery life, and specialized software development platforms to support wearables.
9. *Sensor fusion:* Combining data from different sources can improve accuracy. Data from two sensors is better than data from one. Data from lots of sensors is even better.
10. *Sensor hubs:* Developers will increasingly experiment with sensor hubs for IoT devices, which will be used to offload tasks from the application processor, cutting down on power consumption and improving battery life in the devices.
11. *Big data:* "Things" (sensors, chips, computers) will produce even more data than we have now, taxing our already complex enterprise information management systems.
12. *Blockchain:* Using blockchain technology, developers can set up a distributed model that does not require trusting every node in the network.
13. *Success stories:* To move from hype to real mass adoption, vendors need to provide customer case studies that show best-of-breed implementations of IoT systems and prove the benefits that await enterprises.
14. *Chief IoT officer:* Expect more senior level execs to be put in place to build the enterprise-wide IoT strategy.
15. *Business processes transformation:* IoT enables new automated sense-and-respond systems, disrupting traditional processes and requiring new skills.
16. *Education needed:* Demand will rise for education and skills training related to IoT systems, how to develop applications, implement them, and what to do with the data collected.
17. *Product design:* Consumers and customers will increasingly expect products to come with embedded sensors that can connect to mobile devices and IoT systems.
18. *Network bandwidth:* As more devices come online, networks will clog and service providers will be in a never-ending battle to increase network capacity.
19. *Vertical clouds:* Aggregating all this big data and acting on its findings will be best achieved by capturing, analyzing, and responding from the cloud. In 2015, expect specialized and vertical IoT cloud services.
20. *Industry partnerships:* Traditional IT vendors will accelerate their partnerships with global telecom service providers, semiconductor vendors, and vertical IoT platform providers.

A 3-year forecast for trends related to the IoT says cloud is the platform of choice for integrating data from machines and IT systems. New security risks will emerge, and wearables will gradually replace smartphones. Intelligent maintenance is probably the most tangible scenario to illustrate what the IoT is, and what it will be capable of: IT systems constantly evaluate sensor data, providing key information about machine components that may need to be replaced. This reduces machine downtimes immensely. Yet the IoT is predicted to permeate almost every industry. Data can be collected from practically any conceivable source, including traffic, sports, and at home. The central challenge of the next few years will be to ensure the security of this new data, and to derive meaningful conclusions from it. The most important trends according to International Data Corporation (IDC) (Schmitz, 2015) are:

- *Cloud as the central platform.* The core challenge for companies will be data blending. Data is generated in a range of formats, but ultimately intended for holistic analyses.

- *New security risks in IT networks.* Within the next two years, 9 out of 10 IT networks will be confronted with new security threats.
- *Increased network proximity.* IDC analysts predict that by 2018, the storage, processing, and analysis of data will take place in close proximity to the networks in question.
- *Need for additional network capacity.* About 50% of all companies will be forced to adapt their IT networks to accommodate immense increases in data traffic.
- *New business models shape infrastructure.* Collaboration with an ecosystem of service providers, and joint development of strategic roadmaps and IoT business models together with partners are only possible if you have a flexible infrastructure.
- *All industries will profit.* The production, transportation, smart city, and apps for end-user industries currently account for more than 50% of all IoT projects.
- *Smart cities blazing the trail.* In 2018, governments will spend more than a quarter of their total budgets on developing IoT-based solutions. There will be cloud-based IoT solutions, and the public sector will increasingly work together with the private sector. From a technical standpoint, the importance of location services, digital maps, and satellite imaging services will play an important role.
- *Industry-oriented platforms will emerge.* Currently, 60% of all IT solutions are proprietary, meaning they are available only to user groups with the corresponding licenses. Developments in the IoT market are predominantly open source solutions. Standardizing data transfer will be essential to guarantee effective interfacing between machines and IT systems.
- *Wearables are the smartphones of tomorrow.* Within the next 5 years, wearables such as smart watches and smart glasses will have permeated the market.

In Harper (2015) opinion, 2015 will bring significant refinement in the application of the IoT in all of its realms—the industrial Internet, mobile, public, and private consumer convenience uses:

- *Improving the industrial Internet.* The industrial Internet sector of the IoT is responsible for its continued expansion for the past several years. The primary application of the industrial Internet is asset management, in which a sophisticated series of real-time and predictive analytics continuously monitors equipment and provides timely information pertaining to diagnostics, maintenance, failures, and more. Additional benefits include selling equipment monitoring capabilities to OEM's and their customers.
- *Throughout the enterprise.* The true potential of the IoT is in the absorption and transmission of machine-generated data to produce machine-generated responses that create action. Outside of the industrial Internet, the sector of the IoT that is most likely to take advantage of this potential in 2015 involves enterprises that rely upon operational intelligence for real-time data to generate revenue. The factor that is increasing their ability to do so, and slowly boosting the adoption rate for the IoT in general, is the growing number of analytics vendors that are specializing in instantaneous analytics on time-sensitive data. Many of these vendors offer tools that issue alerts, generate notes, and inform users about their data and its relevance to business objectives.
- *Mobile.* At present, a relatively modest section of the IoT is predicated on mobile technology. The most eminent of these technologies are centered on smartphones and marketing. The sensors in these devices are accessed by marketing departments, which issue certain promotions based on customer and product location, typically for retail ventures. However, as mobile adoption rates increase due to advances in analytics platforms with greater security and overall functionality, the relevance of mobile technologies for the IoT should diversify its applications.

- *Secure cloud analytics.* Other than integration with other data sets, enterprise-wide deployments of the IoT to achieve business objectives hinge on the confluence of analytics, the cloud, and security concerns.
- *Personal consumerization.* Other than the aforementioned mobile use cases, these include conveniences associated with smart homes such as increased video functionality, smart televisions, and a host of other novelty devices that will probably remain costly and in low demand. More development is likely to be seen in the public sector, in which technologies for smart cities including cameras on law enforcement officials, energy reduction applications regarding lighting and traffic management, and water management will evince themselves but not garner widespread adoption until after 2015.
- *Machine generated responses.* The true potential of the IoT not only involves connectivity between a growing number of devices, but the ability to derive action from that connectivity. In the industrial Internet, that action frequently takes the form of an alert or the call for an end user to take preventative or maintenance measures. The IoT will gain widespread adoption throughout the enterprise when the connectivity between machines is used to generate machine-generated responses that create action. There are some instances in which this functionality is possible, such as marketing tools that base customer offers on their locations and that of products.

These main trends related to the IoT (Chamberlin, 2015; Schmitz, 2015; Harper, 2015) in the built environment can be further analyzed in more detail, which will be done using new strategic choices that companies face (Porter and Heppelmann, 2014) and four key ways the IoT will change the face of modern marketing (Rossi, 2015) as an example.

Porter and Heppelmann (2014) research reveals that in a smart, connected world companies face 10 new strategic choices:

1. Which set of smart, connected product capabilities and features should the company pursue? Smart, connected products dramatically expand the range of potential product capabilities and features. Companies may be tempted to add as many new features as possible, especially given the often low marginal cost of adding more sensors and new software applications, and the largely fixed costs of the product cloud and other infrastructure. But just because a company can offer many new capabilities does not mean that their value to customers exceeds their cost.
2. How much functionality should be embedded in the product and how much in the cloud? Once a company has decided which capabilities to offer, it must decide whether the enabling technology for each feature should be embedded in the product (raising the cost of every product), delivered through the product cloud, or both.
3. Should the company pursue an open or closed system? Smart, connected products involve multiple types of functionality and services, and are often systems encompassing multiple products. A closed-system approach aims to have customers purchase the entire smart, connected product system from a single manufacturer. Key interfaces are proprietary, and only chosen parties gain access. An open system, by contrast, enables the end customer to assemble the parts of the solution—both the products involved and the platform that ties the system together—from different companies. A hybrid approach, in which a subset of functionality is open but the company controls access to full capabilities, occurs in industries like medical devices, where manufacturers support an industry standard interface but offer greater functionality only to customers.

4. Should the company develop the full set of smart, connected product capabilities and infrastructure internally or outsource to vendors and partners? Developing the technology stack for smart, connected products requires significant investment in specialized skills, technologies, and infrastructure that have not been typically present in manufacturing companies. Many of these skills are scarce and in high demand. A company must choose which layers of technology to develop and maintain in-house and which to outsource to suppliers and partners. In utilizing outside partners, it must decide whether to pursue custom development of tailored solutions or license off-the-shelf, best-of-breed solutions at each level. Our research suggests that the most successful companies choose a judicious combination of both.
5. What data must the company capture, secure, and analyze to maximize the value of its offering? Product data is fundamental to value creation and competitive advantage in smart, connected products. But collecting data requires sensors, which add cost to the product, as does transmitting, storing, securing, and analyzing this data. Companies may also need to obtain rights to the data, adding complexity and cost. To determine which types of data provide sufficient value relative to cost, the firm must consider questions such as: How does each type of data create tangible value for functionality? For efficiency in the value chain? Will the data help the company understand and improve how the broader product system is performing over time? How often does the data need to be collected to optimize its usefulness, and how long should it be retained? Companies must also consider the product integrity, security, or privacy risks for each type of data and the associated cost.
6. How does the company manage ownership and access rights to its product data? As a company chooses which data to gather and analyze, it must determine how to secure rights to the data and manage data access. The key is who actually owns the data. The manufacturer may own the product, but product usage data potentially belongs to the customer. There is a range of options for establishing data rights for smart, connected products.
7. Should the company fully or partially disintermediate distribution channels or service networks? Smart, connected products enable firms to maintain direct and deep customer relationships, which can reduce the need for distribution channel partners. Companies can also diagnose product performance problems and failures and sometimes make repairs remotely, reducing reliance on service partners. By minimizing the role of the middlemen, companies can potentially capture new revenue and boost margins. They can also improve their knowledge of customer needs, strengthen brand awareness, and boost loyalty by educating customers more directly about product value.
8. Should the company change its business model? Manufacturers have traditionally focused on producing a physical good and capturing value by transferring ownership of the good to the customer through a sales transaction. The owner is then responsible for the costs of servicing the product and other costs of use, while bearing the risks of downtime and other product failures and defects not covered by warranties. Smart, connected products allow the radical alteration of this long-standing business model. The manufacturer, through access to product data and the ability to anticipate, reduce, and repair failures, has an unprecedented ability to affect product performance and optimize service. Companies can also pursue hybrid models between the extremes of product-as-a-service and conventional ownership, such as product sales bundled with warranty or service contracts, or product sales bundled with performance-based contracts.
9. Should the company enter new businesses by monetizing its product data through selling it to outside parties? Companies may find that the data they accumulate from smart, connected products is valuable to entities besides traditional customers. Companies may also discover

that they can capture additional data, beyond what they need to optimize product value, that is valuable to other entities. In either case, this may lead to new services or even new businesses.

10. Should the company expand its scope? Smart, connected products not only transform existing products but often broaden industry boundaries. Products that have been separate and distinct can become parts of optimized systems of related products, or components of systems. Shifting boundaries mean that companies that have been industry leaders for decades may find themselves playing more of a supporting role in a broader landscape.

These choices involve trade-offs, and each must reflect a company's unique circumstances. The choices are also interdependent. The company's entire set of choices must reinforce one another and define a coherent and distinctive overall strategic positioning for the company (Porter and Heppelmann, 2014).

Without the ability to gather, analyze, and react to customer data in real time, marketers will be unable to fully leverage the IoT. The growth in the number of devices connected to the IoT is set to be explosive, with Cisco and EMC suggesting we'll see between 50 and 200 billion by 2020. The impact of so many devices, network sensors, and wireless access points coming online, and the huge volumes of data subsequently created, will have a dramatic effect on the marketing industry. Here are four key ways the IoT will change the face of modern marketing (Rossi, 2015):

- Sharper campaign lifecycles. The IoT will see a huge range of devices coming online, with anything from a wristwatch to a tractor, a factory pallet, or even a simple network sensor increasingly generating data for companies to capture. The ability to process this data autonomously will truly empower marketers by enabling them to analyze customer feedback in real time and derive actionable insights from multiple data sources. This level of instant response will allow companies to improve the efficiency of marketing campaigns while reducing the cost. With the ability to analyze large data streams at such speed, marketing professionals can judge the success of a campaign within hours and days, rather than weeks or months. Rapid campaign analytics mean that dud-campaigns can be killed, while those with the potential to go viral are given the required backing. Meanwhile, the supply chain needed to deliver on the promises of marketing campaigns can be instantly aligned with the projected level of demand based on customer feedback.
- New and surprising bedfellows. According to EMC, the amount of data being generated annually by an expanding IoT network is predicted to skyrocket to 44 zettabytes by 2020; a staggering number considering that, throughout history, humanity created a total of 2.7 zettabytes until 2013. The sheer volume of data being generated will require strategic partnerships between marketers, data specialists, and IoT enablers, such as telcos, in order to maximize the utility of customer data. A marriage between huge data-gathering organizations and analytical specialists could well resemble the IoT itself in becoming greater than the sum of its parts, at least in terms of value creation.
- A new world of click-through-rates (CTRs). The IoT is likely to spell the end for large swathes of mass-media marketing. Personalized customer data generated by wearables and smart devices in the home will mean that irritating pop-ups and banner adverts with CTRs of less than 1% will be replaced. Instead, we will see more targeted, analytics-powered adverts, as companies identify exactly what their customers want and when they want it. For example, an IoT-enabled kitchen will be able to detect a blocked sink and message the home owner's phone to alert them to the issue. At the same time, it will be

able to suggest several solutions, such as local plumbers or drain unblocking liquids, along with relevant coupons, thereby maximizing the value of each advert. In order to respond to this, marketers will need to use streaming analytics, applied to live, real-time data to create personalized advertising strategies designed to arrive at the moment they will have the greatest impact.

- True 360 degree customer profiles. Marketers are already focusing on building the most complete customer profiles they can using analytics and segmentation. However, when only using data from a small section of customers' lives, such as their shopping habits, this is inevitably limited. New high-volume data streams, coupled with advanced customer analytics and personalization algorithms means 360-degree profiles will be dramatically improved.

Analytics is already helping companies connect with customers on a new level by knowing their favorite brands, habits, and what offers they might find most interesting. With the IoT set to create never-before-seen quantities of data providing countless insights into customer behavior, marketers have a golden opportunity to instantly offer individual customers what they need, when they need it, creating a major competitive advantage (Rossi, 2015).

17.6 Internet of Things start-ups

It is necessary to comprehensively analyze and make rational decisions at micro-, meso- and macrolevels in the effort to boost competitiveness and avoid start-up failure. This must not only encompass economic, political, legal, and institutional types of decisions but also other qualitative aspects of start-up and related management issues, including cultural (Kosoff, 2015; Kotak, 2015), social (Kim and Shin, 2016; Ning et al., 2015; Rau et al., 2015; Atzori et al., 2012), ethical (Bradley et al., 2015; Weber, 2013; Hanson, 1998), psychological (Lamprinakos et al., 2015; Zhuge, 2011), trust (Bao and Chen, 2012; Bao et al., 2013; Xiong et al., 2011; Yan et al., 2014), and other such aspects. For example, the group of experts on the IoT has analyzed in subgroups several issues of importance in connection with the IoT, namely architecture, identification, privacy and security, standards, governance, as well as ethics. In the public consultation, responses supported the inclusion of ethical elements into the IoT debate, in particular elements such as personal identity, autonomy of individuals, user consent, fairness, and social justice. Obviously the difficulty consists in the search of finding an agreement regarding the level of ethical standards that might be different according to industrial and societal environments. A special part of the questionnaire looked at procedural issues in ethics, opening the debate of what measures would have to be adopted in order to properly take into account the ethical aspects in the design and development of the IoT. Civil society representatives have been skeptical in assuming that an ethical charter or another form of self-regulation would be sufficient to cover the needs in ethics since respect by IoT providers could not be enforced. Therefore, many respondents have called for regulatory oversight and governance (Weber, 2013). Examination of the best practices of these start-ups reveals several key steps new ventures can take to make ethics a distinguishing mark of the start-up's culture (Hanson, 1998):

- Ethical start-ups recognize the ethical dilemmas that surround them in the first few months. The pressures to cut ethical corners are great in a start-up. How much puffery do you use in presenting your idea to venture capitalists? How do you divide stock ownership and options fairly among the founding team and later hires? How reliable does a product have to be before you ship it? How creative can you be in your accounting when the value of your stock is so sensitive to a stumble? When a deal falls through, how quickly do you tell your board and your funders? How generous can you afford to be in employee benefits in the early days?
- Ethical start-ups make ethics a core value of the enterprise. Start-up founders have discovered that they must explicitly embrace doing business ethically to counter the temptations to fudge various standards. Ethics should appear in business plans, in company mission statements, and in all other company documents.
- The ethical entrepreneur finds early opportunities to make his or her ethical commitment real. A Silicon Valley entrepreneur who took over a months-old company that refused to send faulty financial data to the venture capitalists, over the objections of his new team. "You just don't do business that way," reflects the entrepreneur, who enjoys both financial success and a superb reputation today. He communicated clearly from that day the ethical standards he and the company would follow.
- The ethical entrepreneur anticipates the ethical tensions in day-to-day decisions. As business plans are written and product capabilities are described, the ethical tension between the truthful and the "hopeful" is inevitable. As a start-up tries to attract top talent, there is an unavoidable ethical tension in determining how rosy a picture to draw for the prospect. The ethically thoughtful entrepreneur anticipates these tensions and talks about them with the team before the situations are confronted. In later years of a company's life, this practice will become more formal "ethics training."
- The ethical entrepreneur welcomes ethical questions and debates. Some situations cannot be anticipated, and the ethical entrepreneur must always keep an open door so that new ethical issues can be worked out. Even the willingness to take time to discuss and resolve tough ethical dilemmas gives the signal that ethics is important in the start-up.
- The ethical entrepreneur is watchful about conflicts of interest. It is hard to single out one area of particular ethical concern in start-ups because there are so many of importance. However, the world of high-tech start-ups emphasizes partnerships, strategic alliances, and "virtual relationships." These arrangements are rife with opportunities for conflicts of interest where an entrepreneur or start-up employee can line his or her own pockets to the detriment of the organization. An early and consistent stand against questionable conflicts of interest is an important dimension of a start-up ethics effort.
- The ethical entrepreneur talks about the ethical values all the time. The frantic pace of start-ups and their rapid growth create short memories and a staff that is often very new to the enterprise. Only by continually articulating the ethical commitment can the entrepreneur be sure the members of the organization, particularly new hires, understand the ethical commitment and know it is real.
- The ethical entrepreneur weeds out employees who do not embrace the ethical values of the company. Hiring is among the most important strategic steps a start-up takes. Inevitably, the venture will hire some individuals who believe financial success, perhaps just personal financial success, is the only value. The ethical entrepreneur is on the lookout for "teammates" who do not share the company's values and weeds them out before they can do damage to the reputation or culture of the firm.

- The ethical entrepreneur looks for opportunities to engage the company in the community. The start-up's preoccupation with meeting product and financial goals and with its own growth can lead to blindness about anything other than personal gain. Ethical entrepreneurs find ways to engage the team in community service and to emphasize the continuing importance of the team's family relationships.
- The ethical entrepreneur takes stock occasionally. Just as the entrepreneur must keep an eye on the start-up's cash flow and produce a balance sheet periodically, so he or she must also take stock of the company's commitment to its ethics and other values.
- The ethical entrepreneur renews the commitment to ethical behavior. Companies change as they grow. The most pressing ethical dilemmas of a $10 or $100 million company differ from those of a fledgling start-up. Ethical values and the commitment to ethical behavior must be recast and recommunicated periodically, preparing the company and its employees to deal with the ethical dilemmas currently faced.

Specific interesting business opportunities and recommendations for increasing start-up efficiency and avoiding start-up failure and minimizing its consequences in defined areas are presented next.

Small business website www.Start-ups.co.uk has identified the most interesting business opportunities over the next 12 months. Growth areas and the best businesses to start in 2014 included niche price comparison marketplace, boutique care home, virtual assistant, independent BYOD consultant, personal trainer, and analytics (Wallis, 2014).

The IoT and related issues (cloud–client computing, machine learning, and big data) are among the main start-up trends that entrepreneurs should look to exploit in 2015 (Kosoff, 2015):

- The IoT. The solution to our society's culture of planned obsolescence—where you constantly have to replace old or broken devices because they're not built to last forever—is the IoT. The IoT could change things here, and create a new culture of repair.
- If you're a small family-owned restaurant that can't afford to constantly upgrade equipment or fix things, you can answer a whole new set of questions with the IoT: Is that freezer working extra hard because someone left the door open, or because its compressor is about to fail and you're about to lose $6000 in food?
- Machine learning and big data. Where business intelligence before was about past aggregates (How many properties have we sold in Kentucky?), it will now demand predictive insights (How many properties will we sell in Kentucky?). An important implication of this is that machine learning will not be an activity in and of itself—it will be a property of every application. There will not be a standalone function.
- Cloud–client computing. We have more processing power in our hands today through smartphones than we did in large computers decades ago. So why shouldn't some of this processing move out of the cloud and back into the end point, into the phone? Doing processing locally has its advantages. For instance, the cost of an end-point CPU and memory is a 1000 times cheaper than the cost of CPU and memory in the server. And in many places around the world, connectivity and transmission costs are sometimes far more expensive than the device.

As Internet connectivity gets embedded into every aspect of our lives, investors, entrepreneurs, and engineers are rushing to cash in. According to Gartner, IoT vendors will earn more than $309 billion by 2020. However, most of those earnings will come from services. Gartner also estimates that by 2020, the IoT will consist of 26 billion devices. All of those devices, Cisco believes, will end up dominating the Internet by 2018. In less time than it takes to earn a college degree (much less time these days), machines will communicate over the Internet a lot more than people do. With the IoT space in full gold-rush mode, it evaluated more than 70 start-ups to find the best that look poised to help shape the future of IoT in the built environment. These IoT are listed in alphabetical order and are not ranked (Vance, 2014):

- Chui. Combine facial recognition with advanced computer vision and machine learning techniques to turn faces into universal keys. Home and business security systems are expensive, generate tons of false alarms, and really cannot be called intelligent. Chui's facial recognition technology replaces keys, passwords, or codes, allowing you to disarm a security system with facial recognition. Chui emphasizes that our faces are unique, universal, and nontransferable. Your features cannot be hacked, nor can they be spied on. For businesses, or even homes with service people stopping by regularly, Chui also keeps track of who is coming and going, documenting visitors, and time-stamping their visits. Chui ensures you have complete control over who is entering your home through real-time notifications and the ability to engage in live conversations with visitors.
- Enlighted. Lighting is an immense cost to building operators, as well as a large factor for employee comfort. Lighting consumes 25–40% of commercial office building electricity and is a major factor in worker comfort and productivity, yet more than 90% of existing buildings have little more than a wall switch. Enlighted has created people-smart sensors that gather real-time environmental data and analytics at each light fixture within a building, while networking those sensors in a way that delivers even more value to building owners and operators. Enlighted's sensors not only control lights, but also monitor light levels, temperature, occupancy, and power consumption for the 100 square feet of floor space directly beneath each of them. That means that each sensor knows not only what its individual light is doing, but also what it should be doing. In other words, instead of relying on the vagaries of wireless networking architecture to carry data back to a central control platform, come up with preferred control options, and then send those commands back to light fixtures, Enlighted's nodes can handle most tasks on their own. The app detects heat and movement within conference rooms, so that employees can easily look on an app to see which conference rooms are occupied. This cuts down time employees constantly spend circling an office building looking for a space to hold a meeting. Enlighted is also working on using its sensor data for other things, like integrating their data with building HVAC systems to fine-tune heating and cooling, or to respond to utility demand response calls to shed power use when the grid is under stress.
- Revolv. Revolv unifies control of your smart home via a smartphone or tablet app. The best bet for consolidation is the smartphone, which has already become the de facto command-and-control center of our connected lives. As more and more connected home products hit the market, managing them all means either remote-control sprawl will worsen until it's akin to kudzu along a southern highway, or we'll consolidate management via smartphones. Consumers, however, are already showing signs of app fatigue and long for simplicity. They are confused by the onslaught of connected products, each with different (often incompatible) wireless technologies and control interfaces. Revolv intends to solve this issue by unifying consumers' connected devices through one simple app that helps smart home products work together. Revolv was purpose-built to take the complexity out of connecting, controlling, and automating a smart home. It also has the capability to automate devices used in daily routines through time, place, and sensor triggers. For instance, with its GeoSense technology, Revolv can automatically activate (or shut down) connected devices when the user reaches a certain geo-radius to and from their home. Revolv incorporates seven radios

in its hub and speaks 10 different wireless languages, giving it the ability to support the most popular brand-name smart devices available today.
- TempoDB. TempoDB provides a cloud-based sensor data analytics backend for IoT and M2M. The opportunities to reduce cost or increase revenues by leveraging IoT are well understood, but the challenges in making sense of the data generated by sensors are unlike any big data challenges yet encountered. TempoDB argues that without purpose-built, integrated sensor analytics solutions, businesses will struggle to realize the tremendous promise of a connected and measured world. TempoDB provides proprietary real-time sensor data monitoring and analytics engines, which are integrated into its custom data store, to ensure performance and security as users scale up to (potentially) millions of devices and sensors.

To mitigate the chances of failure, here are four start-up best practices to help you along (Boss, 2014):

- Maximize your resources. In this day and age, obtaining information isn't the challenge—it's sifting through the endless streams of data, insights, and email exchanges that soak up more time than the analyses themselves. Meanwhile, trying to navigate the rough waters of information overload wastes precious time. Know when you need to tackle an issue and when it makes more sense to delegate. Building a better business with outsourcing tools such as Amberly or Zapier is a great way to keep the grind wheel churning when your energy reserves are low.
- Leverage your network. There are five types of people in your network: family, friends, uppers, downers, and influencers. Family and friends speak for themselves, and may also serve as mentors or role models. Uppers is a term for generally positive people. Their evil twins, however, are downers, who are the human toxins that act as social hand grenades because of their negativity. Finally, there are influencers. This latter group is key because they are the ones who can connect you with others and others to you. Remember also that everybody is a mix of the aforementioned types. Nobody is just an influencer or just a friend—everybody knows something worthwhile. The key is to know when to leverage the right people at the right time.
- Build a learning culture. Culture is a product of the people, processes, and systems that define the company's name and all that it stands for. You don't change a culture without changing behavior, and you only change behavior at scale through the systems and processes that guide the company's daily execution. The only bridge to start-up sustainability is how willing you are, as a leader and a company, to adapt to changes in the environment. Success or failure is only determined by where you stop.
- Have an MVP. No, not an individual. A minimum viable product is the least amount of product or service you can bring to market while achieving two objectives: maximizing value to the customer and minimizing costs. Good judgment only comes from experience, and experience typically comes from bad judgment. The toughest lessons to learn are usually the most costly in terms of resources and capital, so the best practice for you is the one that keeps your business unique.

Below are seven tips that are the bedrock for any successful start-up venture (Little, 2014):

1. Fiercely pursue the most impactful thing. While your many to-do's—emails, meetings, people, opportunities—all seem important, none of them will matter unless you nail down the most impactful activities. Be honest with yourself about what tasks only you can do and

what tasks others could do. Many start-ups fail because they cannot allocate their limited time to the right things.

2. Dare to be remarkable. The question to ask about everything related to your business—your product, your marketing, your interactions with customers—is, Is this remarkable? Your offering should be surprising to people, attention grabbing, and should break through all the noise. Being remarkable in all aspects of your business is the ultimate competitive advantage.
3. Start with the "cool kids." When you start with the right core of team members, the rest of the band falls in line.
4. Out-teach the competition. Sales is the lifeblood of any company and you should have at least one phenomenal salesman at your company's core.
5. Don't be afraid to move the boat. When customers seem hard to catch, it might be because you are fishing in the wrong spot. Be humble enough to relocate your focus, and do it before you've burned through your capital. Opportunity costs are exaggerated early in a company's life cycle, so it's important to be nimble and willing to adjust on the fly.
6. Make sure your medicine actually cures the disease. Entrepreneurs often stop innovating once customers are interested and willing to pay for a solution. If you want to last for the long haul, be honest about whether or not your medicine actually cures the disease. Customers today are more informed and connected than any other time in history, and your company will ultimately be exposed if it does not deliver long-lasting solutions.
7. Experience is everything. Be fanatical about every little detail of the customer experience. This means spending the money to hire the best user and customer-experience designers you can find. Your first customers will serve an immensely valuable role as brand ambassadors if you treat them right.

Here are the top five mistakes made in home automation products and what can be learned from them (Kotak, 2015):

- Mistake 1: We were neither experts nor target users of the product that we were building. We had never used the existing home automation products in our homes. We were not experts in the IoT sector. Had we been users of existing smart switches, we would have known that the incremental value that our product was offering was quite low. Had we been experts in IoT, we would have known how to price hardware and the difficulties in building it. By avoiding this mistake, you can avoid a lot of other mistakes that happen as a result of this one. Lesson: Work on something where you are either an expert or a top user. If not, become an expert/top user. Homejoy founder Adora Cheung, for example, worked as a professional cleaner to understand the business.
- Mistake 2: We did not do the due diligence on the idea before we started building the product. We did not understand the market and competition well enough. We also did not figure out the persona of our customer—and whether that customer was looking for the value that we were providing. Lesson: Make a thorough list of hinge-breaking assumptions for your market, product, and competition. Hinge-breaking assumptions are those that can make or break your company. Rank them according to probability of the assumption being wrong and of subsequent risk to the company. Start validating from the top while building as little as possible.
- Mistake 3: We let sunk-cost bias affect our decisions about pivoting. It was not that we were clueless about the problems in our product. We had doubts in our minds. In a start-up, you almost always have doubts. But we had built so much. Is it okay to voice your doubts and bring the team morale down? Build a culture of transparency in your company. Encourage dissent among cofounders and deal with it objectively.

- Mistake 4: We were trying to do everything for everybody. We were making switches that could automate your lights, fans, air conditioners, and water heaters. We would have tried to automate your TV, fridge, oven, and car as well had it been feasible to do so. We were pitching power savings as well as luxury. This made the product and the pitch very complicated. Lesson: As a start-up, your resources are limited. So it is always better to identify and solve one problem very well instead of solving numerous problems in a mediocre way.
- Mistake 5: We underestimated hardware. Building a successful start-up is hard. Building a hardware start-up is 10 times harder. Building a prototype is the easiest part of building a hardware start-up. The real challenge comes in product design, production engineering, manufacturing, distribution, and marketing/sales. Also, hardware product validation and iteration cycles are much longer than software ones. Getting funding is relatively difficult. Managing cash flow is hard because you have to pay your vendors months before you get paid by your customers. Lesson: Understand what you are getting into if you are starting a hardware company and plan accordingly. Get experienced people on your team.

Acknowledgments

The chapter was written in the course of the TEMPUS project, "Reformation of the Curricula on Built Environment in the Eastern Neighbouring Area"; COST Action TU1204, "People Friendly Cities in a Data Rich World"; and "Collaborative Reformation of Curricula on Resilience Management with Intelligent Systems on Open Source and Augmented Reality."

References

Ashton, K., 2009. That "Internet of Things" thing. In the real world, things matter more than ideas. RFID Journal. http://www.RFIDjournal.com/articles/view?4986 (accessed 02.09.15.).

Atzori, L., Iera, A., Morabito, G., Nitti, M., November 14, 2012. The Social Internet of Things (SIoT) – when social networks meet the Internet of Things: concept, architecture and network characterization. Computer Networks 56 (16), 3594–3608.

Bao, F., Chen, I., 2012. Trust management for the Internet of Things and its application to service composition. In: Proceedings of the IEEE International Symposium on World of Wireless, Mobile and Multimedia Networks (WoWMoM), 2012, pp. 1–6.

Bao, F., Chen, I., Guo, J., 2013. Scalable, adaptive and survivable trust management for community of interest based Internet of Things systems. In: Proceedings of the IEEE eleventh International Symposium on Autonomous Decentralized Systems (ISADS), 2013, pp. 1–7.

Boss, J., 2014. 4 Best Practices to Avoid Start-Up Failure. http://www.entrepreneur.com/article/239348.

Bradley, D., Russell, D., Ferguson, I., Isaacs, J., MacLeod, A., White, R., April 2015. The Internet of Things – the future or the end of mechatronics. Mechatronics 27, 57–74.

Chamberlin, B., 2014. Internet of Things (IoT): Important Enabling Technologies. http://ibmcai.com/2014/10/16/internet-of-things-IoT-important-enabling-technologies/ (accessed 03.09.15.).

Chamberlin, B., 2015. Twenty Internet of Things Trends to Watch in 2015. http://ibmcai.com/2015/01/27/twenty-internet-of-things-trends-to-watch-in-2015/ (accessed 09.09.15.).

Clarke, A., Steele, R., 2014. Local processing to achieve anonymity in a participatory health e-Research system. Procedia – Social and Behavioral Sciences 147, 284–292.

Crawford, S., 2014. Chicago Is Your Big (Friendly) Brother. http://www.bloombergview.com/articles/2014-06-19/chicago-is-your-big-friendly-brother (accessed 09.09.15.).

Decentlab, 2015. Wireless monitoring systems in the field of civil engineering. Storchenbruecke. http://www.decentlab.com/storchenbruecke (accessed 09.09.15.).

Digiesi, S., Facchini, F., Mossa, G., Mummolo, G., Verriello, R., 2015. A cyber - based DSS for a low carbon integrated waste management system in a smart city. IFAC-PapersOnLine 48 (3), 2356–2361.

Ding, L.Y., Zhou, C., Deng, Q.X., Luo, H.B., Ye, X.W., Ni, Y.Q., Guo, P., 2013. Real-time safety early warning system for cross passage construction in Yangtze Riverbed Metro Tunnel based on the Internet of Things. Automation in Construction 36, 25–37.

Elmaghraby, A.S., Losavio, M.M., 2014. Cyber security challenges in smart cities: safety, security and privacy. Journal of Advanced Research 5 (4), 491–497.

Enevo, 2015. Waste Collection for Smart Cities. Enevo One. http://cdn.enevo.com/wp/wp-content/uploads/2013/05/enevo_2015_eng_web.pdf (accessed 09.09.15.).

Fielding, R.T., 2000. Architectural Styles and the Design of Network-Based Software Architectures (Dissertation). University of California.

Friess, P., 2012. Towards dynamism and self-sustainability. In: Smith, I.G. (Ed.), The Internet of Things. New Horizons, Halifax, pp. 12–21. Chapter 1.

Friess, P., Woysch, G., Guillemin, P., Gusmeroli, S., Sundmaeker, H., Bassi, A., Eisenhauer, M., Moessner, K., 2012. Europe's IoT Strategic Research Agenda 2012. In: Smith, I.G. (Ed.), The Internet of Things. New Horizons, Halifax, pp. 22–118. Chapter 2.

Handsome, C., 2015. Oxford Flood Network. http://oxfloodnet.co.uk/ (accessed 10.09.15.).

Hanson, K.O., 1998. Implementing Ethics Strategies in Organizations. Monograph. Society of Management Accountants of Canada.

Harper, J., 2015. Trends for the Internet of Things. http://www.dataversity.net/2015-trends-internet-things/ (accessed 07.09.15.).

Henze, M., Hermerschmidt, L., Kerpen, D., Häußling, R., Rumpe, B., Wehrle, K., 2015. A comprehensive approach to privacy in the cloud-based Internet of Things. Future Generation Computer Systems. http://dx.doi.org/10.1016/j.future.2015.09.016.

Höller, J., Tsiatsis, V., Mulligan, C., Karnouskos, S., Avesand, S., Boyle, D., 2014. From Machine-to-machine to the Internet of Things. Academic Press, 352 pp.

IERC definition, 2014. W3C Semantic Sensor Network Incubator Group Report. http://www.w3.org/2005/Incubator/ssn/XGR-ssn-20110628/ (accessed 12.04.14.).

Jaffe, E., 2014. New York-based placemeter is turning disused smartphones into big data. http://www.citylab.com/tech/2014/08/the-view-from-your-window-is-worth-cash-to-this-company/375471/ (accessed 12.09.15.).

Jamil, M.S., Jamil, M.A., Mazhar, A., Ikram, A., Ahmed, A., Munawar, U., 2015. Smart environment monitoring system by employing wireless sensor networks on vehicles for pollution free smart cities. Procedia Engineering 107, 480–484.

Kim, T., Shin, D., February 2016. Social platform innovation of open source hardware in South Korea. Telematics and Informatics 33 (1), 217–226.

Kim, Y., Lee, J., Lee, H., 2014. Life-log data-based window opening and closing for individual customized services in symbiosis houses. Procedia Environmental Sciences 22, 247–256.

Kosoff, M., 2015. 16 Start-Up Trends that Will Be Huge in 2015. Business Insider. http://www.entrepreneur.com/article/242154.

Kotak, Y., 2015. 5 reasons Why My IoT Start-Up Failed. http://venturebeat.com/2015/06/16/5-reasons-why-my-iot-start-up-failed/.

Lamprinakos, G.C., Asanin, S., Broden, T., Prestileo, A., Fursse, J., Papadopoulos, K.A., Kaklamani, D.I., Venieris, I.S., July 1, 2015. An integrated remote monitoring platform towards telehealth and telecare services interoperability. Information Sciences 308, 23–37.

Lau, H., Dwight, R., 2011. A fuzzy-based decision support model for engineering asset condition monitoring – a case study of examination of water pipelines. Expert Systems With Applications 38, 13342–13350.

Lenz, C.E., 2014. Internet of Things: Six Key Characteristics. https://designmind.frogdesign.com/2014/08/internet-things-six-key-characteristics/ (accessed 03.09.15.).

Libelium, 2014. Top 50 Internet of Things Applications. http://www.libelium.com/top_50_iot_sensor_applications_ranking/ (accessed 03.09.15.).

Little, J., 2014. Don't Want to Screw up Your Start-Up? Follow These 7 Rules. http://www.entrepreneur.com/article/239309.

Littman, M., Kortchmar, S., 2014. Internet of Things. The Path to a Programmable World. Footnote. http://footnote1.com/the-path-to-a-programmable-world/ (accessed 04.09.15.).

Louchez, A., 2013. The Internet of Things – Machines, Businesses, People, Everything. ITU News. https://itunews.itu.int/En/4291-The-Internet-of-things-Machines-businesses-people-everything-.note.aspx (accessed 03.09.15.).

Lu, J., Ruan, D., Zhang, G., 2010. A special issue on intelligent decision support and warning systems. Knowledge-Based Systems 23 (1), 1–2.

Luvisi, A., Lorenzini, G., 2014. RFID-plants in the smart city: applications and outlook for urban green management. Urban for Urban Green 13 (4), 630–637.

M2M, 2015. Machine-to-Machine Communication. Adaptive Modules. http://www.adaptivem2m.com/wireless-communication/m2m-machine-to-machine-communication.htm.

McFarland, M., 2014. Five Interesting Things that Result From Gobs of Foot Traffic Data. http://www.washingtonpost.com/news/innovations/wp/2014/06/16/five-interesting-things-that-result-from-gobs-of-foot-traffic-data/ (accessed 09.09.15.).

Minneapolis, 2015. About MIMMI. http://www.minneapolis.org/mimmi/about-mimmi (accessed 09.09.15.).

Morley, M.S., Bicik, J., Vamvakeridou-Lyroudia, L.S., Kapelan, Z., Savic, D.A., 2009. Neptune DSS: a decision support system for near-real time operations management of water distribution systems. In: Proceedings of the 10th International Conference on Computing and Control for the Water Industry, CCWI 2009, pp. 249–255.

Ning, H., Liu, H., Ma, J., Yang, L.T., Huang, R., 2016. Cybermatics: cyber–physical–social–thinking hyperspace based science and technology. Future Generation Computer Systems 56, 504–522.

Park, T., Kim, H., 2013. A data warehouse-based decision support system for sewer infrastructure management. Automation in Construction 30, 37–49.

Patterson, T., 2014. Data Surveillance Centers: Crime Fighters or "Spy Machines?". CNN. http://edition.cnn.com/2014/05/26/tech/city-of-tomorrow-video-data-surveillance/index.html.

Phillips-Wren, G., Mora, M., Forgionne, G.A., Gupta, J.N.D., 2009. An integrative evaluation frame-work for intelligent decision support systems. European Journal of Operational Research 195 (3), 642–652.

Placemeter, 2015. About Placemeter. https://www.placemeter.com/about (accessed 12.09.15.).

Porter, M.E., Heppelmann, J.E., 2014. How Smart, Connected Products Are Transforming Competition. Harvard Business Review. https://hbr.org/2014/11/how-smart-connected-products-are-transforming-competition (accessed 08.09.15.).

Pyayt, A., Lang, B., Mokhov, I., Ozhigin, A., 2010. UrbanFlood, an early-warning system for climate induced disasters. In: Conference 5th Annual Conference of the International Network of Storm Surge Barrier Managers St. Petersburg, July 5–7, 2010.

Rau, P.P., Huang, E., Mao, M., Gao, Q., Feng, C., Zhang, Y., August 2015. Exploring interactive style and user experience design for social web of things of Chinese users: a case study in Beijing. International Journal of Human-Computer Studies 80, 24–35.

Ribeiro, R., Loureiro, D., Barateiro, J., Smith, J.R., Rebelo, M., Kossieris, P., Gerakopoulou, P., Makropoulos, C., Vieira, P., Mansfield, L., 2015. Framework for technical evaluation of decision support systems based on water smart metering: the iWIDGET case. Procedia Engineering 119, 1348–1355.

Rossi, B., 2015. 4 Key Internet of Things Analytics Trends for Marketing Pros. http://www.information-age.com/it-management/strategy-and-innovation/123459628/4-key-internet-things-analytics-trends-marketing-pros (accessed 07.09.15.).

Schmitt, C., Kothmayr, T., Ertl, B., Hu, W., Braun, L., Carle, G., 2014. TinyIPFIX: an efficient application protocol for data exchange in cyber physical systems. Computer Communications (In press, corrected proof, Available online 12.06.14.).

Schmitz, A., 2015. IDC Trends in 2015: How the Internet of Things Is Progressing. http://news.sap.com/idc-trends-2015-internet-things-progressing/ (accessed 07.09.15.).

Sempere-Payá, V.-M., Santonja-Climent, S., 2012. Integrated sensor and management system for urban waste water networks and prevention of critical situations. Computers, Environment and Urban Systems 36 (1), 65–80.

Shrouf, F., Miragliotta, G., 2015. Energy management based on Internet of Things: practices and framework for adoption in production management. Journal of Cleaner Production 100, 235–246.

Suñer, D., Malgrat, P., Leitćo, P., Clochard, B., 2008. COWAMA (Coastal Water Management) integrated and real time management system of urban water cycle to protect the quality of bathing waters. In: 11th International Conference on Urban Drainage, Edinburgh, Scotland, United Kingdom, August 31– September 5, 2008, pp. 1–10.

Tibken, S., 2015. Samsung, SmartThings and the Open Door to the Smart Home (Q&A). http://www.cnet.com/news/smartthings-ceo-on-samsung-being-open-apples-homekit-and-more-q-a/ (accessed 08.09.15.).

Van Rijmenam, M., 2015. Where Does the Internet of Things Come From? https://datafloq.com/read/where-does-the-internet-of-things-come-from/524 (accessed 02.09.15.).

Vance, J., 2014. 10 Hot Internet of Things Start-Ups. CIO. http://www.cio.com/article/2602467/consumer-technology/10-hot-internet-of-things-start-ups.html.

Vermesan, O., Friess, P., Guillemin, P., Gusmeroli, S., Sundmaeker, H., Bassi, A., Jubert, I.S., Mazura, M., Harrison, M., Eisenhauer, M., Doody, P., 2011. Internet of Things Conceptual Framework. http://www.internet-of-things-research.eu/pdf/IoT_Cluster_Strategic_Research_Agenda_2011.pdf (accessed 03.09.15.).

Vilajosana, I., Dohler, M., 2015. 19-Machine-to-machine (M2M) communications for smart cities. Machine-to-machine (M2M) Communications 355–373.

Wallis, I., 2014. Top 14 New Start-Up Trends for 2014. The Telegraph. http://www.telegraph.co.uk/finance/businessclub/10577944/Top-14-new-start-up-trends-for-2014.html.

Wan, S., Lei, T.C., 2009. A knowledge-based decision support system to analyze the debris-flow problems at Chen-Yu-Lan River, Taiwan. Knowledge-Based Systems 22 (8), 580–588.

Wang, Y., Zhang, W., Engel, B.A., Peng, H., Theller, L., Shi, Y., Hu, S., 2015. A fast mobile early warning system for water quality emergency risk in ungauged river basins. Environmental Modelling and Software 73, 76–89.

Weber, R.H., August 2013. Internet of things – governance quo vadis? Computer Law and Security Review 29 (4), 341–347.

Wikipedia, 2015. Built environment. https://en.wikipedia.org/wiki/Built_environment (accessed 12.09.15.).

Xiong, L., Zhou, X., Liu, W., 2011. Research on the architecture of trusted security system based on the Internet of Things.. In: Proceedings of the International Conference on Intelligent Computation Technology and Automation (ICICTA), 2, pp. 1172–1175.

Xu, G., Huang, G.Q., Fang, J., 2015. Cloud asset for urban flood control. Advanced Engineering Informatics 29 (3), 355–365.

Yan, Z., Zhang, P., Vasilakos, A.V., June 2014. A survey on trust management for Internet of Things. Journal of Network and Computer Applications 42, 120–134.

Yang, Y., Tan, W., Li, T., Ruan, D., 2012. Consensus clustering based on constrained self-organizing map and improved Cop-Kmeans ensemble in intelligent decision support systems. Knowledge-Based Systems 32, 101–115.

Zhong, R.Y., Peng, Y., Fang, J., Xu, G., Xue, F., Zou, W., Huang, G.Q., 2015. Towards Physical Internet.

Zhuge, H., April 2011. Semantic linking through spaces for cyber–physical–socio intelligence: a methodology. Artificial Intelligence 175 (5–6), 988–1019.

App programming and its use in smart buildings

18

S. Makonin
Simon Fraser University, Burnaby, BC, Canada

18.1 Introduction

Apps provide an interface between the building occupants and the underlying building automation system that is not hardware-based; (eg, office thermostat). Unlike a thermostat, a software app is very malleable and customizable. This act of creating an app versus a hardware device is less rigid and less methodical—features such as the user interface can change in midstream development. It is these properties that allow apps the potential for creating a user experience that allows occupants to use building systems successfully. Further, a well-developed app can inform the building system about a user's actions and location within a building. A smart building management system may even understand the occupant's intentions and proactively perform actions.

18.1.1 Motivating app start-ups

As we become more and more a culture of connected people, apps play an important part in accessing and interpreting the world around us. Companies whose sole focus is on the creation of apps are profitable. App companies require less startup costs than those who need to create hardware. As an app company, the focus is more of how to interconnect existing hardware and communications technologies into an app that is easy to use, yet powerful enough to be of use in a person's daily life. Interconnecting hardware from different companies is a valuable marketable advantage an app company can have.

Rather than focusing on what sensors to create, an app company can focus on the analysis of data and filter that data into a set of useful app functions and properties that create value for the app and motivation for a user to buy the app. For example, a mobile app that can track your movement and turn lights on automatically in a room if you are the first to enter, or turn them off if you are the last to leave. What is important to note is that this type of automation, or what I have called, in the past, *smart interventions* (Makonin et al., 2013a), need to be designed with great thought and care.

Well-designed apps sell and make a profit—more than ever before. January 2015 Apple announced "[t]o date, App Store developers have earned a cumulative $25 billion from the sale of apps and games" (Apple, 2015). This amount is for iOS apps and is up from $15 billion (Apple, 2014) from the previous year as shown in Fig. 18.1. Apple has not released earnings for OSX apps. However, OSX apps tend to be higher priced compared to their iOS counterparts.

Start-Up Creation. http://dx.doi.org/10.1016/B978-0-08-100546-0.00018-2

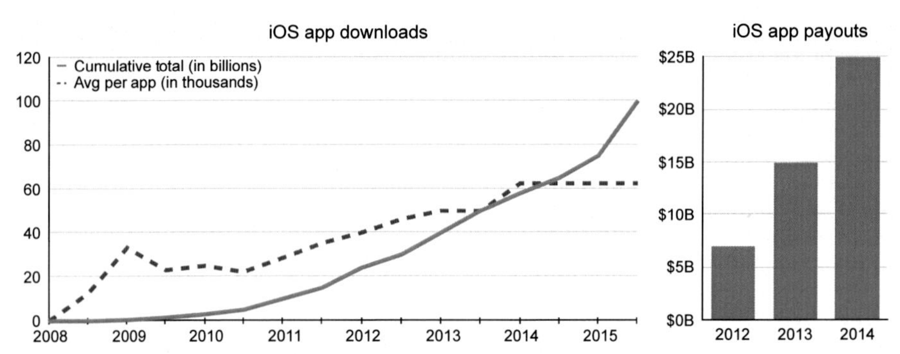

Figure 18.1 Cumulative developer community payouts from 2013 to 2015 (Apple, 2013, 2014, 2015) and app downloads data from 2008 to 2015 (see Wikipedia, 2015).

18.1.2 From building automation to smart buildings

Before 2003, building management system providers such as Johnson Controls, Siemens, Honeywell, and others have developed and distributed building automation products and services that were closed and proprietary (Mitchell, 2005). Building managers needed to choose between one of these companies and were locked-in to their products and services. Apps only existed on a building control server, which was locked away in a communications closet or server room. Only the building manager had access to these apps. If the occupants of a building found it too cold they would need to call the building manager and ask for the heat to be raised. If this was during the weekend and the building manager was not reachable then the occupants would have to work in an uncomfortable environment.

With the standardization of the communication protocols (BACnet, LonTalk, etc.) starting in 2003, building managers could start to choose different building automation products and services. This allowed them to pick the best sensors and actuators that suited their building's needs. Apps were still very much an unaddressed issue. Little existed in the way of innovation in this area. App advancement remained on the central building control server. Such advancements centered around the storage and analysis for sensor and actuator data.

After 2010 with a well-established iPhone and App Store, the iPad was introduced. The idea of the *Internet of Things* (IoT) (Want et al., 2015) also started to become a focal point—all sensors are web accessible. The advent and popularity of mobile devices created new app markets. Apps for smart buildings and building management systems is one of these new markets. There is opportunity for third-party app developers to create apps that allow occupants to have more control over their environment. There is also the opportunity for advancement in innovation as more nimble start-ups create niche apps that reside out of the thinking paradigm of these older, monolithic corporations. There is also the possibility that these corporations will outright buy start-ups with innovative apps rather than building their own apps. These acquisition scenarios are quite prevalent in the software industry. As of April 16, 2015, Google alone has acquired over 180 companies and has spent "$17 billion on hardware, software, and ad-tech companies in the last two years" (D'Onfro, 2014).

18.1.3 Remaining chapter organization

This chapter will discuss the various types of apps you might consider for your smart building management system (Section 18.2). Additionally, we discuss the importance of software methodologies, in particular discussing evolutionary delivery, and give a primer on what type of back-end platform you might want to consider using and what app-development environments (Section 18.3). We finish with conclusions (Section 18.4).

18.2 Types of apps

When we hear the word app, we might think of applications that run on our computer, smartphone, or tablet. We might not think of having apps that run on our thermostat, TV, digital photo frame, and such. In fact, apps that run in these unexpected places can enhance the value of a smart building management system. Here we explore apps in different places.

18.2.1 General building system anatomy

Fig. 18.2 shows a network/systems diagram of a generic smart building management system. This will form the basis for the discussion of this chapter. Not all the components in the diagram are necessary for a smart building management system but are there to show the basic setup of a building system. Devices that can run apps are shown (in the diagram) to create a visual context.

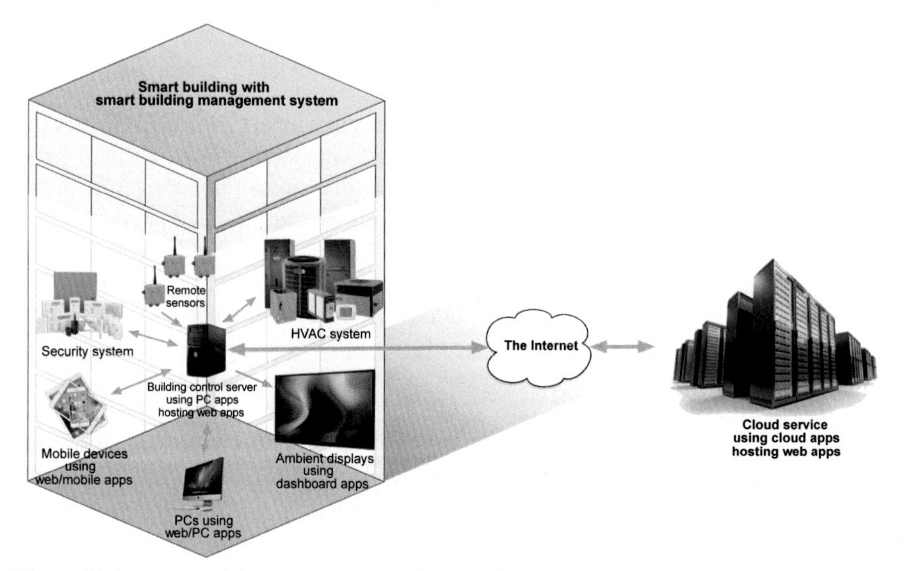

Figure 18.2 A network/systems diagram of a generic smart building management system.

Given a generic smart building management system, there are sensors that need to be monitored and actuators that need to be controlled by a *building control server* that often runs locally in the building. The control server can monitor simple sensors and communicate with more complex systems such as security and HVAC to monitor and control them. The control server has various apps that allow for the configuration of the smart building management system. Configuration can include scripts that automatically run on a schedule, as well as unscheduled events such as the handling of alerts when the system encounters an exception. The building control server may also host web apps that mobile devices and computers can use to access the smart building management system. There may also be ambient displays that have dashboards to inform occupants about the building. All the devices within the building communicate with the building control server either through a wired or wireless connection over many different communication protocols. The control server may also be a connection via the Internet to a cloud service that performs smart building management system operations remotely. In this case, the control server may simply be a gateway for data acquisition and remote command and control.

18.2.2 Native apps

Native apps are apps that are built to run directly on a personal computer (PC) or mobile device. These apps need to be installed on every PC and mobile device that needs to use the app. These apps are integrated into the operating system of the PC or mobile device and can directly use their features; for example, having an iOS app use Siri for voice-activated commands.

Apps that run on PCs have advantages but also have a number of serious disadvantages. These apps can run faster on a PC through hardware acceleration and having direct access to low-level libraries. Additionally, PC apps can store data locally and process that data locally. You can have apps that can produce very sophisticated visualization in real-time. PC apps can also run inference and prediction algorithms in consort with a visualization, allowing for alerts and notifications to be propagated locally to inform the user of unexpected issues or when an event happens. Locally run apps do not rely on an Internet connection to a cloud server to run successfully. Using a local PC can lessen the need for a significant investment with a cloud-computing infrastructure or the need to rent computing time from cloud-computing service providers (eg, Amazon EC2). It can take a significant amount of time and money to develop initially and then maintain PC apps. Costs are additive for each major operating systems (OS X, Linux, Windows) that is planned to be supported. The final issue to consider is whether the PC needs to be on all the time. Does the PC need to be securely stored and have a fail-safe start-up procedure so that on a power outage it will automatically resource itself to be fully operational.

Mobile apps that run on smartphones and tablets provide a convenient way to access data and services provided by your smart building management system. Building maintenance personnel can control the system for any location that has cellular or WiFi access without the need to be logged into the control server. This provides a faster response to any building occupant inquiries or any unexpected issues or events.

The major limitation of these types of apps is the screen size. As the size of the device's screen gets smaller, the greater the limitation of app functionality, and the app's visual design can be drastically different. Compounding this issue is the wide variety of device screen sizes, which requires thoughtful design and can increase the development time and costs involved in creating native apps for each major mobile platform (eg, iOS, Android, Windows Mobile, Tizen, etc.).

18.2.3 Cloud apps

Cloud apps are applications that run on a server cluster somewhere on the Internet. Server clusters are groups of computers that can swap and share the computational responsibility of executing a task or a request. Until recently, a company interested in having such a service needed a significant investment to build a central computing infrastructure. However, with the advent of cloud service providers, such as Amazon EC2, companies can now rent the amount of computing that is needed, when it is needed, for a small fraction of the cost of creating their own cloud. Cloud apps can be viewed as services that customers can subscribe to. This is similar to going to an app store and buying an app, except the app is installed and run in the cloud. For example, you could have an app that monitors a building HVAC system to detect faults and control the system to run optimally.

There are some issues that need to be considered when using cloud apps. Unlike PC apps that run locally, security and privacy of data need to be considered and may be a sales barrier. Sen (2013) and Zhou et al. (2010) provide a good discussion on security and privacy issues with cloud computing. Cloud apps have the benefit of being easily maintainable as they are run in one place without the need of supporting customers having to upgrade apps on their site.

18.2.4 Web apps

If you do not have the resources to create native apps that support different PC and mobile platforms you might want to consider creating only web apps. Web apps can be hosted from the building control server or on the cloud. They can adapt their visual layout and functionality based on the screen size and device that is showing them. This is often referred to as *mobile friendly* or *mobile optimized* or *responsive web design* (RWD, see Wikipedia, 2015). Issues like speed, local storage, native device features/integration can negatively affect the app's usability. However, having a single codebase that is PC- and mobile OS-independent prevents your app from having to be updated when a new version of every OS is released. You are free to release new versions of your web app without requiring all users to upgrade to the new version (Montecuollo, 2014).

18.2.5 Dashboard apps

Dashboards are apps that graphically display different information and statistics. For example, you might have a chart that shows energy savings over a period of time.

Another chart could show the amount of water saved by using a rainwater harvesting system. Dashboard apps run best on panel displays, often called *ambient displays*. Dashboard apps show the status or various metrics of the smart building; in other words, the current state the building is in.

Large ambient displays can be hung in the lobby of a building to show occupants sustainability stats (eg, energy savings). Smaller panels can be digital picture frames that sit on a desk and incorporate touch to allow a user to see different dashboards or statistics. Placement is an issue to consider when using large ambient displays. Things to consider are visibility and readability. Displays need to be placed where they are easy to see, not obstructed by other objects, and not placed where the sun can be reflected off the screen. If displays are placed high above, they need to be angled slightly downward; graphics and text need to be large enough so that occupants and visitors can easily read what is displayed on the dashboard. Information of dashboards should be compelling and motivating that convey a number of simple overview messages or charts.

Each message or chart is often referred to as a dashboard widget. In terms of visual layout, each widget occupies a rectangular space that may be different from other widgets. Good dashboard apps would allow a user to place different widgets in different places on the display without widgets overlapping each other. To enforce alignment of the widgets you might divide the display's real-estate into a grid of columns and rows and have widgets snap into alignment using these grid coordinates.

18.2.6 Ambient devices

The idea of ambient displays can be extended different forms of feedback: eco-visualizations and eco-feedback devices. These devices try to convey high-level information in a more indirect or subliminal way. Often these devices blend into the surrounding building environment and architecture.

Eco-visualization devices usually focus on conveying information as art. For example, abstract art can be used to visualize resource consumption information where the ambient display acts as a live canvas. When consumption of a particular resource increases, the art changes and its look changes dramatically by adding or removing artifacts and colors (Makonin et al., 2011). More pictorial art can be used as well. For instance, a 24-h radial clock chart can be depicted as an abstract pattern or flower with colors representing different appliance energy usage for a 24-h period (Makonin et al., 2012).

Alternately, if you are able to design electronic devices you may want to create eco-feedback devices to convey high-level information as well. For example, you can have an orb-like device that changes color based on how much energy a building is using in real-time (Makonin et al., 2013). Such devices can easily alert occupants of undesirable conditions. For instance, the same orb-like device could pulse red when a building is consuming an above-average amount of power during a period when energy prices (per kWh) are highest. This can create a sense of urgency.

How to design devices that effectively convey information is still a debated topic and forms a large research area within the human–computer interaction (HCI) research community (Bartram, 2015). If designed right, these devices create unique ways to convey information in an intuitive way using principles of visual design (Ware, 2008).

18.2.7 Agent apps

Some apps that run on the building control server may have machine-learning algorithms to infer and predict building state and occupant activity. For example, an energy savings agent could learn to detect when the building is unoccupied (Makonin and Popowich, 2012) and set the HVAC system to heat or cool less. This would allow for a reduction in energy consumption and a saving on the power bill. Such apps can reduce the payback period when purchasing a smart building management system.

Part of the app development process is to test every feature in each app that is developed. For more sophisticated apps, such as agent apps, this is even more important. The machine-learning algorithms within the app need to be tested to measure how accurately they can infer and predict. To do this, it is important to use datasets that contain data recorded from real-world buildings. High accuracy scores on synthesized data do not translate well to the real world. Measuring the accuracy of your agent app requires real-world data. A key point to remember is that agent apps that do not perform accurately will cause occupants to lose confidence in the app's ability to work properly and ultimately reflect poorly on the *smarts* of a building management system.

18.2.8 Other issues to consider

Apps that display data in an analytical way need to be mindful of best practices for visual design. This is often called *visual analytics*. Visual analytics is defined as the science of analytical reasoning facilitated by interactive visual interfaces (Thomas and Cook, 2006). How to create an interface that allows users to drill-down from summary information to details, what data to show/hide at different levels, and what charts are appropriate to use (and when to use them) are key considerations when designing data exploration apps. Visual analytics is a necessary part of any smart building management system because facility personnel need to monitor the building and analyze its data to ensure that it runs optimally. Sun et al. (2013) provide a good survey on this topic.

When designing apps, such as dashboards, make sure the options are provided that allow for customization of visual layout and what information and charts are displayed. Make sure that apps communicate securely and reliably with the smart building management system's local area network. Apps that require intelligence need to be highly accurate. Apps that run reliably, accurately, and securely create trust and satisfaction between the occupants, maintenance personnel, and the system. Satisfied customers become advocates of the smart building management system.

18.3 Methodologies for creating apps

Perhaps one of the most critical decisions for any software development project is choosing the appropriate software development methodology, of which there are many. Some methodologies such as Unified Modeling Language (Booch et al., 2005) require quite a bit of overhead in terms of person-power, software, time, and knowledge of that methodology to be followed successfully. While other methodologies like Evolutionary Prototyping (McConnell, 1996, p. 147) are light and flexible, that actually works quite well for very small development teams that want to get a product to market quickly. To begin with, if you have or have not been a software developer, in solo or in a team, you should review what methodology will suit you, your team, and the goals of your app. McConnell (1996) provides a good review of many different types of software development methodologies you can use.

18.3.1 Creating apps using evolutionary delivery

Evolutionary delivery (see Fig. 18.3 and Table 18.1) is one of the software development methodologies that I like to use for a small team of developers (2–3 people) where a focus on the visual interface and user experience is essential. McConnell (1996, pp. 425–432) discussed this methodology in great detail. One of the key market differentiators of a smart building management system would be the apps that are an interface between the users and the building system. If an end-user interface (app, device, or otherwise) is cumbersome and difficult to use, or dumbed-down, and it does not provide the right features, users will become frustrated and in some cases not use the apps that interface with the building system.

Envision app concept (see Fig. 18.3) refers to having an overall idea of what you want your app to do. One question you might ask is *whether you want one large app or a number of smaller apps to sense and control the smart building management system*? Another important question might be *what type of devices you want apps to run on*? You may decide that you want a building dashboard app that shows the status of

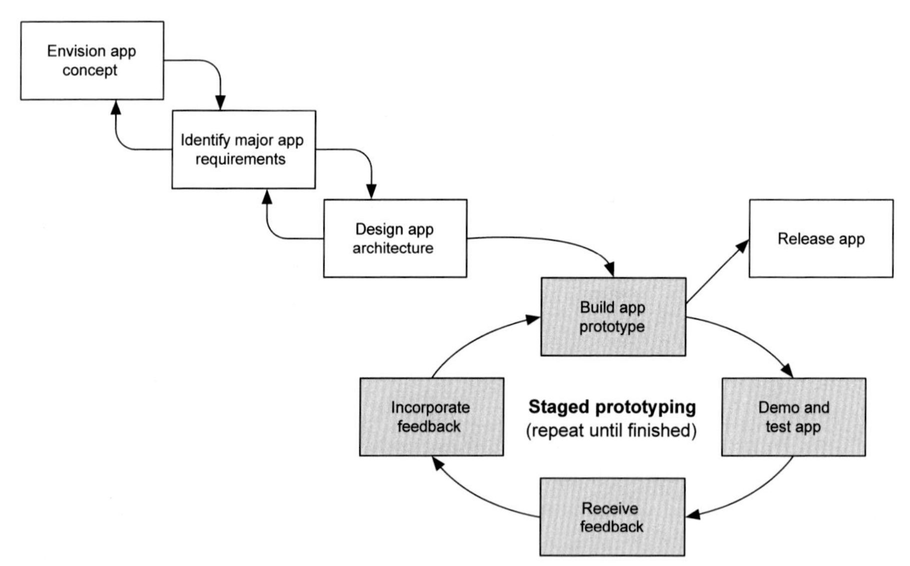

Figure 18.3 The evolutionary delivery methodology.
Adapted from McConnell, S., 1996. Rapid Development: Taming Wild Software Schedules. Microsoft Press, Redmond, WA, USA, p. 151.

Table 18.1 **The evolutionary delivery capability ratings**

Evolutionary delivery capability	Rating
Can work with poorly defined app requirements	★★★★☆
Can work with poorly defined app architecture	★★★★★
Finished app requires very little bug fixing	★★★★☆
Easily modify the app over its lifetime with new features	★★★★★
Ability to manage unforeseen risks during app development	★★★☆☆
Can follow tight app development schedule	★★★☆☆
Has low resource/person overhead to create app	★★★☆☆
Can allow for change in the app's features and visual design	★★★★☆
Ability to test and demo the app before apps fully built	★★★★★
Ability for managers to visually see app's progress	★★★★★
Significant investment in understanding methodology to build the app	★★☆☆☆

Adapted from McConnell, S., 1996. Rapid Development: Taming Wild Software Schedules. Microsoft Press, Redmond, WA, USA, p. 151.

the smart building that can run on mobile devices and ambient displays while a local website app would be used to control the smart building management system and configure various dashboards.

Identify major app requirements (see Fig. 18.3) will start to be identified as you solidify your app concept. Write these requirements down in a list and rank them in priority. This feature list will provide an app development direction and will be used to aid in the design of the app architecture. This list will also help decide what features will be added to each staged prototype. Not all features need to be identified; new features may be found during staged prototyping.

Design app architecture (see Fig. 18.3) is the most important step in the evolutionary delivery methodology. An ill-defined architecture will result in a failure to successfully develop an app. The architecture determines how the app is structured and how features are added to existing prototypes. For instance, adding new features to the previously staged prototype should not cause you to rewrite existing features unless those features operated contrary to what was expected during demonstration and testing. Along with a basic structure, architecture should also plan for common system libraries that need to be created and used by the app's different features or even among your different apps. The use of common code reduces app testing time and increases app reliability. For example, there may be a common set of functions (or libraries) to store and retrieve data from the smart building management system's database. These common libraries can be developed once and reused to enable faster and more reliable app development. Common libraries are not user-facing and will not change due to user feedback. However, there could be changes that speed up application performance. As long as the functions in these libraries do not change their name, their parameters, or the data they return, they can be changed independently of the user interface code and the user interface code should not have to be modified or rewritten.

Staged prototyping (see Fig. 18.3) is essentially an iterative process where you continually add features to your app a stage at a time. In each stage you would build an app prototype with a number of features. Once the prototype is complete you demo and test the app by having different users (managers, quality assurance analysts, a select group of customers) use the features. Once they have used the app you would receive feedback from the users on what they thought of the features, their suggestions, and any bugs they might have found. You would then incorporate the feedback into the prototype and add new features and functionality to create a new prototype.

Release app (see Fig. 18.3) happens when one of the following occurs: you need to release the app due to time or budget constraints, the app has all the planned features integrated into it, or the app is considered complete enough to be released. There may be a point where you have enough features for a version 1.0 of your app and adding more features would only delay its release. Adding other features can be delayed and released in future versions of the app. Additionally, fixing all software bugs is not possible. Attempting to do this is called *gold plating* and should be avoided.

18.3.2 Front-end and back-end app development

Part of the creation of an overall software architecture will involve deciding where the *smarts* live. Inference and prediction algorithms are a driving part of making a building smart. Models and parameters for these algorithms need to be learned and stored. Having an app on a mobile phone where there are limited memory and computational resources to run data intensive algorithms is not a well-designed idea. It would be better to have algorithms that build models run on the building control server or the cloud. The *cloud* could mean a server cluster of many computers that can run algorithms in parallel and in shorter periods of time. However, that is not to say that some algorithms cannot run on a mobile app. In fact, it may be advantageous to have some of the *smarts* run within the mobile app especially when it comes to providing alerts or decisions that require low latency or need to run if the Internet is down. One such example might be deciding what light level to turn the lights on in the middle of the night so as not to interrupt the circadian rhythms of a person (Makonin et al., 2013a).

18.3.3 How to collect and store data

When can an app be mobile or need to run on a PC or server? This will depend on its operation. Data intensive apps—ones that aggregate, summarize, and report on building data—should run on a computer that is close to the data such as the building control server or the cloud. Apps that display the resulting reports or statistics (dashboard apps) should run on mobiles and ambient displays. How data is stored is of importance as well. Real-time data that may be used in dashboards would be better stored in tuple databases (eg, Memcached, MongoDB), while data used for reporting would be better stored in a relational database such as MySQL, PostgreSQL, or MS SQL Server.

Reliable data collection is needed to have accurate reporting and dashboard statistics. It is important to have a reliable building control server that has redundant and fail-safe hardware. Off-site data replication may also be a service that can store a backup copy of a building data on the cloud. If the control server should fail and need to be replaced, then the data can be restored onto a new control server.

18.3.4 Ubiquitous sensor platforms and Internet of Things

Today there exist many ubiquitous platforms that facilitate communication between sensors and apps. Such platforms have been recently coined the *Internet of Things*. IoT platforms will allow a smart building management system to use sensors and actuators for different components and devices to create an environment where the focus of app development can become an important issue for system usability. Table 18.2 lists and compares some of the widely use IoT platforms currently available.

Table 18.2 Listing of different Internet of Things platforms

Internet of Things platform	Platform	Application program interface type	Source code
Apple HomeKit	iCloud hosted	OSX and iOS SDK, and SoC	Not released
Carriots	Hosted	App and embedded SDK	Not released
Comcast Xfinity	Hosted	Closed	Not released
DeviceHive	Any OS, local or hosted	Custom platform	On Github
Google cloud platform	Hosted	APIs and libraries	Not released
iDigi device cloud	Hosted	Embedded modules	Not released
KAA	Any OS, local	Server and embedded SDK	On Github
Kura	Any OS, local	App and embedded API	On Github
Nimbits	Any OS, local or hosted	Java library	On Github
Samsung smart home service	Hosted	Closed	Not released
SiteWhere	Any OS, local	REST API	Partly on Github
Spacebrew	Any OS, local or hosted	WebSockets	On Github
Temboo	Hosted	SDK and Arduino	Not released
Zetta	Any OS, local	REST API and WebSockets	On Github

18.3.5 *App development environments*

The IoT platform chosen to be used in a smart building management system can affect what development environment is needed for developing apps. Creating apps and using existing IoT platforms can dictate what development environment you use. For example, apps that run on Apple computers using OSX and mobile devices using iOS will require you to use the Xcode environment. There are alternatives to using Xcode but among these alternatives there can be a trade-off between cost and functionality. Apps that run on Microsoft Windows computers and mobile devices use the Visual Studio development environment. You can develop apps in Java using the Eclipse development environment. Apps programmed in Java or Python can run in OSX, Linux, and Windows. A Java-based development environment called Processing focuses on creating visualization-based apps and is OS independent. Developing Android-based mobile apps tends to use Android Studio. For web apps, you can markup documents in HTML and CSS in any text editor. However, it does help to have a text editor that supports *colored syntax highlighting*, such as the open source text editor Brackets or TextWrangler.

18.4 Conclusions

As we become more and more a culture of connected people, apps play an important part in accessing and interpreting the world around us. App companies are profitable and require less start-up costs than those companies who need to create hardware. App companies that focus on interconnecting different hardware and communication technologies create powerful apps that focus on data analysis and visualization. These apps can interface with building systems in different ways that can better the occupant's living environment and daily life.

This chapter has focused on app development and has given an overview of many of the issues that need to be taken under consideration when developing apps for any smart home or smart building. The way you choose to develop apps and how they interact with building occupants and the smart building management systems will depend on how experienced your teams of software designers and developers are. Often times, app development choices that are made depend on the prior experiences of your software team. In the chapters to follow, specific app case studies will be discussed to further help understand various issues that apps face.

References

App Store (iOS), October 27, 2015. In Wikipedia, the Free Encyclopedia. Retrieved 19:17, October 29, 2015, from: https://en.wikipedia.org/w/index.php?title=App_Store_(iOS)&oldid=687741665.

Apple Inc., Press release by, 2013. App Store Tops 40 billion Downloads with Almost Half in 2012 (online) Available at: https://www.apple.com/ca/pr/library/2013/01/07App-Store-Tops–40-Billion-Downloads-with-Almost-Half-in–2012.html (accessed 22.10.15.).

Apple Inc., Press release by, 2014. App Store Sales Top $10 billion in 2013 (online) Available at: https://www.apple.com/ca/pr/library/2014/01/07App-Store-Sales-Top−10-Billion-in−2013.html (accessed 22.10.15.).

Apple Inc., Press release by, 2015. App Store Rings in 2015 with New Records (online) Available at: https://www.apple.com/ca/pr/library/2015/01/08App-Store-Rings-in−2015-with-New-Records.html (accessed 22.10.15.).

Bartram, L., 2015. Design challenges and opportunities for eco-feedback in the home. IEEE Computer Graphics and Applications 35 (4), 52−62.

Booch, G., Rumbaugh, J., Jacobson, I., 2005. Unified Modeling Language User Guide, second ed. Addison-Wesley Professional, Boston.

D'Onfro, J., 2014. Google Has Spent More on Acquisitions than Its Top Five Rivals Combined, Business Insider (online) Available at: http://www.businessinsider.com/google-acquisition-spending−2014−1 (accessed 29.10.15.).

Makonin, S., Bartram, L., Popowich, F., 2013a. A smarter smart home: case studies of ambient intelligence. IEEE Pervasive Computing 12 (1), 58−66.

Makonin, S., Kashani, M., Bartram, L., 2012. The affect of lifestyle factors on eco-visualization design. In: Proceedings of Computer Graphics International (CGI).

Makonin, S., Pasquier, P., Bartram, L., 2011. Elements of consumption: an abstract visualization of household consumption. In: Smart Graphics, LNCS, vol. 6815. Springer, Berlin Heidelberg, pp. 194−198.

Makonin, S., Popowich, F., 2012. Home occupancy agent: occupancy and sleep detection. GSTF Journal on Computing 2 (1), 182−186.

Makonin, S., Popowich, F., Moon, T., Gill, B., 2013b. Inspiring energy conservation through open source power monitoring and in-home display. In: Proceedings of the 2013 IEEE Power and Energy Society General Meeting.

McConnell, S., 1996. Rapid Development: Taming Wild Software Schedules. Microsoft Press, Redmond, WA, USA.

Mitchell, R.L., 2005. The Rise of Smart Buildings, Computer World (online) Available at: http://www.computerworld.com/article/2568757/networking/the-rise-of-smart-buildings.html (accessed 22.10.15.).

Montecuollo, M., January 29, 2014. Native or web-based? selecting the right approach for your mobile app. UX Magazine. Article No. 1179, from: https://uxmag.com/articles/native-or-web-based-selecting-the-right-approach-for-your-mobile-app.

Responsive web design. In Wikipedia, the Free Encyclopedia, October 2, 2015. Retrieved 18:52, October 4, 2015, from: https://en.wikipedia.org/w/index.php?title=Responsive_web_design&oldid=683783673.

Sen, J., 2013. Security and Privacy Issues in Cloud Computing. ArXiv e-prints, arXiv: 1303.4814 (cs.CR).

Sun, G.D., Wu, Y.C., Liang, R.H., Liu, S.X., 2013. A survey of visual analytics techniques and applications: state-of-the-art research and future challenges. Journal of Computer Science and Technology 28 (5), 852−867.

Thomas, J.J., Cook, K.A., 2006. A visual analytics agenda. IEEE Computer Graphics and Applications 26 (1), 10−13.

Want, R., Schilit, B.N., Jenson, S., 2015. Enabling the internet of things. Computer 1, 28−35.

Ware, C., 2008. Visual Thinking for Design. Morgan Kaufmann, Burlington, MA, USA.

Zhou, M., Zhang, R., Xie, W., Qian, W., Zhou, A., 2010. Security and privacy in cloud computing: a survey. In: Proceedings of the Sixth International Conference on Semantics Knowledge and Grid (SKG), pp. 105−112.

Apps for smart buildings: a case study on building security

I. Chatzigiannakis
Sapienza University of Rome, Rome, Italy

19.1 Introduction

The smart home and building market has become an extremely popular topic since 2014, and is expected to explode by 2020. Homes, and the way we live and behave in them, have changed dramatically in the past 10 years. There is an explosion of creative ideas happening around the home—a vital structure of our everyday lives. The smart home of the future aspires to help people's busy lifestyles by organizing the environmental aspects and managing the residential energy, while providing a healthy and secure feeling. To fulfill this vision, a variety of products have entered the market, delivering a range of innovative services to homeowners using a variety of intelligent, connected devices.

A key aspect for technologies and concepts wishing to be transformed into successful products, is to understand the consumers' needs. Ericsson ConsumerLab (2015) conducted a study in order to understand people's behaviors and values and provide insights on consumer trends. The study points out that the user's most dominant need in connected home services is the safety of their family and property. The ability to monitor who is entering home and when, or faults in the electricity or water supply would be comforting. Having a connected home would give families a high level of control over what goes on in their house—both while being there or when away.

A number of related services are currently trying to address this need: surveillance cameras, smart locks for locking and unlocking doors, motion detector notifications, and fire and flood alerts. These are already available products that enable homeowners to manage their property remotely through a smartphone application. Does this constitute an increased level of security? Or does remote control entail risks, as it involves insecure data exchanges between the home system and the smartphone? Is it possible there could be an increase in the available options for intruders, by offering a new way in, through the communication network of the house? There is so much that smart homes can do to make a house more secure (eg, remote-controlled door locks, apps that turn lights on and off) but an Internet-connected home may actually be less secure.

In this chapter we look into smart home products which will increase building security. We start by examining the available networking technologies and the need for secure and encrypted data exchange and storage. We look into specific product cases that could expose a building to new types of cyber threats. We also look into products that introduce new mechanisms to reinforce network security, thus reducing the risks of cyber intrusion.

Start-Up Creation. http://dx.doi.org/10.1016/B978-0-08-100546-0.00019-4

19.2 Networking technologies for smart homes

A smart home is understood as an integration system, which takes advantage of a range of techniques such as computers, network communication, as well as synthesized wiring to connect all indoor subsystems that attach to home appliances and household electrical devices as a whole (Ricquebourg et al., 2006). Under this perspective, the development of smart home technology during the past years can be described by examining the networking technologies used to interconnect home devices.

Network technologies is the term used to describe familiar cables that, to a large extent, already exist, both inside and outside the house: telephone cables, TV cables, and the power supply network. It also refers to the extra infrastructure that may or may not exist yet: computer cables and lower voltage cables. It refers to infrared (IR) and radio frequent communication (RF). Essentially, we refer to technologies that effectively centralize house management and services, providing all-round functions for internal information exchange, while keeping in touch with the outside world.

Traditional suppliers' products mainly contain a protocol, which allows communication between the products, remote control, and central control by the resident. Ideally, all different manufacturers' products should communicate via the same protocol. In practice, however, this is not the case. New platforms have been introduced that unify products from different manufacturers, yet there is a lot of work that needs to be done to establish uniform network access.

19.2.1 Wired legacy systems

A homebus is a physical wire, a special low-voltage cable, that is used to transfer signals within the house via a certain protocol. Generally, two types of homebus systems can be distinguished: the systems with fixed, built-in intelligence and the systems that can be programmed via a PC. A special version of homebus systems is the powerline system that transfers signals, within the house, through the power cables. Among the power line systems, the X10 is the oldest. Improvements on this protocol resulted in the so-called A10 system, which has a growing popularity for use in both newly built and existing houses, because of its low price.

All media differ in their properties and have both advantages and disadvantages. The right selection is also a question of cost. As a general rule, higher data speeds lead to higher installation costs. However, the demand for bandwidth or bit rate strongly depends on the application. In general, for control data transmission, a bit rate of some kbps is sufficient. This holds for most of the smart home components (sensors, actuators, control, and visualization units). However, for telecommunication purposes (mainly video communication), the bit rate exceeds the megabit range.

Apart from X10, there are other standards available for setting up a homebus. Here we report some dominant technologies:

- EIB (European Installation Bus) is an open standard widely used in Europe. EIB is available for power line, signal cable, and radio. The single cable version is currently the most widely used in smart homes.

- KNX is a new standard resulting from an amalgamation of three European bus standards, with EIB being one of them. KNX is expected to replace EIB in the near future. KNX fully complies to the EN 50090 series, the European Standard for Home and Building Electronic Systems.
- LON (Local Operating Network) is a proprietary standard, used for energy-control, steering machinery, and access control systems in industry and larger buildings. The standard is mostly known for power line signaling, but also supports signal cables, coaxial cables, radio, and fiber optical transmission.
- BACnet is a standard developed in the United States for the control of functions in larger buildings, but has so far not been observed in European smart homes. BACnet is supposed to easily communicate with the EIB.

Regardless of the technology used, a common denominator of all these technologies is the fact that each product requires cables. As the number of components installed increases, so does the need for additional wires. The Ericsson ConsumerLab (2015) study on consumer trends clearly indicates that families consider too many visible cables annoying and space consuming. In some cases, consumers hire technicians to set up or install their entertainment centers, just to ensure that wires are hidden when a TV is mounted on the wall.

19.2.2 Wireless 802.11 (WiFi)

Wireless technologies are an alternative to laying out long wires across the house. New component installation is done with increased flexibility—clearly a desired property for a smart home product. However, wireless technologies also bring an important drawback. Safety and security cannot reach the levels obtained with wired networks, deterministic response times are not possible, and RF emissions by nearby devices might cause some user concern. However, it is clear that in many cases the advantages overcome the drawbacks and a wireless network has become the most feasible alternative for home automation.

Originally wireless smart home networks were based on protocols specifically designed for this purpose but currently, due to the huge penetration of computer and telecom wireless networks, it seems that this is no longer the case. Many new smart home products are based on the 802.11 family of protocols, reducing the need for installing additional equipment and enabling the smart devices to become Internet-connected with minimum effort.

In 2015 the WiFi Alliance for Smart Home was setup, to leverage WiFi's 16-year legacy of interoperability, industry-standard security, and great user experience for enabling the smart homes of the future. The key point of the alliance is that companies focus on WiFi for the smart home because it is a mature, standards-based technology ecosystem, it is deployed globally across a range of devices, and it can accommodate whole-home ranges.

19.2.3 Wireless 802.15.4 (ZigBee/Z-Wave)

The benefits of setting up a wireless network for smart homes are very clear, however WiFi does not provide an ideal environment. After years of research on wireless sensor networks and wireless personal area networks, the IEEE 802.15 Wireless Personal Area Working Group introduced a new networking standard for low-power devices designed for seamless integration into everyday life. It can be viewed as a low-power WiFi version and is ideally suited for connecting embedded devices to the Internet with extremely long battery life requirements.

Today, organizations use IEEE 802.15.4 enabled microcontrollers to effectively deliver solutions for a variety of areas including consumer electronic device control, energy management and efficiency, home and commercial building automation, as well as industrial plant management. Two variations of the 802.15.4 protocol have been used for developing the majority of smart home products.

ZigBee's name illustrates the mesh networking concept, since messages from the transmitter zigzag like bees, looking for the best path to the receiver. While Z-Wave uses a proprietary technology for operating its system, ZigBee's platform is based on the standard set by IEEE for wireless personal networks. This means any company can build a ZigBee-compatible product without paying licensing fees for the technology behind it, which may eventually give ZigBee an advantage in the marketplace. Like Z-Wave, ZigBee has fully functional devices (or those that route the message) and reduced function devices (or those that do not).

Z-Wave uses a source routing algorithm to determine the fastest route for messages. Each Z-Wave device is embedded with a code, and when the device is plugged into the system, the network controller recognizes the code, determines its location, and adds it to the network. When a command comes through, the controller uses the algorithm to determine how the message should be sent. Because this routing can take up a lot of memory on a network, Z-Wave has developed a hierarchy between devices: Some controllers initiate messages, and some are slaves, which means they can only carry and respond to messages.

ZigBee/Z-Wave and 802.15.4-based wireless sensor networks have been studied extensively in the context of developing smart systems for building security. The ability to distribute smart-sized devices within the building and easily establish a communication network is very important for advanced monitoring of building security. The use of ZigBee technology in combination with other networking technologies (eg, GSM/GPRS) can significantly contribute in developing smart building security systems that achieve fast-rate, low-cost, low-power wireless network communications. In Liu (2014), a typical application that relies on the CC2430 ZigBee wireless radio component is developed that provides real-time acquisition in the home environment temperature, humidity, three tables, IR, smoke, the parameters of the gas, fire, theft alarm, and home appliances. Alarms related to appliances operation, in combination with traditional sensor alarm systems, are forwarded to smartphone applications in order to achieve a flexible, convenient home security monitoring. In Huanga et al. (2011) the ability to position small-scale devices at specific locations in the home environment is utilized to develop a system for building electrical safety. The primary focus of this system is power consumption monitoring, remote control of appliances, overload protection, and energy management. These examples demonstrate the cost-effectiveness of ZigBee-based monitoring and protection systems.

Developing systems for smart homes that rely on small-factor embedded devices that communicate over a 802.15.4 wireless network raises significant security and trust issues. In many cases, these petit computers may need to exchange crucial information that needs to remain private. Moreover, as these embedded devices are distributed throughout the building, in some cases they can be easily acquired by an intruder that is capable of tampering with their hardware components. Several studies exist, such as Goodspeed (2009), that demonstrate how to extract keys from ZigBee hardware and thus intrude the wireless network. These studies point that without appropriate hardware, key secrecy should not be the foundation of the ZigBee product's security architecture.

Adaptic cryptographic algorithms for improving the cyber security of embedded wireless systems is a difficult task due to resource limitations. In Baumgartner et al. (2010) a hardware-agnostic implementation of an elliptic curve cryptosystem is provided. The main advantage is that it uses much smaller keys than conventional, discrete logarithm-based cryptosystems (an 160-bit key in an elliptic curve cryptosystem provides equivalent security with a 1024-bit key in a conventional cryptosystem). This fact makes elliptic curves an excellent approach for significantly improving the levels of cyber security, given the limited resources of the devices. Research results have shown that public-key cryptography based on elliptic curves is feasible to be used in sensor networks (Gura et al., 2004; Malan et al., 2004).

Another approach toward reinforcing the cyber security of the wireless network is to incorporate zero knowledge proofs (ZKPs) as a cryptographic tool for protecting a user's privacy. A ZKP involves two entities, a prover and a verifier. It allows the prover to demonstrate knowledge of a secret while revealing no information whatsoever of use to the verifier in conveying this demonstration of knowledge to others (Menezes et al., 1996). Up to now, although a wide variety of ZKPs of this category has been proposed (Smith, 2005) there exists very few actual implementations regarding resource constrained devices. Among the very few such systems available is the one presented in Chatzigiannakis et al. (2011), where the application of ZKP is studied for the security and privacy empowerment of wireless 802.15.4 networks consisting of low-constrained devices.

19.2.4 Wireless 802.15.6 (BLE)

In 2014, LOCKITRON[1] started a crowd-funding campaign to create a WiFi-enabled smart lock, one that could connect instantly to a WiFi network and also be controlled remotely over the Internet. After more than a year invested in developing the product, the company announced it was pulling the plug on the original device and would soon begin shipments of a next-generation Bluetooth-only smart lock called the Bolt. The company also created a separate WiFi-to-Bluetooth bridge accompanying the Bolt, in order to enable remote connectivity via the Internet. The migration to a Bluetooth low-energy (BLE)-only solution was justified by LOCKITRON because

[1] https://lockitron.com/.

Table 19.1 Matrix networking technology versus device

Networking technology	Motion	Door	Camera	Lock	Presence	Smoke
Wired	X	X	X			X
802.11 (WiFi)			X	X		X
802.15.4 (ZigBee)	X	X		X	X	X
802.15.6 (BLE)				X		

fundamentally, they could not get around how power hungry WiFi is, in nonideal circumstances and the substandard customer experience this creates. By switching to BLE and providing a bridge, they eliminated the hard problem of WiFi power management and gave users instantaneous remote control over the Internet.

This story clearly depicts the motivation behind BLE. As the name implies, this new version of Bluetooth provides considerably increased power-efficiency allowing devices to run off a tiny battery for long periods. Although the standard was introduced in 2010, within only a few years, there are numerous wireless devices offering a painless path to control lights, temperature, household appliances, window and door locks, security systems, and more. This new version of Bluetooth provides a completely new approach in terms of cyber-securing the network. The previous version of Bluetooth was known to contain a number of security vulnerabilities that could lead to exposure of encryption keys (Hager and MidKiff, 2003).

Another benefit of Bluetooth networking is the fact that it is supported by the vast majority of the mobile phones that are in use today. It is therefore easy to use the resident's smartphone to provide a simple interface with the smart devices installed in the premises. Moreover, Bluetooth allows for greater localization accuracy compared to WiFi, due to its more limited range. It is also easier and safer to set up and operate, due to the inherent features in Bluetooth's design (Antoniou et al., 2012) (Table 19.1).

19.3 The vulnerability of wireless networks: a case of cyber-security threat

Each time a new smart home product enters the market, consumers are concerned about the cyber-security levels of the wirelessly connected device like smart cameras and thermostats—devices that are always connected awaiting a remote control. Such connected devices may become an easy target for hackers because they may lack basic security measures. Indeed, as companies (small and established ones) rush to incorporate Internet of Things technologies to produce appealing products, security may not be prioritized as highly as it should.

A particular case that raises a number of issues is the case of LIFX, a smart light bulb created in September 2012, through crowd funding. LIFX is considered a characteristic success story for the smart home industry. A so-called hardware premium, it offers remotely programmable LED light bulbs that can be controlled through a smartphone application. These bulbs are sold at a premium, and are priced around 10 times higher than a compact fluorescent bulb (Wired, 2012). From a consumer perspective, a key driver for buying the product is the novelty in remote-controlling the devices, the ability to combine their operation, and the wide range of colors and brightness provided.

Looking into the technical details of the networking aspects of this specific product operation, there is the use of two wireless technologies in order to optimize both the communication between the bulbs as well as the ease of usage. In more detail, one of the bulbs assumes the role of the controller and connects to the home WiFi network, thus providing a very simple way to communicate with smartphones as well with the Internet so that the bulbs can be operated remotely. In parallel, all the bulbs set up an 802.15.4 6LoWPAN mesh network in order to exchange configuration and control commands. The 802.15.4 wireless technology is designed to keep power consumption at low levels and also operate at adequate levels in environments with high interference (eg, when video is transmitted over the WiFi network, or a microwave oven operates in close proximity).

In June 2014, Context released a detailed report where it demonstrated how researchers, after studying the encryption algorithms, keys and initialization vectors, as well as the mesh network protocol, managed to inject packets into the 802.15.4 mesh network. These packets allowed them to capture the WiFi authentication details and decrypt the credentials. The retrieval of the information did not require any prior authentication and did not cause any alert or trace of a cyber attack.

Context reported that hacking into the light bulb was certainly not trivial but would be within the capabilities of experienced cyber criminals. In the case of LIFX these vulnerabilities were overcome relatively quickly and easily. Very shortly after the release of the report, LIFX issued a firmware update that solved the problem. In fact, prior to Context's report, this vulnerability was totally unexplored (and possibly unknown), most likely due to the complexity of the equipment and reverse engineering required. It should also be noted that since this attack was on the 802.15.4 6LoWPAN wireless mesh network, the attacker had to be within wireless range (ie, within at least 30 m) of a vulnerable LIFX bulb to perform this attack, severely limiting the practicality for exploitation on a large scale.

The hack of devices like the LIFX bulbs is just a sample of the possible security issues related to IoT components. In some cases the vulnerabilities are fundamental to the design of the products. Therefore, it is critical to consider their security at the earliest stages of the design phase and incorporate it throughout the development of the product.

19.4 Reinforcing the security of wireless communications: the case of smart locks

Door locks are probably the first group of hardware devices that have attracted serious developers and have become a smart product. Located at a natural entry point, they are arguably improving the overall user experience for a secure and comfortable smart home. Virtually all smart lock products offer a very straightforward installation process: the smart lock simply replaces the current deadbolt lock, making it a very attractive product even for customers who do not wish to employ a professional technician.

Most products look like normal locks (ie, they offer the option to use a normal key) but provide electronic methods for replacing the need of a physical key. Others give a more futuristic look to your door by requiring either a touch to initiate the authentication procedure or in some cases like the AUGUST Smart Lock,[2] the wireless connection is used to detect if the owner is standing in front of the door and unlocks your door for you. They let you send electronic keys to your friends that work only during times that you specify. Some will even connect to your larger home automation system, telling your smart thermostat when you're away so it can enter into its energy-saving mode.

Regardless of the different features provided, existing products rely upon the built-in protective features of BLE (also known as 4.0, or SMART). BLE built-in security features claim to use the same security protocols as those in online banking. What that essentially means is that anytime your smartphone is communicating with your smart lock, a 128-bit AES encryption cryptomechanism is used. Although 128-bit AES encryption is considered the lowest level of encryption used by government agencies in the United States, it is still good enough for real-life scenarios. Keep in mind that the highest level of encryption used by government agencies in the United States requires 192- or 256-bit AES encryption.

An important aspect of BLE security is how each user or digital key is being authenticated. Unlike Bluetooth version 2.1, the latest version does not actually require two devices to initially pair with each other. So the more common Bluetooth-based attacks like Bluejacking, Bluesnarfing, or Bluebugging are a nonissue. In addition, BLE is introducing adaptive frequency-hopping, which splits encrypted data and transmits it across the 2.4 GHz spectrum.

In order to further improve the security of wireless communication, UNIKEY[3] totally replaces the BLE's security protocol and introduces a public-key infrastructure system to authenticate users. In their product, every communication between the smartphone and smart lock is a unique transaction. So even if someone were able to overhear the wireless communication containing the key, they wouldn't be able to use it again. As a second step to further improve security, UNIKEY provides a wireless system on both sides of the door, thus letting the smart lock understand if the user is on the inside or the outside. Essentially UNIKEY eliminates any false unlocks.

A different approach is followed by AUGUST Smart Lock.[4] Unlike UNIKEY's product, it handles communication between the lock and a permissioned smartphone over Bluetooth. However, the smart lock never communicates with a cloud server, or even the Internet. Authentication is handled through the smartphone and the

[2] http://www.august.com/.
[3] http://www.unikey.com/.
[4] http://www.august.com/.

accompanying AUGUST smartphone application. AUGUST owners can also grant permission or deny access through the website.

These examples demonstrate how companies are trying to address the concerns of consumers regarding cyber security. Although some scenarios exist, where wireless communication security could be breached, such scenarios are very hard to reproduce.

19.5 The need for secure data exchange and storage

Clearly the ability to control a house while being away has to be done in a way that it does not put the residents or the house in danger. We need to keep in mind that the ability to check the status of the house security system from a smartphone and remote control the devices involves the exchange of data between the house and the smartphone application through the Internet. Is the data encrypted throughout all the system's layers? Is the information related to the credentials for controlling the devices stored properly on the smartphone application? How easy is to crack the application code and decode the communication protocols? Such questions are critical while connecting more home elements with the Internet, as the home environment is exposed to new attack vectors that increase the risk of cyber threats.

We also need to keep in mind that a cyber-security breach may lead intruders into understanding the behavior of the residents and their patterns (eg, when they are home, and when away on vacation) putting their safety and privacy at risk. It also becomes possible for someone to turn on certain devices in order to reduce the physical security of the house and thus decrease the difficulty of entering the house. For example consider a possibly "exotic" scenario, where an intruder can turn on the air conditioning during winter, trying to freeze and break pipes while the homeowners are away on vacation, without the intruder ever stepping foot in the house. Indeed the possibilities are really only limited by the imagination and determination of the attacker.

In March 2015, Synack conducted a thorough benchmarking of 15 smart home devices ranging from cameras to home automation controllers (Synack, 2015). The researchers examined the scenario where the smartphone of one of the home residents is stolen or an attacker has a window of opportunity where he or she can control the smartphone for a short period of time (eg, for 5 min). Interestingly, they identified many products whose smartphone applications store the corresponding passwords in plaintext or left behind nonexpiring session credentials that would give an attacker indefinite access to a user's device.

The benchmarking report conducted by Synack also investigated the case where the user uses the smartphone application to remotely control the house from a public WiFi network or a WiFi network that is controlled by the attacker. The malicious adversary is able to overhear the data transmitted over the wireless medium and potentially gain critical information. In fact, multiple products were identified to exchange critical information such as passwords in plaintext. Especially when a connection is made over public WiFi networks, information exchanged by these products can be easily collected even by nonexpert attackers.

In the first case investigated, it is clear that product designers should make sure that all communication must use bidirection encryption. In the second case, the operation of the smart home and the overall security of the product becomes a responsibility of the user. Such an approach is clearly wrong. Products need to incorporate security as part of the design process.

19.6 The need for innovative approaches to handle data generated: the case of smart cameras

Camera sensors are certainly the most traditional equipment used to establish high levels of security for a building. The closed-circuit TV (CCTV) has become the de-facto standard for video surveillance on a specific place. Being a wired-only technology, CCTV requires a certain level of wiring especially when we wish to secure a large building.

Most security cameras are made for indoor use, so they are not weatherproof. However, some high-end products can be also used in outside environments. Some other products offer the ability to use the camera in the dark. These cameras feature night vision, so you get a clear picture of your surroundings in even the poorest light conditions. Another important feature is the ability to rotate the viewing angle of the camera in order to provide a better view of broader scenes.

During the past years, a large number of new products have been made available, replacing the wired network for transmitting camera signals, with wireless technologies based on the 802.11 family of protocols. The so-called smart cameras can be remotely accessible via the Internet allowing the building owner to view a live streaming from the camera wherever in the world he or she may be. This is a great step forward, since wireless networking essentially eliminates the need for wires, thus heavily reducing the total costs of CCTV camera installation.

The use of cameras within a home environment facilitates a broad range of services from illegal intrusion to resident care (Demiris and Hensel, 2008). Camera sensors provide a rich source of information about the home environment with the advantage of a noncontact sensor that is convenient for both living and installation. Hence, video surveillance becomes a practical solution for smart home. However, a downside of camera sensors is that they require human analysis of the streaming in order to identify a potential alert. We also need to consider that the number of cameras that can be inspected at any given time by an individual is limited. Clearly, a large number of camera feeds challenge the cognitive capabilities of the human operator.

A variety of software-based surveillance systems have been proposed that combine a systematical architecture and algorithm pipeline for intelligent video analysis for smart homes (Zhang et al., 2015). Developing such systems entails various types of challenges. Consider that in indoor environments it is common to face sudden illumination changes. Robust behavior representation models need to be developed in order to extract meaningful context information.

Recently, new hardware products are reaching the market, combining different sensing technologies (eg, motion sensors) with image processing techniques embedded on the hardware that are capable of identifying events of certain significance. These products are able to provide notifications at specific times when human intervention is required, in order to infer possible critical events. They also offer the ability to record while the motion sensor is detecting movement in the viewport of the camera. This means the home can capture any activity the moment it is happening, rather than trawl through hours of footage.

Some newer products such as the NETATMO smart camera[5] introduce more advanced image processing techniques for doing face extraction. The owner is able to tag faces, attach names, and characterize them as family or friends. The camera is capable of recognizing individual family members and send immediate notifications depending on the characterization of the people. This is a step forward from sending alerts (with video feed) when a motion is detected, and reduces the number of false positive alerts generated by the camera.

The main challenge for such new products is the homeowner's ability to access the video feeds remotely. The storage of the video (especially if it is at high quality) requires significant storage capacity. Furthermore, if the storage is kept locally (ie, within the building premises) then the Internet connectivity might cause long delays for accessing the video feeds remotely (eg, ADSL is not the ideal solution for this usage scenario). If there is a need for a 24/7 monitoring service, the critical point is whether the number of alerts that require the owners (remote) intervention can be restricted. Essentially these two features lead to a new contract-based service model that resides on the cloud.

As products for smart homes generate large volumes of sensor data, this creates the need to provide mechanisms that will assist the user in handling large volumes. Interestingly, for many start-ups, the ability to capture data, and analyze and sell the findings offers a potential monetization model.

Young companies such as CANARY[6] and WITHINGS[7] offer a monthly fee for using cloud-based resources to store video feeds up to 30 days and storing alarms and notifications to specific time instances (also known as bookmarks). This approach leads to extremely competitive monthly fees.

19.7 Smart home products: a fragmented landscape

As smart home technology has advanced, traditional home security products are increasingly being ported over analog to digital controls that offer expanded functionality and improved wireless connectivity, including integration with mobile technologies (Tankard, 2015). Numerous hardware products are already available in the market, which can be used to set up a smart building environment with emphasis on

[5] https://www.netatmo.com/en-US/product/camera.
[6] http://canary.is.
[7] http://www.withings.com.

securing the property. Unfortunately, the smart home market is a fragmented sector with a multitude of devices and services, many operating in their own bubble of protocols and software interfaces. As the smart home becomes a standard for residences across the world, these devices and services will not only need to work together, but will be incentivized by providing a cooperative interface.

Once again, the drawback of having a broad range of devices that are not able to form a common communication network and offer integrated services becomes an opportunity for new products. SMARTTHINGS[8] is a typical example of a home automation system that creates an ecosystem of products by supporting and combining different wireless networking technologies. The company via its hardware controller, the SmartThings Hub, offers users smart ways to control their home by combining products from different hardware manufacturers. They are positioned in such a way that they profit from both end customers as well as other platform users. Platform users pay the promoter for listing and the promoter also gets a share whenever a product is sold to the end customer on the platform. The company delivers a platform that helps unlock the synergy possibilities that were previously hidden (Forbes, 2014).

Like SmartThings, other companies such as Revolv, Nexia Home Intelligence, and Vera Control also offer a monitoring and controlling hub that allows users to control their connected devices from a single point. These companies offer users the ability to set simple triggers and actions that work with many types of sensors and devices. These triggers can be activated manually, randomly, or on a schedule.

19.7.1 Motion detectors

The most common sensors for securing buildings are motion sensors, also known as occupancy or vacancy sensors. For 30 years, occupancy sensors relied primarily on passive infrared (PIR) technology to infer the presence of people by detection motion. Other technologies include ultrasonic or a combined multisensing technology for increased accuracy. They can be positioned at walls, mounted on the ceiling or placed at a wall switch for both indoor and outdoor use.

A recent technological development for boosting the accuracy of occupancy sensors is the use of image processing. Unlike conventional PIR motion sensors used to infer occupancy, image processing occupancy sensors uses sequential image subtraction techniques for extracting and analyzing motion-dependent occupancy content. This new approach detects human presence at an increased range while reducing false positives and negatives produced by conventional sensors.

19.7.2 Door open/close sensors

Detection sensors for door and window entry are also among the first warning mechanisms introduced in building security. They offer a straightforward method for securing the perimeter of the building. Most available products can detect an open

[8] http://smartthings.com/.

door or window based on a magnetic sensor, while some are using a tilt sensor or a contact plunger. In most wired security systems, magnetic sensors are considered a standard for monitoring the status of several doors and windows throughout the building.

The most common type of a door sensor connection (magnetic) is via a wire. As wires significantly increase the installation costs, the new wave of smart door sensors is switching to wireless technologies. Currently, the 802.15.4 wireless medium is the most common choice for connecting door sensors located in different places within the building, in a network. The low-power requirements of the particular networking technologies allow battery-operated sensors to last longer, thus removing the need for wires. A wireless door sensor can be easily installed and nonintrusively take up the role of building security.

Some products attempt to use the door sensor as a point for providing multisensing capabilities. As an example consider the Door/Window sensor by PHILIO[9] that includes temperature, illumination, and motion sensors.

19.7.3 Presence sensors

The presence sensor lets you receive updates about a person or pet coming and going. It also works for objects, such as cars. It enables the security system to detect people and pets approaching a specified area of the building (eg, front door) and trigger different actions (eg, automatically lock/unlock doors). They can also be placed in children's backpacks and provide notifications when they arrive home from school.

An alternative to carrying a small presence sensor is to use the smartphone as a presence sensor. This approach utilizes the GeoFencing technology available for almost all smartphone operating systems that relies on the 802.11 Wi-Fi and the GPS module of the phone. Essentially it provides a coarse-grained detection of users when they are approaching or leaving the building.

19.8 Conclusions and future trends

In the recent past, many smart home products have entered the market, from connected toothbrushes to appliances and HVAC. A leading segment of the smart home market is that of home security; consumers report that home security is among their top priorities when selecting the next product to acquire. Personal and family security is a key driver in smart home adoption for the majority of the consumers (Icontrol, 2015). The vast majority of consumers agree that security is one of the most important reasons to purchase a smart home system. This justifies the fact that smart cameras and smart locks are among the most popular devices. Still, it is evident there is a chasm between the early adopters and mainstream consumers before products achieve mass adoption.

[9] http://www.philio-tech.com/products/PST02-A.pdf.

At the same time, current smart home products and services create a complex network of components that hugely complicates issues for information security. Cyber threats are realized at several levels: targeting hardware devices, they can be carried out over the communication medium or even the smartphone remote controlling application. Many cases have demonstrated the significant security risks involved with this nascent technology (Proofpoint, 2014). It is evident that overcoming these challenges and addressing the consumer concerns is crucial for developing successful smart home products.

References

Antoniou, A., Theodoridis, E., Chatzigiannakis, I., Mylonas, G., 2012. Using Future Internet Infrastructure and Smartphones for Mobility Trace Acquisition and Social Interactions Monitoring. In: The Future Internet, Volume 7281 of the Series Lecture Notes in Computer Science. Springer-Verlag, pp. 117–129.

Baumgartner, T., Chatzigiannakis, I., Fekete, S.P., Koninis, C., Kröller, A., Pyrgelis, A., 2010. Wiselib: a generic algorithm library for heterogeneous sensor networks. In: 7th European Conference on Wireless Sensor Networks, EWSN 2010, Lecture Notes in Computer Science, 5970, pp. 162–177.

Chatzigiannakis, I., Pyrgelis, A., Spirakis, P.G., Stamatiou, Y.C., October 17–22, 2011. Elliptic curve based zero knowledge proofs and their applicability on resource constrained devices. In: IEEE 8th International Conference on Mobile Adhoc and Sensor Systems, MASS 2011, IEEE Computer Society 2011, pp. 715–720.

Demiris, G., Hensel, B.K., 2008. Technologies for an Aging Society: A Systematic Review of "Smart Home" Applications. Yearbook of Medical Informatics, pp. 33–40.

Ericsson ConsumerLab, 2015. Connected Homes.

Forbes, July 2014. SmartThings Wants to Eviscerate the Home Insurance Industry.

Goodspeed, T., 2009. Extracting Keys from Second Generation ZigBee Chips. Black Hat USA, Las Vegas.

Gura, N., Patel, A., Wander, A., Eberle, H., Shantz, S.C., 2004. Comparing elliptic curve cryptography and RSA on 8-bit CPUs. In: Cryptographic Hardware and Embedded Systems (CHES 2004), Lecture Notes in Computer Science, LNCS 3156. Springer-Verlag, pp. 119–132.

Hager, C.T., MidKiff, S.F., 2003. An analysis of bluetooth security vulnerabilities. In: Wireless Communications and Networking, 2003. WCNC 2003. 2003 IEEE, vol. 3. IEEE.

Huanga, L.-C., Changa, H.-C., Chena, C.-C., Kuo, C.-C., June 2011. A ZigBee-based monitoring and protection system for building electrical safety. Energy and Buildings, Elsevier 43 (6), 1418–1426.

Icontrol, 2015. State of the Smart Home.

Liu, Z.-Y., 2014. Hardware design of smart home system based on ZigBee wireless sensor network. AASRI Conference on Sports engineering and computer Science (SECS 2014), AASRI Procedia, Elsevier 8, 75–81.

Malan, D.J., Welsh, M., Smith, M.D., 2004. A public-key infrastructure for key distribution in tinyos based on elliptic curve cryptography. In: 2nd IEEE International Conference on Sensor and Ad Hoc Communications and Networks (SECON 2004), pp. 71–80.

Menezes, A.J., Vanstone, S.A., Oorschot, P.C.V., 1996. Handbook of Applied Cryptography. CRC Press, Inc., Boca Raton, FL, USA.

Proofpoint, January 2014. Proofpoint Uncovers Internet of Things (IoT) Cyberattack.
Ricquebourg, V., Menga, D., Durand, D., Marhic, B., Delahoche, L., Loge, C., December 2006. The smart home concept: our immediate future. In: 1st IEEE International Conference on E-learning in Industrial Electronics, pp. 23–28.
Smith, W., 2005. Cryptography Meets Voting.
Synack, March 14, 2015. Home Automation Benchmarking Results.
Tankard, C., March 2015. How secure is your building? Network Security, Elsevier 2015 (3), 5–8.
Wired, September 2012. App-controlled LIFX Bulbs Reinvent the Humble Household Light.
Zhang, J., Shan, Y., Huang, K., February 2015. ISEE Smart Home (ISH): smart video analysis for home security. Neurocomputing, Elsevier 149, 752–766. Part B, 3 Pages.

Index

'*Note*: Page numbers followed by "f" indicate figures and "t" indicate tables.'

J

K

L

M

Printed in the United States
By Bookmasters